Construction Estimating Using Excel

Construction Estimating Using Excel

THIRD EDITION

Steven J. Peterson
MBA, PE

330 Hudson Street, NY NY 10013

Vice President, Portfolio Management: Andrew Gilfillan
Portfolio Manager: Tony Webster
Editorial Assistant: Lara Dimmick
Senior Vice President, Marketing: David Gesell
Field Marketing Manager: Thomas Hayward
Marketing Coordinator: Elizabeth MacKenzie-Lamb
Director, Digital Studio and Content Production: Brian Hyland
Managing Producer: Cynthia Zonneveld
Managing Producer: Jennifer Sargunar
Content Producer: Faraz Sharique Ali
Content Producer: Nikhil Rakshit
Manager, Rights Management: Johanna Burke
Operations Specialist: Deidra Smith
Cover Design: Cenveo Publisher Services
Cover Art: Shutterstock
Full-Service Project Management and Composition: R. Sreemeenakshi/SPi Global
Printer/Binder: Edward Brothers
Cover Printer: Phoenix Color
Text Font: SabonLtPro

Copyright © 2018, 2012, 2007. Pearson Education, Inc. All Rights Reserved. Manufactured in the United States of America. This publication is protected by copyright, and permission should be obtained from the publisher prior to any prohibited reproduction, storage in a retrieval system, or transmission in any form or by any means, electronic, mechanical, photocopying, recording, or otherwise. For information regarding permissions, request forms, and the appropriate contacts within the Pearson Education Global Rights and Permissions department, please visit www.pearsoned.com/permissions/.

Acknowledgments of third-party content appear on the appropriate page within the text.

Unless otherwise indicated herein, any third-party trademarks, logos, or icons that may appear in this work are the property of their respective owners, and any references to third-party trademarks, logos, icons, or other trade dress are for demonstrative or descriptive purposes only. Such references are not intended to imply any sponsorship, endorsement, authorization, or promotion of Pearson's products by the owners of such marks, or any relationship between the owner and Pearson Education, Inc., authors, licensees, or distributors.

Microsoft®, Microsoft Windows®, and Microsoft Office® are registered trademarks of the Microsoft Corporation in the United States and other countries. This book is not sponsored or endorsed by or affiliated with the Microsoft Corporation.

Library of Congress Cataloging-in-Publication Data

Names: Peterson, Steven J.
Title: Construction estimating using Excel / Steven J. Peterson, MBA, PE.
Description: Third edition. | Boston : Pearson Education, [2018] | Includes index.
Identifiers: LCCN 2016042595| ISBN 9780134405506 | ISBN 0134405501
Subjects: LCSH: Building—Estimates—Data processing. | Microsoft Excel (Computer file)
Classification: LCC TH437 .P357 2018 | DDC 692/.50285554—dc23 LC record available at https://lccn.loc.gov/2016042595

27 2023

ISBN 10:   0-13-440550-1
ISBN 13: 978-0-13-440550-6

DEDICATION

*This book is dedicated to Bill Thiede,
who taught me how to estimate.*

PREFACE

The second hand of the clock swept rapidly toward the 12, signifying that it was 15 minutes to 1. Fifteen minutes until the bid I was working on was due. I had 10 minutes until our secretary would call for the final price, giving her just enough time to write the number on the bid forms, seal the bid, and turn it in. The bid had been tallied and was ready to go, with the exception of one blank line. The words "Asphalt Paving" along with the blank space to the right glared at me. I had yet to receive a number from my asphalt contractor, who had promised me that he would get me a bid. I called his phone number. No answer! There was only one thing left for me to do. I had to do an asphalt takeoff myself, and I had only 8 minutes left. Quickly I grabbed my scale, turned it to match the scale at the bottom of the drawing (1/8″ = 1′), and began to determine the quantity of asphalt needed for the parking lot. Cutting corners where I could, scribbling numbers on a piece of scratch paper, and punching the numbers on my calculator, I arrived at the square footage of the asphalt. It wasn't exact, but it was close. I then multiplied the square footage by the estimated cost per square foot for the asphalt. The phone rang as I added my estimated cost for the asphalt to the bid. It was our secretary calling for the bid. I gave her the number, and she read it back to me.

I relaxed for the next 20 minutes, waiting for our secretary to call back with the bid results. The phone rang again and it was our secretary. We had won the bid. As I was enjoying the thrill of victory, I glanced at the plans and the words "Warning: Half Size Drawings" jumped off the page. The thrill of victory turned into a large lump in my stomach as I realized why I had won the bid. I had forgotten to double all my measurements as I measured the asphalt, so I had only 25% of the money I needed to purchase the asphalt for the project. Quickly, and carefully this time, I took off the asphalt again. Indeed, I only had a quarter of the asphalt I needed, and it would take all the profit in the job and then some to pay for it.

By working hard during the buyout process, I was able to get the project back on budget. I did this by finding local subcontractors and suppliers for the items I had bid myself who would do the work for less than I had budgeted. In the end, we made a fair profit on the job, but that does not make up for the money we left on the table because of my error.

The job of an estimator is to forecast as accurately as possible the likely cost of a construction project, as well as the required amounts of materials, labor, and equipment necessary to complete the work. Because of the high degree of uncertainty in estimating these items, estimating is more of an art than a science.

There are three key pillars that support the success of an estimator. The first is an understanding of the construction process—how things are built and how the pieces are put together. The second is an understanding of the fundamental principles of estimating. Estimates that are not based on these fundamental principles are nothing more than wild guesses. The third pillar is experience. For a person to become a good estimator, it is required that he or she develops good judgment on how to apply estimating principles. Will Rogers reportedly said, "Good judgment comes from experience, and a lot of that comes from bad judgment."[1] The pillar of experience can only be acquired through practice.

The purposes of this book are to: (1) give beginning estimators an understanding of the fundamental principles of estimating, (2) provide beginning estimators with practical estimating experience, (3) give beginning estimators a basic understanding of how to use spreadsheets, such as Microsoft Excel,[2] to increase their estimating productivity and reduce errors, and (4) give experienced estimators another view on estimating and a chance to improve their estimating skills.

---

[1] Will Rogers as posted on http://www.brainyquote.com/quotes/quotes/w/willrogers411692.html; accessed August 5, 2016.

[2] Microsoft® Excel is a registered trademark of the Microsoft Corporation in the United States and other countries.

This book is divided into five sections. The first section introduces the reader to estimating. The second section introduces the quantity takeoff and teaches the reader how to determine the quantity of materials needed to complete the project. The third section shows how to put costs to the estimate. The fourth section shows how to finalize the bid, incorporate the estimate into the schedule, and buy out the project, and also discusses bidding ethics. The fifth section teaches how to tap into the power of computer spreadsheets—specifically Microsoft® Excel—and demonstrates how spreadsheets can be used to automate estimating functions. It also provides a chapter on estimating methods used to prepare conceptual and preliminary estimates. To provide practice in estimating, three drawing sets are included—a residential garage, a residence, and a retail building.

I hope that you will take the time to carefully read each chapter, work the problems at the end of each chapter, and prepare estimates for the drawings sets included with the book. Doing this will help you begin to gain the experience needed to be a successful estimator.

Feedback on this book can be submitted at: stevenjpeterson9@gmail.com

I would like to thank the following people for providing reviews of the manuscript: Dennis Dorward (Pennsylvania College of Technology), Kristen Gundvaldson (Southeast Technical Institute), Darwin Olson (Anoka Technical College), Mark Steinle (Casper College), James William White (Indiana University–Purdue University Indianapolis), and Manoochehr Zoghi (California State University, Fresno).

STUDENT LEARNING OUTCOMES

During the past few years, higher education has been moving to outcome-based learning, which requires accredited programs to measure their students' ability to meet the required outcomes. Currently, in the United States, there are four accreditation standards for construction management and construction engineering program, which are as follows: (1) American Council for Construction Education (ACCE); (2) ABET—Engineering Accreditation Commission, for construction engineering; (3) ABET—Engineering Technology, for construction engineering technology; and (4) ABET—Applied Science, for construction management. Although each of these standards are different, they all focus on three general outcomes, which can be summarized as follows. Construction management/engineering students should be able to:

- Prepare construction cost estimates. This book includes both a residential and a commercial set of plans. Each chapter, where applicable, includes homework problems related to these plans. By completing the homework problems associated with either the residential plans or the commercial plans, the students will prepare a complete construction estimate. See the instructor's guide for more information on which problems relate to preparing these estimates.
- Effectively communicate in writing. Chapter 24 covers writing a scope of work and Chapter 28 covers writing proposals and communicating by e-mail. The material in these chapters can be used in part to meet the outcome related to written communication.
- Understand ethics as it relates to estimating. Chapter 31 addresses ethic as it relates to estimating and can be used in part to meet the outcome related to ethics.

NEW TO THIS EDITION

The following is a list of key changes that have been made to this edition.

- The Excel contents have been updated to Excel 2016, the latest version of Excel available at the time of this revision.
- A number of changes have been made to ensure that the material in this book aligns with the ACCE and ABET student learning outcomes.
- The material and equipment costs and labor rates have been updated to reflect the costs at the time of this revision.
- Chapters 33 and 34 have been combined into a single chapter and a chapter covering estimating methods used to prepare conceptual and preliminary estimates has been added as Chapter 34.
- The use of data from RSMeans (along with six sample pages from their books) has been added.
- A discussion of the Fair Labor Standards Act, which covers the labor laws dealing with non-exempt employees, has been added.
- A discussion of the Davis–Bacon Act, which covers the pay rates for employees on projects funded by the federal government, has been added.
- The problem sets from Chapters 21 and 22 have been expanded.
- An appendix with sample equipment costs has been added.

DOWNLOAD INSTRUCTOR RESOURCES FROM THE INSTRUCTOR RESOURCE CENTER

To access supplementary materials online, instructors need to request an instructor access code. Go to www.pearsonhighered.com/irc to register for an instructor access code. Within 48 hours of registering, you will receive a confirming e-mail including an instructor access code. Once you have received your code, locate your text in the online catalog and click on the Instructor Resources button on the left side of the catalog product page. Select a supplement, and a login page will appear. Once you have logged in, you can access instructor material for all Pearson textbooks. If you have any difficulties accessing the site or downloading a supplement, please contact customer service at: http://support.pearson.com/getsupport

DISCLAIMER

Microsoft and/or its respective suppliers make no representations about the suitability of the information contained in the documents and related graphics published as part of the services for any purpose. All such documents and related graphics are provided "as is" without warranty of any kind. Microsoft and/or its respective suppliers hereby disclaim all warranties and conditions with regard to this information, including all warranties and conditions of merchantability, whether express, implied or statutory, fitness for a particular purpose, title and non-infringement. In no event shall Microsoft and/or its respective suppliers be liable for any special, indirect or consequential damages or any damages whatsoever resulting from loss of use, data or profits, whether in an action of contract, negligence or other tortious action, arising out of or in connection with the use or performance of information available from the services.

The documents and related graphics contained herein could include technical inaccuracies or typographical errors. Changes are periodically added to the information herein. Microsoft and/or its respective suppliers may make improvements and/or changes in the product(s) and/or the program(s) described herein at any time. Partial screen shots may be viewed in full within the software version specified.

The sample Excel spreadsheets in this book are to provide the reader with examples of how Excel may be used in estimating, and as such, are designed for a limited number of estimating situations. Before using the spreadsheets in this book, the reader should understand the limits of the spreadsheet and carefully verify that the spreadsheets: (1) are applicable to his or her estimating situation and (2) produce an acceptable answer. The reader assumes all risks from the use and/or performance of these spreadsheets. The drawings included with this text are for educational purposes only and are not to be used for construction.

BRIEF CONTENTS

PART I
INTRODUCTION TO ESTIMATING 1

Chapter 1
THE ART OF ESTIMATING 2

Chapter 2
OVERVIEW OF THE ESTIMATING AND BIDDING PROCESS 11

Chapter 3
INTRODUCTION TO EXCEL 21

PART II
THE QUANTITY TAKEOFF 71

Chapter 4
FUNDAMENTALS OF THE QUANTITY TAKEOFF 72

Chapter 5
CONCRETE 84

Chapter 6
MASONRY 107

Chapter 7
METALS 115

Chapter 8
WOODS, PLASTICS, AND COMPOSITES 126

Chapter 9
THERMAL AND MOISTURE PROTECTION 153

Chapter 10
OPENINGS 172

Chapter 11
FINISHES 181

Chapter 12
FIRE SUPPRESSION 192

Chapter 13
PLUMBING 196

Chapter 14
HEATING, VENTILATION, AND AIR-CONDITIONING (HVAC) 201

Chapter 15
ELECTRICAL 209

Chapter 16
EARTHWORK 220

Chapter 17
EXTERIOR IMPROVEMENTS 236

Chapter 18
UTILITIES 242

PART III
PUTTING COSTS TO THE ESTIMATE 249

Chapter 19
MATERIAL PRICING 250

Chapter 20
LABOR PRODUCTIVITY AND HOURS 255

Chapter 21
LABOR RATES 264

Chapter 22
EQUIPMENT COSTS 276

Chapter 23
CREW RATES 283

Chapter 24
SUBCONTRACT PRICING 286

Chapter 25
MARKUPS 290

Chapter 26
PRICING EXTENSIONS 296

Chapter 27
AVOIDING ERRORS IN ESTIMATES 309

PART IV
FINALIZING THE BID 313

Chapter 28
SUBMITTING THE BID 314

Chapter 29
PROJECT BUYOUT 321

Chapter 30
THE ESTIMATE AS THE BASIS OF THE SCHEDULE 325

Chapter 31
ETHICS 328

PART V
ADVANCED ESTIMATING 331

Chapter 32
CONVERTING EXISTING FORMS 332

Chapter 33
CREATING NEW FORMS 352

Chapter 34
OTHER ESTIMATING METHODS 381

Appendix A
REVIEW OF ESTIMATING MATH 393

Appendix B
SAMPLE JOB COST CODES 406

Appendix C
SAMPLE LABOR PRODUCTIVITY RATES 411

Appendix D
SAMPLE EQUIPMENT COSTS 417

Appendix E
MODEL SCOPES OF WORK 419

Appendix F
GLOSSARY 425

Appendix G
INDEX OF DRAWING SETS 432

INDEX 433

CONTENTS

PART 1

INTRODUCTION TO ESTIMATING 1

Chapter 1

THE ART OF ESTIMATING 2

The Estimator 2

Types of Estimates 3

Bid Package 4

Estimating Tools 7

Computerized Estimating 8

Conclusion 9

Problems 9

References 10

Chapter 2

OVERVIEW OF THE ESTIMATING AND BIDDING PROCESS 11

Planning the Bid 11

Pre-Bid-Day Activities 12

Bid-Day Activities 15

Post-Bid Activities 17

Information Flow 18

Conclusion 20

Problems 20

Chapter 3

INTRODUCTION TO EXCEL 21

Conventions Used in This Text 21

Workbook Management 21

Working with Worksheets 25

Entering Data 30

Formatting Worksheets 35

Writing Formulas 50

Basic Functions 53

Printing 59

Testing Spreadsheets 65

Conclusion 68

Problems 69

PART II

THE QUANTITY TAKEOFF 71

Chapter 4

FUNDAMENTALS OF THE QUANTITY TAKEOFF 72

Performing a Quantity Takeoff 72

Work Packages 73

Communication with the Field 74

Counted Items 76

Linear Components 77

Sheet and Roll Goods 78

Volumetric Goods 80

Quantity-From-Quantity Goods 80

Waste 81

Building Information Modeling (BIM) 81

Conclusion 83

Problems 83

Reference 83

Chapter 5

CONCRETE 84

Forms 84

Reinforcing 85

Concrete 85

Spread Footings 86

Columns 87

Continuous Footings 90

Foundation Walls 95

Beams 97

Slab on Grade 97

Raised Slabs 99

Stairs 100

Sample Takeoff for the Residential Garage 101

Conclusion 104

Problems 104

Chapter 6

MASONRY 107

Block and Structural Brick Walls 108

Brick Veneer 111

Conclusion 112

Problems 112

Chapter 7

METALS 115

Types of Structural Steel 115

Common Shapes for Structural Steel 115

Beams, Girders, and Columns 116

Joists and Joist Girders 118

Metal Deck 120

Steel Trusses 120

Stairs and Handrail 121

Miscellaneous Steel 122

Conclusion 122

Problems 122

Chapter 8

WOODS, PLASTICS, AND COMPOSITES 126

Floor Systems 126

Walls 131

Roof Systems 137

Finish Carpentry 143

Cabinetry and Countertops 143

Sample Takeoff for the Residential Garage 143

Conclusion 145

Problems 145

Reference 152

Chapter 9

THERMAL AND MOISTURE PROTECTION 153

Waterproofing and Dampproofing 153

Building Paper and Vapor Barriers 154

Insulation 157

Exterior Insulation Finish System 159

Shingle Roofs 160

Siding, Soffit, and Fascia 162

Membrane Roofing 165

Sample Takeoff for the Residential Garage 165

Conclusion 168

Problems 168

References 171

Chapter 10

OPENINGS 172

Doors 172

Windows 173

Commercial Storefront 174

Glazing 175

Hardware 176

Sample Takeoff for the Residential Garage 176

Conclusion 177

Problems 177

Reference 180

Chapter 11

FINISHES 181

Metal Stud Partitions 181

Gypsum Board 182

Tile 184

Suspended Acoustical Ceilings 184

Wood and Laminate Floors 185

Sheet Vinyl 186

Vinyl Composition Tile 187

Rubber Base 187

Carpet and Pad 188

Paint 188

Sample Takeoff for the Residential Garage 188

Conclusion 189

Problems 189

Chapter 12

FIRE SUPPRESSION 192

Conclusion 194

Problems 194

Chapter 13

PLUMBING 196

Water Supply 196

Drain-Waste-and-Vent System 196

Fixtures and Equipment 199

Conclusion 199

Problems 199

Chapter 14

HEATING, VENTILATION, AND AIR-CONDITIONING (HVAC) 201

Residential HVAC Systems 201

Commercial HVAC Systems 204

Conclusion 206

Problems 206

Chapter 15

ELECTRICAL 209

Residential Wiring 209

Commercial Wiring 210

Sample Takeoff for the Residential Garage 213

Conclusion 216

Problems 216

Chapter 16

EARTHWORK 220

Characteristics of Soils 220

Swell and Shrinkage 220

Geometric Method 223

Average-Width-Length-Depth Method 224

Average-End Method 224

Modified-Average-End Method 225

Cross-Sectional Method 226

Comparison of Methods 231

Backfill 231

Soils Report 232

Sample Takeoff for the Residential Garage 232

Conclusion 233

Problems 233

Chapter 17

EXTERIOR IMPROVEMENTS 236

Asphalt and Base 236

Site Concrete 237

Landscaping 237

Sample Takeoff for the Residential Garage 238

Conclusion 239

Problems 239

Chapter 18

UTILITIES 242

Excavation 242

Bedding 245

Utility Lines 247

Backfill 247

Conclusion 247

Problems 247

PART III

PUTTING COSTS TO THE ESTIMATE 249

Chapter 19

MATERIAL PRICING 250

Shipping Costs 251

Sales Tax 252

Storage Costs 252

Escalation 252

Conclusion 254

Problems 254

Chapter 20

LABOR PRODUCTIVITY AND HOURS 255

Factors Affecting Labor Productivity 255

Historical Data 257

Field Observations 257

National Reference Books 260

Labor Hours 260

Conclusion 262

Problems 262

Reference 263

Chapter 21

LABOR RATES 264

Billable Hours 264

Wages 264

Fair Labor Standards Act 264

Davis-Bacon Act 265

Labor Contracts 265

State and Local Employment Laws 266

Cash Equivalents and Allowances 267

Payroll Taxes 268

Unemployment Insurance 268

Workers' Compensation Insurance 269

General Liability Insurance 270

Insurance Benefits 270

Retirement Contributions 270

Union Payments 271

Other Benefits 271

Annual Costs and Burden Markup 271

Conclusion 273

Problems 273

References 275

Chapter 22

EQUIPMENT COSTS 276

Depreciation and Interest 276

Taxes and Licensing 277

Insurance 277

Storage 277

Hourly Ownership Cost 277

Tires and Other Wear Items 278

Fuel 278

Lubricants and Filters 279

Repair Reserve 279

Leased Equipment 279

Rented Equipment 279

Conclusion 281

Problems 281

Chapter 23

CREW RATES 283

Conclusion 285

Problems 285

Chapter 24

SUBCONTRACT PRICING 286

Request for Quote 286

Writing a Scope of Work 287

Historical 288

Bid Selection 288

Conclusion 288

Problems 289

Chapter 25

MARKUPS 290

Building Permits 290

Payment and Performance Bonds 291

Profit and Overhead 292

Conclusion 294

Problems 294

Reference 295

Chapter 26

PRICING EXTENSIONS 296

Detail Worksheet 296

Material Costs 296

Labor Costs 297

Equipment Costs 297

Total Cost 298

Summary Worksheet 298

Sample Estimate: The Residential Garage 301

Conclusion 308

Problems 308

Chapter 27

AVOIDING ERRORS IN ESTIMATES 309

List Cost Codes 309

Spend More Time on Large Costs 309

Prepare Detailed Estimates 309

Mark Items Counted During the Quantity Takeoff 310

Double-Check All Takeoffs 310

Include Units in Calculations 310

Automate with Spreadsheets 310

Use Well-Tested and Checked Formulas 310

Double-Check All Calculations 311

Perform Calculations in Two Ways 311

Drop the Pennies 311

Have Someone Review the Estimate 311

Review Each Cost Code as a Percentage of the Total Cost 311

Check Unit Costs for Each Cost Code 311

Compare Costs to Those for Another Project 311

Allow Plenty of Time 312

Conclusion 312

Problems 312

PART IV

FINALIZING THE BID 313

Chapter 28

SUBMITTING THE BID 314

Bid Submission With Standardized Documents 314

Writing a Proposal 314

Writing a Business Letter 317

Letter Formats 318

Writing E-Mails 318

Conclusion 320

Problems 320

Chapter 29

PROJECT BUYOUT 321

Subcontracts 321

Purchase Orders 323

Contracts for Materials 324

Conclusion 324

Problems 324

Chapter 30

THE ESTIMATE AS THE BASIS OF THE SCHEDULE 325

Estimating Durations 326

Sample Durations: The Residential Garage 326

Conclusion 327

Problems 327

Chapter 31

ETHICS 328

Work Ethic 328

Bidding Practices 328

Loyalty to Employer 329

Ethical Dilemmas 329

Conclusion 329

Problems 329

Reference 330

PART V

ADVANCED ESTIMATING 331

Chapter 32

CONVERTING EXISTING FORMS 332

Creating the Layout 332

Adding Formulas 333

Automating with Macros 335

Testing the Worksheets 340

Adding Error Protection 341

Conclusion 351

Problems 351

Chapter 33

CREATING NEW FORMS 352

Planning New Forms 352

Setting Up the Spreadsheet 356

Series 358

Naming Cells 361

Adding Dropdown Boxes 362

Referencing Worksheets in a Formula 366

Concatenate 367

Lookup and Vlookup 371

Proposals 376

Conclusion 379

Problems 379

Chapter 34

OTHER ESTIMATING METHODS 381

Design Process 381

Delivery Methods 382

Project Comparison Method 382

Square-Foot Estimating 386

Assembly Estimating 389

And Beyond 391

Conclusion 391

Problems 391

Reference 392

Appendix A

REVIEW OF ESTIMATING MATH 393

Lengths 393

Scaling 394

Pythagorean Theorem 395

Areas 396

Volumes 402

Conversion Factors 404

Appendix B

SAMPLE JOB COST CODES 406

Appendix C

SAMPLE LABOR PRODUCTIVITY RATES 411

Appendix D

SAMPLE EQUIPMENT COSTS 417

Appendix E

MODEL SCOPES OF WORK 419

Footings and Foundations 419

Framing 419

Finish Carpentry 420

Drywall 420

Floor Coverings 421

Painting and Staining 421

Fire Sprinklers 422

Plumbing 422

HVAC 423

Electrical 423

Earthwork and Utilities—Roads and Parking Lots 424

Landscaping 424

Appendix F

GLOSSARY 425

Appendix G

INDEX OF DRAWING SETS 432

INDEX 433

PART ONE

INTRODUCTION TO ESTIMATING

Chapter 1 The Art of Estimating
Chapter 2 Overview of the Estimating and Bidding Process
Chapter 3 Introduction to Excel

In this section, you will be introduced to the art of estimating, the estimating process, and Microsoft Excel 2016. These chapters will prepare you for the rest of the book. For those of you who struggle with math or just want to review the mathematical principles used in estimating, a review of estimating math is given in Appendix A.

CHAPTER ONE

THE ART OF ESTIMATING

In this chapter you will be introduced to the art of estimating. You will learn what it takes to become a good estimator, the components of a bid package, and the tools used by an estimator. Finally, you will be introduced to computerized estimating.

Estimating is the process of determining the expected quantities and costs of the materials, labor, and equipment for a construction project. The goal of the estimating process is to project, as accurately as possible, the estimated costs for a construction project, as well as the required amount of materials, labor, and equipment necessary to complete the work. The American Institute of Architects defines the work as "the construction and services required by the Contract Documents . . . and includes all other labor, materials, equipment and services provided . . . by the Contractor to fulfill the Contractor's obligations."[1] The work is often referred to as the *scope of work*.

Estimating plays a key role in the operation of construction companies. Accurate estimates are needed for a company to be successful in the bidding process while maintaining a reasonable profit margin. If the estimates are too high, the company may starve to death because of the lack of work. If the estimates are too low, the company may lose money and go bankrupt. The estimator is constantly walking a fine line between bidding too low and too high.

THE ESTIMATOR

The estimator is the person responsible for preparing the cost estimates. Large companies may employ an estimating department with one or more full-time estimators. The estimating department is often charged with preparing all of the company's estimates. In smaller companies, the project managers or the company's owner may be responsible for preparing the estimates. Regardless of their job title, employees who are responsible for preparing estimates are estimators. The estimator is responsible for seeing that the estimate accurately determines the quantity of materials, labor, and equipment and incorporates the costs to complete the required scope of work along with a reasonable profit while remaining competitive with other firms in the market. Great skill is required to balance the need to be competitive with the need to be profitable.

To be a good estimator a person must possess the following skills:

- An estimator must have a sound understanding of the construction methods, materials, and the capacities of skilled labor. Because of the great variety of work, it is impossible for an estimator to be versed in all forms of construction. Therefore, an estimator must specialize in one or more areas of construction. An estimator may specialize in a subcontractor trade such as electrical, mechanical, or excavation. An estimator may also specialize in an area of construction such as residential, tenant finish, or highway construction. Because the methods of construction, the preferred materials of construction, and the skill of labor vary from market to market, estimators must also specialize in a specific market area, such as a state or region.

- An estimator must possess the basic skills needed to determine the quantities of materials, labor, and equipment necessary to complete a project. This requires the estimator to read blueprints, understand the design that the architect or engineer has specified, and determine the quantities needed to complete the project. Because much of an estimator's time is spent working with quantities, estimators must have strong mathematical skills. A review of estimating math is given in Appendix A.

- An estimator must be a good communicator, both verbally and in written form. Part of an estimator's job is to obtain pricing from vendors and subcontractors. To do this, the estimator must convince vendors and subcontractors to bid on projects, communicate what pricing is needed, and—when the company has won the job—communicate the responsibilities of the vendors and subcontractors in the form of purchase orders and subcontracts provided to the vendor, subcontractor, superintendent, and so forth. Estimators may also be required to present estimates to owners—which requires good presentation skills—or prepare proposals—which requires good writing skills.
- An estimator must possess strong computer skills. Much of today's estimating is performed using computer software packages, such as Excel, takeoff packages such as On-Screen Takeoff, and estimating software packages. Estimators also need to be able to prepare contracts, proposals, and other documents using a word processing program.
- An estimator must be detail-oriented. Estimators must carefully and accurately determine the costs and quantities needed to complete the project. Simple mistakes—such as forgetting that the drawings are half-scale or not reading the specifications carefully enough to realize that an unusual concrete mix has been specified—can quickly turn a successful project into an unprofitable job and, in extreme cases, bankrupt a company.
- An estimator must have the confidence to quickly prepare takeoffs and make decisions under pressure. Bid days are hectic. Many vendors and subcontractors wait until the last hour to submit their bids. As the time of the bid closing approaches, the estimator must compare and incorporate new pricing as it is received and fill in missing pricing with limited time and information at his or her disposal. Making a bad decision or failing to make a decision under pressure can cause the company to lose the bid or take an unprofitable job.
- Finally, an estimator must have a desire for constant improvement. One way to do this is to get involved with a professional origination such as the American Society of Professional Estimators (ASPE). Other ways include studying other estimating books and attending seminars. Much of a company's success or failure rides on the abilities of the estimator to obtain profitable work.

Because of the high degree of uncertainty in estimating costs, estimating is more of an art than a science. As with any art, only by practicing can one become a good estimator. One would not expect to become a good pianist after a few lessons. It takes practice. Likewise, becoming a good estimator takes practice. But practice is not enough. Practicing bad estimating skills will only turn a person into a bad estimator. Truly, practice makes permanent.

To be a good estimator, a person must practice using good estimating skills; and for a person to practice good estimating skills, he or she must have a sound understanding of the fundamental principles of estimating. This understanding can only be obtained by studying the art of estimating.

If you want to become a good estimator, it is very important to study and practice estimating. The following is a list of things that you can do as you read this book to become a good estimator:

- Carefully read each chapter. The chapters provide you with explanations of the basic principles of estimating.
- As you read each example problem, check the math with a calculator. This will help you to gain a greater understanding of the estimating principles.
- Many of the quantity takeoff problems can be solved using the five methods (counted items, linear components, sheet and roll goods, volumetric goods, and quantity-from-quantity goods) found in Chapter 4. Be sure that you completely understand each of these methods. As you read each chapter, keep a list of which of the five methods to use for each problem type and how each problem type differs from the general methods covered in Chapter 4.
- When reading the example problems that are based upon the garage drawings that accompany this book, refer to these drawings and see if you can get a similar quantity of materials. Minor differences will occur due to difference in rounding.
- Complete all of the computer exercises and sidebars. Take existing sidebars and customize them to your estimating situation. This will help you to become proficient with Excel.
- Work all of the problems at the end of the chapters. Have your instructor or a fellow estimator look over your solutions. These problems have been provided to give you practice in solving simple estimating problems.
- Prepare complete estimates for the Johnson Residence and West Street Video projects listed in Appendix G. Have your instructor or a fellow estimator look over your estimate. (The loose project drawings are provided in a separate package shrink-wrapped to this text.)

In addition to these things, practice estimating whenever you get the chance. Volunteer to help an estimator with his or her estimate. Ask questions of fellow estimators. Remember, learning to estimate takes time.

TYPES OF ESTIMATES

There are three common types of estimates. They are the conceptual estimate, preliminary estimate, and final or detailed estimate.

The conceptual estimate is an estimate prepared while the project is still in a conceptual state. The conceptual estimate is used to study the feasibility of a project or to compare two potential design alternatives (for example, a concrete structure versus a steel structure or three stories versus four stories). These estimates are based on a description of the project or on very limited drawings and as such are the least accurate type of estimate.

The preliminary estimate is an estimate prepared from a partially completed set of drawings. A preliminary estimate is often performed when the drawings are 35% to 50% complete and is used to check to see if the proposed design is on budget and to identify changes to the design that need to be made to meet the budget. Preliminary estimates may be performed any time before the bid. Preliminary estimates are more accurate than conceptual estimates because more information about the design is available.

Chapter 34 covers three estimating methods—project comparison, square-foot, and assembly estimating—that may be used to prepare conceptual and preliminary estimates.

Final or detailed estimates are used to prepare bids and change orders, order materials, and establish budgets for construction projects. They are prepared from a complete or nearly completed set of drawings and are the most accurate type of estimate.

Sometimes shortcut methods (methods which produce a close, but less accurate answer) are used to prepare bids in order to bid more projects in less time. This may save time during the bidding process, but can cause cost-control problems for the company. When shortcut methods are used to order materials, the wrong materials or wrong quantities are often delivered to the site, which results in delays and increased construction costs. When a shortcut method produces a quantity within a few percent of the correct answer, it is hard to determine if a quantity overrun is due to poor material use in the field or inaccuracy in the estimating process. Similarly, when a quantity underrun occurs, it is unclear if it is due to good material use in the field or inaccuracy in the estimating process. For cost-control purposes, it is important to have an accurate quantity and cost estimate for all items, including an estimate for both unavoidable and avoidable waste. Unavoidable waste is waste that is a result of the difference between design dimensions and the size of materials. For example, when carpeting a nine-foot by nine-foot room with carpet that comes in 12-foot-wide rolls, you will have a three-foot-wide strip of waste unless it can be used elsewhere. Avoidable waste is waste that can be avoided by good use of materials in the field. It is not uncommon to have one framing crew use 5% more materials to frame the same house as another crew. The extra materials used are avoidable waste. Avoidable waste also includes materials that are damaged or destroyed on the job.

The primary focus of this book is on preparing accurate final or detail estimates, although many of these principles apply to conceptual and preliminary estimates.

BID PACKAGE

Estimates are prepared from bid packages. The bid package defines the scope of work for the construction project. A well-developed bid package includes a set of plans and a project manual.

The plans graphically show building dimensions and where different materials are used. The process of converting the building dimensions and details into estimated quantities is known as the quantity takeoff. The estimator will prepare the bid from the quantity takeoff. Typically, the plans are organized as follows: civil plans, architectural plans, structural plans, mechanical and plumbing plans, and electrical plans. The project manual provides a lot of information for the bidder. The typical project manual includes the following items:

Invitation to Bid: The purpose of the invitation to bid is to invite bidders to bid on the project and give a bidder enough information to decide whether he or she wants to bid on the project. Often, public agencies are required to invite all qualified bidders to bid on the work. This is typically done by posting the invitation to bid in a few predetermined locations (for example, the local library and the public agency's office) and by printing it in the local newspaper. The invitation to bid is also placed at the front of the project manual. The invitation to bid provides a project description, contact information for the owner and design professional (architect or engineer), bid date, restrictions on bidders, expected price range, and the expected duration of the project.

Bid Instructions: Bid instructions are instructions that must be followed to prepare a responsive bid. The goal of the bid instructions is to help the bidders provide a complete bid with all the necessary documents (for example, bid bond and schedule of values). Bidders may be disqualified for not following these instructions, which may result in a bidder losing the job even though he or she was the lowest bidder.

Bid Documents: Public agencies and owners typically use standard forms for the submission of the bid. These forms may include bid forms, bid-bond forms, a schedule of values, and contractor certifications. A sample bid form is shown in Figure 1-1. The bid bond is discussed in the next paragraph. The schedule of values breaks the bid into smaller portions and is used to determine and evaluate the amount of the progress payments. A sample schedule of values is shown in Figure 1-2. Estimators should read these documents carefully and comply fully with their requirements. Bidders can be disqualified for not submitting all of the required bid documents.

Bonds: Bonds include the bid bond, the payment bond, and the performance bond. Bonds are issued

---

Bid Form

Owner:
West Street Video
John M. Smith, President
P.O. Box 1256
Ogden, Utah 84403

Project:
West Street Video
4755 S. West Street
Ogden, Utah 84403

Dear Sirs:

Having carefully examined the bid documents including the plans, specifications, and other related documents; visited the proposed site of the work; and being familiar with other conditions surrounding construction of the proposed project including the availability of material and labor, the undersigned proposes to furnish all labor, material, equipment, supplies, tools, transportation, services, licenses, fees, permits, sales tax, and so forth required by the bid documents for the sum of

$_____dollars ($_____).
The undersigned also agrees to complete the work in 150 calendar days.

We acknowledge the following addenda: _____

Enclosed is a _____ (bond or check), as required, in the sum of 5% of the bid.

This bid shall remain good for 60 days after the bid opening.

Respectfully Submitted:

SEAL (if a Corporation)

Company _____

By _____

Address _____

_____

License No. _____

Date _____

FIGURE 1-1 Bid Form

by sureties. Bid bonds are provided by the contractor at the time of the bid. The bid bond guarantees that the contractor—should he or she be the lowest bidder—will sign the contract and provide the payment and performance bonds. The payment and performance bonds are provided when the contract for the work is signed. The payment bond guarantees that the vendors, subcontractors, and labor will be paid for the work they perform on the project. In the event that the vendors, subcontractors, and labor are not paid on the project, the surety will step in and make the necessary payments. The performance bond guarantees that the contractor will complete the construction project. In the event that the contractor fails to complete the project, the surety will step in and complete the project. Providing payment and performance bonds can increase the costs of the project by 1% to 2%.

Contract: Public agencies and owners often use standard contracts. Estimators should read the contract carefully and include any costs associated with meeting the terms of the contract in their bid.

General Conditions: Public agencies and owners often use standard general conditions. General conditions affect the cost to complete the work. The general conditions identify the relationships among the owner, design professionals, and the contractor and addresses provisions that are common to the entire project; for example, how changes in the scope of work are to be processed or the need for cleanup. Estimators should read the general conditions carefully and include any costs associated with the general conditions in their bid.

Special Conditions: Special conditions are additional conditions that apply to this specific project. Like the general conditions, the special conditions affect the cost to complete the work. Estimators should read the special conditions carefully and include any costs associated with the special conditions in their bid.

Schedule of Values

General Requirements	$ _____
Grading and Excavation	$ _____
Utilities	$ _____
Asphalt	$ _____
Site Concrete	$ _____
Landscaping	$ _____
Concrete	$ _____
Masonry	$ _____
Rough Carpentry	$ _____
Finish Carpentry	$ _____
Cabinetry and Countertops	$ _____
Insulation	$ _____
Roofing	$ _____
Doors and Windows	$ _____
Finishes	$ _____
Plumbing	$ _____
Fire Sprinklers	$ _____
HVAC	$ _____
Electrical	$ _____
Total	$ _____

FIGURE 1-2 Schedule of Values

Technical Specifications: Technical specifications identify the quality of materials, installation procedures, and workmanship to be used on the project. The specifications also specify the submittal and testing requirements for individual building components. Estimators must read these carefully and understand their implications. Bidding the wrong specification can lead to losing the bid unnecessarily or winning a bid that you do not have sufficient funds to complete. The specifications are typically organized according to the MasterFormat.[2] The 2004 edition expanded the MasterFormat from 16 divisions to 50 divisions (00 to 49).[3] The 2016 MasterFormat is organized as follows:

Division 00: Procurement and Contracting Requirements
Division 01: General Requirements
Division 02: Existing Conditions
Division 03: Concrete
Division 04: Masonry
Division 05: Metals
Division 06: Wood, Plastics, and Composites
Division 07: Thermal and Moisture Protection
Division 08: Openings
Division 09: Finishes
Division 10: Specialties
Division 11: Equipment
Division 12: Furnishings
Division 13: Special Construction
Division 14: Conveying Equipment
Division 21: Fire Suppression
Division 22: Plumbing
Division 23: Heating, Ventilating, and Air Conditioning (HVAC)
Division 25: Integrated Automation
Division 26: Electrical
Division 27: Communications
Division 28: Electronic Safety and Security
Division 31: Earthwork
Division 32: Exterior Improvements
Division 33: Utilities
Division 34: Transportation
Division 35: Waterway and Marine Construction
Division 40: Process Interconnections
Division 41: Material Processing and Handling Equipment

Division 42: Process Heating, Cooling, and Drying Equipment

Division 43: Process Gas and Liquid Handling, Purification, and Storage Equipment

Division 44: Pollution and Waste Control Equipment

Division 45: Industry-Specific Manufacturing Equipment

Division 46: Water and Waste Equipment

Division 48: Electrical Power Generation

Other Inclusions: Other documents such as a soils report and environmental inspections may be included in the project manual. The soils report describes the soil conditions and the water table at the construction site, which greatly affect the excavation costs. Estimators should take this information into account when preparing an estimate for a project.

Included by Reference: Other documents may be included in the specifications by reference. Documents included by reference are not physically attached to the project manual but are treated as if they were. Documents that are often included by reference include standard agency specifications and other readily available standards (for example, American Society for Testing and Materials [ASTM] and American National Standards Institute [ANSI] standards). Estimators need to understand the implication of these standards because they affect how the work is performed or are used to determine compliance with a specification.

ESTIMATING TOOLS

The estimator has a number of tools at his or her disposal to help prepare the estimate. They range from simple paper forms to powerful estimating computer programs. An estimator's tools include the following:

Architect's and Engineer's Scales: Plans are prepared at a reduced scale. Architect's and engineer's scales allow the estimator to measure dimensions from the plans. The architect's scale typically includes the following scales: $1'' = 1'$, $1/2'' = 1'$, $1/4'' = 1'$, $1/8'' = 1'$, $3/16'' = 1'$, $3/32'' = 1'$, $3/4'' = 1'$, $3/8'' = 1'$, $1 1/2'' = 1'$, and $3'' = 1'$. The engineer's scale typically includes the following scales: 1:10, 1:20, 1:30, 1:40, 1:50, and 1:60. Architect's and engineer's scales are shown in Figure 1-3.

Plan Measurer: A plan measurer or wheel measures the length of a line by rolling the wheel of the measurer along the item to be measured. A plan measurer has the advantage of being able to follow curved lines more easily than an architect's or engineer's scale. Analogue plan measurers

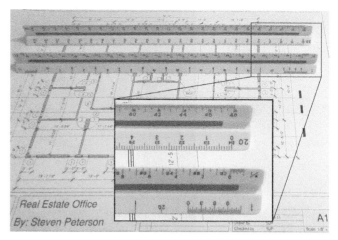

FIGURE 1-3 Architect's Scale (bottom) and Engineer's Scale (top)

have fixed scales, such as $1/2'' = 1'$, $1/4'' = 1'$, and $1/8'' = 1'$; digital plan measurers have more scales available and often include the ability to create custom scales. A digital plan measurer is shown in Figure 1-4.

Digitizer: A digitizer consists of an electronic mat and stylus or puck. The digitizer can measure distances and areas. The digitizer may operate as a stand-alone tool or be connected to a software package. Digitizers have an unlimited number of scales and can perform mathematical functions on distances and areas. A digitizer is shown in Figure 1-5.

Takeoff Packages: The digitizer is being replaced by takeoff packages, such as On Center Software's On-Screen Takeoff and Trimble's Paydirt that allows the user to determine the quantity of material needed from a digital set of drawings. These software packages make determining the length, perimeter, area, and volume quick and easy. On-Screen Takeoff will even allow the user to determine the length of the perimeter, number of tiles, and length of grid needed for a dropped acoustical ceiling by having the estimator trace the

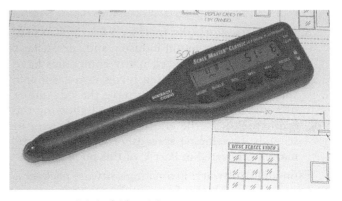

FIGURE 1-4 Digital Plan Measurer

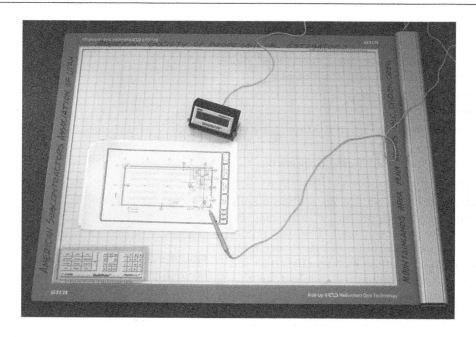

FIGURE 1-5 Digitizer

perimeter of the ceiling. These software packages can be integrated with spreadsheet or other estimating packages.

Building Information Modeling Software: An up and coming trend in the architectural, engineering, and construction (AEC) community is to design buildings using building information modeling (BIM) software, which allows the building to be designed in 3D. Many BIM software packages, such as Autodesk's Revit®, allow estimators to extract quantities from the model.

iPad and Plan Tables: Often the plans and project manuals are available in a digital format, such as pdf file. iPads and larger format plan tables are becoming common tools for the estimator.

Calculator and Paper Forms: In the days before computers, most estimates were prepared by using paper estimating forms and calculators. Some estimates are still prepared in this way, particularly estimates that are prepared in the field; however, with iPads, laptop computers, and portable printers, even paper estimates in the field are being replaced by computerized estimates.

Spreadsheets: With the increased use of computers in the construction industry and spreadsheet programs such as Excel, many companies have converted their estimating forms into computer spreadsheets. Spreadsheets have the advantage of automating the calculations and allowing the estimator to see the effects of minor changes in seconds. Developing spreadsheets for use in estimating is covered in Part V of this book.

Estimating Software: Estimating software packages, such as Trimble WinEst, combines the advantages of a computer spreadsheet with a database. This automates the estimating even further. Estimating software has the additional advantage of being capable of taking off assemblies as a single item. For example, an estimator can take off a wall assembly that includes the top and bottom plates, the studs, insulation inside the wall, and the finish material on both sides of the wall as a single component. Estimating software also allows the estimator to print the bid information in different formats for different uses with little or no setup.

COMPUTERIZED ESTIMATING

Nothing has revolutionized estimating as much as the advent of computers along with spreadsheets, estimating software, and takeoff packages. If used correctly, computers can reduce the time needed to prepare an estimate and decrease the errors in the estimate. If used improperly, they can increase the number of errors in an estimate and decrease the usefulness of the estimate. There are two dangerous mistakes estimators make when using estimating packages.

The first mistake is to turn the thinking over to the computer so the estimator becomes simply a means of entering data into the software package. This happens when the estimator determines the quantity of a given component and enters it into the computer without giving any thought to job conditions and design requirements that may require this component to be handled in a different way than it is usually handled. For example, a hollow-metal door frame for a pair of doors that must be delivered to the tenth floor of an office building up a small elevator must be constructed in pieces and fabricated on site, whereas most door frames are constructed as a single member. Computers are good for performing repetitive tasks quickly without error. They are good for handling the mindless and boring tasks such as totaling a column for the umpteenth time. Estimators can make a minor change to an estimate and the computer will calculate the changes to the estimate almost instantaneously.

Another danger of just entering data into the computer is that computer spreadsheets and formulas in estimating packages are developed for a limited number of circumstances based on a set of assumptions made by the writer of the spreadsheet or formula. Whenever the estimator uses the spreadsheet or formula on a situation that is outside the conditions anticipated by the developer, the estimator may get an inaccurate estimate from the spreadsheet or formula. To protect against this, the estimator must have an understanding of the limits and design of the spreadsheet or formula and must make sure the spreadsheet's or formula's response is reasonable.

The second mistake is to create a new spreadsheet or formula and use it without properly testing it. After creating a spreadsheet or formula, the developer should test the spreadsheet or formula to make sure it is working properly, not only on the situation that it was developed for, but on other conceivable situations. In addition, the developer should try to make the spreadsheet or formula mess up and then build in ways to prevent other users from making the same mistakes by building in error-checking procedures.

Excel is the most popular estimating package. In a survey performed by the American Society of Professional Estimators, 29% of the respondents reported using Excel as their estimating software, 22% reported using Timberline (now Sage Estimating), and 25% reported using another estimating software package such as WinEst (by Trimble) and HeavyBid (Heavy Construction System Specialists, Inc.).[4] Users of estimating packages commonly augment the software package with Excel worksheets to assist in the quantity takeoff and other support functions.

Spreadsheets have the advantage of being inexpensive. Spreadsheet software can be purchased for about $100 and is often included along with other standard applications—such as word processing—that are sold as a software package for use in offices. In addition to being inexpensive, spreadsheets are easily adapted to the existing style and estimating procedures of the company. A company that uses paper forms can easily create a look-alike form in the spreadsheet and let the software perform the mundane and tedious calculations. Finally, spreadsheets are easy to create. With a little training and effort, anyone can develop spreadsheets for estimating. Developing spreadsheets for estimating in Excel is covered in Part V of this book.

Estimating software packages are powerful computer software applications that have been developed specifically for estimating. There are a number of packages available, with some packages having been designed for building construction and others for heavy and highway construction that involves large amounts of earthwork. Estimating software packages have the advantage of automating the takeoff process and decreasing the time it takes to prepare an estimate by combining a spreadsheet with a database. The database contains a list of standard items along with their cost, labor productivity, labor rates, equipment costs, and formulas used to calculate the quantities and costs of the individual items. Estimating packages often allow the user to create assemblies—a group of items that are needed to create a component such as a wall—and take off the assembly in a single step. Another feature of estimating packages is that the data can be easily manipulated and printed in different formats.

For example, one can print the costs by line item with the sales tax appearing at the bottom of the report or one can have the sales tax allocated to the individual line items. All these features come at a price. Estimating software packages are more expensive than spreadsheets—often costing thousands of dollars. In addition to the dollar costs, they require a large time investment to set up and maintain the database. The pricing in the database must be kept current with market pricing, materials must be set up, formulas must be written and tested, and assemblies must be created. Companies must perform large amounts of estimating to justify the cost and time commitment involved in using an estimating software package.

CONCLUSION

Successful estimates are the lifeblood of a construction company. Winning bids while maintaining a good profit margin is necessary for a construction company to succeed. Accurate quantities and costs are needed for strong cost controls. To be a good estimator, a person must study and practice sound principles of estimating. Computers have greatly changed the way estimators prepare estimates. Computers and software packages, if used properly, can increase the productivity of the estimator while decreasing errors. Estimating software includes spreadsheet packages such as Excel, estimating software such as WinEst and takeoff packages such as On-Screen Takeoff.

PROBLEMS

1. Define estimating.
2. Define work or scope of work.

3. What role does estimating play in the success of a construction company?
4. What is the role of an estimator?
5. What skills are required to be a good estimator?
6. Why is practice important for an estimator?
7. What role does the bid package play in estimating?
8. What tools are available to estimators?
9. What are the advantages and disadvantages of using spreadsheets for estimating?
10. What are the advantages and disadvantages of using estimating packages, such as Timberline, for estimating?

REFERENCES

1. *General Conditions of the Contract for Construction*, American Institute of Architects, AIA Document A201-1997, p. 9.
2. For more information on the MasterFormat™ see http://www.csinet.org.
3. Divisions 15 to 20, 24, 29, 30, 36 to 39, 46, 47, and 49 are reserved for future use.
4. American Society of Professional Estimators, 2003 ASPE Cost Estimating Software Survey, available at http://www.aspenational.com/cgi-bin/coranto/viewnews.cgi?id=EpZkZkupuyHEPGialt&tmpl-fullstory&style=fullstory; accessed February 4, 2004.

CHAPTER TWO

OVERVIEW OF THE ESTIMATING AND BIDDING PROCESS

In this chapter you will be introduced to the estimating and bidding process and gain an understanding of the steps needed to complete an estimate. The purpose of this chapter is to give you an overview of the process so that as you read the subsequent chapters you will have a better idea of how they fit into the estimating and bidding process.

Once the bid package is complete, a detailed estimate is needed to establish a budget for the project and, in the case of competitive bidding, selection of the contractor. This chapter looks at the process of preparing a detailed estimate for a competitive bid project. Preparing a preliminary estimate during the design process or preparing an estimate for a change order follows a similar process. Preparing an estimate for a competitive bid project can be divided into four phases: planning the bid, pre–bid-day activities, bid-day activities, and post-bid activities. An overview of these steps is shown in Figure 2-1.

PLANNING THE BID

Planning the bid consists of identifying the scope of work for the project, preparing a schedule for the bidding process, and assigning tasks to members of the estimating team.

Scope of Work: During the planning stage the estimator needs to identify the general scope of work covered by the bid package and prepare a work breakdown structure (WBS) for the estimate. A WBS is a systematic way of dividing the project's scope of work into smaller, more manageable scopes of work aligned with the specific trades that will be preforming the work. Often, this is done by reviewing the bid package and identifying which of the company's cost codes are needed for the project and what pricing needs to be obtained to complete the estimate. Many companies have standard summary sheets, like the one shown in Figure 2-2, where the estimator marks the cost codes in the left column that he or she needs pricing for to complete the estimate. This pricing may come from suppliers or subcontractors or may be estimated in-house. In addition to identifying the cost codes, the estimator needs to identify any items that may be difficult to find a price for (for example, an indoor pistol range and other specialty equipment) and any special items (such as a bid bond, submission of a prequalification package, or attendance at a pre-bid meeting) that are required of the bidders. Appendix B contains a list of items that are typically found in the cost codes shown in Figure 2-2.

Scheduling the Bid: Once the scope of the project has been determined, the estimator must schedule the tasks necessary to complete the bid such that the bid is completed in time to be submitted. If there is insufficient time to complete the bid, the estimator must get additional help, eliminate some of the estimating steps, or decide not to bid on the project.

Assigning Tasks: If the estimating team consists of more than one person, once the scope of work has been identified and a schedule prepared for the bid package, the estimator must delegate the bidding tasks to other members of the team and communicate the deadlines for the completion of these tasks to the team members. For example, for a bid that requires the use of in-house crews and requires a schedule to be submitted as part of the bid, the estimator may assign the estimating

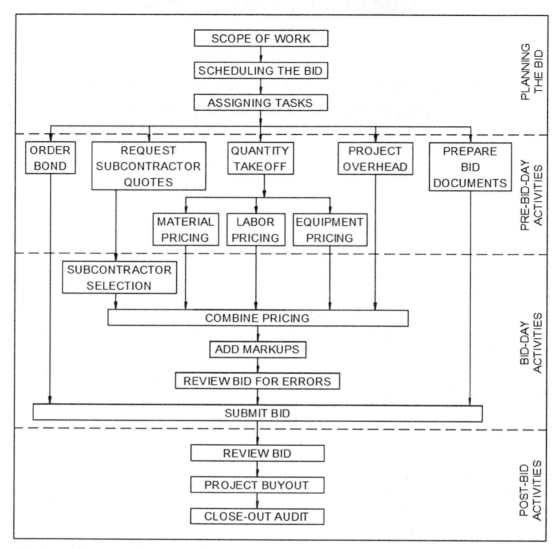

FIGURE 2-1 The Estimating Process

of the in-house work to the person responsible for the in-house crews and the preparation of the schedule to the company's scheduler.

PRE-BID-DAY ACTIVITIES

The pre–bid-day activities are activities or tasks that should be completed before the day of the bid. The activities include ordering the bid bond, requesting quotes from subcontractors, completing the quantity takeoff, obtaining material pricing from vendors, pricing in-house labor, pricing equipment, determining project overhead, and preparing the bid documents.

Order Bond: One of the first activities that should be done is to order a bid bond from the surety company. This gives the surety time to prepare the bid bond, obtain the necessary signatures on the bond, and mail it to the contractor. By ordering the bond early the contractor eliminates the need to make a special trip to the surety's office to pick up the bond. The estimator also finds out early in the bidding process if there will be any problem getting the bond before spending a lot of time preparing the estimate. This is important if the engineer's or architect's estimate is near the contractor's bonding limits.

Request Subcontractor Quotes: Another step that must be completed early in the bid process is identifying which subcontractors need to be invited to bid on the project and requesting quotes from these subcontractors. Starting this process early is important because subcontractors need to find the time to look at the plans and project manual and prepare their estimates. When dealing with a limited number of plan sets it can be quite a challenge to give all of the subcontractors a chance to look at the plans within the allowed time for the bid. When dealing with a limited number of plan sets, the goal of the estimator should be to get one

Code	Description	Materials	Labor	Equipment	Subcontract	Total
01-000	**GENERAL REQUIREMENTS**					
01-300	Supervision					
01-500	Temporary Utilities					
01-510	Temporary Phone					
01-520	Temporary Facilities					
01-700	Clean-Up					
02-000	**EXISTING CONDITIONS**					
02-400	Demolition					
03-000	**CONCRETE**					
03-200	Rebar					
03-300	Footing and Foundation—Labor					
03-310	Footing and Foundation—Concrete					
03-320	Slab/Floor—Labor					
03-330	Slab/Floor—Concrete					
03-340	Concrete Pump					
03-400	Precast Concrete					
03-500	Light-weight Concrete					
04-000	**MASONRY**					
04-200	Masonry					
05-000	**METALS**					
05-100	Structural Steel					
05-200	Joist and Deck					
05-500	Metal Fabrications					
05-900	Erection					
06-000	**WOOD, PLASTICS, AND COMPOSITES**					
06-100	Rough Carpentry					
06-110	Lumber					
06-120	Trusses					
06-200	Finish Carpentry					
06-210	Wood Trim					
06-400	Cabinetry and Counter Tops					
06-410	Counter Tops					
07-000	**THERMAL AND MOISTURE PROTECTION**					
07-100	Waterproofing					
07-200	Insulation					
07-210	Rigid Insulation					
07-220	Stucco					
07-400	Siding					
07-500	Roofing					
07-600	Sheet Metal					
07-700	Roof Specialties					
07-710	Rain Gutters					
07-800	Fireproofing					
07-900	Caulking and Sealants					
08-000	**OPENINGS**					
08-100	Metal Doors and Frames					
08-110	Wood Doors					
08-300	Overhead Doors					
08-400	Store Fronts					
08-500	Windows					
08-700	Hardware					
08-800	Glass and Glazing					

FIGURE 2-2a The Summary Worksheet

subcontractor to bid in each category of work before getting second and third bids. By doing this the contractor has a bid for each category of work rather than having multiple bids in one category and no bid in another category. In addition, the contractor should try to get bids on the high-dollar items first because these items carry the most risk should the contactor have to bid these items himself or herself. Preparing requests for subcontractor quotes is discussed in Chapter 24.

Quantity Takeoff: After lining up subcontractors and arranging for the bid bond, the next most important task is to prepare a quantity takeoff for the work to be performed in-house. The quantity takeoff is necessary to prepare material, labor, and equipment costs (unless the vendors are doing the quantity takeoff for their work) for the estimate. If the vendors are preparing quantity takeoffs for their work, the estimator should request a quote from these vendors at the same time he or she requests quotes from subcontractors. Often vendors who prepare a quantity takeoff for contractors do not guarantee that their quantities are accurate, and the estimator should prepare his or her own quantity takeoff to make sure there are sufficient quantities of materials to complete the work. If time allows, the estimator should prepare quantity takeoffs for subcontractor work where there is a concern that the subcontractor's bid may not be received in time to complete the bid. The quantity takeoff is discussed in Part II (Chapters 4 through 18) of this title.

Code	Description	Materials	Labor	Equipment	Subcontract	Total
09-000	**FINISHES**					
09-200	Drywall					
09-210	Metal Studs					
09-300	Ceramic Tile					
09-500	Acoustical Ceilings					
09-600	Flooring					
09-700	Wall Coverings					
09-900	Paint					
10-000	**SPECIALTIES**					
10-100	Signage					
10-200	Toilet Partitions					
10-210	Toilet and Bath Accessories					
10-400	Fire Extinguishers and Cabinets					
11-000	**EQUIPMENT**					
11-300	Appliances					
12-000	**FURNISHINGS**					
12-200	Window Treatments					
14-000	**CONVEYING EQUIPMENT**					
14-200	Elevators					
21-000	**FIRE SUPPRESSION**					
21-100	Fire Sprinklers					
22-000	**PLUMBING**					
22-100	Plumbing					
23-000	**HVAC**					
23-100	HVAC					
26-000	**ELECTRICAL**					
26-100	Electrical					
27-000	**COMMUNICATIONS**					
27-100	Communications					
31-000	**EARTHWORK**					
31-100	Clearing and Grubbing					
31-200	Grading and Excavation					
32-000	**EXTERIOR IMPROVEMENTS**					
32-100	Asphalt					
32-110	Site Concrete—Labor					
32-120	Site Concrete—Concrete					
32-130	Rebar					
32-300	Fencing					
32-310	Retaining Walls					
32-320	Dumpster Enclosures					
32-330	Signage					
32-340	Outside Lighting					
32-900	Landscaping					
33-000	**UTILITIES**					
33-100	Water Line					
33-300	Sanitary Sewer					
33-400	Storm Drain					
33-500	Gas Lines					
33-700	Power Lines					
33-800	Telephone Lines					
	SUBTOTAL					
	Building Permit					
	Bond					
	SUBTOTAL					
	Profit and Overhead Markup (%)					
	Profit and Overhead					
	TOTAL					

FIGURE 2-2b The Summary Worksheet

Materials Pricing: Estimators often send their quantity takeoffs to vendors and request that the vendors provide pricing for the materials identified in the quantity takeoff. This requires the quantity takeoff to be completed in time to send the list of needed materials to the vendors, have the vendors price the materials, receive a quote from the vendors, and incorporate the pricing into the estimate. Materials pricing is discussed in Chapter 19.

Labor Pricing: Once the quantity takeoff has been completed, the estimator can determine which crews are to be used on the in-house work, the associated productivity rates for the work, hourly crew costs, and the cost of the labor to perform the in-house work. Labor productivity is discussed in Chapter 20, determining labor rates is discussed in Chapter 21, and determining labor cost for a crew made up of different worker classes is discussed in Chapter 23.

Equipment Pricing: Once the quantity takeoff has been completed, the estimator can determine the equipment needs of the project and determine the

OVERHEAD CHECKLIST

Project Supervision:
- ☐ Project Manager
- ☐ Assistant Project Manager
- ☐ Superintendent
- ☐ Project Engineer
- ☐ Project Estimator
- ☐ Project Accountant
- ☐ Project Secretary
- ☐ Detailer/Draftsman

General Labor:
- ☐ Labor Foreman
- ☐ Laborer

Engineering:
- ☐ Engineering
- ☐ Surveying

Testing and Inspections:
- ☐ Soils Report
- ☐ Soil Testing and Inspections
- ☐ Concrete Testing and Inspections
- ☐ Rebar Inspections
- ☐ Masonry Testing and Inspections
- ☐ Welding Testing and Inspections
- ☐ Other Testing and Inspections

Safety:
- ☐ First Aid Kits
- ☐ Hard Hats
- ☐ Safety Equipment
- ☐ Safety Rails, Toe Boards, etc.
- ☐ Other Safety Supplies

Temporary Offices and Facilities:
- ☐ Office Trailer
- ☐ Mobilize/Demobilize Trailer
- ☐ Trailer Skirts
- ☐ Utility Hookups for Trailer
- ☐ Storage Bins

Temporary Utilities:
- ☐ Temporary Telephone
- ☐ Temporary Power
- ☐ Temporary Heating
- ☐ Temporary Water
- ☐ Temporary Lights
- ☐ Temporary Toilets

Barriers and Enclosures:
- ☐ Temporary Fencing and Gates
- ☐ Pedestrian Canopies
- ☐ Temporary Barricades
- ☐ Dust Partitions
- ☐ Weather Protection

Traffic Control:
- ☐ Flaggers
- ☐ Barricades
- ☐ Flares and Lights

Rental Equipment and Tools:
- ☐ Scaffolding
- ☐ Personnel Lift
- ☐ Forklift
- ☐ Crane
- ☐ Other Equipment and Tools

Page 1

FIGURE 2-3a Overhead Checklist

cost of the equipment. This may include obtaining rental pricing or pricing company-owned equipment. Pricing equipment is discussed in Chapter 22 and Chapter 23.

Project Overhead: Before the bid day, the estimator should prepare an estimate for project overhead. Project overhead includes all costs that are associated with the project but cannot be identified with a specific component of the project's construction. The project overhead should be prepared by preparing a quantity takeoff for the project overhead and putting costs to the quantities just as is done with material estimates. An overhead checklist is shown in Figure 2-3.

Prepare Bid Documents: Finally, the bid documents need to be prepared and the appropriate signatures need to be obtained. Any incomplete items that must be filled in on the day of the bid, such as the bid price and the prices for the schedule of values, should be marked with Post-It notes to minimize the possibility of forgetting to fill in a blank on the day of the bid.

BID-DAY ACTIVITIES

The bid-day activities are activities or tasks that are performed predominantly on the day of the bid, although some of these activities may be started before the bid day. This is because most subcontractors wait until a few hours before the bid deadline to submit their bids. As a result, bid days can be quite hectic. Bid-day activities include selecting subcontractors; combining the labor, material, equipment, subcontractor, and overhead pricing; incorporating markups; reviewing the bid for errors; and submitting the bid.

Subcontractor Selection: When receiving bids from a subcontractor, it is important for the estimator to verify that the subcontractor has included the entire scope of work for the items that the subcontractor is bidding. Any items that are missing must be added to the estimate for it to be complete and to avoid having an incomplete estimate. In addition, the estimator must make sure the bids from the subcontractors used in the estimate do not include items that are included in other subcontractor's

```
                        OVERHEAD CHECKLIST

Project Supplies:                    Subsistence:
    ☐ Office Supplies                    ☐ Temporary Housing
    ☐ Office Furnishings                 ☐ Maid Service
    ☐ Fax/Copy Machines                  ☐ Air Fare
    ☐ Computers                          ☐ Hotel Rooms
    ☐ Software                           ☐ Rental Car and Fuel
    ☐ Drinking Water                     ☐ Meals and Expenses
    ☐ Radios
    ☐ Blue Print Reproduction        Final Inspection and Close Out:
    ☐ Construction Photos                ☐ Warranty
    ☐ Postage/Fed Express                ☐ Damage Repairs
                                         ☐ Maintenance Manuals
Security:
    ☐ On-site Guards                 Bonds and Insurance:
    ☐ Security Patrols                   ☐ Prime Contract Bonds
                                         ☐ Subcontractor Bonds
Access and Parking:                      ☐ Fire Insurance
    ☐ Access Road Construction           ☐ Earthquake/Flood
    ☐ Access Road Maintenance            ☐ Increased Liability
    ☐ Temporary Parking                  ☐ Builder's Risk

Cleanup:                             City Fees and Permits:
    ☐ Cleanup Labor                      ☐ Plan Check Fee
    ☐ Trash Removal and Disposal         ☐ Building Permit Fee
                                         ☐ Grading or Excavation Permit
Signage:                                 ☐ Sewer Assessment Fee
    ☐ Project Signs                      ☐ Water Meter Fee
    ☐ Safety Signs                       ☐ Storm Drainage Fee
    ☐ Other Signs                        ☐ Power Company Charges
                                         ☐ Gas Company Charges
                                         ☐ Telephone Company Charge
                                         ☐ Fire Water Connection Fee
                                         ☐ Business License

                              Page 2
```

FIGURE 2-3b Overhead Checklist

bids or their bids. Including an item twice inflates the estimate, and the contractor risks losing the bid because of poor estimating practices. When bids for a bid item are available from more than one subcontractor, the best subcontractor must be selected. Selection may be made based on completeness of the scope of work, price, and past experience with the subcontractor. When a subcontractor phones in a quote, the estimator should take careful notes as to what the subcontractor is bidding. This information should include the bidder and contact information, project name, what the bidder is bidding, addends received, specific exclusions, whether freight and sales tax are included, bid price, the date and time the bid was received, and who took the bid. The best way to do this is to fill out a phone quote sheet such as the one shown in Figure 2-4. Today, most phone quotes have been replaced by faxed or e-mailed quotes; however, they should contain the same basic information. If any of this information is missing, the subcontractor should be contacted and the information obtained. Subcontractor pricing is discussed in Chapter 24.

Combining Pricing: As the pricing becomes available, the estimator should add the pricing to the bid. The pricing includes pricing for labor, materials, equipment, subcontractor, and overhead. The estimator should exercise caution to be sure that all items necessary to complete the bid are included in the bid and none of the items have been included twice. Combining the pricing should begin before bid day and often continues right up until the bid is submitted. Combining the pricing is discussed in Chapter 26.

Add Markups: On the day of the bid, the estimator should finalize the rates for the markups. Markups are also known as add-ons. Markups include the profit markup, overhead markup, building permit costs, bonding costs, and sales tax. The profit markup is used to provide the owners of the construction company a return on their investment. Often, the profit markup is adjusted based on the competition bidding on the project. As the estimator finds out which plan holders have dropped out of the bidding, the profit margin is often adjusted. The overhead markup is used to cover the company's general overhead costs. The project's specific overhead should not be included in the overhead markup but should be bid as part of the project costs and included in general conditions.

```
                        Phone Quote Sheet

    Project Team:_____

    Project:_____

    Subcontractor:_____

    Contact:_____  Phone Number:_____


    Bid Items:_____

    Bid:_____

    Addends: _____

    Inclusions:_____
    _____
    _____
    _____

    Exclusions:_____
    _____
    _____


    Sales Tax Included: Yes  No      Freight Included: Yes  No

    Taken By:_____     Date:_____
```

FIGURE 2-4 Phone Quote

The bonding costs, and sometimes the building permit costs, are based on the estimated construction costs, which are not known until the bid is completed. This requires the costs for these items to be calculated near the completion of the bid. Sales tax may be added to the individual material costs or to the material costs of the bid as a whole. When added to the material costs of the bid as a whole, the sales tax is treated as a markup and is incorporated into the bid when the material costs are totaled. For cost control purposes, the sales tax must be added to the individual cost codes when preparing the budget for the project. Markups are discussed in Chapter 25.

Review Bid for Errors: As the bid approaches completion, the bid needs to be reviewed for errors. This includes making sure that there is pricing for all of the bid items, comparing the bid to historical costs for similar projects, and having someone else check the estimate for errors. Avoiding errors in estimates is discussed in Chapter 27.

Submit Bid: Finally, the price and other bid-day information should be added to the bid documents, and the bid should be submitted to the contracting office. Care must be taken to accurately enter the pricing and make sure all remaining blanks are filled in. It would be a great waste of time to lose the bid because you forgot to fill in a blank or you wrote the wrong price on the bid document. Submitting the bid is discussed in Chapter 28.

POST-BID ACTIVITIES

The post-bid activities are different for jobs for which the contractor is the low bidder than they are for jobs for which the contractor is not. Here are some of the activities that may occur after the submission of the bid.

Bid Review: After completing the bid, the estimator should review the bid and the bidding process to identify any mistakes that were made and try to find ways to avoid these mistakes in the future. Besides identifying mistakes, the estimator should look at what works and identify ways to make sure that he or she does the same things on future bids. One of the characteristics of a successful estimator is that he or she is committed to continuous improvement. To do this, estimators must learn from their mistakes and repeat those actions that made them successful.

Project Buyout: If the estimator wins the bid, the estimator must buyout the project. The project buyout is the process of hiring subcontractors and procuring materials and equipment for the construction project. During the buyout the estimator must prepare contracts for the subcontractor's work and purchase orders for the materials used on the project and communicate this information to the subcontractors, suppliers, and the general contractor's project management team (the superintendent, project manager, and so forth). During the buyout process the estimator should identify any errors in the estimate and identify ways to avoid the errors in the future. The estimator should also update the pricing in the pricing database as needed at this time. The project buyout is discussed in Chapter 29.

Close-Out Audit: The project close-out audit consists of reviewing all activities completed during the construction process, including the estimating process, after the project is complete. As part of this audit, the estimator should review the actual costs to complete the project and compare the costs against the original estimate. The goal is to find ways to improve the accuracy of the estimate. The estimator should look at areas where his or her pricing is too high or low so that his or her future estimates will reflect the actual costs, updating the pricing in the pricing database as needed. For the close-out audit to be useful, the actual costs must be accounted for in the same cost categories as they were included in the estimate. For this to happen a company must have a set of standardized job cost codes that identifies the job cost and spells out what is included in each cost code. This needs to be done in a formal document. Sample job cost codes are found in Appendix B.

INFORMATION FLOW

Throughout the remainder of the book, there are a number of spreadsheets that are used to prepare a bid for a construction project. These include the two spreadsheets in Chapter 3 and the Quick Tips that are spread throughout many of the chapters. Before proceeding, it is important that the reader have an understanding how the information flows from one spreadsheet to another as a bid is prepared. The information flow is shown in Figure 2-5.

The **Wage Calculation** (Quick Tip 21-1) spreadsheet is used to determine the labor cost per hour for a specific class of worker and includes the wages paid to the employee, fringe benefits (vacation, health insurance, etc.), and other

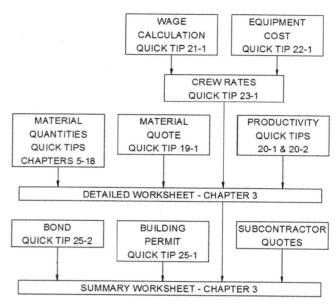

FIGURE 2-5 Information Flow

labor burden costs (Social Security, Medicare, unemployment taxes, workers' compensation insurance, etc.).

The **Equipment Cost** (Quick Tip 22-1) spreadsheet is used to determine the hourly ownership and operation costs of equipment and includes the purchase price, taxes, insurance, fuel, lubrication, tires, repairs, and so on.

The **Crew Rates** (Quick Tip 23-1) spreadsheet is used to determine the cost per labor hour for a crew that consists of different classes of workers and types of equipment. The hourly wage rates from the Wage Calculation spreadsheet and the hourly equipment costs from the Equipment Cost spreadsheet feed into the Crew Rates spreadsheet, as showing in Figure 2-6.

The **Material Quantities** spreadsheet (numerous Quick Tips from Chapters 5–18) are used to determine the quantity of materials needed for a specific building component. Other methods may be used to determine the needed material quantities.

The **Material Quote** (Quick Tip 19-1) spreadsheet is used to obtain pricing for the needed materials.

The **Productivity** (Quick Tips 20-1 and Quick Tips 20-2) spreadsheets are used to determine the productivity of a construction crew. Other methods, such as historical data, may be used to determine the productivity rates.

The **Detailed Worksheet** (from Chapter 3) is used to determine the materials, labor, and equipment costs for work bid in house. The material quantities come from the quick tips in Chapters 5 through 18 or other estimating methods. The unit cost for the materials comes from the Material Quote spreadsheet or suppliers quotes. The labor hours per unit comes from the Productivity spreadsheet or other data. The labor cost per labor hour and equipment cost per labor hour comes from the Crew Rates spreadsheet. This information flow is shown in Figure 2-7.

The **Bond** (Quick Tip 25-2) spreadsheet is used to estimate the cost of the payment and performance bonds.

Overview of the Estimating and Bidding Process 19

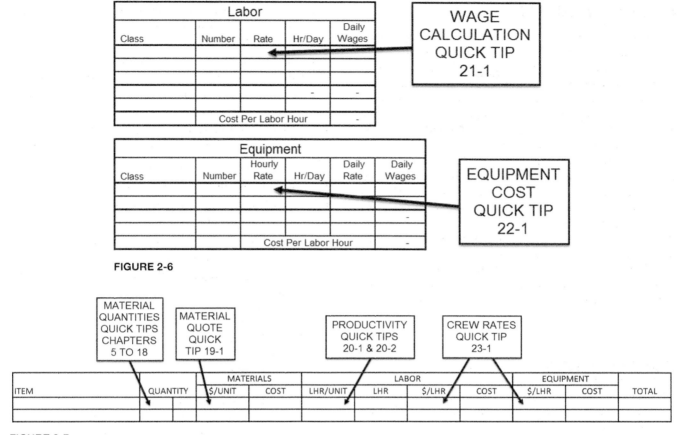

FIGURE 2-6

FIGURE 2-7

The **Building Permit** (Quick Tip 25-1) spreadsheet is used to estimate the cost of the building permit.

Subcontractor Quotes include pricing from subcontractors. Although there is not a spreadsheet for subcontractor quotes, they provide important information needed to complete the bid.

The **Summary Worksheet** (from Chapter 3) is used to summarize the bid by cost code and prepare the final estimate. The material, labor, and equipment cost (by cost code) come from the Detailed Worksheet. The subcontractor cost (by cost code) comes from the subcontractor quotes. The building permit cost comes from the Building Permit spreadsheet; and the bond cost comes from the Bond spreadsheet. This information flow is shown in Figure 2-8.

When a bid is properly prepared, another construction manager should be able to trace the information from the summary sheet back to its original source.

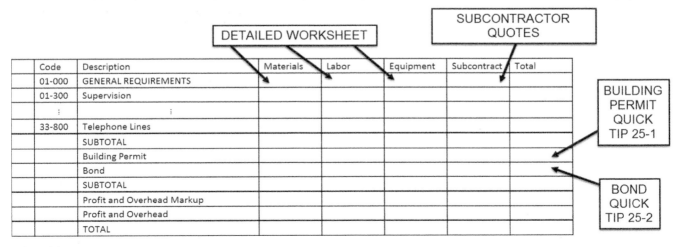

FIGURE 2-8

CONCLUSION

There are a number of activities that must be completed to complete the bid. The first group of activities comprises planning the bid. This includes defining the scope of work, scheduling the bidding activities, and assigning the activities to the members of the estimating team. The second group of activities comprises the pre–bid-day activities. These include ordering the bond; requesting subcontractor quotes; preparing the quantity takeoff; pricing materials, equipment, and labor; preparing a project overhead budget; and preparing the bid documents. The third group of activities comprises the bid-day activities. These include selecting the subcontractors, combining all of the prices together into the bid, adding the markups, reviewing the bid for errors, and submitting the bid. The fourth group of activities comprises the post-bid activities. These include reviewing the bid and, if one is successful in winning the bid, buying the project out and performing a project close-out audit at the end of the project.

PROBLEMS

1. What are some of the things you can do if there is insufficient time to complete the estimate before the bid is due?
2. Why should the bond be ordered early in the bid process?
3. Why should subcontractor quotes be requested early in the bidding process?
4. Why should you get one bid in each area you are going to subcontract out before getting a second bid?
5. Why should you get a bid on the high-dollar subcontractor items first?
6. Why should you prepare a quantity takeoff for materials even when the supplier is preparing a quantity takeoff?
7. What is one way you can prevent the possibility of forgetting to fill in a blank on the bid forms that needs to be filled in on bid day?
8. Why is it important for an estimator to verify that a subcontractor's bid has a complete scope of work?
9. What are some of the criteria that may be used to select subcontractors?
10. What should be recorded when taking a telephone quote from a subcontractor or material supplier?
11. Why should you use phone quote sheets?
12. Identify five different types of markups.
13. What is the overhead markup used for?
14. Define the term buyout.
15. At what two points in a project should an estimator update the pricing database?
16. Using the information in Figure 2-2 and Appendix B, under which cost code would the following materials be found:
    a. Rebar for a sidewalk
    b. Bathroom sinks
    c. Vinyl siding
    d. Metal stairs
    e. Wood base
    f. Waterline from the street
17. Using the summary sheets shown in Figure 2-2 and the descriptions of the costs codes found in Appendix B, determine which cost codes will need pricing to complete the bid for the Johnson Residence given in Appendix G.
18. Prepare a schedule for completing the estimate for the Johnson Residence given in Appendix G.
19. Using the summary sheets shown in Figure 2-2 and the descriptions of the costs codes found in Appendix B, determine which cost codes will need pricing to complete the bid for the West Street Video project given in Appendix G.
20. Prepare a schedule for completing the estimate for the West Street Video project given in Appendix G.

CHAPTER THREE

INTRODUCTION TO EXCEL

In this chapter you will be introduced to the basic operation of Excel. This includes managing workbooks and worksheets, entering data, formatting spreadsheets, writing formulas, using a few of the basic Excel functions, printing, and testing worksheets. Those who are just learning Excel should carefully read this chapter and complete all of the exercises, whereas those who are experienced with Excel are encouraged to quickly skim this chapter and work the exercises to pick up additional knowledge on how to get the most out of Excel. Exercises 3-1 through 3-5, 3-8, and 3-9 are used as the starting point for the exercises in Chapter 32.

Excel is a powerful spreadsheet program that can be used to perform mathematical calculations. An Excel file is known as a workbook and has an .xlsx or .xlsm (for files that contain macros) extension at the end of the file name indicating its file type. Files created with Excel 2003 or older have an .xls extension. The Windows operating system has the option not to display the file type; therefore, you may not see the file type on your computer display.

The screenshot figures in this book are based on Excel 2016 running on Windows 7 Enterprise with the Windows 7 theme. Figure 3-1 shows the typical layout for Excel 2016. Your layout may look different based on the selected theme, version of the operating system, version of Excel, width of your screen, and settings used on your computer. As the width of Excel decreases, large buttons are replaced with smaller buttons, which are stacked on the ribbon; therefore, your ribbon may not look the same as the screenshots used in this text. If you have customized the ribbon it will look different.

CONVENTIONS USED IN THIS TEXT

The following conventions are used in this text to describe actions that must be performed to set up a worksheet or change the settings for a worksheet:

Clicking, unless otherwise noted, means left clicking.
Bold words in the text represent buttons, tabs, icons, radio buttons, items in dropdown boxes, and check boxes that must be selected and keys that must be typed to complete the exercises.
Italic words in the text represent text or numbers that must be entered in a spreadsheet cell or text box to complete the exercises.
The <> brackets signify that the user must enter data represented by the text between the brackets, such as a file name. The appropriate data rather than the text should be entered between the brackets. For example, if you are instructed to type <your name> and your name is John Smith, you would type *John Smith*.
The shorthand notation "select **File>Alignment>Center** button" means that the user should select the **Center** Button from the **Alignment** Group from the **File** tab.
The shorthand "**Ctrl+N**" means that the user should type the **N** key while holding down the **Ctrl** key.

WORKBOOK MANAGEMENT

Workbook management consists of creating new workbooks, opening existing workbooks, and saving changes to workbooks. Let's look at creating new workbooks.

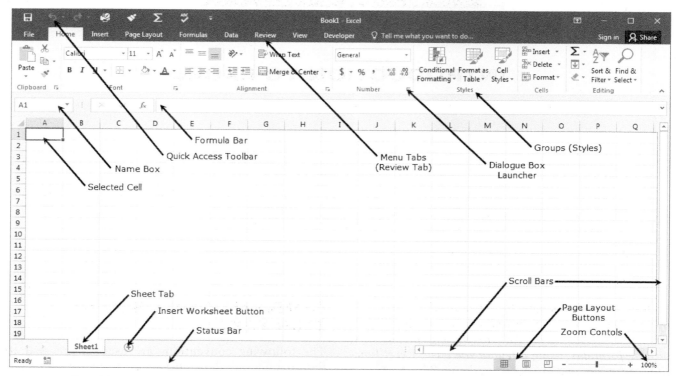

FIGURE 3-1 Excel Layout

Creating Workbooks

A new workbook is created by clicking on the **Excel** icon shown in Figure 3-2.

A new workbook can be created while working in an existing Excel workbook by typing **Ctrl+N** to bring up a blank workbook or by selecting the **File** tab, clicking on **New** in the left pane, and double clicking on **Blank workbook** (shown in Figure 3-3).

Opening Workbooks

Existing workbooks can be opened by double clicking on the file name in the Windows folder. Workbooks can also be opened from Excel by typing **Ctrl+O** or selecting the **File** tab, clicking on **Open** in the left pane, and double clicking on **This PC** to bring up the Open dialogue box shown in Figure 3-4. From the Open dialogue box, the user selects the folder where the file is stored in the left side of the dialogue box and opens the file by (1) double clicking on the file name or (2) clicking on the file name and clicking the **Open** button. Recently used files may be opened by selecting the **File** tab, clicking on **Open** in the left pane, selecting **Recent,** and double clicking the file from the list of recently opened files (see in Figure 3-5).

Saving Workbooks

Workbooks are saved by clicking on the **Save** button on the Quick Access toolbar, by typing **Ctrl+S**, or by selecting the **File** tab and clicking on **Save** in the left pane. The first time a file is saved, any one of these actions will take the user to the File tab where the user can double click on **This PC** to bring up the Save As dialogue box shown in Figure 3-6. From the Save As dialogue box the user selects the folder where the file is to be stored on the left side of the dialogue box, types the file name in the File name: dropdown box, and clicks on the **Save** button.

An existing workbook is saved under a different file name by selecting the **File** tab, clicking on **Save As** in the left pane, and double clicking on **This PC** to bring up the Save As dialogue box shown in Figure 3-6. The workbook is saved with a different file name in the same manner that the user saved a workbook the first time.

Now that you know the basics of spreadsheet file management, you are ready to create your first spreadsheet. The following exercise will take you step by step through the process. You are encouraged to take the time to create the spreadsheet on your computer. This spreadsheet is used as a starting point for the subsequent exercises in this chapter and the exercises in Chapter 32.

FIGURE 3-2 Excel Icon

Introduction to Excel 23

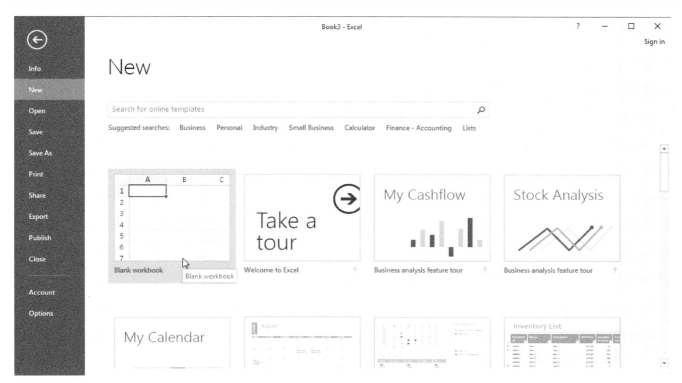

FIGURE 3-3 Adding a New Workbook

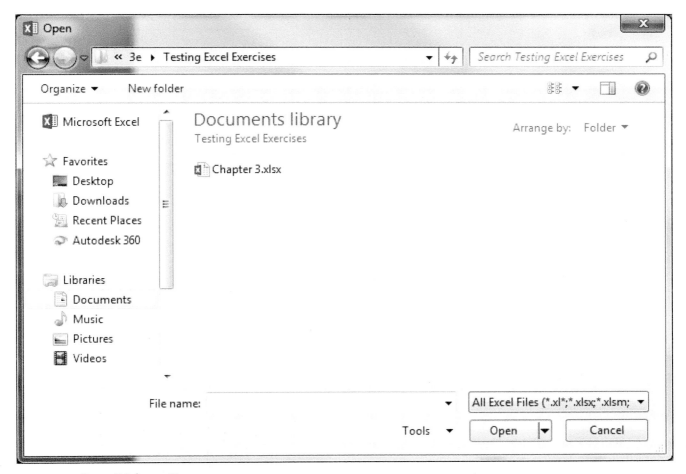

FIGURE 3-4 Open Dialogue Box

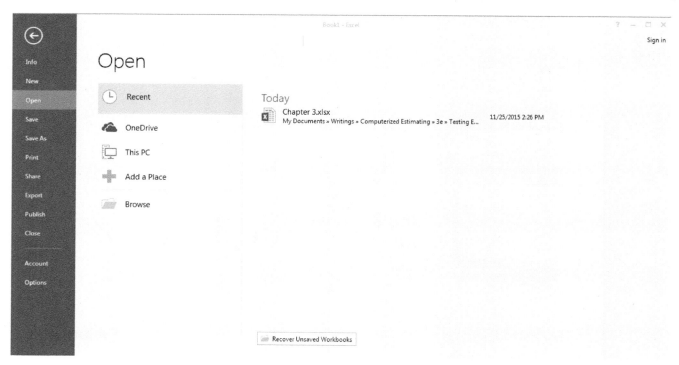

FIGURE 3-5 Opening Recent Files

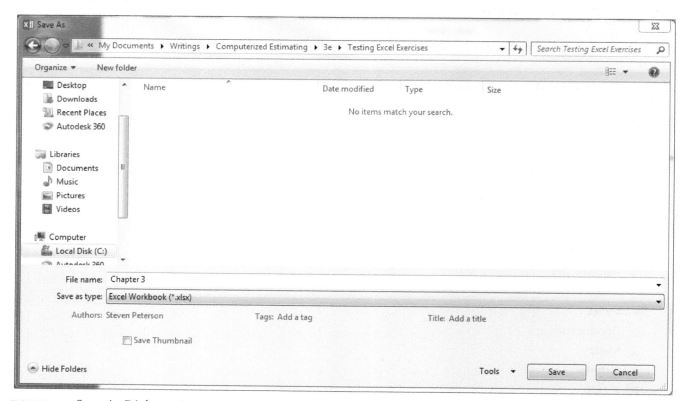

FIGURE 3-6 Save As Dialogue Box

Exercise 3-1

First, you will create a folder where you will store the exercises from this book using the following steps:

1. Open the Documents folder on your hard drive. If you would like to store the documents in another folder or drive, open the folder or drive of your choice rather than the Documents folder.

2. Select the **New Folder** button (shown in Figure 3-7) to create a new folder. The default name of the folder should be "New Folder" and if there is a folder already named "New Folder," the new folder

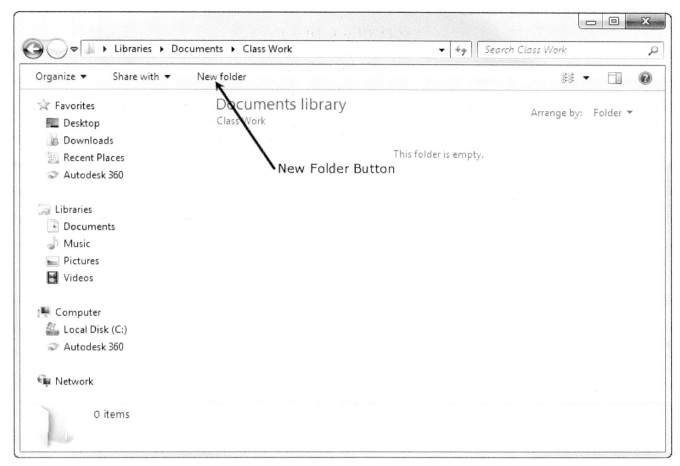

FIGURE 3-7 New Folder Button

has a number in parentheses included as part of its name.

3. The folder's existing name should be highlighted. If it is not, right click on the newly created folder and select **Rename** from the popup menu to highlight the folder's name.
4. Type *Excel Exercises* and press the **Enter** key to change the name of the folder to Excel Exercises.
5. Double click on **Excel Exercises** to open the folder.

Next, create and save a workbook using the following steps:

6. Open a new workbook in Excel by clicking on the Excel icon (shown in Figure 3-2) on your desktop or by selecting it from the Windows' start menu and double clicking on **Blank workbook** (shown in Figure 3-3).
7. Save the workbook by selecting the **File** tab, clicking on **Save** in the left pane, double clicking on **This PC** to bring up the Save As dialogue box, selecting the **Excel Exercises** folder in the right pane, and typing *Chapter 3* in the File Name: text box. The Save As dialogue box should appear as it does in Figure 3-6.
8. Click the **Save** button to save the workbook.

WORKING WITH WORKSHEETS

An Excel workbook contains one or more worksheets. The worksheets are marked by tabs at the bottom of the screen. You may move from one worksheet to another by clicking on these tabs. By default, new workbooks contain one worksheet named Sheet1. You can add, delete, rename, and change the order of the worksheets. Let's begin by looking at adding worksheets.

Adding Worksheets

Blank worksheets are added to the workbook in three ways. First, a blank worksheet may be added to the right of the existing sheets by clicking on the **New sheet** button located to the right of the sheet tabs (see Figure 3-1). Second, a blank worksheet may be added to the left of the active worksheet by selecting the **Home** menu tab, selecting the arrow below the Insert button in the Cells group, and selecting **Insert Sheet** from the popup menu as shown in Figure 3-8. If the Home menu tab is already displayed, you need not select it. Third, a blank worksheet may be added to the left of a worksheet by right clicking on the tab to the right of the location where the worksheet is to be inserted and selecting **Insert . . .** from

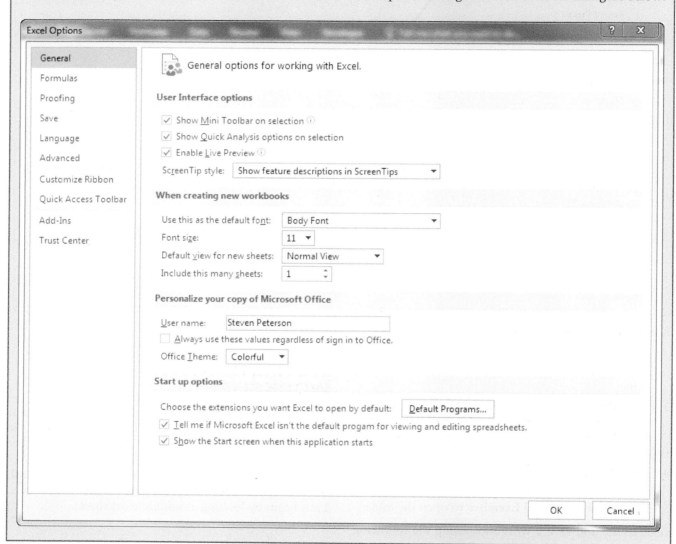

> **EXCEL QUICK TIP 3-1**
> **Setting the Number of Worksheets in a New Workbook**
>
> The number of worksheets that appear when a new workbook is created is changed by clicking on the **File** tab, clicking on **Options** to bring up the Excel Options dialogue box, selecting **General** in the left pane, and under the When creating new workbooks heading in the right pane choosing the number of sheets to be included in the Include this many sheets: list box. The General tab of the Excel Options dialogue box is shown in the figure below:

the popup menu (shown in Figure 3-9) to bring up the Insert dialogue box (shown in Figure 3-10). From the Insert dialogue box the user selects **Worksheet** and clicks the **OK** button to insert the worksheet.

Copying Worksheets

A worksheet is copied by (1) right clicking on the worksheet tab and selecting **Move or Copy . . .** from the popup menu (shown in Figure 3-9) or (2) by selecting the **Home** menu tab, selecting the arrow below the Format button in the Cell group, and selecting **Move or Copy Sheet . . .** from the popup menu as shown in Figure 3-11 to bring up the Move or Copy dialogue box, shown in Figure 3-12. The user checks the **Create a copy** check box and selects the location where the new sheet is to be inserted in the Before sheet: list box. If the worksheet is to be the last worksheet in the workbook, the user must select (**move to end**) as the location of the worksheet. The user can copy the worksheet to another workbook or a new workbook by selecting the desired workbook in the To book: dropdown box. To copy the worksheet to an existing workbook, the workbook must be open so it will appear in the To book: dropdown box.

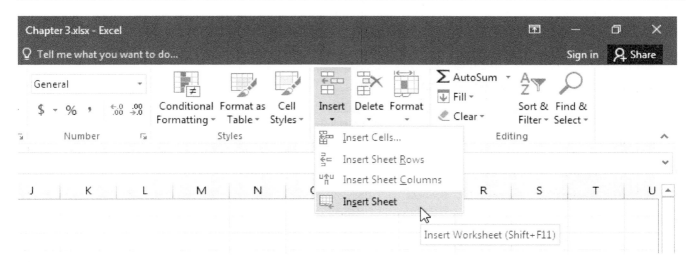

FIGURE 3-8 Home: Insert Popup Menu

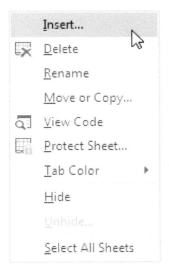

FIGURE 3-9 Popup Menu

Deleting Worksheets

Worksheets can be deleted by right clicking the tabs of the unneeded worksheets and selecting **Delete** from the popup menu (shown in Figure 3-9) or by selecting the **Home** menu tab, selecting the arrow below the Delete button in the Cells group, and selecting **Delete Sheet** from the popup menu as shown in Figure 3-13. If the worksheet contains data, a warning dialogue box (see Figure 3-14) will appear warning the user that data exists and giving the user the opportunity to cancel the delete. To delete the worksheet, click on the **Delete** button. To cancel the delete, click on the **Cancel** button.

Renaming Worksheets

Worksheets are renamed in three ways: (1) by right clicking the tab of the worksheet to be renamed and selecting **Rename** from the popup menu shown in Figure 3-9, (2) by double clicking on the tab, or (3) by selecting the **Home** menu tab, selecting the arrow below the Format button in the Cells group, and selecting **Rename Sheet**

> **EXCEL QUICK TIP 3-2**
> **Selecting Multiple Cells or Worksheets**
>
> The Shift and Ctrl keys are used to select multiple cells or worksheets at a time.
>
> The Shift key is used to select contiguous blocks of cells or worksheets. For example, Cells A12 through B13 are selected by clicking on Cell A12, holding down the **Shift** key, and clicking on Cell B13. Cells A12, A13, B12, and B13 will now be selected. The user must hold down the Shift key after he or she has selected Cell A12, because the selection will begin at the cell the cursor is in when the Shift key is held down.
>
> The Ctrl key is used to select noncontiguous cells. For example, Cells A15 and B16 are selected by clicking on Cell A15, holding down the **Ctrl** key, and clicking on Cell B16. Cells A15 and B16 are now selected.
>
> Both the Shift and Ctrl keys may be used to select a group of cells. For example, Cells A12 through B13, Cell A15, and Cell B16 are selected by clicking on Cell A12, holding down the **Shift** key, clicking on Cell B13, releasing the **Shift** key, holding down the **Ctrl** key, and clicking on Cells A15 and B16.

from the popup menu as shown in Figure 3-11. The name of the tab will then be highlighted and can be edited as one would edit text in a word processor. After changing the name of the worksheet, the user must click on any of the cells or press the **Enter** key to save the name change.

Organizing Worksheets

The order of the worksheet tabs can be changed by holding down the left mouse button and dragging the worksheet to its new location. When the user holds down the left mouse button, a little black arrow will appear showing where the worksheet's tab will be located (as shown in Figure 3-15). When the user releases the left mouse button, the worksheet is moved to the new location.

28 CHAPTER THREE

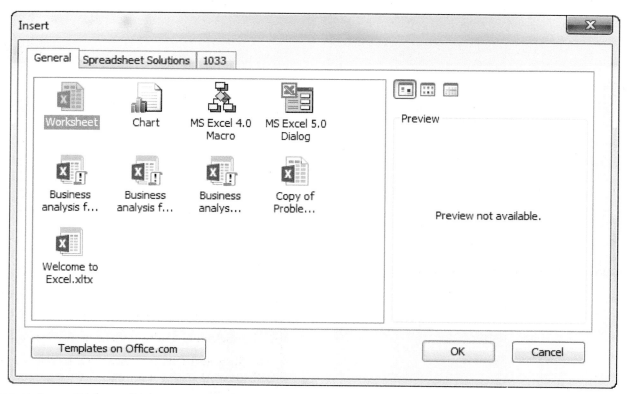

FIGURE 3-10 Insert Dialogue Box

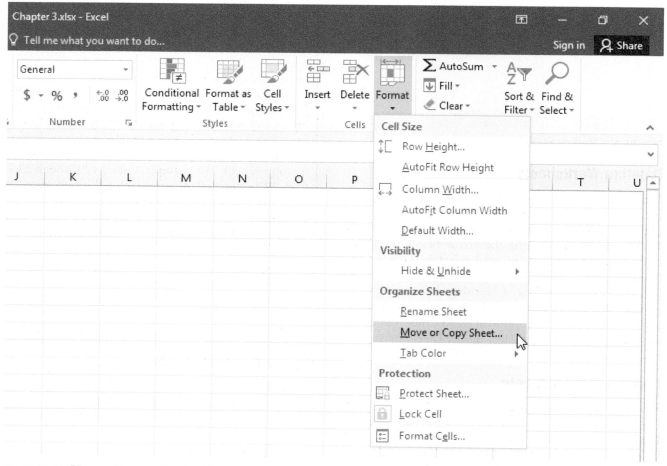

FIGURE 3-11 Home: Format Popup Menu

worksheet is created rather than moving the existing worksheet.

Exercise 3-2

In this exercise you will modify the file you created in Exercise 3-1 to contain two worksheets named "Summary" and "Detail." Begin by changing the number of worksheets to two, using the following steps:

1. Make sure that Chapter 3, the workbook created in Exercise 3-1, is open. If it is not open, open it by selecting the folder where the file is kept and double clicking on **Chapter 3**.

2. Insert additional worksheet by clicking on the **New sheet** button located to the right of the sheet tab. Add a second sheet in the same manner.

3. Delete all but two of the worksheets by right clicking the tabs of the unneeded worksheets and selecting **Delete** from the popup menu (shown in Figure 3-9).

Next, change the names of the worksheets to "Summary" and "Detail" using the following steps:

4. Change the name of the right worksheet to "Summary" by double clicking on the tab and, with the worksheet name highlighted, typing *Summary*.

5. Change the name of the left worksheet to "Detail" by right clicking the tab of the worksheet, selecting **Rename** from the popup menu (shown in Figure 3-9), and, with the sheet name highlighted, typing *Detail*.

6. Move the cursor to Cell D1 by clicking on Cell D1.

Next, move the Detail worksheet so it is on the right (the end of the workbook) and save the workbook using the following steps:

7. Right click on the **Detail** worksheet tab and select **Move or Copy . . .** from the popup menu (shown in Figure 3-9) to bring up the Move or Copy dialogue box.

FIGURE 3-12 Move or Copy Dialogue Box

Alternatively, a worksheet can be moved by (1) right clicking on the worksheet tab and selecting **Move or Copy . . .** from the popup menu (shown in Figure 3-9) or (2) by selecting the **Home** menu tab, selecting the arrow below the Format button in the Cell group, and selecting **Move or Copy Sheet . . .** from the popup menu, as shown in Figure 3-11, to bring up the Move or Copy dialogue box, shown in Figure 3-12. The user must select the location where the worksheet is to be moved in the Before sheet: list box. If the worksheet is to be the last worksheet in the workbook, the user must select (**move to end**) as the location of the worksheet. The user can move the worksheet to another workbook or a new workbook by selecting the desired workbook in the To book: dropdown box. To move the worksheet to an existing workbook, the workbook must be open so it will appear in the To book: dropdown box. If the user checks the **Create a copy** check box, a copy of the

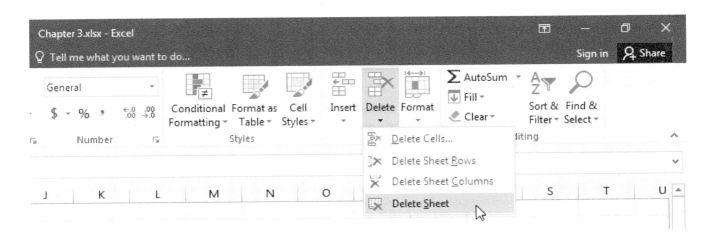

FIGURE 3-13 Home: Delete Popup Menu

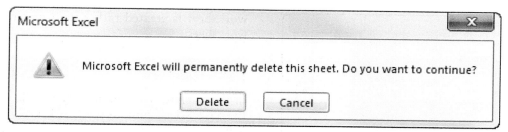

FIGURE 3-14 Warning Dialogue Box

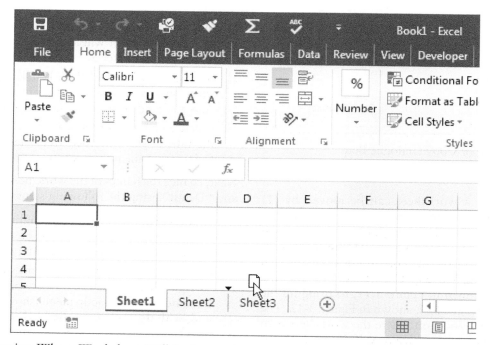

FIGURE 3-15 Location Where Worksheet Will Be Inserted

8. Select (**move to end**) in the Before sheet: list box. The Move or Copy dialogue box should look like Figure 3-12.
9. Click the **OK** button to complete the move. The Detail worksheet should be on the right and the Summary worksheet should be on the left.
10. Save the workbook by clicking on the **Save** button on the Quick Access toolbar or typing **Ctrl+S**.

ENTERING DATA

Data is entered in the worksheet by clicking on the cell and typing the data. When the Enter key is typed, the cursor will move to the next cell. Data is moved from one cell or group of cells to another cell or group of cells by cutting and pasting the data from one cell to another cell. When the same data is used in multiple cells, the data may be copied and pasted from one cell to a number of cells. Pictures can also be added to worksheets. Let's begin by looking at cutting and pasting data.

Cut and Paste

Data is moved by cutting the data from a cell or group of cells to the clipboard and then pasting the data to a new cell. The data is cut by selecting the cell or cells containing the data to be cut and (1) selecting **Home>Clipboard>Cut** button (shown in Figure 3-16), (2) right clicking on one of the selected cells and selecting **Cut** from the popup menu (shown in Figure 3-17), or (3) typing **Ctrl+X**. A moving dashed line will then appear around the cells that have been cut and placed on the clipboard.

The data must be pasted to the desired cells before you perform any other action. If you perform another action before pasting the data, the data on the clipboard is left in the original cell and the user must cut the data again before pasting. The data is pasted to a cell or cells by selecting the upper-left cell of the paste location and (1) selecting **Home>Clipboard>Paste** button (shown in Figure 3-18) or (2) by typing **Ctrl+V**. Alternatively, the user may right click on the upper-left cell and select the **Paste** button from the popup menu (shown in Figure 3-19). The data will then be moved from the original cell or cells to the new cell or cells.

EXCEL QUICK TIP 3-3
Move after Typing Enter

After you press the Enter key, Excel moves to a neighboring cell. The user can set Excel to move to the left, right, up, or down when the Enter key is typed. The default direction is down. To set the direction, select the **File** tab, click on the **Options** button to bring up the Excel Options dialogue box, select **Advanced** in the left pane to display the Advanced options for working with Excel in the right pane, and under the Editing options selecting **Left, Right, Up,** or **Down** in the Direction: dropdown box directly under the After pressing Enter, move selection check box. To prevent the cursor from moving, uncheck the **After pressing Enter, move selection** check box. The Excel Options dialogue box is shown in the figure below:

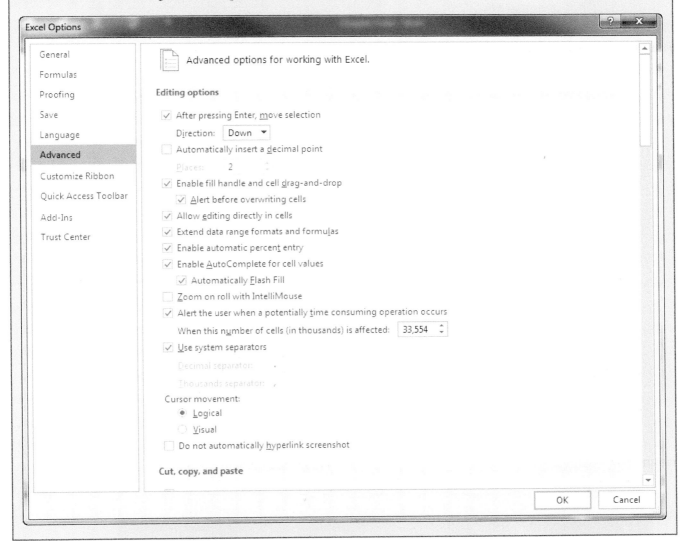

Copy and Paste

Data is copied from a cell or group of cells to multiple cells or groups of cells by copying the data in a cell or cells to the clipboard and pasting the data to new cells. The data is copied by selecting the cell or cells containing the data to be copied and (1) selecting the **Home>Clipboard>Copy** button (shown in Figure 3-20), (2) right clicking on one of the selected cells and selecting **Copy** from the popup menu (shown in Figure 3-17), or (3) by typing **Ctrl+C**. A moving dashed line will then appear around the cells that have been copied and placed on the clipboard.

The data must be pasted to the desired cells before performing any other action. The data is pasted using the same process as was used to paste data after cutting it from a cell with one exception: you may paste the data multiple times until you perform another action.

When copying, the user has the option to use the paste special command, which allows the user to copy only certain properties of the cell. The paste special

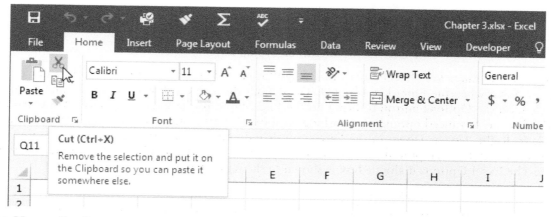

FIGURE 3-16 Home: Cut Button

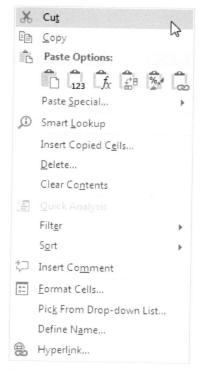

FIGURE 3-17 Popup Menu

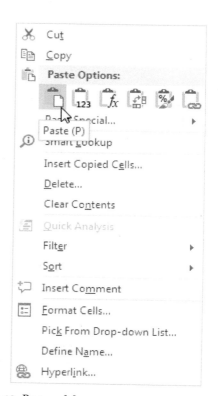

FIGURE 3-19 Popup Menu

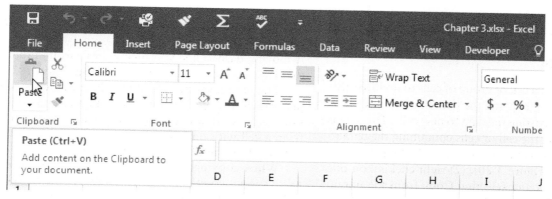

FIGURE 3-18 Home: Paste Button

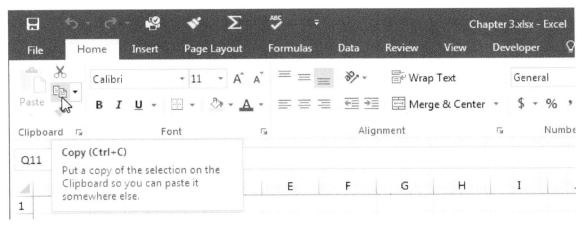

FIGURE 3-20 Home: Copy Button

FIGURE 3-21 Paste Special Popup Menu

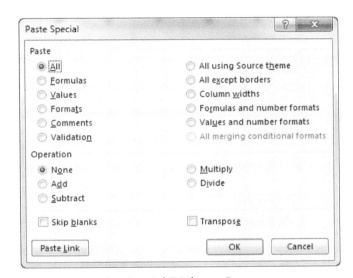

FIGURE 3-22 Paste Special Dialogue Box

command is used by right clicking on a cell and placing the cursor over **Paste Special . . .** from the popup menu (shown in Figure 3-17) or by selecting the **Home** menu tab and clicking on the arrow below the Paste button to bring up the Paste Special popup menu (shown in Figure 3-21) and selecting the attributes to paste. From this popup menu, the user may (from top left) Paste, Formulas, Formulas & Number Formatting, Keep Source Formatting, No Borders, Keep Source Column Widths, Transpose, Values, Values & Number Formatting, Values & Source Formatting, Formatting, Paste Link, Picture, and Linked Picture.

Clicking on **Paste Special . . .** from the either of the popup menus will open the Paste Special dialogue box (shown in Figure 3-22). From the Paste Special dialogue box the user may select to paste All (this performs the same function as the paste command), Formulas, Values, Formats, Comments, Validation, All using Source theme, All except borders, Column widths, Formulas and number formats, or Values and number formats. Once the user selects the attributes that he or she wants to paste, the user must click the **OK** button to complete the paste. For example, if the user selects **Values** from the Paste Special dialogue box, only the values of the cell will be pasted instead of any formulas; whereas, if **All except borders** is selected, the contents of the cell and all of its attributes except the borders will be copied.

Adding Pictures

Pictures are added to the worksheet by selecting **Insert>Illustrations>Picture** button (shown in Figure 3-23) to bring up the Insert Picture dialogue box (shown in Figure 3-24). From the Insert Picture dialogue box, the user selects the folder where the file is stored in the left pane and opens the file by double clicking on the file name or clicking on the file name and clicking the **Insert** button.

In the next few exercises you will modify the Detail worksheet you created in Exercise 3-2 to look like Figure 3-25.

Exercise 3-3

In this exercise you will modify the file from Exercise 3-2 by adding data to the cells. Begin by adding data to individual cells using the following steps:

1. Make sure that Chapter 3, the workbook modified in Exercise 3-2, is open. If it is not open, open it by selecting the folder where the file is kept and double clicking on **Chapter 3**.

34 CHAPTER THREE

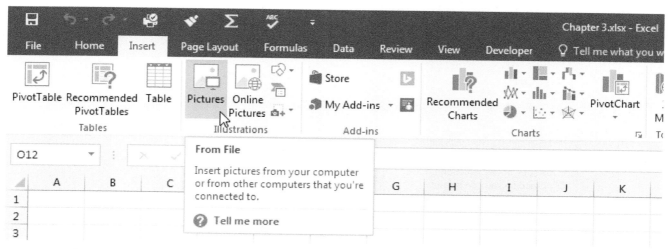

FIGURE 3-23 Insert: Picture Button

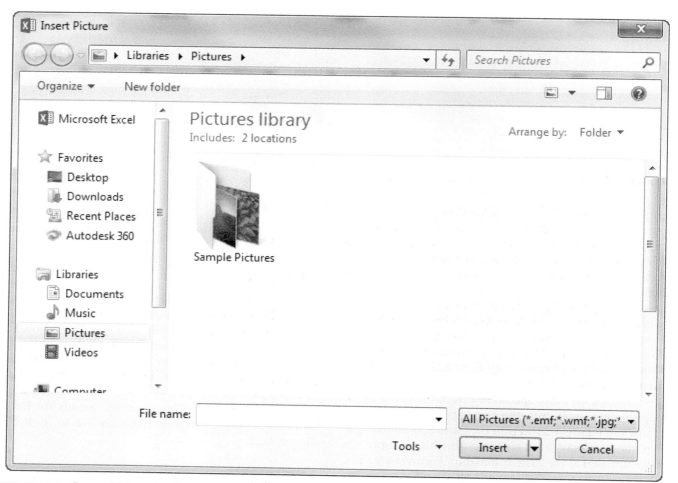

FIGURE 3-24 Insert Picture Dialogue Box

A	B	C	D	E	F	G	H	I	J	K	L
			MATERIALS			LABOR			EQUIPMENT		
ITEM	QUANTITY		$/UNIT	COST	LHR/UNIT	LHR	$/LHR	COST	$/LHR	COST	TOTAL

FIGURE 3-25 Detail Worksheet

TABLE 3-1 Data for Cells in Exercise 3-3

Cell	Data
G1	LABOR
J1	EQUIPMENT
A2	ITEM
B2	QUANTITY
D2	$/UNIT
E2	COST
F2	LHR/UNIT
G2	LHR
H2	$/LHR
J2	$/LHR
L2	TOTAL

2. Make sure the worksheet named Detail is the active worksheet by clicking on the **Detail** tab.
3. Begin setting up the spreadsheet by typing *MATERIALS* in Cell A1.
4. Type the text shown in Table 3-1 into the specified cells.

Next, move the data from Cell A1 to Cell D1 using the following steps:

5. Right click on Cell A1 and select **Cut** from the popup menu (shown in Figure 3-17) to cut the contents of the cell to the clipboard.
6. Right click on Cell D1 and select the Paste button below Paste Options: from the popup menu (shown in Figure 3-19) to paste the contents of the clipboard to Cell D1. Cell A1 should now be blank and "MATERIALS" should be in Cell D1.

Next, copy the data from Cell E2 to Cells I2 and K2 using the following steps:

7. Right click on Cell E2 and select **Copy** from the popup menu (shown in Figure 3-17) to copy the contents of the cell to the clipboard.
8. Select Cell I2 and select **Home>Clipboard>Paste** button (shown in Figure 3-18) to paste the contents of the clipboard to Cell I2.
9. Select Cell K2 and type **Ctrl+V** to paste the contents of the clipboard to Cell K2. "COST" should now appear in Cells E2, I2, and K2.
10. Save the workbook by clicking on the **Save** button on the Quick Access toolbar or typing **Ctrl+S**.

FORMATTING WORKSHEETS

Formatting allows the user to create a specific look for worksheets and makes the worksheets easier to read. Common formatting includes setting the column widths, formatting and aligning the data in the cells, adding borders around a cell or group of cells, setting how much of the worksheet is visible on the computer screen, and adding footers and headers. Let's begin by looking at the column widths.

Column Widths

The width of the columns can be changed in three ways. First, the user may place the cursor over the right side of a column's heading and hold down the left mouse button to drag the right side of the column to the desired width. When the cursor is over the right side of the column heading, the cursor changes to a vertical line with arrows pointing to the right and left. Place the cursor in the location shown in Figure 3-26 to adjust

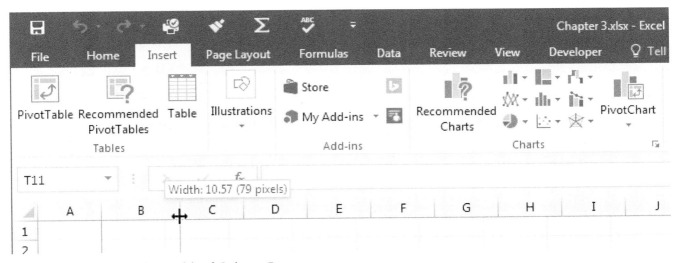

FIGURE 3-26 Changing the Width of Column B

FIGURE 3-27 Column Width Dialogue Box

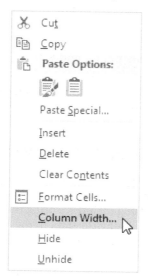

FIGURE 3-28 Popup Menu

FIGURE 3-30 Format Popup Menu

the width of Column B. Second, the user may double click on the right side of the column's heading to have Excel automatically change the width of the column to the minimum width necessary to show the data in the widest cell within the column. Double clicking on the location shown in Figure 3-26 will make this column width adjustment to Column B. Third, the user may set the width using the Column Width dialogue box shown in Figure 3-27 . To bring up the Column Width dialogue box, the user right clicks on the heading for the column and selects **Column Width . . .** from the popup menu (shown in Figure 3-28) or selects **Home>Cells>Format** button shown in Figure 3-29, and selects **Column Width . . .** from the popup menu (shown in Figure 3-30). The column width is then entered into the Column width: text box on the Column Width dialogue box and the **OK** button is selected to change the width of the column.

Row Heights

Row heights are adjusted in the same way as column heights except that the user selects row rather than column.

Cell Formatting

Data in the cells is divided into two types: text (alphanumeric) and numeric data. Both numeric and text data can

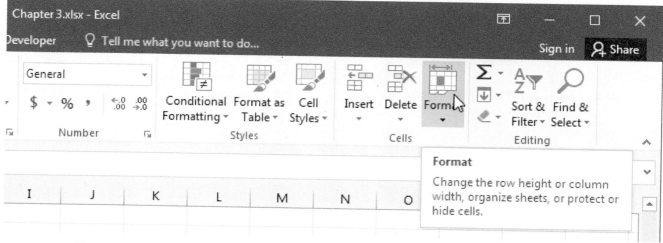

FIGURE 3-29 Home: Format Button

be aligned left, aligned right, or centered, and the font can be changed. The data is aligned left by selecting the desired cells and selecting **Home>Alignment>Align Left** button shown in Figure 3-31. The data is centered by selecting the desired cells and selecting the **Home>Alignment>Center** button shown in Figure 3-32. The data is aligned right by selecting the desired cells and selecting the **Home>Alignment>Align Right** button shown in Figure 3-33.

The font style for the data is changed by selecting the desired cells, selecting the **Home** menu tab, and then selecting the desired font from the Font dropdown box in the Font group. The Font dropdown box is shown in Figure 3-34.

The font size for the data is changed by selecting the desired cells, selecting the **Home** menu tab, and then selecting the desired font size from the Font Size dropdown box in the Font group. The Font Size dropdown box is shown in Figure 3-35.

The font is bolded by selecting the desired cells and (1) selecting the **Home>Font>Bold** button shown in Figure 3-36 or (2) by typing **Ctrl+B**.

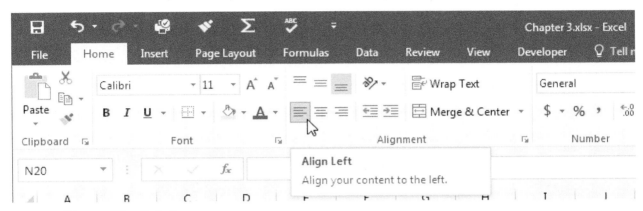

FIGURE 3-31 Home: Align Left Button

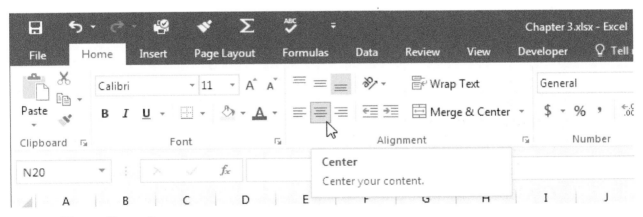

FIGURE 3-32 Home: Center Button

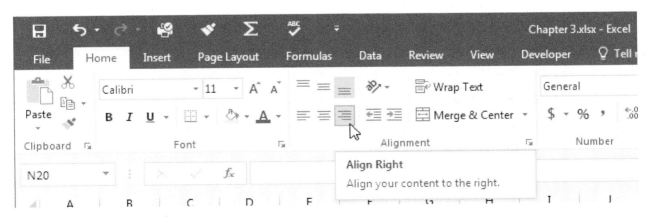

FIGURE 3-33 Home: Align Right Button

Alternatively, both the font style and font size can be changed and the font bolded from the Format Cells dialogue box, shown in Figure 3-37, which is opened by selecting the desired cells and performing one of the following actions: (1) by typing **Ctrl+1** and selecting the **Font** tab, (2) right clicking on one of the selected cells, selecting **Format Cells . . .** from the popup menu (shown in Figure 3-17), and selecting the **Font** tab, or (3) selecting the **Home** menu tab and selecting the **Format Cells: Font Dialogue Box Launcher** (shown in Figure 3-38) located at the lower-right corner of the Font group. From the Font: combo box, the user selects the font; from the Font style: combo box, the user can bold the font; and from the Size: combo box, the user selects the font size. The user must complete the changes by clicking on the **OK** box to close the Format Cells dialogue box.

For very wide text, the text may be wrapped to allow the text to occupy more than one line. This is done by selecting the cells in which the text is to be wrapped and selecting the **Home>Alignment>Wrap Text** button shown in Figure 3-39.

At times the user may want to have text cover a number of rows or columns, as is the case when creating headings. This is done by selecting the cells to be combined and selecting the **Home>Alignment>Merge & Center** button shown in Figure 3-40. In addition to merging the cells, the Merge & Center button centers the data. For the merge and center to be successful, the selected cells must be in the shape of a rectangle. If more than one of the cells contains data, only data from the upper-left cell is retained and the remaining data is lost. If data is to be lost, Excel will provide the user with the warning dialogue

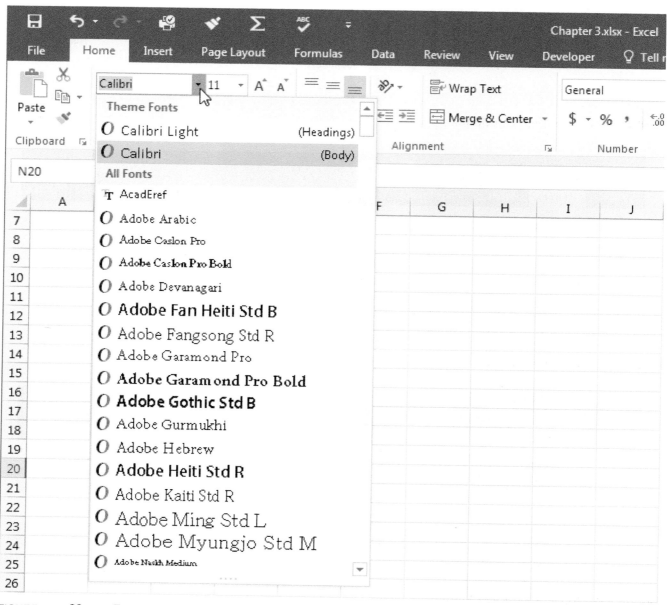

FIGURE 3-34 Home: Font Dropdown Box

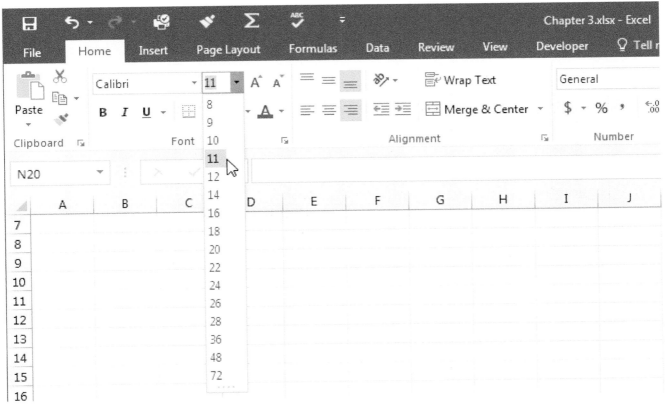

FIGURE 3-35 Home: Font Size Dropdown Box

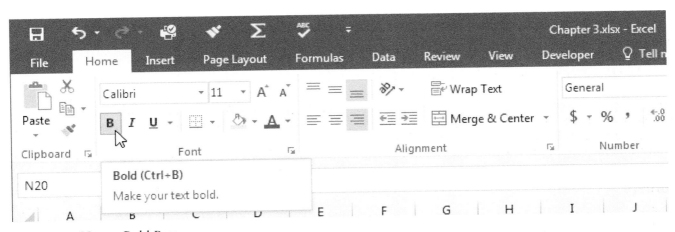

FIGURE 3-36 Home: Bold Button

box shown in Figure 3-41 before proceeding. If the user wants to continue, he or she must click on the **OK** button.

To unmerge cells, select the merged cells and click the **Merge & Center** button. The data from the cells will be placed in the upper-left cell.

With numeric data, the user often wants to control how the data is displayed. For estimating the user often wants to display the numbers as dollars and cents with commas between the thousands and hundreds. An easy way to do this is to select the desired cells and select **Home>Number>Comma Style** shown in Figure 3-42. The Comma Style button will format the data to an accounting style with two decimal places shown, negative numbers will be placed in parentheses, and a zero will be shown as a dash.

Numbers can be expressed as a percentage by selecting the desired cells and selecting the **Home>Number>Percentage Style** button shown in Figure 3-43.

The number of decimal places is increased by selecting the desired cells and selecting **Home>Number>Increase Decimal** button shown in Figure 3-44. The number of decimal places is decreased by selecting the desired cells and selecting **Home>Number>Decrease Decimal** button shown in Figure 3-45.

Other formatting styles, such as date and time, are selected from the Number tab of the Format Cells dialogue

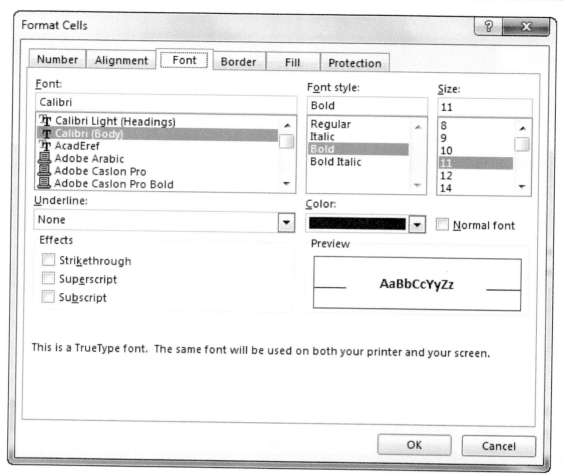

FIGURE 3-37 Font Tab of the Format Cells Dialogue Box

box (shown in Figure 3-46), which is open by (1) right clicking on one of the selected cells, selecting **Format Cells...** from the popup menu (shown in Figure 3-17), and selecting the **Number** tab or (2) selecting the **Home** menu tab and selecting the **Format Cells: Number** Dialogue Box Launcher at the lower-right corner of the Number group.

Borders

Borders are drawn around cells or groups of cells using the Borders button, located in the Font group on the Home menu tab. The Borders button is shown in Figure 3-47. The user may use the most recently used border by clicking on the left side of the button.

The user can select from the 12 preset borders shown in Figure 3-48 by selecting the small arrow on the right side of the Borders button to bring up the Borders popup menu. After selecting a preset border type, the popup menu will disappear and the Border button will change to show the most recently used preset border. To erase all borders associated with a cell, the user selects the **No Border** from the Borders popup menu shown in Figure 3-48.

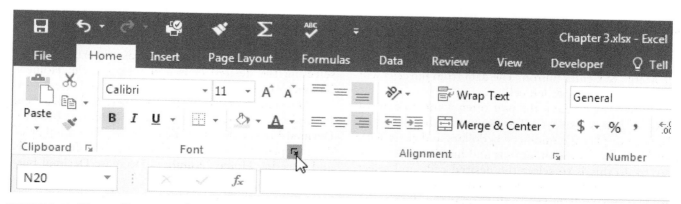

FIGURE 3-38 Home: Format Cells: Font Dialogue Box Launcher

Introduction to Excel 41

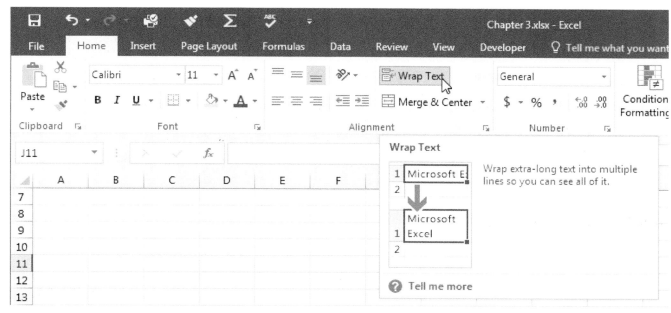

FIGURE 3-39 Home: Wrap Text Button

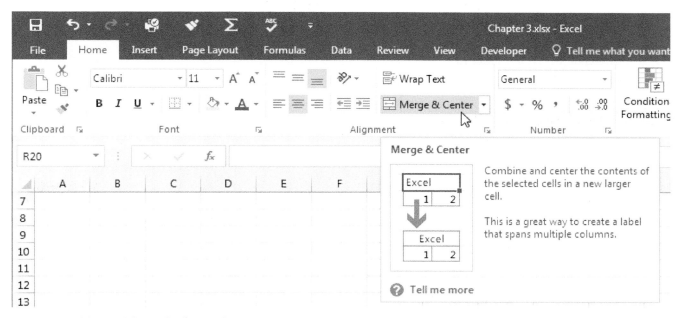

FIGURE 3-40 Home: Merge & Center Button

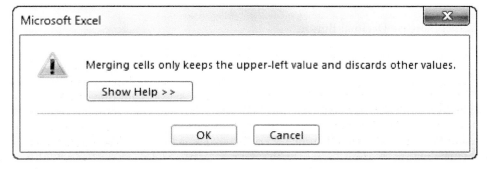

FIGURE 3-41 Warning Dialogue Box

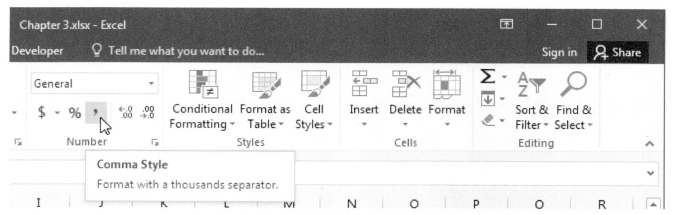

FIGURE 3-42 Home: Comma Style Button

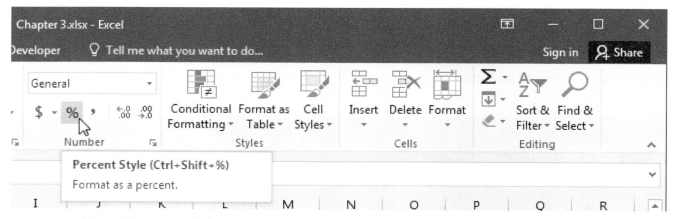

FIGURE 3-43 Home: Percentage Style Button

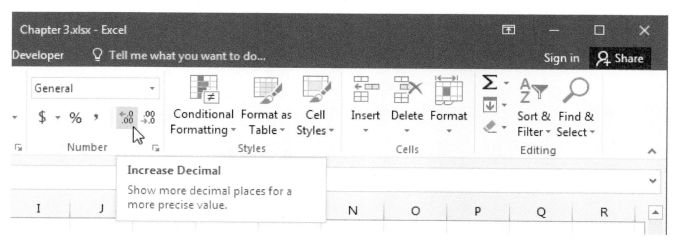

FIGURE 3-44 Home: Increase Decimal Button

Zooming

The extent of the spreadsheet shown on the computer screen is controlled by the Zoom slider located in the lower-right hand corner. The user may zoom in and out by sliding the control knob as shown in Figure 3-49 or by clicking on the Plus or Minus buttons located to the right and left of the slider. When clicking on the Plus or Minus buttons the zoom progresses in increments of 10%. From the Zoom dialogue box (shown in Figure 3-50), the user can select from one of the standard zoom percentages, select a custom zoom percentage, or zoom to fit the width or height of a selected number of cells. The Zoom dialogue box is opened by clicking on the Zoom level located to the right of the Zoom Slider (shown in Figure 3-51) or by selecting the **View>Zoom>Zoom** button (shown

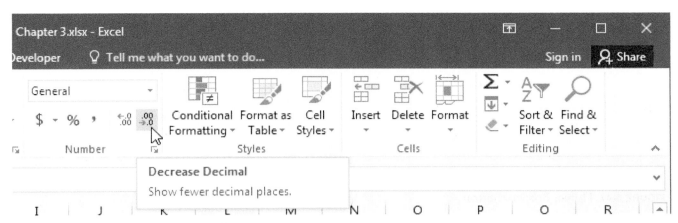

FIGURE 3-45 Home: Decrease Decimal Button

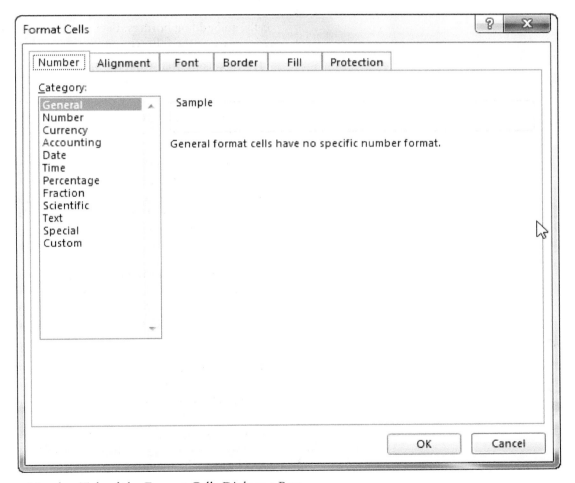

FIGURE 3-46 Number Tab of the Format Cells Dialogue Box

in Figure 3-52). By selecting a number of rows or columns and selecting **Fit selection** from the Zoom dialogue box the spreadsheet will zoom to match the width of the selected columns or the height of the selected rows. This may also be done by selecting the rows or columns and selecting the **View>Zoom>Zoom to Selection** button (shown in Figure 3-53). It is important to understand that changing the zoom will not change the scale at which the spreadsheet will be printed. This is done by setting the print scale, which is discussed later in this chapter.

44 CHAPTER THREE

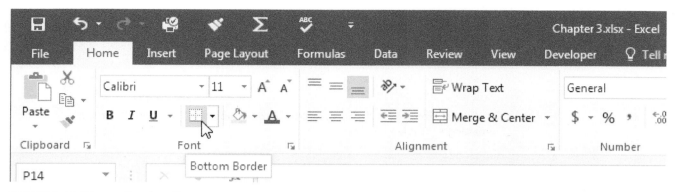

FIGURE 3-47 Home: Borders Button

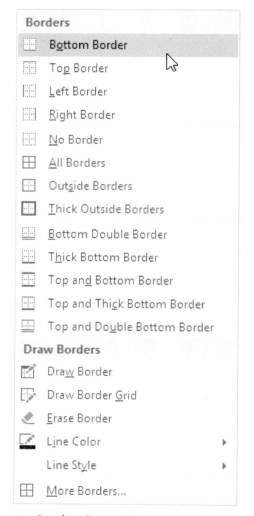

FIGURE 3-48 Borders Popup Menus

Exercise 3-4

In this exercise you will modify the file from Exercise 3-3 by changing the column widths, merging and formatting cells, and adding borders. Begin by changing the column widths using the following steps:

1. Make sure that Chapter 3, the workbook modified in Exercise 3-3, is open.

2. Change the width of Column A to 25 by right clicking on the letter A at the top of Column A, selecting **Column Width . . .** from the popup menu (shown in Figure 3-28) to bring up the Column Width dialogue box (shown in Figure 3-27), enter 25 in the Column Width: text box, and click the **OK** button.

3. Change the width of Column B to 6, that of Column C to 4, and that of Columns D through L to 10 by using the same procedures. Hint: Columns D through L can be changed at once by selecting Columns D through L using the Shift key, right clicking on any one of the highlighted columns, and selecting **Column Width . . .** from the popup menu to bring up the Column Width dialogue box. You can now change the width of all of the columns at once.

Next, format the cells using the following steps:

4. Center the text in Cells D2 through L2 by selecting Cells D2 through L2 using the Shift key and clicking **Home>Alignment>Center** button (shown in Figure 3-32).

5. Center the text in Column C by selecting the heading of Column C and clicking the **Home>Alignment>Center** button.

FIGURE 3-49 Zoom Slider

6. Change the formatting for columns E, I, K, and L to the accounting style by selecting Column E and holding down the **Ctrl** key while selecting Columns I, K, and L to highlight these columns. With the columns highlighted, click **Home>Number>Comma Style** button (shown in Figure 3-42) to change the formatting to the accounting style and then reduce the number of decimal points to zero by clicking **Home>Number>Decrease Decimal** button (shown in Figure 3-45) twice.

7. Change the formatting for columns D, F, G, H, and J to the comma style as done previously; however, do not change the number of decimal points.
8. Merge Cells A1 and A2 by highlighting Cells A1 and A2 and clicking **Home>Alignment>Merge & Center** button (shown in Figure 3-40).
9. Left justify the text in the merged cells by clicking **Home>Alignment>Align Left** button (shown in Figure 3-31).
10. Merge Cells B1 through C2 by highlighting Cells B1 through C2 and clicking **Home>Alignment>Merge & Center** button.
11. Merge Cells D1 and E1 in the same manner.
12. Merge Cells F1 through I1.
13. Merge Cells J1 and K1.
14. Merge Cells L1 and L2.

Next, you will underline the headings you have created using the following steps:

15. Underline Row 2 by highlighting Cells A1 through L2 and clicking the **Bottom Border** button (shown in Figure 3-47) in the Font group on the Home menu tab. If another border is shown in the Borders button, select the correct border by clicking on the small arrow to the right of the Borders button and selecting the **Bottom Border** from the Borders popup menu (shown in Figure 3-48).

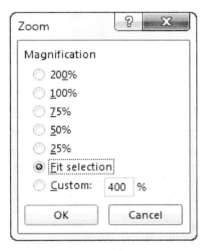

FIGURE 3-50 Zoom Dialogue Box

FIGURE 3-51 Zoom Level

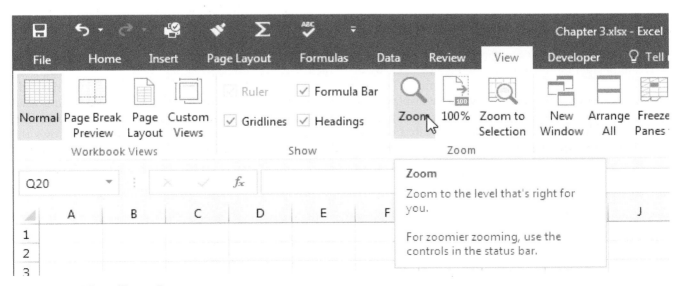

FIGURE 3-52 View: Zoom Button

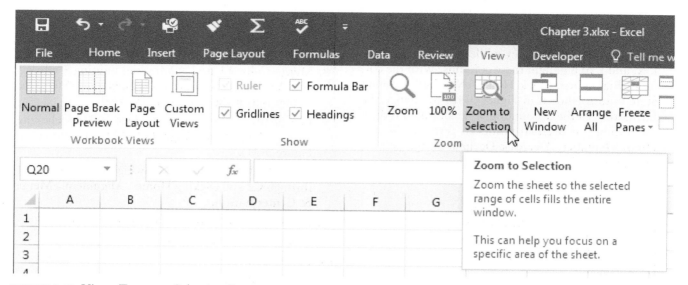

FIGURE 3-53 View: Zoom to Selection Button

Next, set the zoom so you can see Columns A through L using the following steps:

16. Select Columns A through L by clicking on Column A and hold down the **Shift** key while clicking on Column L.

17. Click on the **Zoom level** (shown in Figure 3-51) at the right of the Status bar to bring up the Zoom dialogue box (shown in Figure 3-50), select **Fit selection**, and click the **OK** button to close the Zoom dialogue box. Columns A through L should cover the entire width of the spreadsheet.

18. Save the workbook by clicking on the **Save** button on the Quick Access toolbar or typing **Ctrl+S**.

The first two rows of your spreadsheet should be the same as the first two rows of the spreadsheet in Figure 3-25.

Headers and Footers

Custom headers and footers can be added to each spreadsheet. The headers and footers may include text, data from the workbook and computer, or graphics. To edit headers and footers, the worksheet must be displayed in Page Layout view. The Page Layout view is displayed by clicking on the **Page Layout** button located on the Status bar (shown in Figure 3-54) or by clicking the **View>Workbook Views>Page Layout** button (shown in Figure 3-55). Once the spreadsheet is in Page Layout view, the user can click on the header or footer to edit the header or footer. Alternately, the user may click the **Insert>Text>Header & Footer** button (shown in Figure 3-56) to switch to the Page Layout view and open the header and footer for editing.

The headers and the footers are divided into three areas: left (for left-justified elements), center (for centered elements), and right (for right-justified elements). A sample header is shown in Figure 3-57. It is important to note that elements in any of these areas are not limited to their areas. They simply identify how the element is aligned. Text is typed in the headers or footers and the font is changed in the same manner as the font in the cells is changed. When the header or footer has been selected, the Header & Footer Tools: Design menu tab becomes visible. Elements are added to the header and footers by clicking on the **Header & Footer Tools: Design** tab and selecting the appropriate element from the Header & Footer Elements group. The elements of the Header & Footer Elements group, along with an explanation of their use, are shown in Figure 3-58.

The Page Number button is used to automatically number the pages. The Number of Pages button is used to display the total number of pages in the worksheet. This, along with the Page Number button, allows the user to number the pages "1 of 5," "2 of 5," and so

FIGURE 3-54 Page Layout Button

Introduction to Excel   47

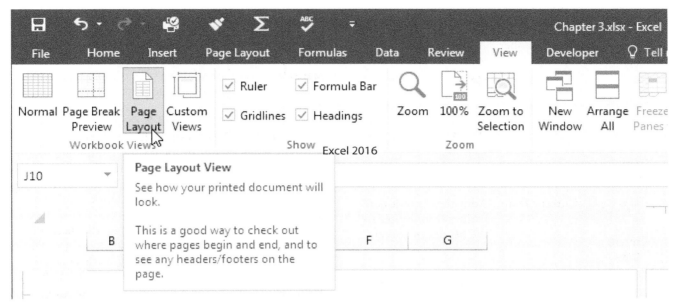

FIGURE 3-55 View: Page Layout Button

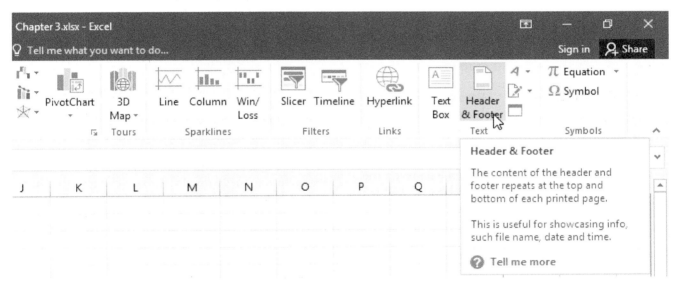

FIGURE 3-56 Insert: Header & Footer Button

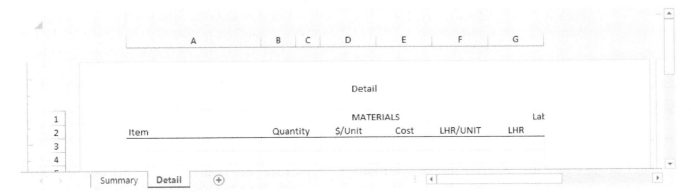

FIGURE 3-57 Sample Header

Icon	Name	Use
Page Number	Page Number	Adds the page number to each page of the header or footer.
Number of Pages	Number of Pages	Adds the total number of pages in the header or footer.
Current Date	Current Date	Adds the current date from the computer to the header or footer.
Current Time	Current Time	Adds the current time from the computer to the header or footer.
File Path	File Path	Adds the file name and path (location of the folder) to the header or footer.
File Name	File Name	Adds the file name to the header or footer.
Sheet Name	Sheet Name	Adds the name on the worksheet tab to the header or footer.
Picture	Picture	Brings up the Picture dialogue box that is shown in Figure 3-24, which allows pictures to be added to the header or footer.
Format Picture	Format Picture	Allows inserted pictures to be formatted.

FIGURE 3-58 Elements Available to Headers and Footers

forth. By numbering the pages in this manner, it is easy to see if pages at the end of the document are missing.

The Current Date button adds the current date from the computer to the header or footer. The Current Time button adds the current time from the computer to the header or footer. The date and time will be automatically updated when the worksheet is printed. By putting the date and time on each worksheet, it is easy to determine which printed version of the worksheet is the most current. This is very useful when preparing bids and the user wants to make sure he or she is using the most current worksheet.

The File Path button adds the file name and path (the location of the folder) to the header or footer. This is useful when trying to locate the computer file for a worksheet that has been printed. The File Name button adds just the file name to the header or footer. The Sheet Name button adds the name on the worksheet tab to the header or footer.

The Insert Picture button brings up the Insert Picture dialogue box, which allows the user to add a picture, such as the company's logo, to the worksheet. The Insert Picture dialogue box is shown in Figure 3-24. Pictures are added to the headers and footers in the same manner as pictures are added to the worksheet. The Format Picture button brings up the Format Picture dialogue box, which allows inserted pictures to be resized, rotated, and cropped. The Format Picture dialogue box is shown in Figure 3-59.

The user may move from the header to the footer by clicking the **Header & Footer Tools: Design>Navigation>Go to Footer** button (shown in Figure 3-60) and from the footer to the header by clicking the **Header & Footer Tools: Design>Navigation>Go to Header** button (shown in Figure 3-61).

Once the header and footers have been set up, the user clicks on a cell in the worksheet to close the header or footer and return to editing the worksheet. The user may return to the Normal view by clicking on the **Normal** button located on the Status bar (shown in Figure 3-62) or by clicking the **View>Workbook Views>Normal** button (shown in Figure 3-63).

Exercise 3-5

In this exercise you will modify the file from Exercise 3-4 by adding headers and footers to the Detail worksheet. In the header you will add the sheet name. In the footer you will add the date and time (which will record the date and time the spreadsheet is printed), the file name, the page number, and a place for the preparer and reviewer to initial. Begin by adding a header using the following steps:

1. Make sure that Chapter 3, the workbook created in Exercise 3-1, is open.
2. Click on the **Insert>Text>Header & Footer** button (shown in Figure 3-56) to edit the header.
3. Place the cursor in the center section and click the **Header & Footer Tools: Design>Header & Footer Elements>Sheet Name** button. The Header should look like Figure 3-57. When the cursor is

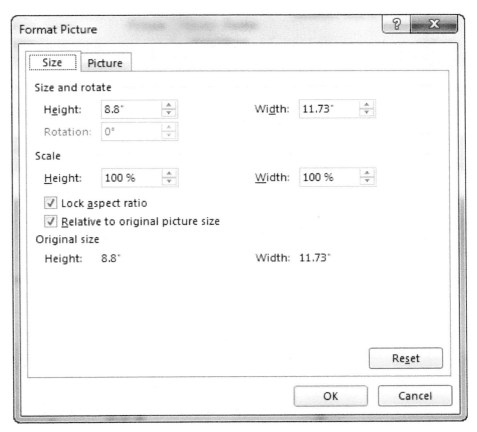

FIGURE 3-59 Format Picture Dialogue Box

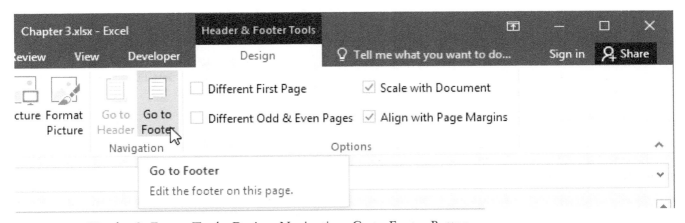

FIGURE 3-60 Header & Footer Tools: Design: Navigation: Go to Footer Button

moved from the center section of the header, the &[Tab] will be replaced with the name of the tab, in this case *Detail*.

Next, add a footer using the following steps:

4. Right click the **Header & Footer Tools: Design>Navagation>Go to Footer** button (see Figure 3-60) to move to the footer.
5. Place the cursor in the right section and type *Prepared by:* followed by pressing the **Underline** key 10 times.
6. Press the **Return** key and type *Checked by:* followed by pressing the **Underline** key 10 times.
7. Place the cursor in the center section and click the **Header & Footer Tools: Design>Header & Footer Elements>Page Number** button.
8. Following "&[Page]" type *of*, including a space before and after the *of*.
9. Click the **Header & Footer Tools: Design>Header & Footer Elements>Number of Pages** button and press the **Enter** button to move it up one line.

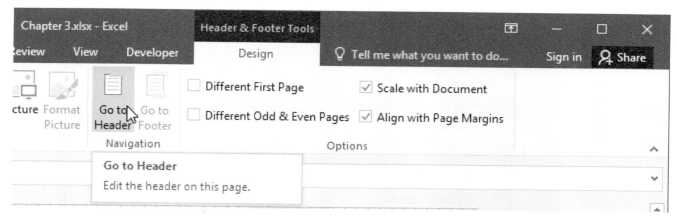

FIGURE 3-61 Header & Footer Tools: Design: Navigation: Go to Header Button

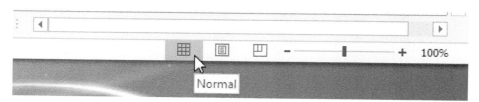

FIGURE 3-62 Normal Button on the Status Bar

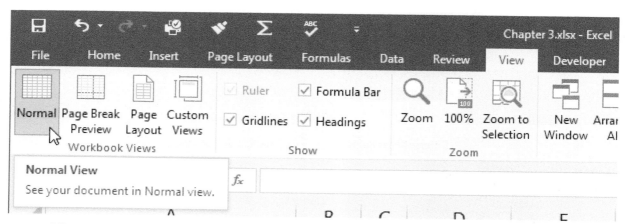

FIGURE 3-63 View: Normal Button

10. Place the cursor in the left section, click the **Header & Footer Tools: Design>Header & Footer Elements>Current Date** button, type the **Space** bar, and click the **Header & Footer Tools: Design > Header & Footer Elements > Current Time** button.
11. Press the **Enter** key to move to the following line and click the **Header & Footer Tools: Design>Header & Footer Elements>File Name** button.
12. The Footer should look like Figure 3-64. When the cursor is moved from the left section of the footer, the &[Date] &[Time] will be replaced with the current date and time and the &[File] will be replaced with the file name.
13. Click on any cell to return to the worksheet.
14. Click the **View>Workbook Views>Normal** button (shown in Figure 3-63) to return to the Normal view.
15. Save the workbook by clicking on the **Save** button on the Quick Access toolbar or typing **Ctrl+S**.

WRITING FORMULAS

Formulas allow Excel to perform simple or complex calculations. The formulas may contain mathematical calculations and functions. In this segment, the writing of formulas is discussed. Formulas may consist of

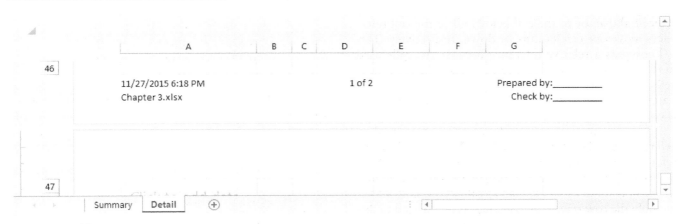

FIGURE 3-64 Footer

numbers, variables, and operators. Numbers are typed in from the key board and do not change. Cells within the worksheets that are referenced by the formula are known as variables. Variables are entered by typing the cell reference or by selecting the cell with the mouse while writing the formula.

Operators

Excel allows the use of the following operators:

- = Equals Sign
- + Plus Sign
- − Minus Sign
- * Multiplication Sign or Asterisk
- / Division or Backslash
- (Opening Parentheses
-) Closing Parentheses
- ^ Caret

For Excel to recognize the data being entered into a cell as a formula, the formula must begin with an equal sign, a plus sign, or a minus sign. If the user enters a plus or minus sign, Excel will add an equal sign before the plus or minus sign.

For each opening parenthesis there must be a closing parenthesis. As the formula is entered, Excel will color code the parentheses in pairs. Should the user forget to have an equal number of opening and closing parentheses, Excel will warn the user of the problem and suggest a correction.

The caret character allows the user to raise a value to a power. For example, the mathematical function X^Y is written in Excel as follows:

$$X\wedge Y$$

The formula 4^2 is equal to 4 squared or 16. To find the square root of a number, raise the number to the 0.5 power as follows:

$$X\wedge 0.5$$

Order of Operation

Excel performs the calculation of a formula in the following order:

First, the calculations enclosed by the parentheses are performed, starting with the innermost pair and working out to the outermost pair of parentheses.
Second, the variable or number raised to a power is performed.
Third, multiplication and division are performed.
Finally, addition and subtraction are performed.
When two calculations have the same priority, they are performed left to right.

For example, Excel calculates the following formula by adding 2 and 1 together to get 3, then squaring the 3 to get 9, adding 7 to the 9 to get 16, dividing the 16 by 4 to get 4, and, finally, adding 1 to get 5:

$$= ((2 + 1)\wedge 2 + 7)/4 + 1$$
$$= (3\wedge 2 + 7)/4 + 1$$
$$= (9 + 7)/4 + 1$$
$$= 16/4 + 1$$
$$= 4 + 1$$
$$= 5$$

Editing Formulas

Existing formulas are edited by selecting the cell (clicking on the cell) and editing the formula in the Formula bar. When building complex formulas, it is a good idea to build formulas in steps, making sure each step of the formula works before adding the next step.

Absolute and Relative References

By default, when a cell is copied into the cell to the right, all of the cell references move one cell to the right. Likewise, when a cell is copied to the cell to the left, to the cell above, or to the cell below, all of the cell references move one cell left, up, or down, respectively. This is known as a relative reference because the cells used in the formula maintain the same relative position with cells containing the formula. Figure 3-65 shows what happens when Cell B2 contains a relative reference and is copied to Cells A1 through C3.

The dollar sign ($) used in conjunction with a cell reference allows the user to fix a portion of the cell's reference as it is copied. For example, the use of a cell reference of $H22 fixes the column portion of the reference to the H column as it is copied to other columns, allowing only the row reference to change as it is copied to different rows. The column portion of the reference would then be an absolute reference and the row portion would be relative. Figure 3-66 shows what happens when Cell B2 contains an absolute-column, relative-row reference and is copied to Cells A1 through C3.

Likewise, the use of a cell reference of H$22 fixes the row portion of the reference to the twenty-second row as it is copied to other rows, allowing only the column reference to change as it is copied to different columns. The row portion of the reference would then be an absolute reference and the column portion would be relative. Figure 3-67 shows what happens when Cell B2 contains a relative-column, absolute-row reference and is copied to Cells A1 through C3.

The use of a cell reference of $H$22 would fix both the row and the column, preventing the row and column reference from being changed as the cell is copied. The entire reference would then be an absolute reference. Figure 3-68 shows what happens when Cell B2 contains an absolute reference and is copied to Cells A1 through C3.

The user may move between the absolute and relative reference types by highlighting the reference to be changed and pressing the **F4** key. This cycles the reference from relative, to both the column and rows being absolute, to only the row being absolute, to only the

	A	B	C
1	=G21	=H21	=I21
2	=G22	=H22	=I22
3	=G23	=H23	=I23

FIGURE 3-65 Copying a Relative Reference

	A	B	C
1	=$H21	=$H21	=$H21
2	=$H22	=$H22	=$H22
3	=$H23	=$H23	=$H23

FIGURE 3-66 Copying an Absolute-Column, Relative-Row Reference

	A	B	C
1	=G$22	=H$22	=I$22
2	=G$22	=H$22	=I$22
3	=G$22	=H$22	=I$22

FIGURE 3-67 Copying a Relative-Column, Absolute-Row Reference

	A	B	C
1	=$H$22	=$H$22	=$H$22
2	=$H$22	=$H$22	=$H$22
3	=$H$22	=$H$22	=$H$22

FIGURE 3-68 Copying an Absolute Reference

column being absolute, and back again to a relative reference.

Exercise 3-6

In this exercise you will explore absolute and relative references. Begin by completing the following steps:

1. Begin by opening a new workbook. If Excel is not open, open it. If Excel is open, type **Ctrl+N** to open a new workbook.
2. Type the data shown in Table 3-2 into the indicated cells.

Next, create a relative reference using the following steps.

3. Type = B2 into Cell B6.
4. Copy Cell B6 to Cells A5 through C7 by right clicking on Cell B6 and selecting **Copy** from the popup menu (shown in Figure 3-17), selecting Cells A5 through C7, and clicking on the **Paste** button from the Home tab (shown in Figure 3-18).

Cells A5 through A7 should contain the numbers shown in Figure 3-69.

Next, create an absolute reference using the following steps:

5. Type = $B$2 into Cell B10.
6. Copy Cell B10 to Cells A9 through C11.

Because it is an absolute reference, the formula in all of the cells is the same as the formula entered into Cell B10; therefore, the value in all of the cells is the same as the value in Cell B2 or 5. Cells A9 through C11 should appear as they do in Figure 3-69.

Next, create a formula with an absolute row reference using the following steps:

7. Type = B$2 into Cell B14.
8. Copy Cell B14 to Cells A13 through C15.

TABLE 3-2 Data for Example 3-6

Cell	Data
A1	1
A2	2
A3	3
B1	4
B2	5
B3	6
C1	7
C2	8
C3	9

Because the row portion of the reference is an absolute reference, it does not change. The value for Cells A13 through A15 should be 2, the value for Cells B13 through B15 should be 5, and the value for Cells C13 through C15 should be 8. Cells A13 through C15 should appear as they do in Figure 3-69.

Next, create a formula with an absolute column reference using the following steps:

9. Type = $B2 into Cell B18.
10. Copy Cell B18 to Cells A17 through C19.

Because the column portion of the reference is an absolute reference, it does not change. The value for Cells A17 through C17 should be 4, the value for Cells A18 through C18 should be 5, and the value for Cells A19 through C19 should be 6. Cells A17 through C19 should appear as they do in Figure 3-69.

11. Close the workbook without saving it.

BASIC FUNCTIONS

In addition to mathematical equations, computer functions can also be included in the formulas. Functions are entered by (1) typing the functions, (2) selecting the **Formulas** menu tab, selecting the function category from the Function Library group, and selecting the function from the popup menu, or (3) by clicking on the **Insert Function** button (shown in Figure 3-70) on the formula bar to bring up the Insert Function dialogue box (shown in Figure 3-71). This section covers the rounding functions (including the CEILING and FLOOR functions), the SUM function, the AVERAGE function, and the IF function. Additional functions are available from the Insert Function dialogue box.

From the Function dialogue box the user may make a function available for selection by selecting All or the function's category from the Or select a category: dropdown box. Alternatively, the user may perform a key word search by typing key words in the Search for a function: text box and clicking on the **Go** button. When the function is shown in the Select a function: list box, the user double clicks on the function or selects the function and clicks on the **OK** button to bring up the Function Arguments dialogue box. The Function Arguments dialogue box for the ROUND function is shown in Figure 3-72. Each function has its own Function Arguments dialogue box.

The Function Arguments dialogue box prompts the user for the information necessary to complete the function. The user enters the requested information and clicks the **OK** button to close the Function Arguments dialogue box and create the function.

Rounding

Excel provides the user three standard rounding functions: (1) ROUND, which rounds to the closest number; (2) ROUNDDOWN, which rounds the number down; and (3) ROUNDUP, which rounds the number up. Each of these functions requires the user to indicate the number or cell containing the number to be rounded (number) and the number of digits to round to (num_digits), in the following format:

=ROUND(<number>,<num_digits>)

The user does not need to capitalize the word "round" because Excel will automatically capitalize the word "round" when the formula is completed. Rounding to a positive number of digits represents the number of digits

	A	B	C
1	1	4	7
2	2	5	8
3	3	6	9
4			
5	1	4	7
6	2	5	8
7	3	6	9
8			
9	5	5	5
10	5	5	5
11	5	5	5
12			
13	2	5	8
14	2	5	8
15	2	5	8
16			
17	4	4	4
18	5	5	5
19	6	6	6

FIGURE 3-69 Spreadsheet for Exercise 3-6

54 CHAPTER THREE

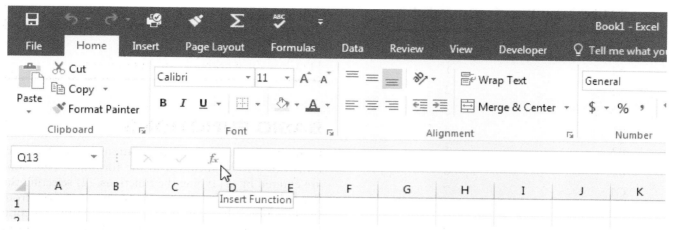

FIGURE 3-70 Insert Function Button

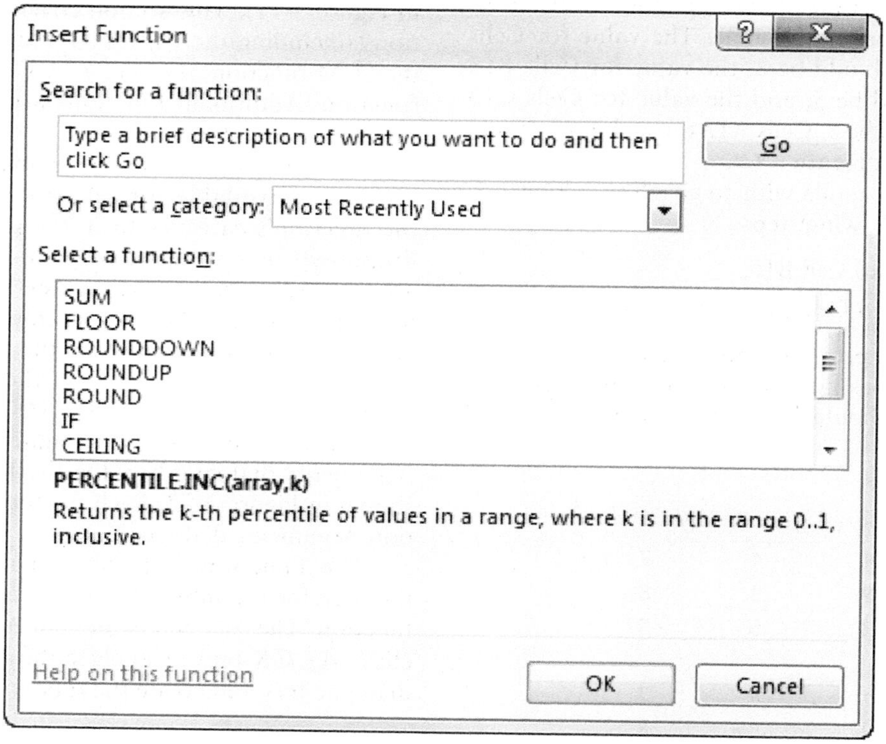

FIGURE 3-71 Insert Function Dialogue Box

on the right side of the decimal point. For example, if the number of digits equals 2, the function rounds to hundredths. The function

$$=ROUND(123.228,2)$$

rounds 123.228 to 123.23. Figure 3-72 shows how this function would be set up using the Function Arguments dialogue box for the ROUND function. Rounding to a negative number of digits represents the number of digits on the left side of the decimal point. The function

$$=ROUND(123.228,-2)$$

rounds 123.228 to 100.00. The ROUNDUP function always rounds the number up and the ROUNDDOWN function always rounds the number down. The function

$$=ROUNDUP(123.228,2)$$

rounds 123.228 up to 123.23, whereas the function

$$=ROUNDDOWN(123.228,2)$$

rounds 123.228 down to 123.22.

The three standard rounding functions can only be used to round to decimal places, such as the hundreds, tens, whole numbers, tenths, hundredths, and so forth.

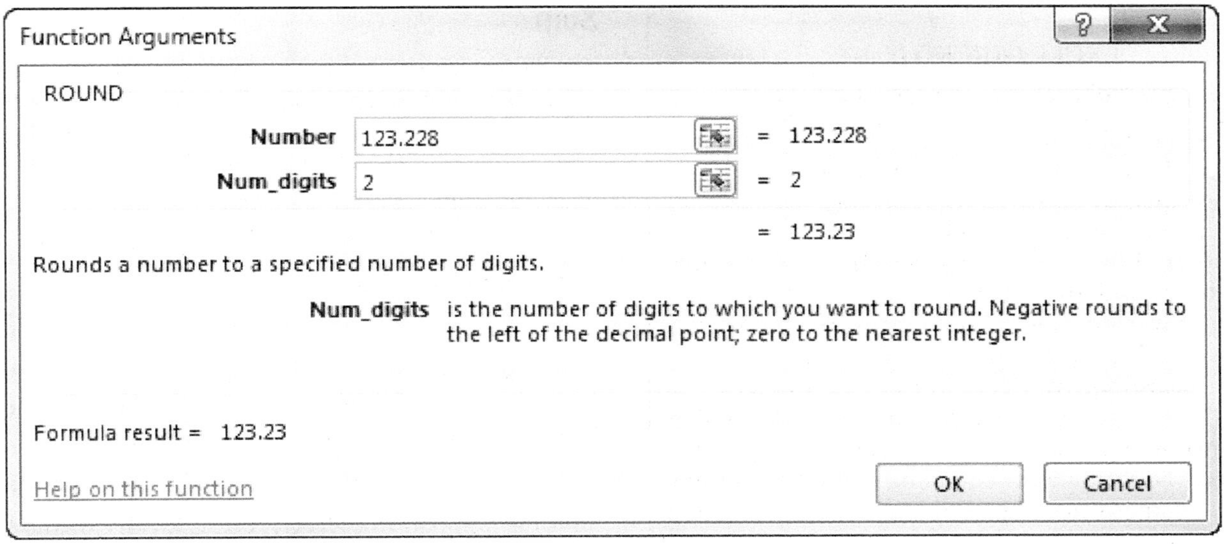

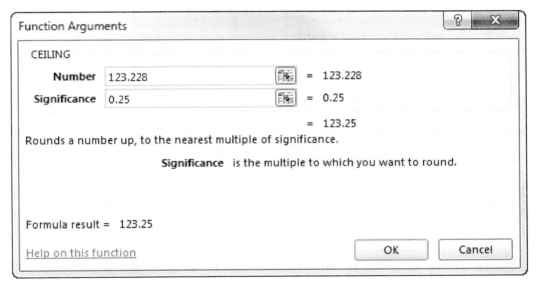

FIGURE 3-72 Function Arguments Dialogue Box for the ROUND Function

FIGURE 3-73 Function Arguments Dialogue Box for the CEILING Function

The rounding functions will not round to quarters of a yard. For this we must use the CEILING or FLOOR functions.

Ceiling

The CEILING function may be used to round up a number to the next multiple of a specified number (referred to by Excel as significance). For example, if the significance is 0.25, the CEILING function will round up to the next 0.25 increment. The CEILING function requires the user to indicate the number or cell containing the number to be rounded (number) and the significance, in the following format:

=CEILING(<number>,<significance>)

For example, the function

=CEILING(123.228,0.25)

rounds 123.228 to 123.25. Figure 3-73 shows how this function would be set up using the Function Arguments dialogue box for the CEILING function.

Floor

The FLOOR function works in the same way as the CEILING function except it rounds down. For example, the function

=FLOOR(123.228,0.25)

rounds 123.228 to 123.00.

EXCEL QUICK TIP 3-4
Rounding to Nearest Quarter Yards

Excel's rounding, CEILING, and FLOOR functions do not allow the user to automatically round to the nearest quarter-cubic-yards because Excel's rounding functions will only round to decimal places (such as hundredths, tenths, tens, hundreds, and whole numbers) and the CEILING and FLOOR functions only round up or down, not to the nearest quarter-cubic-yard. However, the user can round to the nearest quarter-cubic-yards by converting the units to quarter-cubic-yards, rounding the number to a whole number, and converting back to cubic yards. The formula for rounding a number in cubic yards to the nearest quarter-cubic-yard is written as follows:

$$=\text{ROUND}(<\text{number}>*4,0)/4$$

For example, the following function is used to round 123.228 cubic yards to the nearest quarter-cubic-yard:

$$=\text{ROUND}(123.228*4,0)/4$$

The function first multiplies 123.228 cubic yards by 4 to get 492.912 quarter-cubic-yards, rounds 492.912 quarter-cubic-yards to 493 quarter-cubic-yards (the nearest whole number), and then divides 493 quarter-cubic-yards by 4 to get 123.25 cubic yards. Formulas can be written for other fractions of a whole number by using the same concepts.

Sum

The SUM function adds a group of numbers. It is particularly useful when adding a column, row, or block numbers, inasmuch as only the cells containing the first and last number need to be included in the formula. The SUM function is written as follows:

$$=\text{SUM}(<\text{beginning Cell}>:<\text{ending Cell}>)$$

In Excel, the formula

$$=\text{SUM}(C1:C5)$$

adds the numbers in Cells C1, C2, C3, C4, and C5. Noncontiguous cells may be added by placing a comma between the cells references. The formula

$$=\text{SUM}(C1:C5,D17)$$

adds the numbers in Cells C1, C2, C3, C4, C5, and D17. The SUM function is quickly inserted into a cell by clicking the **Formula>Function Library>AutoSum** button shown in Figure 3-74. The AutoSum will place a moving dashed line around the cells that Excel thinks the user wants to sum. If the user wants to include different cells, the user must select the cells. The AutoSum is completed by pressing the **Enter** key.

Average Function

The AVERAGE function allows the computer to average the values in a range of cells. Only the cells containing numbers are included in the average. The AVERAGE function is written as follows:

$$=\text{Average}(<\text{beginning Cell}>:<\text{ending Cell}>)$$

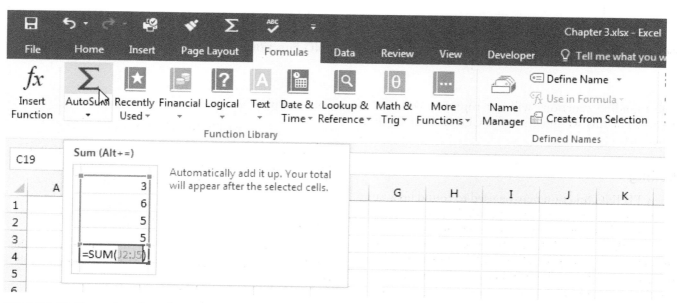

FIGURE 3-74 Formulas: AutoSum Button

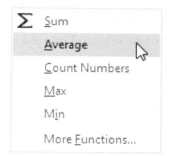

FIGURE 3-75 Average Example

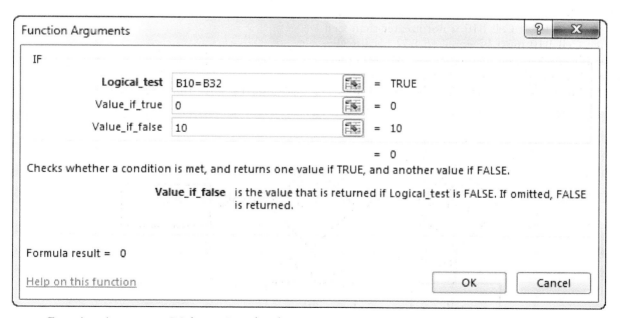

FIGURE 3-76 Popup Menu

In Excel, the formula

$$=\text{Average}(C1:C5)$$

takes the average of the numbers in Cells C1, C2, C3, C4, and C5, ignoring blank cells and cells with text. If Cell C1 is 3; Cell C2 is blank; Cell C3 contains the text "Joe"; Cell C4 is 10; and Cell C5 is 17, the AVERAGE function ignores Cells C2 and C3 and averages Cells C1, C4, and C5 to get 10. This example is shown in Figure 3-75.

The AVERAGE function may be typed directly into a cell or may be quickly inserted into a cell by clicking on the arrow below the AutoSum button (shown in Figure 3-74) in the Function Library group on the Formula menu tab to bring up the popup menu (shown in Figure 3-76) and selecting **Average** from this menu. When selecting Average from the popup menu, Excel will place a moving dashed line around the cells that Excel thinks the user wants to average. If the user wants to include different cells, the user must select the cells. The AVERAGE is completed by pressing the **Enter** key.

IF Function

The IF function allows the computer to select between two responses based on the value found in a specified cell. If the value in the specified cell meets the test requirements, then Excel does one thing, and if it does not meet the test requirement, it does another. The IF function is written as follows:

$$=\text{IF}(<\text{logical\_test}>,<\text{value\_if\_true}>,<\text{value\_if\_false}>)$$

The logical test consists of any statement that is either true or false. The logical test should consist of (1) the cell to be tested, (2) a logical operator that identifies the type of test to perform, and (3) a cell, function, or value to test against. Excel allows for the use of the following logical operators:

\> Greater Than
< Less Than
= Equal to
>= Greater Than or Equal to

FIGURE 3-77 Function Arguments Dialogue Box for the IF Function

<= Less Than or Equal to
<> Not Equal to

For example, the formula

$$=IF(B10=B32,0,10)$$

is equal to 0 if Cell B10 is equal to Cell B32 and is equal to 10 if Cell B10 is not equal to Cell B32. Figure 3-77 shows how this function would be set up using the Function Arguments dialogue box for the IF function.

The value_if_true and the value_if_false may contain numbers, cell references, functions, and text. For example, the formula

$$=IF(B10=B32,B34*B35,SUM(C17:C28))$$

is equal to Cell B34 multiplied by Cell B35 if Cell B10 is equal to B32, and is equal to the sum of Cells C17 through C28 if Cell B10 is not equal to Cell B32.

When using text for the value_if_true or the value_if_false, the text is placed between quotation marks. For example, the formula

$$=IF(B10=B32,\text{"True"},\text{"False"})$$

is equal to "True" if Cell B10 is equal to B32, and is equal to "False" if Cell B10 is not equal to Cell B32.

The user can select among more than two options by creating nested IF functions: placing one IF function inside another IF function, as follows:

$$=IF(B10<B32,-10,IF(B10>B32,10,0))$$

The logic for this function is shown in Figure 3-78. This function will first test to see if Cell B10 is less than Cell B32. If Cell B10 is less than Cell B32, the function is equal to −10. If Cell B10 is not less than Cell B32, the function will test to see if Cell B10 is greater than Cell B32. If Cell B10 is greater than Cell B32, then the function is equal to 10. If Cell B10 is not greater than Cell B32, then Cell B10 must be equal to B32 because we have already checked and know that Cell B10 is not less than Cell B32, and the function is equal to zero.

These functions are used in the following exercise.

Exercise 3-7

In this exercise you will practice using the rounding functions, the SUM function, the AVERAGE function, and the IF function. At the completion of this exercise your spreadsheet should look like Figure 3-79.

1. Begin by opening a new workbook. If Excel is not open, open Excel. If Excel is open, type **Ctrl+N** to open a new workbook.

In the first section you will work with the rounding functions using the following steps:

2. Enter the data in Table 3-3 into the indicated cells.
3. Type 57.3333 in Cell A1. Cells B1 through C4 should have the values shown in Table 3-4.
4. Experiment with the rounding function by changing Cell A1 to different numbers.
5. Now change Cell A1 back to 57.3333.

In this next section you will sum Cells B1 through C4 with the SUM function using the following steps:

6. Select Cell C5, click the **Formula>Function Library>AutoSum** button (shown in Figure 3-74), select Cell B1, hold down the left mouse button, and drag the cursor to Cell C4. A dashed line should

	A	B	C	D
1	57.3333	57.33	57	57.5
2		57.34	58	58
3		57.33	57	57.25
4		57.25	100	56
5		62.65625	501.25	
6				
7				
8				
9				
10		12	Equal to or Greater Than	
11		11	Greater Than	

FIGURE 3-79 Spreadsheet for Exercise 3-7

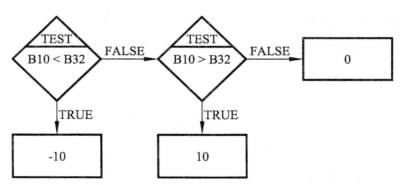

FIGURE 3-78 Nested IF function

TABLE 3-3 Data for ROUND Function

Cell	Data
B1	=ROUND(A1,2)
B2	=ROUNDUP(A1,2)
B3	=ROUNDDOWN(A1,2)
B4	=ROUND(A1*4,0)/4
C1	=ROUND(A1,0)
C2	=ROUNDUP(A1,0)
C3	=ROUNDDOWN(A1,0)
C4	=ROUND(A1,-2)
D1	=CEILING(A1,0.25)
D2	=CEILING(A1,2)
D3	=FLOOR(A1,0.25)
D4	=FLOOR(A1,2)

TABLE 3-4 Results for 57.3333

Cell	Value	Explanation
B1	57.33	Rounded to nearest hundredth
B2	57.34	Rounded up to nearest hundredth
B3	57.33	Rounded down to nearest hundredth
B4	57.25	Rounded to nearest quarter
C1	57	Rounded to nearest whole number
C2	58	Rounded up to nearest whole number
C3	57	Rounded down to nearest whole number
C4	100	Rounded to nearest hundred
D1	57.5	Rounded up to nearest quarter
D2	58	Rounded up to nearest even number
D3	57.25	Rounded down to nearest quarter
D4	56	Rounded down to nearest even number

be around Cells B1 through C4, and =SUM(B1:C4) will appear in Cell C5.

7. Press the **Enter** key to complete the function. Cell C5 should now contain 501.25.

In this next section you will average Cells B1 through C4 with the AVERAGE function using the following steps:

8. Select Cell B5 and type = AVERAGE(B1:C4).
9. Press the **Enter** key to complete the function. Cell B5 should now contain 62.65625 when five decimal places are shown.

In this next section you will create an IF function and a nested IF function using the following steps:

10. In Cell C10, type = IF(A10<A11, "Less Than", "Equal to or Greater Than"). If the value in Cell A10 is less than the value in Cell A11, the words *Less Than* will appear in Cell C10. If Cell A10 is equal to or greater than Cell A11, the words *Equal to or Greater Than* will appear in Cell C10.

11. In Cell C11, type = IF(A10<A11, "Less Than", IF(A10 = A11,"Equal to","Greater Than")). If the value in Cell A10 is less than the value in Cell A11, the words *Less Than* will appear in Cell C11. If Cell A10 is equal to Cell A11, the words *Equal to* will appear in Cell C11. If Cell A10 is greater than Cell A11, the words *Greater Than* will appear in Cell C11.

12. In Cell A10, type *10* and in Cell A11 type *11*. The words *Less Than* should appear in Cells C10 and C11.

13. In Cell A10, type *11*. The words *Equal to or Greater Than* should appear in Cell C10 and the words *Equal to* should appear in Cell C11.

14. In Cell A10, type *12*. The words *Equal to or Greater Than* should appear in Cell C10 and the words *Greater Than* should appear in Cell C11.

15. Close the workbook without saving it.

PRINTING

A worksheet is printed by clicking on the **Quick Print** button on the Quick Access toolbar or by opening the Print dialogue box. By clicking on the **Quick Print** button on the Quick Access toolbar (shown in Figure 3-80), the worksheet is printed without giving the user a chance to change the printing settings—which printer, number of copies, range, and so forth. If the **Quick Print** button does not appear on the Quick Access toolbar, the user may add it by clicking on the arrow at the right of the Quick Access toolbar (shown in Figure 3-81) and selecting **Quick Print** from the popup menu shown in Figure 3-82.

To open the Print dialogue box, shown in Figure 3-83, the user may (1) type **Ctrl+P** or (2) select the **File** tab and click on **Print** in the left pane to bring up the Print dialogue box.

From the Print Dialogue Box, the user can select the printer to be used under Printer and set the properties for the printer by clicking on **Printer Properties** to bring up a dialogue box for the printer. The user can select to print the active sheet, the entire workbook, or only the selected (the highlighted) cells from the first settings dropdown box.

Page Orientation

The orientation of the page is selected by clicking on the **Page Layout**>**Page Setup**>**Orientation** (shown in

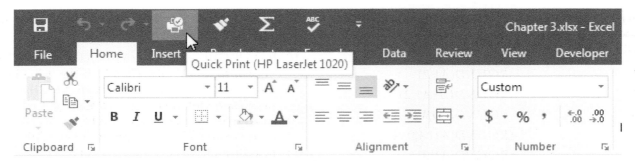

FIGURE 3-80 Quick Print Button

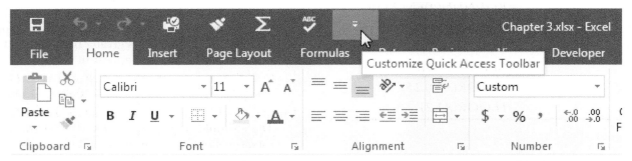

FIGURE 3-81 Customize Quick Access Toolbar Button

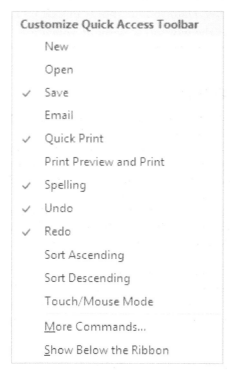

FIGURE 3-82 Customize Quick Access Toolbar Popup Menu

Figure 3-84), and clicking on **Portrait** or **Landscape** from the popup menu.

Repeating Row at the Top of Each Page

Rows from the spreadsheet can be repeated at the top of each page to create column headings. This is done by clicking the **Page Layout>Page Setup>Print Titles** button (shown in Figure 3-85 to bring up the Sheet tab of the Page Setup dialogue box (shown in Figure 3-86), typing the beginning row followed by a colon followed by the ending row in the Rows to repeat at top: text box, and clicking on the **OK** button to close the Page Setup dialogue box. The settings in Figure 3-86 would repeat the first row of the worksheet on every page printed.

Printing Gridlines

To have the gridlines of the workbook print, the user must check the **Gridlines** check box on the Sheet tab of the Page Setup dialogue box (shown in Figure 3-86). Alternately, the user may click on the **Page Layout>Sheet Options>Gridlines: Print** check box (shown in Figure 3-87).

Setting Print Scale

The area of a worksheet that fits onto one page is changed by adjusting the scale in the Scale to Fit group on the Page Layout menu tab (shown in Figure 3-88) or from the Page Break Preview.

The worksheet may be scaled fit on one to nine pages wide by selecting the number of pages wide from the Width: dropdown box shown in Figure 3-88. Similarly, the worksheet may be scaled fit on one to nine pages high by selecting the number of pages high from the Height: dropdown box. To scale the workbook to more than nine pages high or wide, the user must open the Page tab of the Page Setup dialogue box (shown in Figure 3-89) by clicking on the **Page Layout** menu tab and clicking

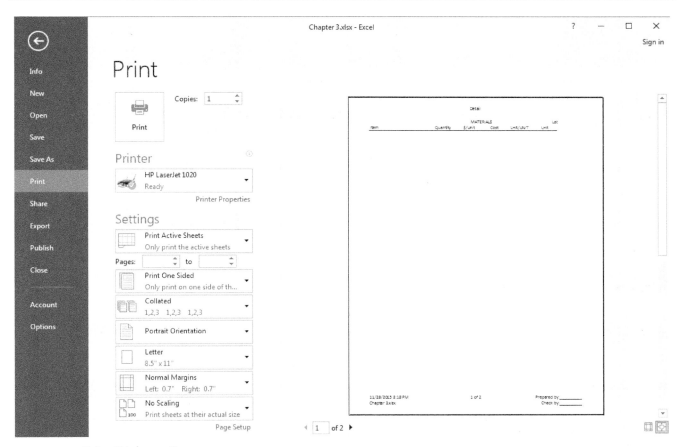

FIGURE 3-83 Print Dialogue Box

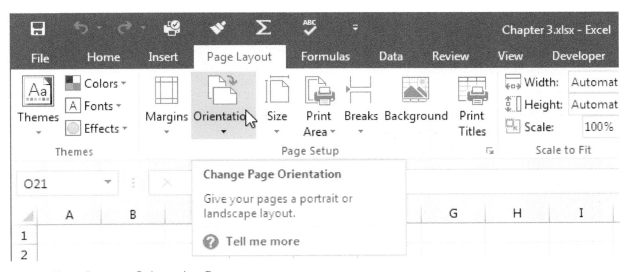

FIGURE 3-84 Page Layout: Orientation Button

on the **Page Setup Dialogue Box Launcher** located at the bottom-right corner of the Scale to Fit group (shown in Figure 3-90). The page width and height are set by clicking **Fit to:** radio button and entering the number of pages wide and high the printout is to be scaled to. Excel will then adjust the scale to match the number of pages specified by the user. The user must click on the **OK** button to close the Page Setup dialogue box.

The printout of the worksheet may be set to a specific scale by clicking on the **Page Layout** menu tab and selecting the scale from the Scale: box. By default the print scale is set to a scale of 100% of the normal size.

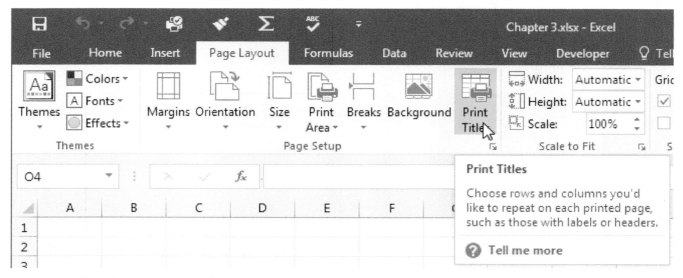

FIGURE 3-85 Page Layout: Print Titles Button

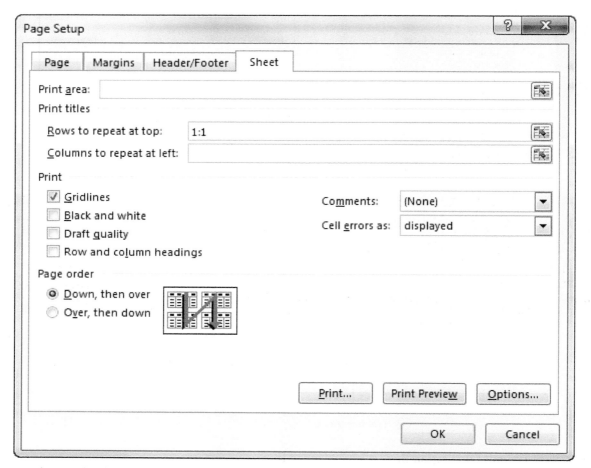

FIGURE 3-86 Sheet Tab of the Page Setup Dialogue Box

Setting Page Breaks

The layout of the page breaks is viewed and adjusted from the Page Break View (shown in Figure 3-91). The Page Break view is displayed by clicking on the **Page Break Preview** button located on the Status bar (shown in Figure 3-92) or by clicking the **View>Workbook Views>Page Break Preview** button in the Workbook Views group (shown in Figure 3-93).

The page breaks are moved by placing the cursor over the page-break line to be moved, holding down the left mouse button while dragging the page-break line to

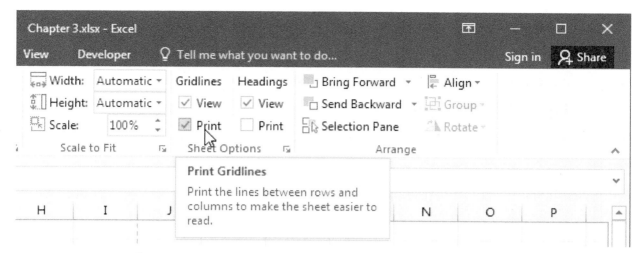

FIGURE 3-87 Printing Gridlines

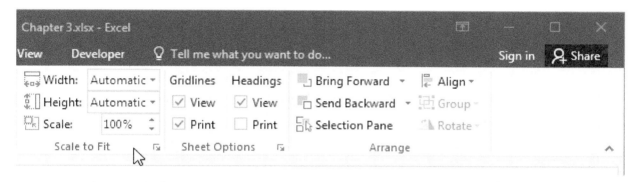

FIGURE 3-88 Page Settings: Scale to Fit Group

the desired location, and releasing the mouse button. When the page breaks are moved they automatically change the scaling shown on the Page tab of the Page Setup dialogue box. The user may return to the Normal view by clicking on the **Normal** button located on the Status bar (shown in Figure 3-62) or by clicking the **View>Workbook Views>Normal** button (shown in Figure 3-63).

Exercise 3-8

In this exercise you will prepare the file from Exercise 3-5 to be printed by changing the orientation to landscape, repeating Row 1 at the top of each page, and changing the scaling to one page wide using the following steps:

1. Make sure that Chapter 3, the workbook modified in Exercise 3-5, is open.

Next, change the page orientation to landscape using the following steps:

2. Click on the **Page Layout>Page Setup>Orientation** button (shown in Figure 3-84), and click on **Landscape** from the popup menu.

Next, set Row 1 to be repeated at the top of each page using the following steps:

3. Click on the **Page Layout>Page Setup>Print Titles** button (shown in Figure 3-85) to bring up the Sheet tab of the Page Setup dialogue box (shown in Figure 3-86), and type *1:1* in the Rows to repeat at top: text box.

4. Check the **Gridlines** check box. The Page Setup dialogue box should appear as it does in Figure 3-86. Click on the OK button to close the Page Setup dialogue box.

Next, you are going to set the spreadsheet so its width matches the width of a piece of paper using the Page Break view. This is done by completing the following steps:

5. Click on the **View>Workbook Views>Page Break Preview** button (shown in Figure 3-93) to bring up the Page Break Preview view. The spreadsheet should look similar to Figure 3-91, with a dashed blue line between columns J and K.

6. Using the cursor, drag the dashed line to the right side of column L. This will reduce the printing size of the detail sheet to one page wide.

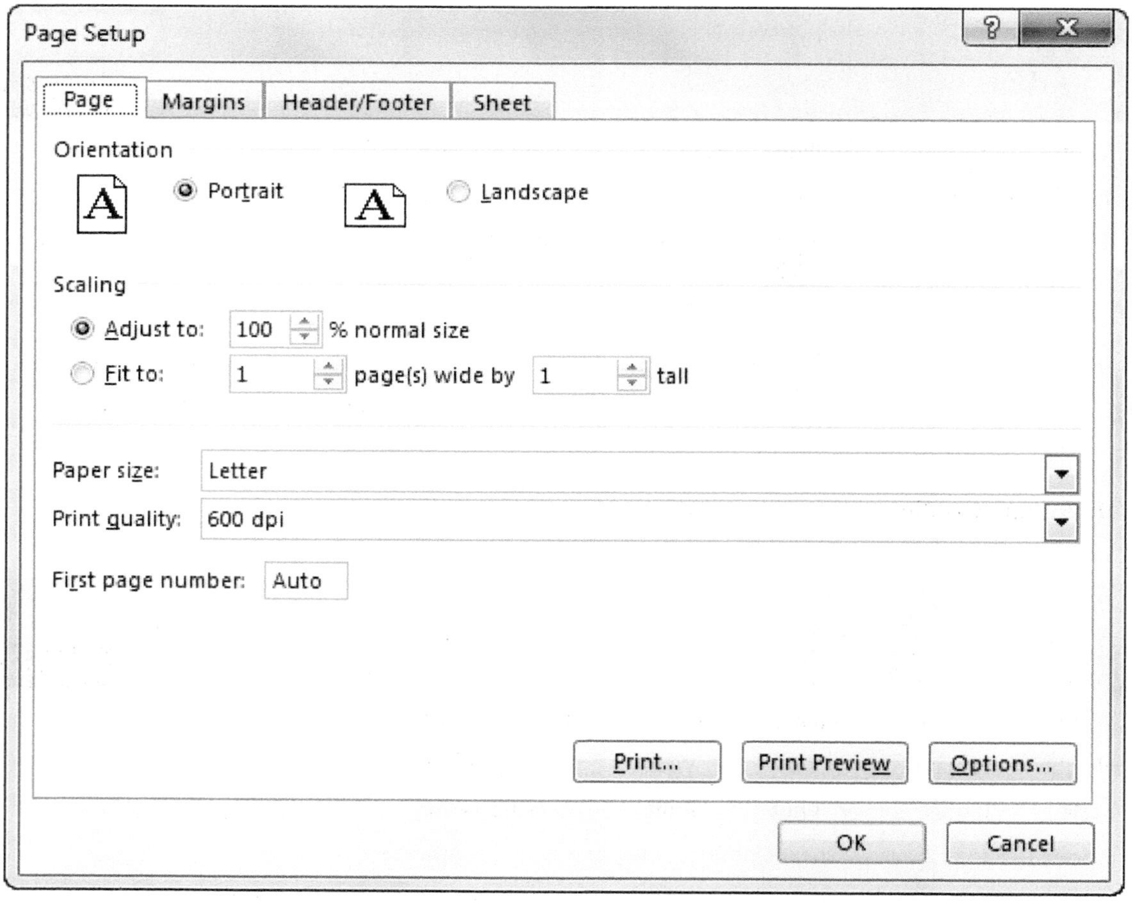

FIGURE 3-89 Page Tab of the Page Setup Dialogue Box

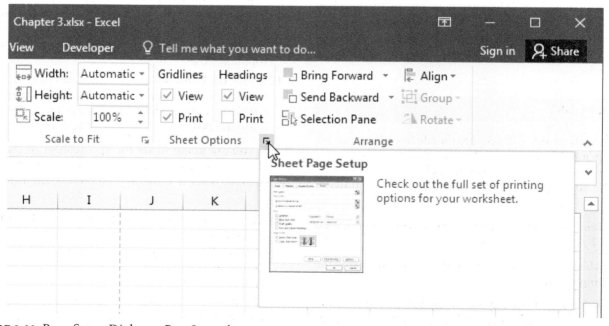

FIGURE 3-90 Page Setup Dialogue Box Launcher

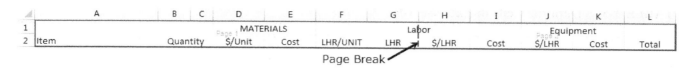

FIGURE 3-91 Page Break Preview

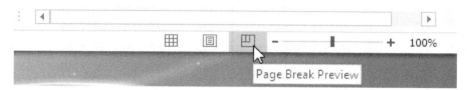

FIGURE 3-92 Page Break Preview Button on the Status Bar

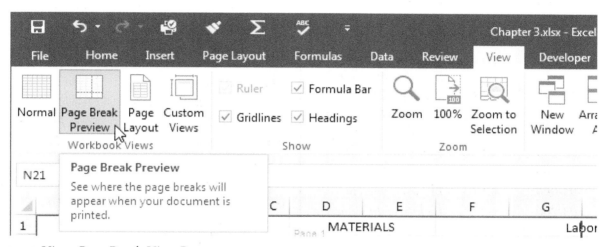

FIGURE 3-93 View: Page Break View Button

7. Click on the **View>Workbook Views>Normal** button in the Workbook Views group (shown in Figure 3-63) to return to the Normal view.
8. Click the **Save** button on the Quick Access toolbar or type **Ctrl+S** to save the workbook.

TESTING SPREADSHEETS

Before using any spreadsheet for estimating, the user must carefully test the spreadsheet to make sure it is performing the calculations correctly. The object is to make the spreadsheet foolproof. This is very important when someone other than the spreadsheet designer is going to use the spreadsheet. Foolproofing a spreadsheet is done by trying to find ways to mess up the spreadsheet and then finding ways to prevent this from happening. This is not as easy as it sounds because there is always someone out there who will come up with a new and creative way to mess things up. No spreadsheet can be completely foolproof; therefore, it is important that the user understand what the spreadsheet does and the limitations of the spreadsheet. It is best if you test the spreadsheet as you go rather than waiting until the spreadsheet is complete. It is much easier to make small changes to the worksheet, solving the problems as you go, than it is to find a small error in a large worksheet.

Exercise 3-9

This exercise, when completed, will be used as the starting point for the exercises in Chapter 32. In this exercise you will setup the Summary Worksheet in the file from Exercise 3-8 using the following steps:

1. Make sure that Chapter 3, the workbook modified in Exercise 3-8, is open.
2. Select the **Summary** tab.

Begin by entering the data into the cells using the following steps:

3. Type the text shown in Table 3-5 into the specified cells.
4. Save the workbook by clicking on the **Save** button on the Quick Access toolbar or typing **Ctrl+S**.

TABLE 3-5 Data for Cells in Exercise 3-9

Cell	Data	Cell	Data	Cell	Data	Cell	Data
B1	Code	C2	GENERAL REQUIREMENTS	B34	06-120	C42	Insulation
C1	Description	C3	Supervision	B35	06-200	C43	Rigid Insulation
D1	Materials	C4	Temporary Utilities	B36	06-210	C44	Stucco
E1	Labor	C5	Temporary Phone	B37	06-400	C45	Siding
F1	Equipment	C6	Temporary Facilities	B38	06-410	C46	Roofing
G1	Subcontract	C7	Cleanup	B40	07-000	C47	Sheet Metal
H1	Total	C9	EXISTING CONDITIONS	B41	07-100	C48	Roof Specialties
B2	01-000	C10	Demolition	B42	07-200	C49	Rain Gutters
B3	01-300	C12	CONCRETE	B43	07-210	C50	Fireproofing
B4	01-500	C13	Rebar	B44	07-220	C51	Caulking and Sealants
B5	01-510	C14	Footing and Foundation—Labor	B45	07-400	C53	OPENINGS
B6	01-520	C15	Footing and Foundation—Concrete	B46	07-500	C54	Metal Doors and Frames
B7	01-700	C16	Slab/Floor—Labor	B47	07-600	C55	Wood Doors
B9	02-000	C17	Slab/Floor—Concrete	B48	07-700	C56	Overhead Doors
B10	02-400	C18	Concrete Pump	B49	07-710	C57	Store Fronts
B12	03-000	C19	Precast Concrete	B50	07-800	C58	Windows
B13	03-200	C20	Light-weight Concrete	B51	07-900	C59	Hardware
B14	03-300	C22	MASONRY	B53	08-000	C60	Glass and Glazing
B15	03-310	C23	Masonry	B54	08-100	C61	FINISHES
B16	03-320	C25	METALS	B55	08-110	C62	Drywall
B17	03-330	C26	Structural Steel	B56	08-300	C63	Metal Studs
B18	03-340	C27	Joist and Deck	B57	08-400	C64	Ceramic Tile
B19	03-400	C28	Metal Fabrications	B58	08-500	C65	Acoustical Ceilings
B20	03-500	C29	Erection	B59	08-700	C66	Flooring
B22	04-000	C31	WOOD, PLASTICS, AND COMPOSITES	B60	08-800	C67	Wall Coverings
B23	04-200	C32	Rough Carpentry	B61	09-000	C68	Paint
B25	05-000	C33	Lumber	B62	09-200	C70	SPECIALTIES
B26	05-100	C34	Trusses	B63	09-210	C71	Signage
B27	05-200	C35	Finish Carpentry	B64	09-300	C72	Toilet Partitions
B28	05-500	C36	Wood Trim	B65	09-500	C73	Toilet and Bath Accessories
B29	05-900	C37	Cabinetry and Counter Tops	B66	09-600	C74	Fire Extinguishers and Cabinets
B31	06-000	C38	Counter Tops	B67	09-700	C76	EQUIPMENT
B32	06-100	C40	THERMAL AND MOISTURE PROTECTION	B68	09-900	C77	Appliances
				B70	10-000	C79	FURNISHINGS
				B71	10-100	C80	Window Treatments
				B72	10-200	C82	CONVEYING EQUIPMENT
B33	06-110	C41	Waterproofing	B73	10-210	C83	Elevators

TABLE 3-5 Data for Cells in Exercise 3-9 (*Contd.*)

Cell	Data	Cell	Data
B74	10-400	C85	FIRE SUPPRESSION
B76	11-000	C86	Fire Sprinklers
B77	11-300	C88	PLUMBING
B79	12-000	C89	Plumbing
B80	12-200	C91	HVAC
B82	14-000	C92	HVAC
B83	14-200	C94	ELECTRICAL
B85	21-000	C95	Electrical
B86	21-100	C97	COMMUNICATIONS
B88	22-000	C98	Communications
B89	22-100	C100	EARTHWORK
B91	23-000	C101	Clearing and Grubbing
B92	23-100	C102	Grading and Excavation
B94	26-000	C104	EXTERIOR IMPROVEMENTS
B95	26-100	C105	Asphalt
B97	27-000	C106	Site Concrete—Labor
B98	27-100	C107	Site Concrete—Concrete
B100	31-000	C108	Rebar
B101	31-100	C109	Fencing
B102	31-200	C110	Retaining Walls
B104	32-000	C111	Dumpster Enclosures
B105	32-100	C112	Signage
B106	32-110	C113	Outside Lighting
B107	32-120	C114	Landscaping
B108	32-130	C116	UTILITIES
B109	32-300	C117	Water Line
B110	32-310	C118	Sanitary Sewer
B111	32-320	C119	Storm Drain
B112	32-330	C120	Gas Lines
B113	32-340	C121	Power Lines
B114	32-900	C122	Telephone Lines
B116	33-000	C123	SUBTOTAL
B117	33-100	C124	Building Permit
B118	33-300	C125	Bond
B119	33-400	C126	SUBTOTAL
B120	33-500	C127	Profit and Overhead Markup
B121	33-700	C128	Profit and Overhead
B122	33-800	C129	TOTAL

Next, format the cells to match those in Figure 2-2 using the following steps:

5. Change the width of column A to 4, that of column B to 8, that of column C to 36, and that of columns D through H to 12.
6. Bold the data in the following Cells: B2, C2, B9, C9, B12, C12, B22, C22, B25, C25, B31, C31, B40, C40, B53, C53, B61, C61, B70, C70, B76, C76, B79, C79, B82, C82, B85, C85, B88, C88, B91, C91, B94, C94, B97, C97, B100, C100, B104, C104, B116, C116, C123, C126, and C129. *Hint.* Use the ctrl key to select multiple cells before changing them to bold.
7. Change the height of Rows 8, 11, 21, 24, 30, 39, 52, 69, 75, 78, 81, 84, 87, 90, 93, 96, 99, 103, and 115 to 5. *Hint.* Use the ctrl key to select multiple rows before changing their height.
8. Select columns A through H and click the **View>Zoom>Zoom to Selection** button, so that Columns A through H cover the entire width of the spreadsheet.
9. Change the formatting for Columns D through H to the accounting style by clicking the **Home>Number>Comma Style** button and then reduce the number of decimal points to zero by clicking the **Home>Number>Decrease Decimal** button twice.
10. Center the text in Column A by selecting Column A and clicking the **Home>Alignment>Center** button.
11. Center the text in Column B.
12. Center the text in Cells D1 through H1.
13. Change the formatting for Cells D127 through H127 to percentage using the **Home>Number>Percent Style** button and then increase the number of decimal points to one by clicking the **Home>Number>Increase Decimal** button.
14. Underline Row 1 by selecting Cells A1 through H1 and clicking the **Home>Font>Bottom Borders** button. If another border is shown in the Borders button, select the correct border by clicking on the small arrow to the right of the Borders button and selecting the **Bottom Border** from the popup menu.
15. Underline Rows 122, 125, and 128 in the same manner.
16. Save the workbook by clicking on the **Save** button on the Standard toolbar or typing **Ctrl+S**.

Next, change the font for Column A so a checkmark (✓) can be placed in the column indicating that the associated cost code is needed for the bid. This is done as follows:

17. Select Column A and select **Webdings** from the Font dropdown box in the Font group on the Home

menu tab. If Webdings is not available, try to find another font that has a check mark or replace the check marks with another symbol, such as an X.

18. Type an *a* in Cell A3, and a check mark (✓) should appear in Cell A3.
19. Delete the contents of Cell A3.

Headers and footers must be set up for each individual sheet. Next, add a header and footer to the Summary worksheet using the following steps:

20. Click the **Insert>Text>Header & Footer** button to edit the header.
21. Place the cursor in the center section of the header and click the **Header & Footer Tools: Design>Header & Footer Elements>Sheet Name** button. The header should look like the header of Figure 3-57, except the sheet name will be *Summary*. When the cursor is moved from the center section of the header, the &[Tab] will be replaced with the name of the tab, in this case *Summary*.
22. Click the **Header & Footer Tools: Design> Navigation>Go to Footer** button.
23. Place the cursor in the right section and type *Prepared by:* followed by pressing the **Underline** key 10 times.
24. Press the **Return** key and type *Checked by:* followed by pressing the **Underline** key 10 times.
25. Place the cursor in the center section and click the **Header & Footer Tools: Design>Header & Footer Elements>Page Number** button.
26. Following "&[Page]" type *of*, including a space before and after the *of*.
27. Click the **Number of Pages** button from the Header & Footer Elements group and press the **Enter** button to move it up one line.
28. Place the cursor in the left section, click the **Header & Footer Tools: Design>Header & Footer Elements>Current Date** button, press the **Space** bar, and click the **Header & Footer Tools: Design>Header & Footer Elements>Current Time** button.
29. Press the **Enter** key to move to the following line and click the **Header & Footer Tools: Design>Header & Footer Elements>File Name** button.
30. The Footer should look like Figure 3-64. When the cursor is moved from the left section of the footer, the &[Date] &[Time] will be replaced with the current date and time and the &[File] will be replaced with the file name.
31. Click on any cell in the workbook.

Next, prepare the worksheet to be printed by repeating the first row on each page, adding gridlines, and setting the page breaks using the following steps:

32. Click on the **Page Layout>Page Setup>Print Titles** button to bring up the Sheet tab of the Page Setup dialogue box, type *1:1* in the Rows to repeat at top: text box, and click on the **OK** button to close the Page Setup dialogue box.
33. Check on the **Page Layout>Sheet Options: Gridlines>Print** check box.
34. Click on the **View>Workbook Views>Page Break Preview** button to bring up the Page Break Preview view.
35. Using the cursor, drag the horizontal page-break line (the dashed line) at the bottom of Page 1 between Rows 60 and 61.
36. Drag the horizontal page-break line for the bottom of Page 2 below Row 129.
37. If there are any vertical page-break lines, drag the page break to the right side of Column H. The Summary Worksheet should now be on two pages.
38. Click on the **View>Workbook Views>Normal** button to return to the Normal view.
39. Click the **Save** button on the Quick Access toolbar or type **Ctrl+S** to save the workbook.

When you print the Summary Worksheet, the worksheet should be the same as Figure 2-2.

When you set up the Summary Worksheet for your company, you should use your company's cost codes rather than the ones used here. The formulas are added to the Summary and Detail Worksheets in Chapter 32.

CONCLUSION

Excel allows the user many ways to customize the look of the spreadsheet and perform calculations. This chapter covered how to manage workbooks and worksheets; enter data; format spreadsheets; write formulas; round numbers using the rounding functions (including CEILING and FLOOR functions); use the SUM, the AVERAGE, and IF functions; set up and print the worksheets; and test worksheets. These skills are needed to complete the Excel Quick Tip worksheets found in the following chapters.

PROBLEMS

1. What is the difference between a workbook and a worksheet?
2. What three ways can be used to save a workbook?
3. How are worksheets added to a workbook?
4. How can a copy of a worksheet be created?
5. How are worksheets renamed?
6. If you were to cut the contents from Cell H15 and then save the worksheet, what would happen to the contents from Cell H15?
7. Besides using the Column Width dialogue box, what are two other ways to change the width of a column?
8. What are two ways to change the font for a cell?
9. What does the Merge & Center button do?
10. How do you add the worksheet name to the header of the worksheet?
11. How do you add the date the worksheet was printed to the header of the worksheet?
12. What is the caret (^) used for?
13. List the calculation steps taken to solve the following formula: $(2\wedge 5 + 4/2)*2 - 1$.
14. Cell C17 was copied to Cells B16 through D18, and the content of Cell C17 is $B5. What are the contents of Cells B16 through D18?
15. What value would Excel calculate for the formula ROUND(122.568,−1)?
16. What value would Excel calculate for the formula ROUNDUP(17.222,1)?
17. When A12 equals 17, what value would Excel calculate for the formula IF(A12>10,"Error","OK")?
18. When A12 equals 25, what value would Excel calculate for the formula IF(A12<0,"Error",IF(A12>20,"Error""OK"))?
19. How are rows of a spreadsheet repeated at the top of every page?
20. What is the page break view used for?

PART TWO

THE QUANTITY TAKEOFF

Chapter 4   Fundamentals of the Quantity Takeoff
Chapter 5   Concrete
Chapter 6   Masonry
Chapter 7   Metals
Chapter 8   Woods, Plastics, and Composites
Chapter 9   Thermal and Moisture Protection
Chapter 10 Openings
Chapter 11 Finishes
Chapter 12 Fire Suppression
Chapter 13 Plumbing
Chapter 14 Heating, Ventilation, and Air-Conditioning (HVAC)
Chapter 15 Electrical
Chapter 16 Earthwork
Chapter 17 Exterior Improvements
Chapter 18 Utilities

In this section, you will be introduced to the quantity takeoff. The purpose of this section is not to provide an all-inclusive list of equations and tables to determine the quantities of every conceivable component on a construction project, but to provide basic quantity takeoff concepts and show how these concepts can be applied to common estimating situations. The basic concepts are covered in Chapter 4.

Chapters 5 through 18 apply these concepts to common estimating situation. Chapters 5 through 18 are grouped by MasterFortmat® divisions and are presented in the order they appear in the 2016 MasterFormat® and the Summary Worksheet from Chapter 3. Chapter 4 should be read first, because it lays the foundation for the quantity takeoff methods used in the following chapters. Some people prefer starting with excavation (Chapter 16) and then proceeding to concrete (Chapter 5). The chapters are written so that the reader may do this if he or she wishes. Not all divisions of the MasterFormat® are covered for example, Divisions 10 through 14 are not covered because they are often a matter of counting the items.

CHAPTER FOUR

FUNDAMENTALS OF THE QUANTITY TAKEOFF

In this chapter you will be introduced to the process of completing the quantity takeoff and the basic methods of determining the quantities for counted items (both nonrepetitive and repetitive members), linear components, sheet and roll goods, volumetric goods, and quantity-from-quantity goods. This chapter also includes a discussion of waste and the use of Building Information Modeling (BIM) to prepare a quantity takeoff. The purpose of this chapter is to introduce you to the basic concepts used in the quantity takeoff, not to make you proficient in them. As you apply these principles in Chapters 5 through 18, you will see how these concepts apply to specific construction components, and your understanding of these concepts will be solidified.

The quantity takeoff is the process whereby the estimator prepares a complete list of all building components that need to be constructed to complete the project. The quantity takeoff may also be called the quantity survey. The quantity takeoff includes components that are incorporated into the building (for example, footings, masonry walls, and windows) as well as jobsite or direct overhead. Jobsite or direct overhead includes all items that are needed to support construction that can be identified with a specific project, such as supervision and temporary facilities.

Commonly used units for the quantity takeoff include feet or lineal feet, square feet, board feet, square yards, cubic yards, acres, each, months, and lump sum. Months are used when the quantity of an item is based on the duration of the project, such as office trailer rental. Lump sum is used when multiple items are combined into a single item for pricing, such as all of the plumbing. The selection of takeoff units is determined by the way the bid is to be priced. For a concrete slab on grade, the concrete is priced in cubic yards, the forming in lineal feet, the labor to pour the concrete in cubic yards, and the labor to finish the slab in square feet or square yards.

PERFORMING A QUANTITY TAKEOFF

While performing the quantity takeoff the estimator needs to determine all the materials, equipment, and labor tasks needed to complete the construction of the project. The most accurate method of doing this is for the estimator to build the construction project in his or her mind while recording the materials, equipment, and labor tasks needed to complete the work.

To demonstrate this, let's look at the construction of the prehung door shown in Figure 4-1. To construct the door, the worker would begin by cutting the two side jambs and top jamb to length. Next, the worker would rabbet the top of the side jambs to receive the top jamb, mortise one of the side jambs for the hinges, mortise the other side jamb for the bolt and strike, and staple the top and side jambs together to complete the door frame. Next, the worker would cut stops for the top and side jambs and staple the stops to the jambs. The worker then would mortise the door slab for the hinges and drill the holes for the door handle. Finally, the worker would fasten the hinges to the door slab and the door frame to complete the prehung door. The door is now ready to install or to be packaged for shipping to the jobsite.

The materials needed to complete the prehung door are shown in Table 4-1. The materials are listed in the order they were used to complete the prehung door.

The labor tasks required to complete the prehung door are shown in Table 4-2. Again they are listed in the order in which they are performed. Often on construction projects, the estimator will combine a number of labor tasks into a single task. In the case of the prehung door, the labor tasks could be combined into a single task entitled *assemble prehung door*.

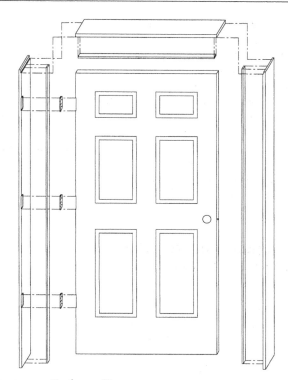

FIGURE 4-1 Prehung Door

TABLE 4-1 **Materials for a 2′6″ × 6′8″ Prehung Door**

Quantity	Item
2 ea	4 5/8″ × 6′10″ Jamb
1 ea	4 5/8″ × 2′6 3/4″ Jamb
2 ea	3/8″ × 1 1/4″ × 6′9″ Stop
1 ea	3/8″ × 1 1/4″ × 2′6″ Stop
1 ea	2′6″ × 6′8″ Door Slab
3 ea	Hinges

TABLE 4-2 **Labor Tasks and Equipment for a 2′6″ × 6′8″ Prehung Door**

Labor Task	Equipment
Cut jambs	Miter saw
Rabbet side jambs	Router
Mortise side jambs	Router
Staple jambs together	Staple gun and compressor
Cut stops	Miter saw
Staple stops to jambs	Staple gun and compressor
Mortise door slab	Router
Drill door slab for handles	Drill
Fasten hinges to jamb and slab	Drill

Finally, let's look at the equipment needed to complete the prehung door. To determine the equipment needed, the estimator must determine the equipment that will be used by the worker while completing the labor tasks. The equipment needed to complete the labor tasks is shown in Table 4-2. To complete the prehung door, the worker will need a miter saw to cut the jambs and stop; a router with the appropriate bits to rabbet the jambs and mortise the jambs and the door slab; a staple gun and compressor to assemble the jambs and stops; and a drill with the appropriate bits to drill the door slab and attach the hinges. This equipment will vary based on the construction techniques used by the worker. For example, if the worker constructs the frame with nails rather than staples, he or she would need a nail gun in place of a staple gun.

This same method is used on a large, complex building. The estimator begins at the bottom of the building and constructs it from the ground up: first the excavation, then the footings, the foundation, and so forth.

WORK PACKAGES

As the estimator performs the quantity takeoff the materials, equipment, and labor tasks must be divided into work packages. A work package is a group of items that are related. At a minimum the quantity takeoff should be broken down by the job cost codes to be used on the job. Sample job cost codes are shown in Appendix B. The job cost codes are used to track the material, labor, and equipment costs on the job, and, as such, a budget will need to be established for each of the job cost codes.

A single job cost code may be broken down into multiple work packages because the materials are ordered at different times. For example, the lumber for a three-story building may be ordered in the following lumber packages: first-floor walls, floor between the first and second floors, second-floor walls, floor between the second and third floors, third-floor walls, and roof. To facilitate the ordering of these materials in lumber packages, the estimate must be grouped by work packages that divide the work into the first-floor walls, the floor between the first and second floors, and so forth.

The work packages also should match the way the work on the project is scheduled. The estimate for the labor hours should be used to determine the durations for the schedule. As a result, the work packages should also match the scheduling tasks. This is discussed in more detail in Chapter 30.

As the estimator develops the work packages, he or she must make sure everything needed to complete the work is included in the work package at the point it will be first installed. For example, if the footing and foundation wall shown in Figure 4-2 are to be included in separate work packages, the footing package would consist of the forms for the footing, the four #4 continuous rebar, the #5 dowels, and the concrete for the footing.

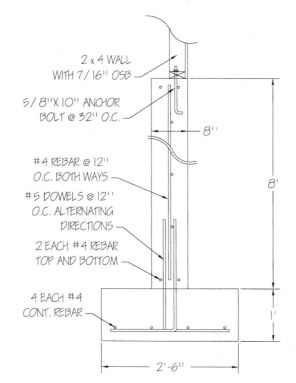

FIGURE 4-2 Wall Section

bottom of the wall, the #4 rebar at 12 inches on center running vertical and horizontal, concrete for the walls, and the anchor bolts. The anchor bolts, although they are often considered part of the lumber package, need to be included in the foundation wall package because they will be installed at the time the foundation wall is poured. The same is the case with seismic straps that are poured into the foundation wall.

One of the great advantages of building the project in the estimator's mind is that he or she can organize the material orders by work packages and collect the data needed to prepare the schedule at the same time he or she prepares the estimate.

COMMUNICATION WITH THE FIELD

To eliminate many potential problems and delays, the estimator must communicate to field employees the planned method of construction and the usage of materials and equipment. If the field employees use different construction methods or different materials or equipment, the materials and equipment ordered based on the estimator's quantity takeoff may be insufficient. This is shown in the following case study.

A project contained a small, slab-on-grade office building with an attic. The construction between the office and the attic consisted of 2×12 wood joists as shown in Figure 4-3.

The #5 dowels, even though they are part of both the footing and foundation, will need to be installed with the footing; therefore, they are part of the footing package. The foundation wall package would consist of the forms for the foundation, the #4 rebar located at the top and

To avoid the expense of ordering 22-foot-long joists, the estimator planned to break the joists along the bearing walls and ordered nine 8-foot joists, ten 10-foot

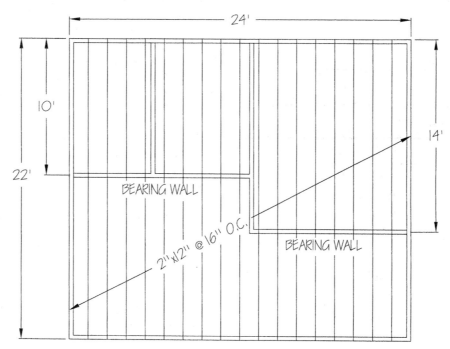

FIGURE 4-3 Floor Design

joists, ten 12-foot joists, and nine 14-foot joists. The estimator planned that the carpenters would use the joists as shown in Figure 4-4.

The carpenters, not knowing how the estimator had bid the materials, decided to add a header as shown in Figure 4-5. The materials needed by the carpenters were nineteen 8-foot joists and twenty-one 14-foot joists. As a result the carpenters needed ten additional 8-foot joists (which they cut from 10-foot joists), and twelve 14-foot joists had to be ordered from the lumber yard, causing additional expense and construction delays. These costs and delays could have been avoided had the estimator provided the carpenters a figure similar to Figure 4-4 to show the planned material usage on the building. This can be accomplished by redlining the framing plan in Figure 4-3.

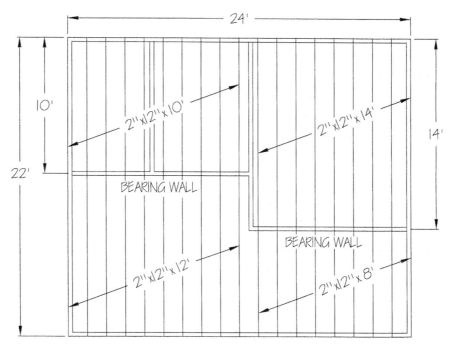

FIGURE 4-4 Estimator's Material Order

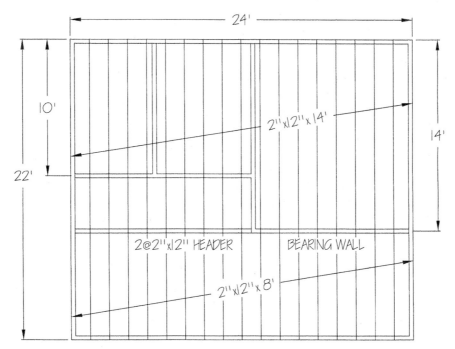

FIGURE 4-5 Field Material Use

As the estimator builds the project in his or her mind, the materials can be grouped into five broad categories: counted items, linear components, sheet and roll goods, volumetric goods, and quantity-from-quantity goods. The principles for estimating each of these broad categories are applied to many different materials. For example, the principles for estimating the number of rolls of building felt needed for a roof are the same principles used to estimate the number of rolls of welded-wire fabric that are needed in a slab. The remainder of this chapter looks at these general principles, which are applied to specific materials in Chapters 5 through 18.

COUNTED ITEMS

The simplest way to determine the quantity of an item is to count the number of items. There are shortcut methods that may be used when the counted items are repetitive members such as studs in a wall. The most common unit for the quantity takeoff of counted items is *each*.

Items that may be estimated as counted items include rebar in spot footings, dowels, anchor bolts, steel lintels, timber brackets, metal stairs, steel beams and columns, steel joists, steel trusses, joist headers, joist hangers, wood columns and beams, wood joists, wood studs, wood trusses, wood rafters, cabinetry, residential doors and windows, hollow-metal doors, hardware, overhead doors, metal studs, many HVAC and plumbing components, electrical fixtures, and landscaping plants.

When counting items it is important to mark the items off as they are counted to avoid counting an item twice. Care also must be exercised to ensure that all of the items are counted. After counting the items one must review the drawings to make sure that no items have been missed. When counting items, care must be taken to avoid grouping similar but different items together. For example, when counting doors in a residence, closet doors, which have a simple passage lockset, must be kept separate from bedroom and bathroom doors, which require a privacy lockset. This slight difference in items, a passage versus a privacy lockset, may only amount to a few dollars in material costs, but can be very costly in both labor costs and construction delays if the wrong materials are delivered to the jobsite.

When counted items are repetitive, as with studs in a wall, time is saved by using a mathematical equation to count the number of studs needed. The following equation is used to calculate the number of respective members over a specified distance and includes a repetitive member at both ends:

$$\text{Number} = \frac{\text{Distance}}{\text{Spacing}} + 1 \qquad (4\text{-}1)$$

where

Number = Number of Counted Items
Distance = Distance over Which the Items Occur
Spacing = Spacing between the Items

The distance and spacing should be measured in the same units (feet or inches), or a conversion factor needs to be included in the equation. The 1 in this equation adds a repetitive member to the beginning of the distance (hereafter referred to as the end condition). This is shown in the following example.

EXAMPLE 4-1

Determine the number of studs needed for a 4-foot-long wall where the studs are spaced at 1 foot on center.

Solution: The number of studs is calculated using Eq. (4-1) as follows:

$$\text{Number} = \frac{4 \text{ ft}}{1 \text{ ft}} + 1 = 5$$

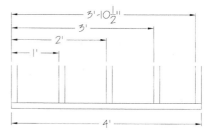

FIGURE 4-6 Stud Placement

Five studs allows for one stud at the left end, one at 1 foot from the end, one at 2 feet from the end, one at 3 feet from the end, and one at the right end, as shown in Figure 4-6. ∎

If a repetitive member is only needed at one end, then Eq. (4-1) can be written as follows:

$$\text{Number} = \frac{\text{Distance}}{\text{Spacing}} \qquad (4\text{-}2)$$

If a repetitive member is not needed at either end, then Eq. (4-1) can be written as follows:

$$\text{Number} = \frac{\text{Distance}}{\text{Spacing}} - 1 \qquad (4\text{-}3)$$

When calculating the number of repetitive members, one must be careful to make sure that the distance and spacing are converted to the same units; otherwise, the number will be off by a significant amount.

When using Eqs. (4-1) through (4-3), one must round up the quantity to the next whole number to have adequate materials, as shown in the following example.

EXAMPLE 4-2

Determine the number of studs needed for a wall 3 feet 4 inches long where the studs are spaced at 1 foot on center.

Solution: This example is the same as Example 4-1, except that the wall is 8 inches shorter. The number of studs is calculated using Eq. (4-1) as follows:

$$\text{Number} = \frac{3 \text{ ft } 4 \text{ in}}{1 \text{ ft}} + 1 = 4.33$$

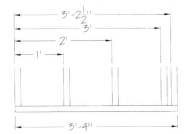

FIGURE 4-7 Stud Placement

Using conventional rules of rounding, one would round this down to 4, leaving the estimate 1 stud short of the needed number of studs. Rounding this up, one gets 5 studs, the number of studs needed to complete this wall. The stud placement for this wall is shown in Figure 4-7.

LINEAR COMPONENTS

Linear components are components that are measured by their length. When a linear component is composed of multiple pieces, the pieces may be attached by butting one piece to another piece or overlapping the pieces as shown in Figure 4-8. Common units for linear components include *lineal feet* (often referred to as *feet*) or *each* when the component is composed of pieces. Items that may be estimated as linear components include footing forms, rebar for continuous footings, steel handrails, wood sills, blocking, top and bottom plates, fascia, soffit, cant strips, wood trim, sheet metal, cabinetry, rubber base, and fencing.

The linear feet of a component is often measured directly off of the plans by using a scale, plan measurer, digitizer, or computer takeoff package; or extracted from a building information model. The number of pieces that are needed for a linear component is calculated using the following equation:

$$\text{Number} = \frac{\text{Length}}{\text{Effective Length}} \quad (4\text{-}4)$$

where

Number = Number of Pieces Needed
Length = Length of Linear Component
Effective Length = Distance between the Start of the Pieces

The effective length is shown in Figure 4-9.

The effective length is equal to the length of the pieces that make up the component less the length of the lap and is written as follows:

$$\text{Effective Length} = \text{Length}_{\text{Piece}} - \text{Lap} \quad (4\text{-}5)$$

where

$\text{Length}_{\text{Piece}}$ = Length of the Pieces That Make up the Linear Component
Lap = Length of the Lapping of the Pieces

When calculating the number of linear components, one must be careful to make sure that the length and spacing are converted to the same units; otherwise, the number will be off by a significant amount.

Substituting Eq. (4-5) into Eq. (4-4), we get the following:

$$\text{Number} = \frac{\text{Length}}{\text{Length}_{\text{Piece}} - \text{Lap}} \quad (4\text{-}6)$$

When the pieces are butt jointed, the lap is zero, and we get the following equation:

$$\text{Number} = \frac{\text{Length}}{\text{Length}_{\text{Piece}}} \quad (4\text{-}7)$$

These equations ignore the end conditions, such as the cover over the end of the rebar, and the fact that unless the linear component makes a complete circle, there is one less lap than there are pieces. In most cases, this provides a slight increase in the materials calculated over the actual materials needed. These effects are shown in the following example.

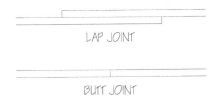

FIGURE 4-8 Connection of Linear Components

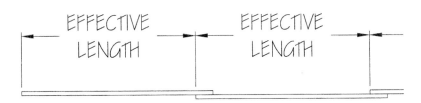

FIGURE 4-9 Effective Length

EXAMPLE 4-3

Determine the number of 20-foot-long pieces of rebar that are needed to complete a 40-foot-long footing. The pieces are spliced with a lap of 2 feet and form a single bar running the length of the footing. Three inches of cover over the rebar is required at each end of the footing.

Solution: The number of pieces is calculated using Eq. (4-6) as follows:

$$\text{Number} = \frac{\text{Length}}{\text{Length}_{\text{Piece}} - \text{Lap}} = \frac{40 \text{ ft}}{20 \text{ ft} - 2 \text{ ft}} = 2.22 \text{ pieces}$$

This is slightly more than the required amount. The first piece will start 3 inches from the end of the footing and end 20 feet 3 inches from the end of the footing. The second piece will start at 18 feet 3 inches, lapping the previous bar by 2 feet, and end at 38 feet 3 inches. The third, and final, piece will start at 36 feet 3 inches and end at 39 feet 9 inches. The third piece is 3 feet 6 inches long or 0.175 (3.5 ft/20 ft) of a 20-foot-long bar.

Linear components should be rounded up to the next whole number to make sure that there are adequate materials.

SHEET AND ROLL GOODS

Sheet and roll goods are components that are measured by their area. When sheet and roll goods are composed of multiple pieces, the pieces are attached by butting or lapping them together in two directions, as shown in Figure 4-10. In other cases, such as with brick, a space is placed between the pieces that is filled with another material, such as mortar. Common units for sheet and roll goods include *square feet, square yards, squares* (100 square feet), or *each* when the component is composed of pieces.

The key difference between sheet goods and roll goods is that the length of rolled goods is much greater than their width. Items that may be estimated as sheet goods include concrete wall forms, face brick, block and structural brick, steel deck, tongue and groove sheeting, oriented strand board (OSB) sheathing, wood decks, wood siding, building paper, ridged insulation, asphalt shingles, membrane roofing, siding, drywall, ceramic tile, acoustical tile, and vinyl composition tile (VCT). Items that may be estimated as roll goods include welded-wire fabric building paper, vinyl, and carpet.

The quantities for sheet and roll goods can be determined by one of two methods: the area method and the row and column method.

Area Method

The area method determines the quantities of materials needed by determining the area to be covered by the materials and dividing it by the area covered by one sheet or roll of the materials as shown in the following equation:

$$\text{Number} = \frac{\text{Area}}{\text{Effective Area}} \qquad (4\text{-}8)$$

where

Number = Number of Pieces Needed
Area = Area to Be Covered
Effective Area = Area Covered by One Piece Including Spacing or Lap

The effective area for butt-jointed pieces equals the length times the width of the piece and is calculated by the following equation:

$$\text{Effective Area} = (\text{Length})(\text{Width}) \qquad (4\text{-}9)$$

The effective area for lapped pieces equals the effective length times the effective width of the piece. The effective length equals the length less one lap joint, just as it did with linear components. The effective width equals the width less one lap joint. The effective area for lapped pieces is calculated by the following equation:

Effective Area
$$= (\text{Length} - \text{Lap}_{\text{Length}})(\text{Width} - \text{Lap}_{\text{Width}}) \quad (4\text{-}10)$$

The laps at the end and sides of a roll or sheet may be different or they may be the same.

The effective area for spaced pieces equals the effective length times the effective width of the piece. The effective length equals the length plus one-half of the spacing around the piece as shown in Figure 4-11.

Therefore, the effective length equals the length plus two times one-half of the spacing around the piece or the length of the piece plus the width of one space. The effective width equals the width plus the width of one space. The effective area for spaced pieces is calculated by the following equation:

Effective Area
$$= (\text{Length} + \text{Space}_{\text{Length}})(\text{Widht} + \text{Space}_{\text{Width}})$$
$$(4\text{-}11)$$

Care must be taken when using the area method because it ignores edge conditions and may underestimate

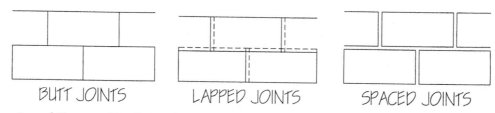

FIGURE 4-10 Lapping of Sheet and Roll Goods

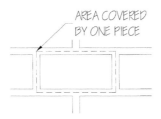

FIGURE 4-11 Effective Area for Spaced Pieces

the quantities needed to complete the work, as seen in the following example.

EXAMPLE 4-4

A 21-inch-high by 48-inch-long wall is constructed of 15 5/8-inch-long by 7 5/8-inch-high block. The mortar joints between the blocks are 3/8 inch thick. How many blocks are needed to construct the wall?

Solution: Using Eq. (4-11), we calculate the effective area of one block as follows:

Effective Area = (15 5/8 in + 3/8 in)(7 5/8 in + 3/8 in)
= (16 in)(8 in) = 128 in²

The area of the wall is calculated as follows:

Area = (Length)(Height) = (48 in)(21 in) = 1,008 in²

The number of blocks is calculated using Eq. (4-8) as follows:

$$\text{Number} = \frac{1{,}008 \text{ in}^2}{128 \text{ in}^2} = 7.87 \text{ blocks}$$

FIGURE 4-12 Block Layout

Using the area method, we would have ordered eight blocks. The blocks would be laid out as shown in Figure 4-12. The first row would require three blocks. The second row would require three blocks, with one of the blocks being cut in half and each half placed on either end of the row. The top row will require three blocks cut to a height of about 5 inches. From this example we see that we would need nine rather than eight blocks because the waste from the cut blocks cannot be spliced together to form a new block. This waste is unavoidable. ∎

Row and Column Method

Example 4-4 introduces us to the other method for calculating the quantities for sheet and roll goods: the row and column method. With the row and column method, one determines the number of rows and the number of columns needed to construct the components. The number of rows is calculated using Eq. (4-12) for lapped components, Eq. (4-13) for butt-jointed components, and Eq. (4-14) for spaced components:

$$\text{Number}_{\text{Rows}} = \frac{\text{Height}}{\text{Height}_{\text{Piece}} - \text{Lap}} \quad (4\text{-}12)$$

$$\text{Number}_{\text{Rows}} = \frac{\text{Height}}{\text{Height}_{\text{Piece}}} \quad (4\text{-}13)$$

$$\text{Number}_{\text{Rows}} = \frac{\text{Height}}{\text{Height}_{\text{Piece}} + \text{Space}} \quad (4\text{-}14)$$

These equations should be familiar because they are similar to those used for linear components. When determining the number of rows, one should always round up, unless the fraction of a row is less than the lap. Rounding up should be to the next whole number, 1/2, 1/3, 1/4, or 1/n (where n is a whole number) fraction of a sheet or roll. Rounding up to 2/5, for example, will give you the wrong quantity because 1/5 of the material will be wasted when you take two pieces that are 2/5 of a sheet; therefore, 2/5 should be rounded up to 1/2, which will properly take the waste into account.

The number of pieces per row (hereafter referred to as columns) is calculated using Eq. (4-15) for lapped components, Eq. (4-16) for butt-jointed components, and Eq. (4-17) for spaced components:

$$\text{Number}_{\text{Columns}} = \frac{\text{Length}}{\text{Length}_{\text{Piece}} - \text{Lap}} \quad (4\text{-}15)$$

$$\text{Number}_{\text{Columns}} = \frac{\text{Length}}{\text{Length}_{\text{Piece}}} \quad (4\text{-}16)$$

$$\text{Number}_{\text{Columns}} = \frac{\text{Length}}{\text{Length}_{\text{Piece}} + \text{Space}} \quad (4\text{-}17)$$

Width may be substituted for Height and Width$_{\text{Piece}}$ for Height$_{\text{Piece}}$ in Eqs. (4-12) through (4-14) for horizontal surfaces such as slabs on grade.

The number of pieces is then determined by multiplying the number of rows by the number of columns using the following equation:

$$\text{Number} = (\text{Number}_{\text{Rows}})(\text{Number}_{\text{Columns}}) \quad (4\text{-}18)$$

EXAMPLE 4-5

Solve Example 4-4 using the row and column method. How many blocks are needed to construct the wall?

Solution: Using Eq. (4-14) to find the number of rows, we get the following:

$$\text{Number}_{\text{Rows}} = \frac{21 \text{ in}}{7\ 5/8 \text{ in} + 3/8 \text{ in}} = \frac{21 \text{ in}}{8 \text{ in}} = 2.625 \text{ rows}$$

Three rows of blocks are needed. Using Eq. (4-17) to get the number of columns, we get the following:

$$\text{Number}_{\text{Columns}} = \frac{48 \text{ in}}{15\ 5/8 \text{ in} + 3/8 \text{ in}} = \frac{48 \text{ in}}{16 \text{ in}} = 3 \text{ columns}$$

Three columns of blocks are needed. Using Eq. (4-18), we calculate the number of blocks needed as follows:

$$\text{Number} = (3 \text{ rows})(3 \text{ columns}) = 9 \text{ blocks}$$

The row and column method calculates the correct number of blocks needed to construct the wall, and it is easy to see that the entire top row of blocks will need to be cut. This allows the estimator to take this into account as he or she calculates the labor for the installation of the block.

VOLUMETRIC GOODS

Volumetric goods are commonly measured in *cubic yards* or *cubic feet*. Items that may be estimated as volumetric goods include concrete, excavation, and asphalt. The quantity of rectangular column-shaped volumetric goods is calculated by multiplying the height by the width by the length as expressed in the following equation:

$$\text{Volume} = (\text{Height})(\text{Width})(\text{Length}) \quad (4\text{-}19)$$

Chapter 16 discusses four additional methods for calculating the volume of excavation. These are the average-width-length-depth method, average-end method, modified-average-end method and cross-sectional method. Appendix A has additional formulas that may be used to calculate the volume of a cylinder, prism, cone, and pyramid.

EXAMPLE 4-6

Determine the volume of a 100-foot by 100-foot concrete slab. The slab is 4 inches thick. Express your answer in cubic yards.

Solution: The volume of the slab is determined by Eq. (4-19) as follows:

$$\text{Volume} = (100 \text{ ft})(100 \text{ ft})\left(\frac{4 \text{ in}}{12 \text{ in/ft}}\right)\left(\frac{1 \text{ yd}^3}{27 \text{ ft}^3}\right) = 123.5 \text{ yd}^3$$

QUANTITY-FROM-QUANTITY GOODS

Quantity-from-quantity goods are estimated based upon the quantity of another item. For example, the number of gallons of paint may be estimated by dividing the area of the surface to be painted by the coverage rate for a gallon of paint. Alternately, roofing nails may be estimated by multiplying the area of the roof in squares (100 square feet) by the average number of nails used per square. Items that may be estimated by the quantity-from-quantity method include mortar, perlite, fasteners, saw blades, waterproofing and dampproofing, blown insulation, stucco, shims, drywall tape and mud, adhesives, grout, tack strip and seaming tape for carpet, paint, and electrical wiring. Quantity-from-quantity estimates are performed using one of the following equations:

$$\text{Quantity} = \frac{\text{Quantity}_{\text{Base}}}{\text{Coverage}} \quad (4\text{-}20)$$

Quantity
$$= (\text{Quantity}_{\text{Base}})(\text{Average Quantity Required}) \quad (4\text{-}21)$$

where

Quantity = Quantity of Material Needed
Quantity$_{\text{Base}}$ = Quantity of the Item that the Estimate is Based Upon
Coverage = Coverage of the Material Needed
Average Quantity Required = Average Quantity Needed to Cover One Unit of Base Material

The coverage and average quantity required may be obtained from historical data, manufacturers' data, or reference books. Estimating the quantity from quantity is shown in the following two examples:

EXAMPLE 4-7

Determine the quantity of paint needed to paint 5,000 square feet of wall. One gallon of paint will cover 300 square feet of wall.

Solution: Using Eq. (4-20), we get the following quantity:

$$\text{Quantity} = \frac{5{,}000 \text{ sf}^2}{300 \text{ ft}^2/\text{gallon}} = 16.7 \text{ gallons}$$

EXAMPLE 4-8

Determine the quantity of roofing nails needed for a 14-square roof. Historically, 340 nails are needed per square.

Solution: Using Eq. (4-21), we get the following quantity:

$$\text{Quantity} = (14 \text{ sq})(340 \text{ nails/sq}) = 4{,}760 \text{ nails}$$

Fundamentals of the Quantity Takeoff

WASTE

Ordering the foregoing quantities may not provide the quantities necessary to complete the construction project because some of the materials are lost because of waste. Unavoidable waste is waste that is the result of not being able to use scrap materials (as we saw in Examples 4-4 and 4-5). Avoidable waste is waste that is due to improper use of materials, lost or damaged materials, and the difference between the actual dimensions and the design dimensions (for example, pouring a 4.5-inch-thick slab where a 4-inch-thick slab is required because of poor grade control). Where possible, unavoidable waste should be included in the original quantity takeoff of the materials as it was in Example 4-5. This is necessary to accurately track and control avoidable waste. A waste factor is often added to the quantities to account for avoidable waste, as well as unavoidable waste that has not already been included. The waste factor is expressed as a percentage of the calculated quantity. The quantity of material needed, including the waste, is calculated using the following equation:

$$\text{Quantity}_{\text{with Waste}} = \text{Quantity}_{\text{without Waste}}\left(1 + \frac{\text{Waste Percentage}}{100}\right) \quad (4\text{-}22)$$

Volume, area, or number may be substituted into Eq. (4-22) for the quantity. The inclusion of waste is shown in the following example.

EXAMPLE 4-9

Determine the volume of concrete for the slab in Example 4-6. Include 10% waste in the calculated volume.

Solution: From Example 4-6, the volume of the slab is 123.5 cubic yards. Using Eq. (4-22), we add the waste as follows:

$$\text{Volume}_{\text{with Waste}} = (123.5 \text{ yd}^3)\left(1 + \frac{10}{100}\right)$$
$$= (123.5 \text{ yd}^3)(1.10) = 135.8 \text{ yd}^3$$

The waste factor should be determined from historical data. Historical data is obtained by comparing the estimated quantity without waste to the actual quantities used on the project, using the following equation:

$$\text{Waste Percentage} = 100\left(\frac{\text{Quantity}_{\text{Used}}}{\text{Quantity}_{\text{Estimated}}} - 1\right) \quad (4\text{-}23)$$

EXAMPLE 4-10

The slab in Example 4-6 required 138 cubic yards of concrete. Determine the actual waste percentage for the slab.

Solution: From Example 4-6, the estimated volume of the slab was 123.5 cubic yards. The waste percentage is calculated using Eq. (4-21) as follows:

$$\text{Waste Percentage} = 100\left(\frac{138 \text{ yd}^3}{123.5 \text{ yd}^3} - 1\right) = 11.74\%$$

Data on the quantities used should be available from the company's accounting system.[i] When performing the quantity takeoff, the estimator should keep an accurate estimate of the quantities before adding avoidable waste. This quantity is to be used to measure the actual quantity of waste. On one apartment project the author reviewed, he found that the number of studs needed to complete identical apartment units varied by almost 10% because of differences in the use of materials.

BUILDING INFORMATION MODELING (BIM)

A building information model—a three-dimensional computer model that contains parametric data about the materials used in construction—can be used to determine the quantity of materials needed to construct a building. Figure 4-13 show the quantity of wall—divided by type—needed to construction a wood-framed office building. This data was extracted from a building information model, which allows the user to list each individual wall or group the walls by type. Building information models also allow the user to extract a material takeoff for a component. Figure 4-14 show the materials required for the walls shown in Figure 4-13, with the materials grouped by type.

Before using a building information model to prepare an estimate, one must be aware of some of the limitations and problems with using a model. These limitations and problems are explained in the following paragraphs.

Method of Determining Quantities: Before relying on the quantities extracted from a building information model, one must understand how the software calculates the quantities. For the software program that was used to generate Figure 4-13 and Figure 4-14, the

<Wall Schedule>			
A	B	C	D
Type	Count	Length	Area
Exterior - Brick on 2x4	4	241' - 0"	1818 SF
Foundation - 8" Concrete	4	241' - 4"	402 SF
Interior - 4 3/4" Wood Partition (1-hr)	28	462' - 10 1/2"	3483 SF

FIGURE 4-13 Quantity of Wall by Type

<Wall Material Takeoff>		
A	B	C
Type	Material: Name	Material: Area
Exterior - Brick on 2x4	Air Barrier - Air Infiltration Barrier	1818 SF
Exterior - Brick on 2x4	Gypsum Wall Board Type X	1818 SF
Exterior - Brick on 2x4	Masonry - Brick	1818 SF
Exterior - Brick on 2x4	Misc. Air Layers - Air Space	1818 SF
Exterior - Brick on 2x4	Vapor / Moisture Barriers - Vapor Retarder	1818 SF
Exterior - Brick on 2x4	Wood - Dimensional Lumber	1818 SF
Exterior - Brick on 2x4	Wood - Sheathing - plywood	1818 SF
Foundation - 8" Concrete	Concrete - Cast-in-Place Concrete	402 SF
Interior - 4 3/4" Wood Partition (1-hr)	Gypsum Wall Board Type X	6966 SF
Interior - 4 3/4" Wood Partition (1-hr)	Wood - Dimensional Lumber	3483 SF

FIGURE 4-14 Material Takeoff for Walls

software subtracts openings when calculating the area. When installing the wood sheathing most of the sheathing cut from the openings will become waste. Given that the exterior walls are 9 feet 1 inch high and assuming that the material in the openings becomes waste, a more accurate quantity for sheathing is calculated as follows:

$$\text{Quantity} = 241' \times 9'1'' = 2{,}189 \text{ ft}^2$$

Should one use the quantity from the model in the bid, the quantity of sheathing would be short by about 17 percent.

Incomplete Models: Quantities can only be extracted for components that have been accurately modeled. For example, the core of the walls in Figure 4-14 have been modeled as solid wood rather than sill plate, studs, and top plates. As such, the extracted quantity is 1,818 square feet of Wood – Dimensional Lumber; and one cannot determine the quantity of sill plate, studs, and top plate from this quantity. However, these quantities can be determined from the length of the exterior walls from the takeoff shown in Figure 4-13. Using a stud spacing of 16 inches on center and adding one stud for each of the four walls, the number of studs for the exterior walls can be calculated as follows:

$$\text{Studs} = 241' \times 12''/16'' + 4 = 185$$

Quantities cannot be obtained from components that have not been modeled. For example, the exterior walls in Figure 4-14 require insulation, which has not been modeled. As such, insulation does not show up in the material takeoff.

Conflicts between the Model and the Construction Drawings: Components may be drawing on the construction drawings that are not included in the model, which creates a conflict between the construction drawings and the model. Typically the contractor will be held to the information on the construction drawings, and if the construction drawings call out for base and paint on the interior side of the walls, the contractor will be obligated to provide base and paint for these walls even though they do not appear in the material list. Another example of a conflict is when an elevation for one bathroom wall is used for multiple walls by marking the elevation tags for the additional walls as being similar to the elevation of the first wall without modeling the components for each wall. Because the components have not been modeled for each wall, only the modeled components from the first wall will show up in the material takeoff.

The estimator should carefully review the contract documents to determine the legal implications of using a building information model to prepare the estimate.

Different Means and Methods: The means and methods needed to install a type of component may change based on its location. For example, installing sheathing on the second floor wall will require the use of scaffolding or other lift equipment, which will not be required on the first floor. If both the first and second floor walls are modeled as the same wall type, the wall types must be separated into groups with similar means and methods if an accurate estimate is to be prepared.

In spite of these limitations and potential problems, building information models can be used to prepare a detailed estimate provided one understands and takes into account these limitations and potential problems. In addition to preparing a detailed estimate, building information models may be used to prepare a conceptual/preliminary estimate where less accuracy is needed or as a rough check of the quantity takeoff obtained by a different estimating method.

CONCLUSION

The quantity takeoff is best performed by the estimator building the project in his or her mind, keeping track of the materials, equipment, and labor tasks needed to complete the project. The estimate should be broken down into work packages that contain related materials and tasks, such as materials that will be ordered together. The planned use of materials, labor, and equipment must be communicated with the field personnel.

The quantity takeoff for materials can be divided into counted items (both repetitive and nonrepetitive), linear components, sheet and roll goods, volumetric goods, and quantity-from-quantity goods. Similar methods are used to determine the quantities in each of these categories.

PROBLEMS

1. What is the best way of performing the quantity takeoff?
2. Why should the estimator communicate the planned material use and construction methods to the field personnel?
3. Determine the number of studs needed for a 75-foot-long wall where the studs are spaced at 1 foot on center.
4. Determine the number of studs needed for a 120-foot-long wall where the studs are spaced at 16 inches on center.
5. Determine the number of 20-foot-long pieces of pipe needed to complete 150 feet of pipe. The pipe is connected with a butt joint.
6. Determine the number of 20-foot-long pieces of rebar needed to complete a 240-foot-long footing. The pieces are spliced with a lap of 18 inches and form a single bar running the length of the footing.
7. An 8-foot-high by 50-foot-long wall is constructed of 15 5/8-inch-long by 7 5/8-inch-high blocks. The mortar joints between the blocks are 3/8 inch thick. Using the area method, determine the number of blocks needed to construct the wall.
8. An 8-foot-high by 50-foot-long wall is faced with 7 1/2-inch-long by 2 1/2-inch-high bricks. The mortar joints between the bricks are 1/2 inch thick. Using the area method, determine the number of bricks needed to face the wall.
9. Use the row and column method to solve Problem 7.
10. Use the row and column method to solve Problem 8.
11. Determine the volume of a 200-foot by 75-foot concrete slab. The slab is 6 inches thick. Express your answer in cubic yards and include 10% waste.
12. Determine the volume of a 20-foot by 15-foot by 4-foot-high concrete footing. Express your answer in cubic yards and include 5% waste.
13. Determine the quantity of paint needed to paint 12,620 square feet of wall. One gallon of paint will cover 250 square feet of wall.
14. Determine the quantity of joint compound needed to finish 3,000 square feet of gypsum-board wall. One gallon of joint compound will cover 200 square feet of wall.
15. Determine the quantity of roofing nails needed for a 27.65 square (square = 100 square foot) roof. Historically, 375 nails are needed per square.
16. Determine the quantity of shims needed to install 22 doors. Historically, 0.75 bundles of shims are needed per door.
17. The slab in Problem 11 required 300 cubic yards of concrete. Determine the actual waste percentage for the slab.
18. The footing in Problem 12 required 47 cubic yards of concrete. Determine the actual waste percentage for the slab.

REFERENCE

1. For more information on construction accounting systems and tracking quantities, see Steven J. Peterson, Construction Accounting and Financial Management, Prentice Hall, Upper Saddle River, NJ, 2005.

CHAPTER FIVE

CONCRETE

In this chapter you will learn how to apply the principles in Chapter 4 to concrete, forms, and reinforcing for continuous and spread footings, walls, square and round columns, beams, slabs on grade, raised slabs, and stairs. This chapter includes sample worksheets that may be used in the quantity takeoff. You are encouraged to set these worksheets up in Excel and learn how to use them. This chapter also includes example takeoffs from the residential garage drawings given in Appendix G. Some readers prefer starting with excavation (Chapter 16) before covering concrete (Chapter 5). If you wish to do so, skip to Chapter 16 and come back to Chapter 5 after completing Chapter 16.

Concrete work consists of three tasks: installation and removal of the forms, installation of the reinforcement (rebar and welded wire fabric), and placement and finishing of the concrete. The installation of the forms and rebar may occur in many different orders. For footings, the forms are placed and then the rebar is placed inside the forms; for columns, the rebar is placed and the forms are placed around the rebar; and for walls, one side of the forms is placed, the rebar is placed, and then the forms are finished. After the rebar is placed and the forms are completed, the concrete is placed and then finished. After the concrete has cured, the forms are removed. With careful planning and scheduling of the concrete work, the forms can be reused many times on the same job.

FORMS

Forms or formwork may be bid by itemizing all of the components of the forms, or they may be bid by the square foot, the lineal foot, or set of forms. When bidding formwork by itemizing all of the components of the forms, the estimator must be familiar with the forming system to be used and build the forms in his or her mind, counting the items as he or she goes.

Estimators will often bid the formwork based on the square footage, the lineal footage, or the set of forms, rather than accounting for each of the items needed. To do this the estimator must have historical data on the equipment, materials, and labor needed to complete a square foot, lineal foot, or set of forms.

Beams, girders, walls, and tall footings (footings greater than 12 inches high) are often bid based on the square footage of forms needed to form the concrete member. When figuring the area, the estimator must make sure to include forms on all sides and the bottom (if needed) of the concrete member. For example, on a foundation wall the estimator will need to include both sides of the wall and the ends of the wall. The area of the forms should be based on the area of the forms, not the area of the wall. Wall forms typically come in 2-, 4-, and 8-foot heights. When forming an 18-inch-high footing or wall, a 2-foot-high form would be used. In this case the forming height would be 2 feet rather than the wall height of 18 inches.

Short footings (footings 12 inches high or less) can be formed by using 2 × 10s and 2 × 12s; therefore, they are often bid based on the lineal footage of forms. Just as in the case of the walls, the estimator needs to be sure to include forms on both sides and ends of the footings.

Columns are often bid based on the square footage (square and rectangular columns only), lineal footage of the column, or set of forms. When prefabricated forms or Sonotube are used, they may be bid based on the set of forms required.

Raised slabs are often bid based on the square footage of the slab, often ignoring the small amount of forms needed around the perimeter. When forming a raised slab, the bottom forms of the slab need to be supported until the slab has cured sufficiently to support its own weight.

When estimating forms, the estimator must determine not only the quantity of forms to be installed, but also the quantity of form material that must be purchased for the project, as it affects the cost of the forms. The quantity of forms that needs to be purchased for a project is based on how many times the forms can be reused, whether the forms can be used on other projects, and the life of the forms. If the concrete for a project can be poured in four separate pours with sufficient time between the pours to allow the concrete to cure, the project will take one-fourth of the forms it would need if the concrete on the project was poured in a single pour. If the forms are available from another project or can be used on a subsequent project, the purchase of the forms may be spread out over many projects, reducing the cost of forms that need to be included in the estimate for the project. However, if a project requires a custom form that will not be reused, the entire purchase of the forms must be included in the estimate. The final factor is how long the forms will last before they wear out or are consumed as they are cut to meet specific dimensions.

REINFORCING

Concrete reinforcing consists of rebar, welded wire fabric, chairs to position the rebar, and so forth. Rebar is designated by bar size, which represents the nominal diameter of the bar in eighths of an inch. For example, a #7 bar has a nominal diameter of 7/8 inch. Rebar is taken off by the lineal foot or by the number of pieces. Small quantities of rebar are often purchased by the piece or by the lineal foot, whereas large quantities of rebar are often purchased by the pound or ton. The weight per foot for common rebar sizes is shown in Table 5-1.

Continuous rebar must be lapped. The lapping is specified by the structural engineer or other design professional. In the event it is not specified by the design professional, the contractor has to meet the lapping requirement in the building code of the jurisdiction where the building is being constructed. Lapping is specified by a minimum length or a number of bar diameters or both. For example, the drawings may require that the rebar be lapped 24 bar diameters or 1 foot, whichever is greater. For a #3 bar, 24 bar diameters is 9 inches (24 × 3/8 in). The bar would have to be lapped the greater of 9 inches or 1 foot; therefore, the bar would have to be lapped 1 foot. For a #6 bar, 24 bar diameters is 18 inches (24 × 6/8 in), and the bar would have to be lapped the greater of 18 inches or 1 foot.

On small jobs the rebar is often fabricated in the field from straight bar. A common length for rebar is 20 feet; however, lengths up to 60 feet are often available. For large jobs, the rebar is often shop fabricated and shipped to the jobsite ready for placement.

Many projects are now specifying epoxy-coated rebar. Typically, epoxy-coated rebar is light green and is cut and bent in a fabrication shop. Epoxy-coated rebar that has been field cut and bent must have all cuts and bends recoated. This will require additional installation time in the field and epoxy paint to coat the rebar. Field cuts and bends should be noted when performing the takeoff so they can be included in the estimate.

Welded wire fabric (WWF) or welded wire reinforcement (WWR) is designated by the spacing of the wires and the wires size, with the longitudinal specifications appearing before the transverse. For example, WWF 6 × 12—W10 × W10 would consist of smooth number 10 wires spaced 6 inches on center longitudinally and smooth number 10 wires spaced 12 inches on center transversely. The W in front of the wire size indicates the wire is smooth; whereas, a D would indicate that the wires are deformed. Welded wire fabric is available in flat sheets or rolls 5 feet wide by 150 feet long. Like continuous rebar, welded wire fabric must be lapped as specified by the design profession or building code. Typically a lap of 8 inches or more is required.

CONCRETE

The specifications for concrete vary based on its use. Concrete used for slabs often has a higher strength than concrete used for footings and foundations. The estimator must read the specification to determine the different types of concrete used on the project. When estimating concrete, the quantities for different types of concrete must be kept separate. Concrete specifications may be performance based, design based, or both. A performance-based specification sets the requirements for the physical properties of the concrete, such as strength and slump. A design-based specification sets limits on the various components used to compose the concrete. Concrete is typically composed of cement, large aggregate, fine aggregate, and water. Cement powder comes in five types. Type I is normal cement, Type II is moderate-sulfate-resistance cement, Type III is high-early-strength cement, Type IV is low-heat-of-hydration cement, and Type V is high-sulfate-resistance cement. In some concrete mixes, part of the cement is replaced by fly ash, a pozzolan. Admixtures may be added to achieve specific physical properties, such as air entrainment. All of these affect the price of the concrete. The estimator needs to

TABLE 5-1 Rebar Sizes and Weights (lb/ft)

Bar Size	Weight	Bar Size	Weight
2	0.167	8	2.670
3	0.376	9	3.400
4	0.668	10	4.303
5	1.043	11	5.313
6	1.502	14	7.650
7	2.044	18	13.600

carefully review the specifications to make sure he or she understands what type of concrete is required to complete the job.

Pricing for concrete is best obtained from the concrete suppliers. It is best to provide the supplier with a copy of the specifications for use in pricing the concrete. Concrete companies often charge extra for small or short loads. Care should be taken to note any short loads during the takeoff process.

During hot weather, ice or chilled water is often added to keep the concrete from overheating. This increases the cost of the concrete, and the estimator must incorporate these costs into the concrete estimate. During the quantity takeoff the estimator needs to identify the need for ice or chilled water.

When concrete is placed in cold weather, precautions need to be taken to ensure that the concrete does not freeze. During cold weather the water added to the concrete is usually heated, adding to the cost of the concrete. In addition, the concrete is often protected by covering the concrete with plastic, straw, or concrete blankets and by providing heat. The need for hot water and concrete protection should be included in the concrete takeoff.

The labor required to place the concrete consists of two components: the placement of the concrete and the finishing of the concrete. The placement of the concrete is bid by the cubic yard; the finishing of the concrete is bid by the square foot or square yard of the finished surface. Often the two components are combined into a single cost for a concrete component of a specified size.

A concrete pump is often used to place concrete that cannot be discharged directly from the concrete truck into its final place. Estimators need to be sure they understand what access is available to the site so they can determine the amount of concrete pumping that will be necessary to complete the work. When pumping concrete, care must be taken to ensure that the concrete mix has been designed for pumping. Not all concrete mixes can be pumped with satisfactory results.

Concrete waste varies based on the type of member being constructed. Beams, columns, walls, and other members that have a small area of contact with the ground require less waste than members that have a high area of contact with the ground, such as slabs and curbs. A typical waste factor for beams, columns, walls, and other members that have a small area of contact with the ground is 5%. A typical waste factor for slabs, curbs, footings, and other members that have a high area of contact with the ground is 10%. Leaving an area 1/2 inch too low increases the amount of concrete needed to construct a 4-inch concrete slab by 12.5%. Careful grading of the subgrade is critical to controlling concrete waste. Actual waste factors should be monitored and compared to estimated quantities to ensure that an appropriate waste factor is being used in the estimates. In addition, extra gravel is often provided to the concrete crew to be used to fill in low spots in an effort to minimize concrete waste.

Let's look at how the equations in Chapter 4 are used to calculate the quantities for some common concrete components.

SPREAD FOOTINGS

The work package for the spread footings includes the concrete for the footings, forms, the horizontal rebar in the footings, and the dowels needed to connect the footings to columns. The ties at the bottom of the column are often included in the rebar order for the spread footings because they are used to properly space the dowels that run between the spread footing and the column. The concrete is treated as a volumetric good, the forms for short footings (12 inches thick or less) are treated as a linear component, the form for tall footings (thicker than 12 inches) are treated as sheet goods and are measured by the square foot, and rebar is treated as a counted item. The quantity takeoff for a spread footing is shown in the following example.

EXAMPLE 5-1

Determine the volume of concrete in cubic yards, the lineal feet of forms, and the rebar needed to complete 10 of the footings shown in Figure 5-1. Provide 3 inches of cover for the rebar in the footings and 2 inches of cover for the rebar in the columns. Include 10% waste in the calculated volume of concrete, and express the volume of concrete in quarter-yard increments. The ties are to be ordered with the column rebar.

Solution: The volume of the concrete is determined as follows:

$$\text{Volume} = (\text{No. of Footings})(\text{Width 1})(\text{Width 2})(\text{Thickness})$$

$$= (10 \text{ ea})\left[3 \text{ ft} + \frac{2 \text{ in}}{12 \text{ in/ft}}\right]\left[3 \text{ ft} + \frac{2 \text{ in}}{12 \text{ in/ft}}\right](1 \text{ ft})$$

$$= (100.3 \text{ ft}^3)\left(\frac{1 \text{ yd}^3}{27 \text{ ft}^3}\right) = 3.71 \text{ yd}^3$$

Add the waste using Eq. (4-22) as follows:

$$\text{Quantity}_{\text{with Waste}}$$
$$= \text{Quantity}_{\text{without Waste}}\left(1 + \frac{\text{Waste Percentage}}{100}\right)$$

$$\text{Volume}_{\text{with Waste}} = (3.71 \text{ yd}^3)\left(1 + \frac{10}{100}\right)$$
$$= (3.71 \text{ yd}^3)(1.10) = 4.08 \text{ yd}^3$$

Rounding up to quarter-yard increments, we get 4.25 cubic yards. The lineal feet of forms are calculated as follows:

Length
$$= (\text{No. of Footings})(\text{Width 1} + \text{Width 2} + \text{Width 1} + \text{Width 2})$$
$$= (\text{No. of Footings})(2)(\text{Width 1} + \text{Width 2})$$
$$= (10 \text{ ea})(2)\left[3 \text{ ft} + \frac{2 \text{ in}}{12 \text{ in/ft}} + 3 \text{ ft} + \frac{2 \text{ in}}{12 \text{ in/ft}}\right] = 127 \text{ ft}$$

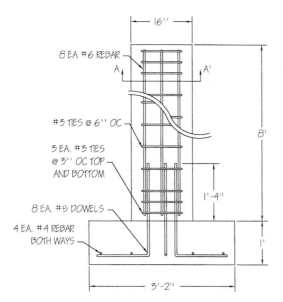
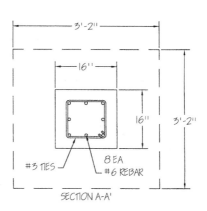

FIGURE 5-1 Footing and Column Details

Each footing will require eight #4 horizontal bars for a total of 80 bars (10 footings × 8 bars/footing). The length of the bars needs to be 3 inches shorter on both ends of the bar than the width of the footings. The bars will be 32 inches (3 ft 2 in − 2 × 3 in) long.

Each footing will require eight #5 dowels for a total of 80 dowels. The length of the long leg of the dowels is the height of the footing less the cover at the bottom of the footing plus the height of the dowel above the top of the footing, or 25 inches (1 ft − 3 in + 1 ft 4 in). The short leg of the dowels will begin 3 inches inside the footing and run 2 inches inside the column. The distance from the edge of the column to the edge of the footing is 11 inches (3 ft 2 in/2 − 16 in/2). The length of the short leg is 10 inches (11 in − 3 in + 2 in).

EXCEL QUICK TIP 5-1
Spread Footing

The volume of concrete and square footage of forms (for footing greater than 12 inches thick) or lineal feet of forms (for footings 12 inches thick or less) needed for a spread footing is set up in a spreadsheet by entering the data and formatting the cells as follows:

	A	B	C
1	No. of Footings	10	ea
2	Width 1	3	ft
3		2	in
4	Width 2	3	ft
5		2	in
6	Thickness	1	ft
7		-	in
8	Concrete Waste	10	%
9	Forms	126.67	ft
10	Volume	4.09	cyd

The following formulas need to be entered into the associated cells:

Cell	Formula
B9	=IF((B6*12+B7)>12,2*B1*(B2+B3/12+B4+B5/12) *Roundup((B6+B7/12)/2,0)*2,2*B1 *(B2+B3/12+B4+B5/12))
B10	=B1*(B2+B3/12)*(B4+B5/12)*(B6+B7/12)/27*(1+B8/100)
C9	=IF((B6*12+B7)>12,"sft","ft")

The formula in Cell B9 uses an IF function to select between square footage of forms for footings over 12 inches thick and linear feet for footings 12 inches thick or less. When calculating the area of forms, Cell B9 uses a ROUNDUP function to round up the form height to 2-foot increments. Cell C9 uses an IF function to select between the units of feet (ft) and square feet (sft).

The data for the spread footings are entered in Cells B1 through B8. The data shown in the foregoing figure is from Example 5-1 and is formatted using the comma style, which replaces zeros with dashes.

COLUMNS

The work package for the columns includes the concrete for the columns, forms, the vertical rebar in columns, and the ties. The dowels, and often the ties at the bottom of the column, are included in the footing package.

The concrete is treated as a volumetric good. The forms for square and rectangular columns are estimated by the square foot, lineal foot, or set of forms. The forms for round columns are estimated by the lineal foot or set of forms. The rebar is treated as a counted item. The quantity takeoff for a square and round column is shown in the following examples.

EXAMPLE 5-2

Determine the volume of concrete in cubic yards, the square feet of forms, and the rebar needed to complete 10 of the columns shown in Figure 5-1. Provide 2 inches of cover for the rebar in the columns. The ties begin 3 inches from the bottom of the column and end 3 inches from the top of the column. Include 5% waste in the calculated volume of concrete and express the volume of concrete in quarter-yard increments.

Solution: The volume of the concrete is determined as follows:

$$\text{Volume} = (\text{No. of Columns})(\text{Width1})(\text{Width 2})(\text{Height})$$
$$= (10 \text{ ea})\left(\frac{16 \text{ in}}{12 \text{ in/ft}}\right)\left(\frac{16 \text{ in}}{12 \text{ in/ft}}\right)(8 \text{ ft})\left(\frac{1 \text{ yd}^3}{27 \text{ ft}^3}\right)$$
$$= 5.27 \text{ yd}^3$$

Add the waste using Eq. (4-22) as follows:

$$\text{Volume}_{\text{with Waste}} = (5.27 \text{ yd}^3)\left(1 + \frac{5}{100}\right) = 5.53 \text{ yd}^3$$

Rounding to quarter-yard increments, we get 5.5 cubic yards. The square feet of forms are calculated as follows:

Area
$$= (\text{No. of Columns})(\text{Width 1} + \text{Width 2} + \text{Width 1} + \text{Width 2})$$
$$\times (\text{Height})$$
$$= (\text{No. of Columns})(2)(\text{Width 1} + \text{Width 2})(\text{Height})$$
$$= (10 \text{ ea})(2)\left(\frac{16 \text{ in}}{12 \text{ in/ft}} + \frac{16 \text{ in}}{12 \text{ in/ft}}\right)(8 \text{ ft}) = 427 \text{ ft}^2$$

Each column will require eight #6 vertical bars for a total of 80 bars (10 columns × 8 bars/column). The length of the bars needs to be 3 inches shorter than the height of the column, or 7 foot 9 inches (8 ft − 3 in) long.

The ties for the column are located in 6-inch increments starting at 3 inches and ending at 7 foot 9 inches. The distance over which the ties occur is 7 feet 6 inches (7 ft 9 in − 3 in). The number of ties is calculated using Eq. (4-1) as follows:

$$\text{Number} = \frac{7 \text{ ft 6 in}}{6 \text{ in}} + 1 = 16 \text{ ties}$$

Two additional ties are needed to meet the requirement for three ties at 3 inches on center at the top and bottom of the column. These ties are located at 6 inches from the bottom and top of the column. The ties will be located at the following distances from the bottom of the column: 3 in, 6 in, 9 in, 1 ft 3 in, 1 ft 9 in, 2 ft 3 in, 2 ft 9 in, 3 ft 3 in, 3 ft 9 in, 4 ft 3 in, 4 ft 9 in, 5 ft 3 in, 5 ft 9 in, 6 ft 3 in, 6 ft 9 in, 7 ft 3 in, 7 ft 6 in, and 7 ft 9 in.

The number of ties needed for each column is 18 (16 + 2), and the number of ties needed for 10 columns is 180 (10 columns × 18 ties/column) ties. The width of the ties will be 2 inches shorter on both sides of the column, or 12 inches (16 in − 2 in − 2 in) square.

EXCEL QUICK TIP 5-2
Rectangular Column

The volume of concrete and area of forms needed for a rectangular column is set up in a spreadsheet by entering the data and formatting the cells as follows:

	A	B	C
1	No. of Columns	10	ea
2	Width 1	-	ft
3		16	in
4	Width 2	-	ft
5		16	in
6	Height	8	ft
7	Concrete Waste	5	%
8	Forms	427	sft
9	Volume	5.53	cyd

The following formulas need to be entered into the associated cells:

Cell	Formula
B8	=B1*2*(B2+B3/12+B4+B5/12)*B6
B9	=B1*(B2+B3/12)*(B4+B5/12)*(B6)/ 27*(1+B7/100)

The data for the column footings is entered in Cells B1 through B7. The data shown in the foregoing figure is from Example 5-2 and is formatted using the comma style, which replaces zeros with dashes.

EXAMPLE 5-3

Determine the volume of concrete in cubic yards, the lineal feet of the Sonotube, and the rebar needed to complete 10 of the columns shown in Figure 5-2. Provide 2 inches of cover for the rebar in the columns. The ties begin 3 inches from the bottom of the column and end 3 inches from the top of the column. Include 5% waste in the calculated volume of concrete, and express the volume of concrete in quarter-yard increments.

Solution: The volume of the concrete is determined as follows:

$$\text{Volume} = (\text{No. of Columns})(\pi)(\text{Diameter}/2)^2(\text{Height})$$
$$= (10 \text{ ea})(\pi)\left(\frac{16 \text{ in}}{(12 \text{ in/ft})(2)}\right)^2(8 \text{ ft})\left(\frac{1 \text{ yd}^3}{27 \text{ ft}^3}\right)$$
$$= 4.13 \text{ yd}^3$$

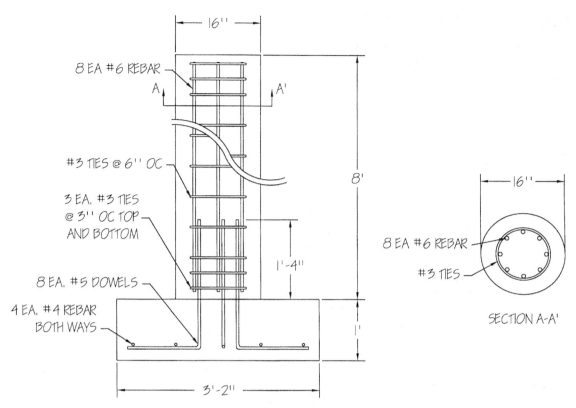

FIGURE 5-2 Column Detail

Add the waste using Eq. (4-22) as follows:

$$\text{Volume}_{\text{with Waste}} = (4.13 \text{ yd}^3)\left(1 + \frac{5}{100}\right) = 4.33 \text{ yd}^2$$

Rounding up to quarter-yard increments, we get 4.50 cubic yards. The lineal feet of forms are calculated as follows:

Length = (No. of Columns)(Height) = (10 ea)(8 ft) = 80 ft

Each column will require eight #6 vertical bars for a total of 80 bars (10 columns × 8 bars/column). The length of the bars needs to be 3 inches shorter than the height of the column, or 7 feet 9 inches (8 ft − 3 in) long.

The ties for the column are located at 6-inch increments starting at 3 inches and ending at 7 feet 9 inches. The distance over which the ties occur is 7 feet 6 inches (7 ft 9 in − 3 in). The number of ties is calculated using Eq. (4-1) as follows:

$$\text{Number} = \frac{7 \text{ ft } 6 \text{ in}}{6 \text{ in}} + 1 = 16 \text{ ties}$$

Two additional ties are needed to meet the requirement for three ties at 3 inches on center at the top and bottom of the column. These ties are located at 6 inches from the bottom and top of the column. The total number of ties needed for each column is 18 (16 + 2). The total number of ties needed for 10 columns is 180 (10 columns × 18 ties/column) ties. The width of the ties will be 2 inches smaller than the diameter of the column on both sides of the column, or 12 inches (16 in − 2 in − 2 in) in diameter.

EXCEL QUICK TIP 5-3
Round Column

The volume of concrete needed for a round column is set up in a spreadsheet by entering the data and formatting the cells as follows:

	A	B	C
1	No. of Columns	10	ea
2	Diameter	-	ft
3		16	in
4	Height	8	ft
5	Concrete Waste	5	%
6	Volume	4.34	cyd

The following formula needs to be entered into Cell B6:

`=B1*(PI()*((B2+B3/12)/2)^2)*B4/27*(1+B5/100)`

In Excel, π (3.141592654) is written as follows: PI()

The data for the column is entered in Cells B1 through B5. The data shown in the foregoing figure is from Example 5-3 and is formatted using the comma style, which replaces zeros with dashes.

CONTINUOUS FOOTINGS

The work package for the continuous footing includes the concrete for the footing, forms, the horizontal rebar in the footings, and the dowels needed to connect the footings to columns. The concrete is treated as a volumetric good, the forms for short footings (12 inches thick or less) are treated as a linear component, the forms for tall footings (thicker than 12 inches) are treated as sheet goods and are measured by the square foot, the dowels are treated as a counted item, and the continuous rebar is treated as a linear component.

Continuous footings are similar to spread footings except that they are much longer than they are wide and they often turn to follow the outline of the building. Continuous footings are handled in a similar manner to spread footings with two exceptions that are a result of the footing turning to follow the outline of the building. First, when measuring the length of continuous footings one must be careful not to count the corners twice. If one were to measure the outside of the footings, the corners would be counted twice as shown in Figure 5-3.

To avoid counting the corners twice, one must begin the measurement for L2 as shown in Figure 5-4.

Second, when calculating the number of dowels, a dowel to account for the end condition must be included for each straight segment of continuous footing. This ensures that there is a dowel located at each corner, end, or intersection.

The quantity takeoff for a continuous footing is shown in the following example.

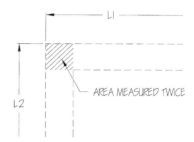

FIGURE 5-3 Double-Counted Corner

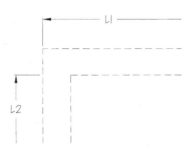

FIGURE 5-4 Proper Way to Measure Corners

EXAMPLE 5-4

Determine the volume of concrete in cubic yards, the lineal feet of forms, and the rebar needed to complete the footings shown in Figures 5-5 and 5-6. Include 8% waste in the calculated volume of concrete, and express the volume of concrete in quarter-yard increments. Provide 3 inches of cover for the rebar in the footings. The continuous rebar is purchased in 20-foot lengths and is lapped 18 inches.

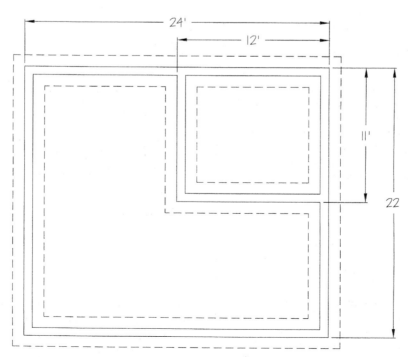

FIGURE 5-5 Foundation Plan

Concrete 91

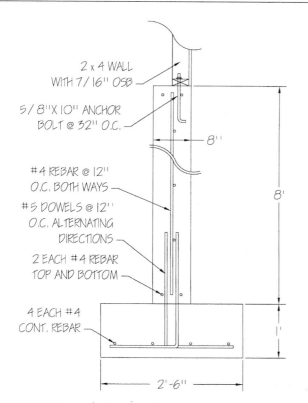

FIGURE 5-6 Foundation Section

Solution: The first step is to find the length of the footings. We begin by dividing the footings into segments as shown in Figure 5-7.

The length of the footing at the top of the foundation plan is equal to the length of the foundation wall plus the length the footing extends beyond the foundation wall at both ends. The footing extends half of the difference between the width of the footing and the width of the wall, or 11 inches (30 in/2 − 8 in/2) beyond the foundation wall as shown in Figure 5-8. The length of the footing is 25 feet 10 inches (24 ft + 11 in + 11 in). Similarly, the length of the footing at the bottom of the foundation plan is also 25 feet 10 inches.

The footing at the left side of the foundation plan begins at the bottom of the top footing and ends at the top of the bottom footing as shown in Figure 5-7. The distance from the exterior of the foundation wall and the bottom of the top footing is 1 foot 7 inches (8 inches + 11 inches) as shown in Figure 5-9. Similarly, the distance from the exterior of the foundation wall and the top of the bottom footing is 1 foot 7 inches. The length of the footing is 18 feet 10 inches (22 ft − 1 ft 7 in − 1 ft 7 in).

Similarly, the length of the right footing is 18 feet 10 inches. The length of the right-center footing is 11 feet 4 inches (12 ft − 1 ft 7 in + 11 in). The length of the top-center footing is 7 feet 10 inches (11 ft − 1 ft 7 in − 1 ft 7 in). The lengths of the footings are shown in Figure 5-10.

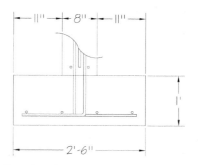

FIGURE 5-8 Footing Distances

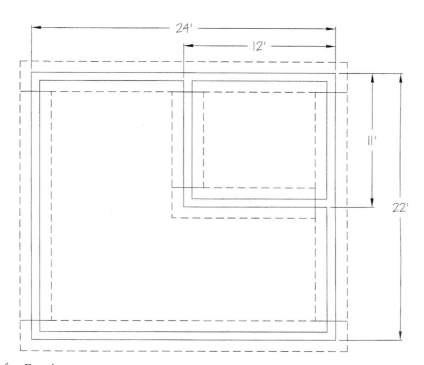

FIGURE 5-7 Segments for Footings

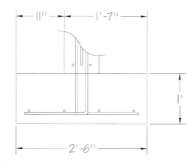

FIGURE 5-9 Footing Distances

The total length of the footing is the sum of the individual lengths and is calculated as follows:

$$\text{Length} = 25 \text{ ft } 10 \text{ in} + 25 \text{ ft } 10 \text{ in} + 18 \text{ ft } 10 \text{ in}$$
$$+ 18 \text{ ft } 10 \text{ in} + 11 \text{ ft } 4 \text{ in} + 7 \text{ ft } 10 \text{ in}$$
$$\text{Length} = 108 \text{ ft } 6 \text{ in} = 108.5 \text{ ft}$$

The volume of the concrete is determined as follows:

$$\text{Volume} = (\text{Length})(\text{Width})(\text{Thickness})$$
$$= (108.5 \text{ ft})(2.5 \text{ ft})(1 \text{ ft})\left(\frac{1 \text{ yd}^3}{27 \text{ ft}^3}\right) = 10.04 \text{ yd}^3$$

Add the waste using Eq. (4-22) as follows:

$$\text{Volume}_{\text{with Waste}} = (10.04 \text{ yd}^3)\left(1 + \frac{8}{100}\right) = 10.84 \text{ yd}^3$$

Rounding up to quarter-yard increments, we get 11 cubic yards. Using the same procedures that were used to determine the length of the footing, we can determine the length of the individual forms. The lengths of the individual forms are shown in Figure 5-11.

The lineal feet of forms is the sum of the individual lengths and is calculated as follows:

$$\text{Length} = 25 \text{ ft } 10 \text{ in} + 23 \text{ ft } 10 \text{ in} + 25 \text{ ft } 10 \text{ in} + 23 \text{ ft } 10 \text{ in}$$
$$+ 9 \text{ ft } 6 \text{ in} + 10 \text{ ft } 4 \text{ in} + 11 \text{ ft } 4 \text{ in} + 8 \text{ ft } 6 \text{ in}$$
$$+ 20 \text{ ft } 10 \text{ in} + 18 \text{ ft } 10 \text{ in} + 8 \text{ ft } 10 \text{ in} + 7 \text{ ft } 10 \text{ in}$$
$$+ 8 \text{ ft } 10 \text{ in} + 7 \text{ ft } 10 \text{ in}$$
$$= 212 \text{ ft}$$

There are four continuous bars in the footings with an average total length of 108.5 feet per bar, for a total length of 434 feet. In addition to this rebar, four 36-inch-long (twice the lap or 2 × 18 inches) L-shaped bars are needed at each intersection for an additional 24 feet (2 intersections × 4 bars/intersection × 3 ft/bar). The L-shaped bars are used to connect the rebar in the exterior footings to the rebar in the interior footings. A total of 458 feet is needed. The number of 20-foot bars is calculated using Eq. (4-6) as follows:

$$\text{Number} = \frac{\text{Length}}{\text{Length}_{\text{Piece}} - \text{Lap}}$$

$$\text{Number} = \frac{458 \text{ ft}}{(20 \text{ ft} - 1.5 \text{ ft})} = 25 \text{ each}$$

The short leg of the #5 dowels will run from a point 3 inches inside the footing to the center of the footing. The short leg will be 12 inches long (30 in/2 − 3 in). The long leg will run from a point 3 inches above the bottom of the footing to 18 inches (the required lap) above the top of the footing. The long leg will be 27 inches long (12 in − 3 in + 18 in). The number of dowels is calculated using the following equation:

$$\text{Number} = \frac{\text{Distance}}{\text{Spacing}} + 6 = \frac{108.5 \text{ ft}}{1 \text{ ft}} + 6 = 115 \text{ each}$$

The 6 in this equation is to account for the end condition for all of the six walls.

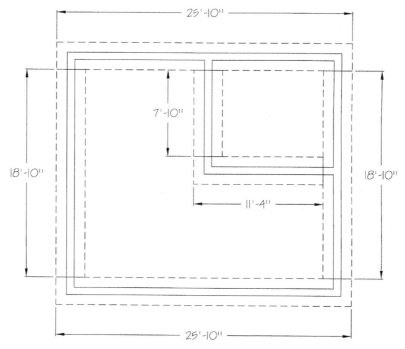

FIGURE 5-10 Footing Lengths

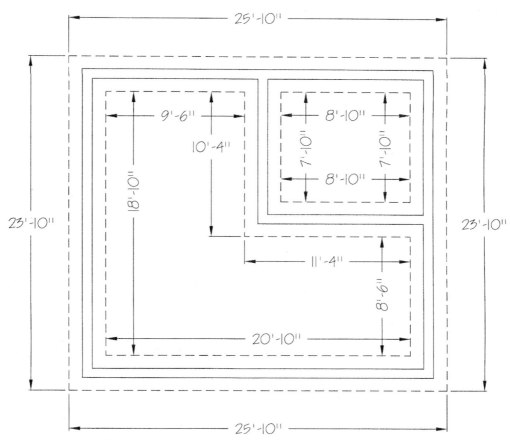

FIGURE 5-11 Lengths of the Footing Forms

Although it is time-consuming to calculate the length of the footing and the length of the side of the footings, this can be rapidly done by using a plan measurer, such as the one in Figure 1-4, a digitizer, such as the one in Figure 1-5, or a takeoff software package. The length of the forms can be approximated by multiplying the footing length by two. Using this method for the preceding example gives the length of the forms as 217 feet rather than the actual length of 212 feet. This method of calculation can overstate or understate the quantity of forms needed.

Another method used to determine the length of a footing is the centerline method. With the centerline method the lengths of the footings are measured along the centerlines of the footings. When measuring around corners, the centerline method double counts one-fourth of the corner and ignores one-fourth of the corner as shown in Figure 5-12. The area double counted offsets the area ignored. When using the centerline method at intersections, such as the one shown in Figure 5-13, the measurement of the intersecting footing must stop at the edge of the other footing to avoid including an area twice. If you measure to the centerline, the shaded area in Figure 5-13 will be measured twice.

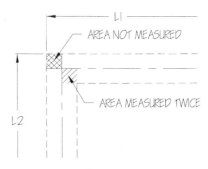

FIGURE 5-12 Measuring Corners

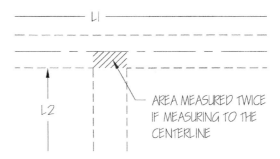

FIGURE 5-13 Measuring Intersections

EXAMPLE 5-5

Determine the volume of concrete in cubic yards for the footing in Example 5-4 using the centerline method.

Solution: The length of the centerline of the footing on the top of the foundation plan is equal to the length of the foundation wall less one-half of the thickness of the foundation wall on both ends, or 23 feet 4 inches (24 ft − 8 in/2 − 8 in/2). The length of the centerline of the footing on the bottom of the foundation plan is the same as that at the top.

The length of the centerline of the footing on the left of the foundation plan is equal to the length of the foundation wall less one-half of the thickness of the foundation wall on both ends, or 21 feet 4 inches (22 ft − 8 in/2 − 8 in/2). The length of the centerline of the footing on the right of the foundation plan is the same as that on the left.

The length of the centerline of the right-center footing is equal to the length of the foundation wall less one-half of the thickness of the foundation wall on the left end less the distance from the outside of the foundation wall to the inside of the footing on the right end. This distance is 19 inches (11 in + 8 in) as shown in Figure 5-8. The length of the footing is 10 feet 1 inch (12 ft − 8 in/2 − 19 in).

The length of the centerline of the top-center footing is calculated in the same way as the right-center footing. Its length is 9 feet 1 inch (11 ft − 8 in/2 − 19 in). The dimensions for the footings are shown in Figure 5-14.

The total length of the footing is the sum of the individual lengths and is calculated as follows:

Length = 23 ft 4 in + 23 ft 4 in + 21 ft 4 in + 21 ft 4 in
  + 10 ft 1in + 9 ft 1 in
  = 108 ft 6 in = 108.5 ft

The volume of the concrete is determined as follows:

$$\text{Volume} = (\text{Length})(\text{Width})(\text{Thickness})$$

$$= 108.5 \text{ ft}(2.5 \text{ ft})(1 \text{ ft})\left(\frac{1 \text{ yd}^3}{27 \text{ ft}^3}\right) = 10.04 \text{ yd}^3$$

Add the waste using Eq. (4-22) as follows:

$$\text{Volume}_{\text{with Waste}} = (10.04 \text{ yd}^3)\left(1 + \frac{8}{100}\right) = 10.84 \text{ yd}^3$$

Rounding up to quarter-yard increments, we get 11 cubic yards. This is the same answer as we got for Example 5-4. ■

EXCEL QUICK TIP 5-4
Continuous Footing

The volume of concrete needed for a continuous footing is set up in a spreadsheet by entering the data and formatting the cells as follows:

	A	B	C
1	Length	108	ft
2		6	in
3	Width	2	ft
4		6	in
5	Thickness	1	ft
6		-	in
7	Concrete Waste	8	%
8	Volume	10.85	cyd

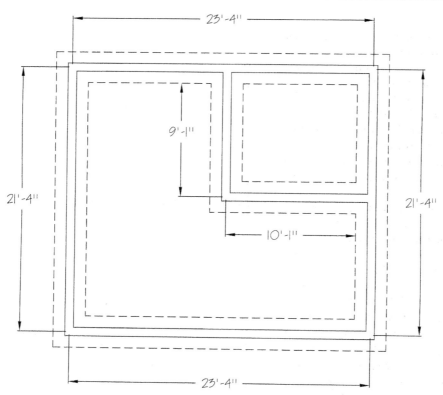

FIGURE 5-14 Footing Dimensions

The following formula needs to be entered into Cell B8:

=(B1+B2/12)*(B3+B4/12)*(B5+B6/12)/27*(1+B7/100)

The data for the footings is entered in Cells B1 through B7. The data shown in the foregoing figure is from Example 5-4 and is formatted using the comma style, which replaces zeros with dashes.

FOUNDATION WALLS

The work package for the foundation walls includes the concrete for the walls, forms, the vertical and horizontal rebar in walls, and anchor bolts. The dowels are included in the footing package. The concrete is treated as a volumetric good. The forms are estimated by the square foot. The vertical rebar is treated as a counted item; the horizontal rebar is treated as a linear good. The same principles used for handling corners of continuous footings apply to foundation walls. Foundation walls may be thought of as narrow, tall, continuous footings. The quantity takeoff for a foundation wall is shown in the following example.

EXAMPLE 5-6

Determine the volume of concrete in cubic yards, the square feet of forms, the rebar, and the anchor bolts needed to complete the foundation wall in Figures 5-5 and 5-6. Include 5% waste in the calculated volume of concrete, and express the volume of concrete in quarter-yard increments. Provide 2 inches of cover for the rebar in the wall and lap the continuous rebar 18 inches. The continuous rebar will be ordered in 20-foot lengths.

Solution: The length of the wall is calculated in the same manner as the lengths of the footings were calculated in Example 5-4. The lengths of the walls are shown in Figure 5-15.

The total length of the wall is calculated as follows:

Length = 24 ft + 20 ft 8 in + 24 ft + 20 ft 8 in
+ 9 ft 8 in + 11 ft 4 in
= 110 ft 4 in

The volume of the concrete is determined as follows:

$$\text{Volume} = (\text{Length})(\text{Width})(\text{Thickness})$$
$$= (110.33 \text{ ft})(8 \text{ ft})\frac{(8 \text{ in})}{(12 \text{ in/ft})}\left(\frac{1 \text{ yd}^3}{27 \text{ ft}^3}\right)$$
$$= 21.79 \text{ yd}^3$$

Add the waste using Eq. (4-22) as follows:

$$\text{Volume}_{\text{with Waste}} = (21.79 \text{ yd}^3)\left(1 + \frac{5}{100}\right) = 22.88 \text{ yd}^3$$

Rounding up to quarter-yard increments, we get 23 cubic yards. The length of the individual forms can be determined using the same procedures that were used to determine the length of the footing forms. The lengths of the individual forms are shown in Figure 5-16.

The lineal feet of forms is the sum of the individual lengths and is calculated as follows:

Length = 24 ft + 22 ft + 24 ft + 22 ft + 11 ft 4 in
+ 10 ft 4 in + 11 ft 4 in + 10 ft 4 in
+ 22 ft 8 in + 20 ft 8 in + 10 ft 8 in
+ 9 ft 8 in + 10 ft 8 in + 9 ft 8 in
= 219 ft 4 in

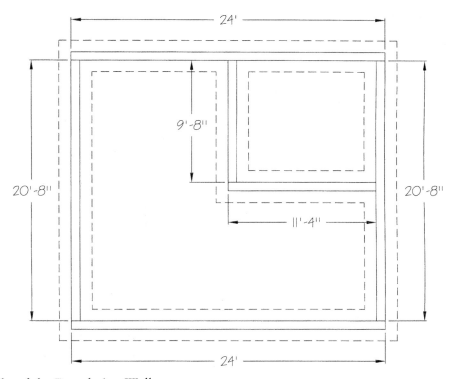

FIGURE 5-15 Lengths of the Foundation Walls

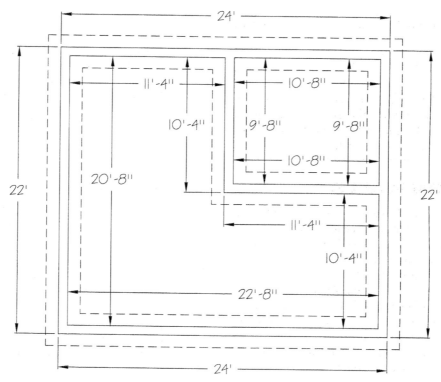

FIGURE 5-16 Lengths of the Foundation Forms

Alternatively, they may be measured using a plan measurer, digitizer, or software takeoff package. The area of the forms is calculated as follows:

Area = (Length)(Height) = (219.33 ft)(8 ft) = 1,755 ft²

The number of continuous bar needed, for one bar every 12 inches, is calculated using Eq. (4-1) as follows:

$$\text{Number} = \frac{8 \text{ ft}}{1 \text{ ft}} + 1 = 9 \text{ each}$$

Because the top and bottom bars are doubled, two additional bars need to be added to the number of bars needed, for one bar every 12 inches. The number of continuous bars is 11 (9 ea + 2 ea). The average length of the continuous bars is 110.33 feet for a total length of 1,214 feet. As was done for the footings, eleven 36-inch L-shaped bars need to be added at each intersection for an additional 66 feet of rebar (2 intersections × 11 bars/intersection × 3 ft/bar). The L-shaped bars are used to connect the rebar in the exterior walls to the rebar in the interior walls. The total length of continuous rebar is 1,280 feet. The number of 20 foot bars is calculated using Eq. (4-6) as follows:

$$\text{Number} = \frac{1,280 \text{ ft}}{(20 \text{ ft} - 1.5 \text{ ft})} = 70 \text{ each}$$

The vertical bars run from the bottom of the wall to 2 inches below the top of the wall, for a length of 7 feet 10 inches. The number of vertical bars is calculated using the following equation:

$$\text{Number} = \frac{\text{Distance}}{\text{Spacing}} + 6 = \frac{110.33 \text{ ft}}{1 \text{ ft}} + 6 = 117 \text{ each}$$

The 6 in this equation is to account for the end condition for all of the six walls. The number of anchor bolts is calculated as follows:

$$\text{Number} = \frac{\text{Distance}}{\text{Spacing}} + 6 = \frac{(110.33 \text{ ft})(12 \text{ in/ft})}{(32 \text{ in})} + 6$$
$$= 48 \text{ each}$$

EXCEL QUICK TIP 5-5
Foundation Wall

The volume of concrete needed for a foundation wall is set up by changing the Cell A5 of the worksheet in Excel Quick Tip 5-4 to *Height* as follows:

	A	B	C
1	Length	110	ft
2		4	in
3	Width	-	ft
4		8	in
5	Height	8	ft
6		-	in
7	Concrete Waste	5	%
8	Volume	22.88	cyd

The data for the foundation wall is entered in Cells B1 through B7. The data shown in the foregoing figure is from Example 5-6 and is formatted using the comma style, which replaces zeros with dashes.

BEAMS

The same principles that were used to determine the quantity of concrete, forms, and rebar for footings, columns, and walls can be used for determining the quantities for beams. The key differences are that forms need to be included for the bottom of the beam and the forms will need to be supported until the concrete has cured. The takeoff for a concrete beam is shown in the following example.

EXAMPLE 5-7

Determine the volume of concrete in cubic yards, the square feet of forms, and rebar needed to complete the beam in Figure 5-17. Only half of the beam is shown in Figure 5-17, with the other half being a mirror image of the one shown. Include 5% waste in the calculated volume of concrete, and express the volume of concrete in quarter-yard increments. Provide 2 inches of cover for the rebar in the beam.

Solution: The volume of the concrete is calculated as follows:

$$\text{Volume} = (\text{Length})(\text{Width})(\text{Thickness})$$
$$= (16 \text{ ft})(1.5 \text{ ft})(1 \text{ ft})\left(\frac{1 \text{ yd}^3}{27 \text{ ft}^3}\right)$$
$$= 0.89 \text{ yd}^3$$

Add the waste using Eq. (4-22) as follows:

$$\text{Volume}_{\text{with Waste}} = (0.89 \text{ yd}^3)\left(1 + \frac{5}{100}\right) = 0.93 \text{ yd}^3$$

Rounding up to quarter-yard increments, we get 1 cubic yard. Forms are needed on the bottom of the beam between the end columns. The area of the forms is calculated as follows:

$$\begin{aligned}\text{Area} &= 2(\text{Area of Side}) + 2(\text{Area of End}) + \text{Area of Bottom} \\ &= 2(16 \text{ ft})(1.5 \text{ ft}) + 2(1 \text{ ft})(1.5 \text{ ft}) + (16 \text{ ft} - 2 \\ &\quad \times 1 \text{ ft } 4 \text{ in})(1 \text{ ft}) \\ &= 64.34 \text{ ft}^2\end{aligned}$$

The #5 dowels should have been ordered with the column. Two #5 bars 5 feet 1 inch long are needed for each end of the beam, for a total of four. Three #5 bars 8 feet 4 inches (4 ft 2 in × 2) are needed for the center of the beam. Six stirrups are needed for each end, for a total of 12. The stirrups will need to be 4 inches less than the dimensions for the beam in order to provide 2 inches of cover. The dimensions for the stirrups are 14 inches high by 8 inches wide.

SLAB ON GRADE

The work package for a slab on grade includes the concrete for the slab, the area to be finished, forms, and rebar or welded wire fabric. The concrete is treated as a volumetric good, the forms are treated as a linear component, rebar is treated as a linear component, and welded wire fabric is treated as a sheet good. The quantity takeoff for a slab on grade is shown in the following examples.

EXAMPLE 5-8

Determine the volume of concrete in cubic yards, the area to be finished, the lineal feet of forms, and number of 20-foot bars of #4 rebar needed to complete a slab on grade 100 feet long by 40 feet wide. The slab is reinforced with #4 rebar at 18 inches on center. Include 12% waste in the calculated volume

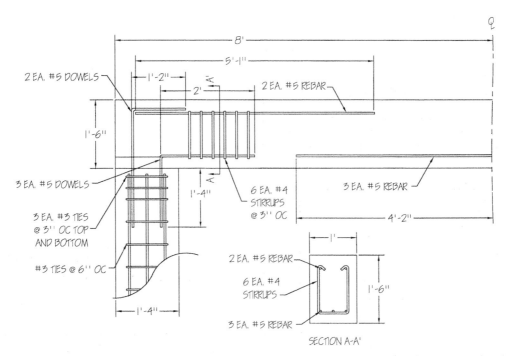

FIGURE 5-17 Beam Section

of concrete, and express the volume of concrete in quarter-yard increments. Provide 36 bar diameters of lap on the rebar.

Solution: The volume of concrete is calculated as follows:

$$\text{Volume} = (\text{Length})(\text{Width})(\text{Thickness})$$
$$= (100 \text{ ft})(40 \text{ ft})(4 \text{ in})\left(\frac{1 \text{ ft}}{12 \text{ in}}\right)\left(\frac{1 \text{ yd}^3}{27 \text{ ft}^3}\right) = 49.38 \text{ yd}^3$$

Add the waste using Eq. (4-22) as follows:

$$\text{Volume}_{\text{with Waste}} = (49.38 \text{ yd}^3)\left(1 + \frac{12}{100}\right) = 55.31 \text{ yd}^3$$

Rounding up to quarter-yard increments, we get 55.50 cubic yards. The area to be finished is calculated as follows:

$$\text{Area} = (100 \text{ ft})(40 \text{ ft}) = 4,000 \text{ ft}^2$$

The length of the forms is calculated as follows:

$$\begin{aligned}\text{Length} &= 2(\text{Length of Side}_1) + 2(\text{Length of Side}_2) \\ &= 2(100 \text{ ft}) + 2(40 \text{ ft}) \\ &= 280 \text{ ft}\end{aligned}$$

The number of 100-foot-long bars needed to cover the slab in the long direction is calculated using Eq. (4-1) as follows:

$$\text{Number} = \frac{40 \text{ ft}}{1.5 \text{ ft}} + 1 = 28 \text{ bars}$$

The number of 40-foot-long bars needed to cover the slab in the short direction is calculated using Eq. (4-1) as follows:

$$\text{Number} = \frac{100 \text{ ft}}{1.5 \text{ ft}} + 1 = 68 \text{ bars}$$

The total length of rebar is calculated as follows:

Length = 28 bars(100 ft/bar) + 68 bars(40 ft/bar) = 5,520 ft

The lap is 18 inches (36 × 0.5 in) or 1.5 feet. The number of 20-foot bars is calculated using Eq. (4-6) as follows:

$$\text{Number} = \frac{5,520 \text{ ft}}{(20 \text{ ft} - 1.5 \text{ ft})} = 299 \text{ each}$$

EXCEL QUICK TIP 5-6
Slab on Grade with Rebar

The volume of concrete and the number of bars of rebar needed for a slab on grade are set up in a spreadsheet by entering the data and formatting the cells as follows:

	A	B	C
1	Slab Length	100	ft
2	Slab Width	40	ft
3	Slab Thickness	4	in
4	Waste	12	%
5	Bar Spacing	18	in
6	Bar Size	4	#
7	Bar Length	20	ft
8	Lap	36	# of dia.
9			
10	Area	4,000	sft
11	Volume	55.31	cyd
12	No. of Bars	299	each

The following formulas need to be entered into the associated cells:

Cell	Formula
B10	=B1*B2
B11	=(B1*B2*(B3/12)/27)*(1+B4/100)
B12	=ROUNDUP((ROUNDUP(B1*12/B5+1,0)*B2+ROUNDUP(B2*12/B5+1,0)*B1)/(B7-B8*(B6/8)/12),0)

The data for the slab is entered in Cells B1 through B8. The data shown in the foregoing figure is from Example 5-8.

EXAMPLE 5-9

Determine the number of 150-foot by 5-foot rolls of welded wire fabric that would be needed for the slab in Example 5-8 if the #4 rebar were replaced by welded wire fabric. The welded wire fabric will need to be lapped 8 inches.

Solution: Run the rolls the long direction in the slab. Using the row and column method, determine the number or rows of welded wire fabric needed to cover the 40-foot width of the slab using Eq. (4-12) as follows:

$$\text{Number}_{\text{Rows}} = \frac{40 \text{ ft}}{(5 \text{ ft} - 0.667 \text{ ft})} = 10 \text{ rows}$$

The number of columns is calculated using Eq. (4-15) as follows:

$$\text{Number}_{\text{Columns}} = \frac{100 \text{ ft}}{(150 \text{ ft} - 0.667 \text{ ft})} = 0.67 \text{ columns}$$

The number of rolls is calculated using Eq. (4-18) as follows:

$$\text{Number} = (\text{Number}_{\text{Rows}})(\text{Number}_{\text{Columns}})$$
$$\text{Number} = (10 \text{ rows})(0.67 \text{ columns}) = 6.7 \text{ rolls}$$

EXCEL QUICK TIP 5-7
Welded Wire Fabric–Reinforced Slab on Grade

The volume of concrete and the number of rolls/sheets of welded wire fabric needed for a slab on grade are set up in a spreadsheet by entering the data and formatting the cells as follows:

Concrete 99

	A	B	C
1	Slab Length	100	ft
2	Slab Width	40	ft
3	Slab Thickness	4	in
4	Waste	12	%
5	Welded Wire Fabric Length	150	ft
6	Welded Wire Fabric Width	5	ft
7	Welded Wire Fabric Lap	8	in
8			
9	Area	4,000	sft
10	Volume	55.31	cyd
11	Welded Wire Fabric	7	each

The following formulas need to be entered into the associated cells:

Cell	Formula
B9	=B1*B2
B10	=(B1*B2*(B3/12)/27)*(1+B4/100)
B11	=ROUNDUP(ROUNDUP(2*B2/(B6-B7/12), 0)/2*B1/(B5-B7/12),0)

The formula for Cell B11 rounds the number of rows up to the nearest half row. The data for the slab is entered in Cells B1 through B7. The data shown in the foregoing figure is from Example 5-9.

RAISED SLABS

The estimating of raised slabs is similar to the estimating of slabs on grade, with two exceptions. First, the thickness of the slab may vary because the bottom of the slab is poured over an intentionally uneven surface (such as a metal deck) and because the structure supporting the slab may sag under the weight of the concrete, making the slab thicker at the center of the spans than near the supports. To account for the uneven surface of the deck, the average slab thickness must be determined.

For the metal deck shown in Figure 5-18, the average thickness is determined by the following equation:

$$T_{Ave} = T + \frac{(W_1 + W_2)(D)}{2S} \quad (5.1)$$

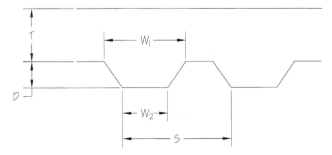

FIGURE 5-18 Metal Deck

where

T = Distance from Top of Slab to Top of Metal Deck
W_1 = Width of the Metal Deck Cell at the Top
W_2 = Width of the Metal Deck Cell at the Bottom
D = Depth of the Metal Deck Cell
S = Spacing of the Metal Deck Cells

EXAMPLE 5-10

A 100-foot by 50-foot by 3-inch-thick slab is poured over the metal deck shown in Figure 5-19. The depth of the slab is measured from the top of the slab to the top of the metal deck. Determine the average thickness of the slab and the number of yards of concrete needed to pour the slab. Add 10% waste.

Solution: The average thickness is determined using Eq. (5-1) as follows:

$$T_{Ave} = 3 \text{ in} + \frac{(4\ 1/2 \text{ in} + 2\ 1/2 \text{ in})(1\ 1/2 \text{ in})}{2 \times 6 \text{ in}} = 3.875 \text{ in}$$

The volume of concrete is calculated as follows:

$$\text{Volume} = (\text{Length})(\text{Width})(\text{Thickness})$$
$$= (100 \text{ ft})(50 \text{ ft})(3.875 \text{ in})\left(\frac{1 \text{ ft}}{12 \text{ in}}\right)\left(\frac{1 \text{ yd}^3}{27 \text{ ft}^3}\right) = 59.80 \text{ yd}^3$$

Add the waste using Eq. (4-22) as follows:

$$\text{Volume}_{\text{with Waste}} = (59.80 \text{ yd}^3)\left(1 + \frac{10}{100}\right) = 65.78 \text{ yd}^3$$

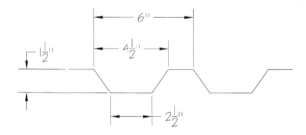

FIGURE 5-19 Metal Deck Dimensions

EXCEL QUICK TIP 5-8
Concrete Slab on Metal Deck

The volume of concrete needed for a concrete slab on a metal deck similar to Figure 5-18 is set up in a spreadsheet by entering the data and formatting the cells as follows:

	A	B	C
1	Slab Length	100	ft
2	Slab Width	50	ft
3	T	3.00	in
4	D	1.50	in
5	W1	2.50	in
6	W2	4.50	in
7	S	6.00	in
8	Waste	10	%
9	Volume	65.78	cyd

> The following formula needs to be entered into Cell B9:
>
> ```
> =B1*B2*((B3+(B5+B6)*B4/
>   (2*B7))/12)/27*(1+B8/100)
> ```
>
> The data for the slab is entered in Cells B1 through B8. The data shown in the foregoing figure is from Example 5-10.

The amount of concrete needed to account for the sag is a function of how much a structure deflects under the weight of the concrete as it is being poured. This additional concrete is treated in the same manner as waste; identify the historical amount of concrete needed to account for the sag as a percentage of the estimated volume of concrete. Typically, sag adds 10% to 15% to the amount of concrete needed to pour the slab.

The second way that estimating raised slabs differs from estimating slabs on grade is that the bottom of the slab must be formed. There are three basic ways to form the bottom of the slab. The first is to form the bottom of the slab with removable forms with a structural support that are removed after the concrete has cured to the point that it can support its own weight. In this case, the cost of the forms and supporting structure is often based on the square footage of the slab. A second way is to use a self-supporting metal deck as the form. In this case, the metal deck and the building structure that supports it remain as part of the building after the construction of the slab is complete. The metal deck and supporting structure are included in the materials needed for the construction of the building. The third way is to use a metal deck that is not capable of supporting the weight of the concrete, which must be supported until the concrete has cured to the point that it can support its own weight, as is the case with composite slabs. When the concrete has cured, the support structure is removed, leaving the metal deck and concrete. In this case the support structure is included in the forming, and the metal deck is included in the materials needed for the construction of the building.

STAIRS

The volume of concrete needed for stairs can be estimated by determining the cross-sectional area of the stairs and multiplying the cross-sectional area by the width of the stairs as shown in the following example.

EXAMPLE 5-11

Determine the volume of concrete needed to construct the stairs in Figure 5-20.

Solution: The cross-sectional area of the stairs can be treated as a parallelogram and seven triangles as shown in Figure 5-21.

The area of the parallelogram equals the length of two of the parallel sides multiplied by the perpendicular distance between the two sizes and is calculated as follows:

$$\text{Area} = (6.5 \text{ in})(5.25 \text{ ft})\left(\frac{1 \text{ ft}}{12 \text{ in}}\right) = 2.84 \text{ ft}^2$$

The area of the triangles is one-half the base times the height multiplied by the number of stairs. The area of the seven triangles is calculated as follows:

$$\text{Area} = (7 \text{ ea})\left(\frac{1}{2}\right)(10 \text{ in})(7 \text{ in})\left(\frac{1 \text{ ft}}{12 \text{ in}}\right)^2 = 1.70 \text{ ft}^2$$

The cross-sectional area is 4.54 square feet (2.84 ft² + 1.70 ft²). The volume is calculated as follows:

$$\text{Volume} = (4.54 \text{ ft}^2)(5 \text{ ft})\left(\frac{1 \text{ yd}^3}{27 \text{ ft}^3}\right) = 0.84 \text{ yd}^3$$

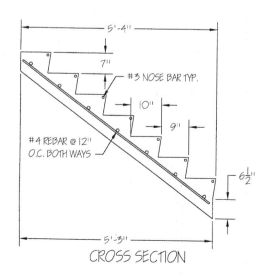
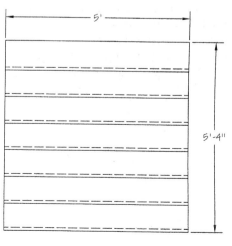

FIGURE 5-20 Stairs

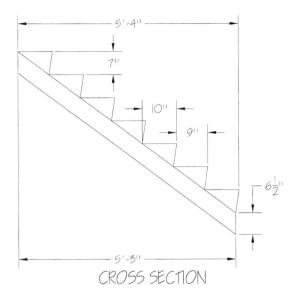

FIGURE 5-21 Stair Cross-Sectional Area

SAMPLE TAKEOFF FOR THE RESIDENTIAL GARAGE

A sample takeoff from a set of plans is shown in the following example.

EXAMPLE 5-12

Determine the concrete, rebar, and forms needed to pour the footings, foundation wall, and interior concrete slab for the residential garage given in Appendix G. Add 5% waste for the footings and foundation wall and 10% waste for the floor slab.

The continuous rebar is to be lapped 18 inches and is ordered in 20-foot lengths.

Solution: The dimensions for the concrete footing are shown in Figure 5-22.

The footing length and volume of concrete needed to pour the footing are calculated as follows:

$$\text{Length} = 3 \text{ ft } 4 \text{ in} + 27 \text{ ft} + 21 \text{ ft } 8 \text{ in} + 27 \text{ ft} + 3 \text{ ft } 4 \text{ in}$$
$$= 82 \text{ ft } 3 \text{ in} = 82.33 \text{ ft}$$
$$\text{Volume} = (\text{Length})(\text{Width})(\text{Thickness})$$
$$= (82.33 \text{ ft})(10 \text{ in})(20 \text{ in})\left(\frac{1 \text{ ft}}{12 \text{ in}}\right)^2\left(\frac{1 \text{ yd}^3}{27 \text{ ft}^3}\right) = 4.24 \text{ yd}^3$$

Add the waste using Eq. (4-22) as follows:

$$\text{Volume}_{\text{with Waste}} = (4.24 \text{ yd}^3)\left(1 + \frac{5}{100}\right) = 4.45 \text{ yd}^3$$

The length of 10-inch-high forms needed for the footing is shown in Figure 5-23.

The length of the forms needed to pour the footing is calculated as follows:

$$\text{Length} = 27 \text{ ft} + 25 \text{ ft} + 27 \text{ ft} + 5 \text{ ft} + 1 \text{ ft } 8 \text{ in} + 3 \text{ ft } 4 \text{ in}$$
$$+ 23 \text{ ft } 8 \text{ in} + 21 \text{ ft } 8 \text{ in} + 23 \text{ ft } 8 \text{ in} + 3 \text{ ft } 4 \text{ in}$$
$$+ 1 \text{ ft } 8 \text{ in} + 5 \text{ ft} = 168 \text{ ft}$$

The footing requires three #4 continuous bars, for a total length of 247 feet (3×82.33 ft). The number of 20-foot bars is calculated using Eq. (4-6) as follows:

$$\text{Number} = \frac{247 \text{ ft}}{(20 \text{ ft} - 1.5 \text{ ft})} = 14 \text{ each}$$

The short leg of the #5 dowels will run from a point 3 inches inside the footing to the center of the footing. The short leg will

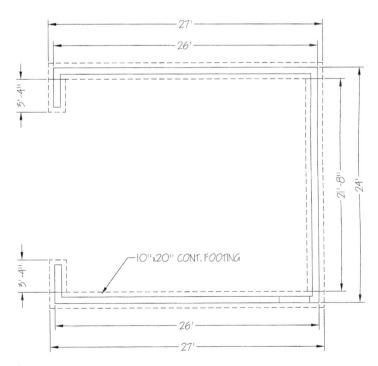

FIGURE 5-22 Length of Footings

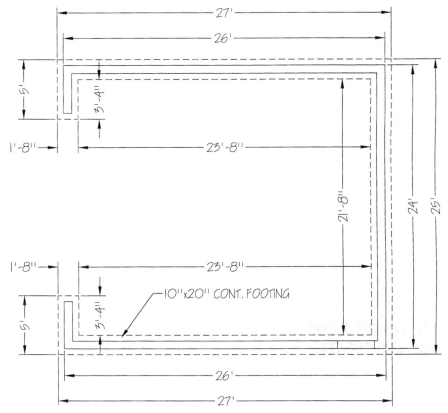

FIGURE 5-23 Length of Footing Forms

be 7 inches long (20 in/2 − 3 in). The long leg will run from a point 3 inches above the bottom of the footing to 2 inches below the top of the 24-inch-high foundation wall. The long leg will be 29 inches long (10 in − 3 in + 24 in − 2 in). The dowels only need to be located in the footing where there is a foundation wall above the footing. The length of the foundation wall is calculated as follows:

Length = 3 ft 4 in + 26 ft + 22 ft 8 in + 26 ft + 3 ft 4 in
= 81 ft 4 in = 81.33 ft

The number of dowels is calculated using the following equation:

$$\text{Number} = \frac{\text{Distance}}{\text{Spacing}} + 5 = \frac{(81.33 \text{ ft})(12 \text{ in/ft})}{16 \text{ in}} + 5 = 66 \text{ each}$$

The 5 in this equation is to account for the end condition for all five of the wall segments. The foundation wall consists of the 2-foot-high by 8-inch-wide wall around the perimeter of the garage and the 14-inch-high by 8-inch-wide frost wall under the garage door. The volume of concrete needed for the foundation wall is calculated as follows:

$$\begin{aligned}\text{Volume} &= (\text{Length})(\text{Width})(\text{Height})\\ &= (81.33 \text{ ft})(2 \text{ ft})(8 \text{ in})\left(\frac{1 \text{ ft}}{12 \text{ in}}\right)\\ &\quad + (16 \text{ ft})(14 \text{ in})(8 \text{ in})\left(\frac{1 \text{ ft}}{12 \text{ in}}\right)^2\\ &= (120.88 \text{ ft}^3)\left(\frac{1 \text{ yd}^3}{27 \text{ ft}^3}\right) = 4.48 \text{ yd}^3\end{aligned}$$

Add the waste using Eq. (4-22) as follows:

$$\text{Volume}_{\text{with Waste}} = (4.48 \text{ yd}^3)\left(1 + \frac{5}{100}\right) = 4.70 \text{ yd}^3$$

Two-foot-high forms are used to form both the foundation wall and the frost wall. The length of the forms needed is calculated as follows:

Length = 26 ft + 24 ft + 26 ft + 24 ft + 24 ft 8 in
+ 22 ft 8 in + 24 ft 8 in + 22 ft 8 in
= 194 ft 8 in

The area of the forms needed is calculated as follows:

Length = (194 ft 8 in)(2 ft) = 390 ft²

Four #4 bars of rebar are needed in the foundation wall and frost wall. The length of the rebar is calculated as follows:

Length = 4(81.33 ft + 16 ft) = 389.32 ft

The number of 20-foot-long bars needed is calculated using Eq. (4-6) as follows:

$$\text{Number} = \frac{389.32 \text{ ft}}{(20 \text{ ft} - 1.5 \text{ ft})} = 21 \text{ each}$$

Eighteen-inch by 18-inch L-shaped #4 dowels are needed between the frost wall and the floor slab. The number of dowels is calculated using Eq. (4-1) as follows:

$$\text{Number} = \frac{(16 \text{ ft})(12 \text{ in/ft})}{16 \text{ in}} + 1 = 13 \text{ each}$$

The length of the wall that needs anchor bolts equals the length of the wall less the width of the 3-foot-wide door. The number of anchor bolts with nuts and washers is calculated using the following equation:

$$\text{Number} = \frac{\text{Distance}}{\text{Spacing}} + 5 = \frac{(81.33 \text{ ft} - 3 \text{ ft})}{2 \text{ ft}} + 5 = 45 \text{ each}$$

In addition, four HPAHD22 hold-downs are needed at the front of the garage. The floor slab extends from the inside of the foundation wall around the perimeter of the garage except at the overhead door, where it extends to the exterior of the wall. At the overhead door the floor slab is thickened. The concrete for the interior slab without the concrete at the overhead door is calculated as follows:

$$\begin{aligned}\text{Volume} &= (\text{Length})(\text{Width})(\text{Thickness}) \\ &= (24 \text{ ft } 8 \text{ in})(22 \text{ ft } 8 \text{ in})(4 \text{ in})\left(\frac{1 \text{ ft}}{12 \text{ in}}\right)\left(\frac{1 \text{ yd}^3}{27 \text{ ft}^3}\right) \\ &= 6.90 \text{ yd}^3\end{aligned}$$

The cross-sectional area of the slab at the overhead door is shown in Figure 5-24. The cross-sectional area is calculated as follows:

$$\begin{aligned}\text{Area} &= (8 \text{ in})(4 \text{ in}) + 0.5(12 \text{ in} + 8 \text{ in})(4 \text{ in}) \\ &= 72 \text{ in}^2 \left(\frac{1 \text{ ft}}{12 \text{ in}}\right)^2 = 0.5 \text{ ft}^2\end{aligned}$$

The volume of concrete at the overhead door is calculated as follows:

$$\begin{aligned}\text{Volume} &= (\text{Area})(\text{Length}) = (0.5 \text{ ft}^2)(16 \text{ ft})\left(\frac{1 \text{ yd}^3}{27 \text{ ft}^3}\right) \\ &= 0.30 \text{ yd}^3\end{aligned}$$

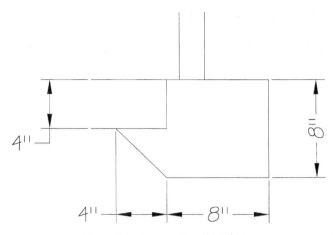

FIGURE 5-24 Cross Section at Overhead Door

The volume of concrete without waste is 7.20 cubic yards (6.90 yd³ + 0.30 yd³). Add the waste using Eq. (4-22) as follows:

$$\text{Volume}_{\text{with Waste}} = (7.20 \text{ yd}^3)\left(1 + \frac{10}{100}\right) = 7.92 \text{ yd}^3$$

The area that needs to be finished is calculated as follows:

$$\text{Area} = (24 \text{ ft } 8 \text{ in})(22 \text{ ft } 8 \text{ in}) + (16 \text{ ft})(8 \text{ in}) = 570 \text{ ft}^2$$

The quantities needed for the garage, grouped by the cost codes in Appendix B, are shown in Table 5-2.

TABLE 5-2 Quantities for Residential Garage

03-200 Rebar		
#4 × 20′ rebar (footings)	14	ea
7″ × 29″ L-dowels (footings)	66	ea
#4 × 20′ rebar (foundation)	21	ea
18″ × 18″ L-dowels (foundation)	13	ea
1/2″ × 10″ anchor bolt (foundation)	45	ea
1/2″ nut (foundation)	45	ea
1/2″ washer (foundation)	45	ea
HPAHD22 hold-downs	4	ea
03-300 Footings and Foundation—Labor		
10″-high footing forms	168	ft
Install continuous rebar 20′	14	ea
Install dowels	66	ea
Pour footings	4.50	yd³
2′-high foundation forms	390	ft²
03-300 Footings and Foundation—Labor		
Install continuous rebar 20′	21	ea
Install dowels	13	ea
Install anchor bolt	45	ea
Pour foundation	4.75	yd³
03-310 Footings and Foundation—Concrete		
3,500-psi concrete (footings)	4.50	yd³
3,500-psi concrete (foundation)	4.75	yd³
03-320 Slab/Floor Labor		
Pour slab	8.00	yd³
Finish slab	570	ft²
03-330 Slab/Floor Concrete		
4,000-psi concrete (slab)	8.00	yd³

CONCLUSION

Concrete consists of the concrete materials, forms, and rebar. The amount of concrete materials needed is determined by the cubic yard. The amount of forms needed may be calculated by the lineal foot, square foot, or each. Forms need to be included on both sides of the concrete member. When calculating the square footage of forms, the area should be based on the height of the forms rather than the height of the concrete member. Short pieces of rebar are determined by the number of pieces, and continuous bars are determined by the number of shorter bars of rebar.

PROBLEMS

1. How many cubic yards of concrete are required to construct a 100-foot-long by 30-inch-wide by 12-inch-thick continuous footing? Include 8% waste. Concrete must be ordered in quarter-yard increments.
2. Determine the length of forms needed for Problem 1.
3. The footing in Problem 1 has four #4 continuous bars. The rebar is lapped 18 inches. Determine the number of 20-foot-long bars that needs to be ordered for the footing.
4. How many cubic yards of concrete are required to construct a 75-foot-long by 24-inch-wide by 10-inch-thick continuous footing? Include 10% waste. Concrete must be ordered in quarter-yard increments.
5. Determine the length of forms needed for Problem 4.
6. The footing in Problem 4 has three #4 continuous bars. The rebar is lapped 21 inches. Determine the number of 20-foot-long bars that needs to be ordered for the footing.
7. How many cubic yards of concrete are required to construct a 145-foot-long by 4-foot-wide by 18-inch-thick continuous footing? Include 5% waste. Concrete must be ordered in quarter-yard increments.
8. Determine the square footage of forms needed for Problem 7.
9. How many cubic yards of concrete are required to construct 11 each 36-inch by 36-inch by 18-inch-thick spread footings? Include 5% waste. Concrete must be ordered in quarter-yard increments.
10. Determine the square footage of forms needed for Problem 9.
11. How many cubic yards of concrete are required to construct 20 each 32-inch by 32-inch by 14-inch-thick spread footings? Include 5% waste. Concrete must be ordered in quarter-yard increments.
12. Determine the square footage of forms needed for Problem 11.
13. How many cubic yards of concrete are required to construct a 250-foot-long by 8-foot-high by 8-inch-thick concrete wall? Include 5% waste. Concrete must be ordered in quarter-yard increments.
14. Determine the square footage of forms needed for Problem 13.
15. How many cubic yards of concrete are required to construct a 59-foot-long by 4-foot-high by 6-inch-thick concrete wall? Include 5% waste. Concrete must be ordered in quarter-yard increments.
16. Determine the square footage of forms needed for Problem 15.
17. How many cubic yards of concrete are required to construct 7 each 16-inch by 16-inch by 10-foot-high concrete columns? Include 5% waste. Concrete must be ordered in quarter-yard increments.
18. Determine the square footage of forms needed for Problem 17.
19. How many cubic yards of concrete are required to construct 5 each 24-inch by 12-inch by 8-foot-high concrete columns? Include 5% waste. Concrete must be ordered in quarter-yard increments.
20. Determine the square footage of forms needed for Problem 19.
21. How many cubic yards of concrete are required to construct 14 each 22-inch-diameter, 12-foot-high concrete columns? Include 5% waste. Concrete must be ordered in quarter-yard increments.
22. How many cubic yards of concrete are required to construct 36 each 12-inch-diameter, 10-foot-high concrete columns? Include 5% waste. Concrete must be ordered in quarter-yard increments.
23. How many cubic yards of concrete are required to construct a 100-foot by 50-foot by 4-inch-thick concrete slab? Include 8% waste. Concrete must be ordered in quarter-yard increments.
24. Determine the length of forms needed for Problem 23.
25. The slab in Problem 23 is reinforced with #4 rebar at 16 inches on center. The rebar is lapped 18 inches. Determine the number of 20-foot-long bars of #4 rebar needed to reinforce the slab.
26. The slab in Problem 23 is reinforced with welded wire fabric, which is lapped 8 inches. Determine the number of 5-foot by 150-foot rolls of welded wire fabric needed to reinforce the slab.
27. How many cubic yards of concrete are required to construct a 75-foot by 40-foot by 5-inch-thick concrete slab? Include 10% waste. Concrete must be ordered in quarter-yard increments.
28. Determine the length of forms needed for Problem 27.

29. The slab in Problem 27 is reinforced with #4 rebar at 24 inches on center. The rebar is lapped 12 inches. Determine the number of 20-foot-long bars of #4 rebar needed to reinforce the slab.

30. The slab in Problem 27 is reinforced with welded wire fabric, which is lapped 10 inches. Determine the number of 5-foot by 10-foot sheets of welded wire fabric needed to reinforce the slab.

31. A 75-foot by 75-foot by 4-inch-thick slab is poured over the metal deck shown in Figure 5-25. The depth of the slab is measured from the top of the slab to the top of the metal deck. Determine the average thickness of the slab and the number of yards of concrete needed to pour the slab. Include 12% waste.

32. How many cubic yards of concrete are needed to construct the continuous footing shown in Figures 5-26 and 5-27? Include 10% waste.

33. How many lineal feet of forms are needed to construct the continuous footing shown in Figures 5-26 and 5-27?

34. Determine the rebar that needs to be ordered to construct the continuous footing shown in Figures 5-26 and 5-27. The rebar is to be lapped 18 inches.

35. How many cubic yards of concrete are needed to construct the spread footings shown in Figures 5-26 and 5-28? Include 10% waste.

36. How many lineal feet of forms are needed to construct the spread footings shown in Figures 5-26 and 5-28?

37. Determine the rebar that needs to be ordered to construct the spread footings shown in Figures 5-26 and 5-28. Provide 3 inches of cover in the footing and 2 inches of cover in the column. The ties are to be ordered with the column rebar.

38. How many cubic yards of concrete are needed to construct the wall shown in Figures 5-26 and 5-27? Add 10% waste.

39. How many square feet of forms are needed to construct the wall shown in Figures 5-26 and 5-27?

40. Determine the rebar that needs to be ordered to construct the wall shown in Figures 5-26 and 5-27. The rebar is to be lapped 18 inches and 2 inches of cover is required.

41. How many cubic yards of concrete are needed to construct the columns shown in Figures 5-26 and 5-28? Add 5% waste.

42. How many square feet of forms are needed to construct the columns shown in Figures 5-26 and 5-28?

43. Determine the rebar that needs to be ordered to construct the columns shown in Figures 5-26 and 5-28. Two inches of cover is required on the ties.

44. Determine the concrete, rebar, and forms needed to pour the footings, foundation wall, and interior concrete slab for the Johnson Residence given in Appendix G. Add 5% waste for the footings and foundation wall and 10% waste for the floor slab. The continuous rebar is to be lapped 18 inches and will be ordered in 20-foot lengths.

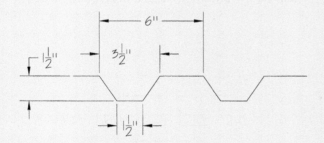

FIGURE 5-25 Metal Deck Dimensions

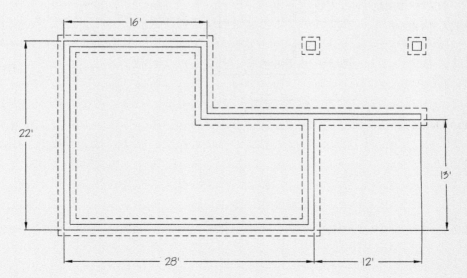

FIGURE 5-26 Footing Plan

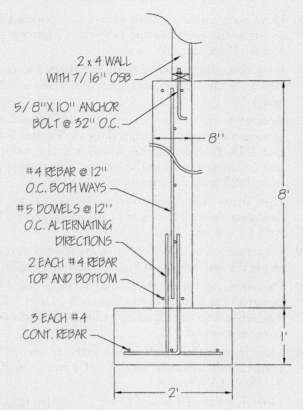

FIGURE 5-27 Wall Section

45. Determine the concrete, rebar, and forms needed to pour the footings and interior concrete slab for the West Street Video project given in Appendix G. Add 5% waste for the footings and 10% waste for the floor slab. The continuous rebar is to be lapped 18 inches and will be ordered in 20 foot lengths.
46. Set up Excel Quick Tip 5-1 in Excel.
47. Set up Excel Quick Tip 5-2 in Excel.
48. Set up Excel Quick Tip 5-3 in Excel.
49. Set up Excel Quick Tip 5-4 in Excel.
50. Set up Excel Quick Tip 5-5 in Excel.
51. Set up Excel Quick Tip 5-6 in Excel.
52. Set up Excel Quick Tip 5-7 in Excel.
53. Set up Excel Quick Tip 5-8 in Excel.
54. Modify Excel Quick Tip 5-2 to allow the height to be entered in feet and inches.
55. Modify Excel Quick Tip 5-3 to allow the height to be entered in feet and inches.
56. Modify Excel Quick Tip 5-6 to allow the slab length and slab width to be entered in feet and inches.
57. Modify Excel Quick Tip 5-7 to allow the slab length and slab width to be entered in feet and inches.
58. Modify Excel Quick Tip 5-8 to allow the slab length and slab width to be entered in feet and inches.

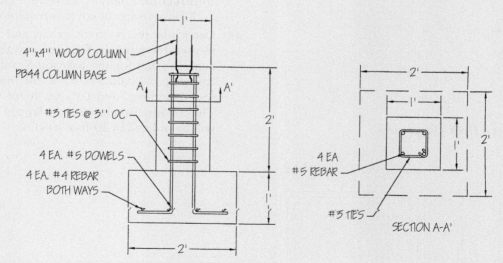

FIGURE 5-28 Column Detail

CHAPTER SIX

MASONRY

In this chapter you will learn how to apply the principles in Chapter 4 to block and structural brick walls and brick veneer. This chapter includes sample spreadsheets that may be used in the quantity takeoff.

For this chapter we divide masonry into two categories: (1) block and structural brick and (2) brick veneer.

Block and structural brick walls are characterized by being self-supporting and are often used as structural walls. The components of block and structural brick walls include the block or brick, reinforcement, mortar, grout, imbeds, insulation used to fill the ungrouted cells, and anchor bolts. Commonly the wall is reinforced with rebar running in the vertical and horizontal directions. The bars running in the vertical direction are commonly spaced at 32 inches on center, and additional bars are often added around openings. This vertical rebar often consists of a single bar (often a #5 bar) placed in a column of cells that are grouted solid. Because the rebar is placed in the cells of the blocks or bricks, the rebar spacing must match the block or brick cell spacing. When placing the blocks, they must be lifted over the top of the rebar and then placed on the blocks in the wall, or the rebar must be stabbed down the cells after the block is placed. This requires that the rebar be ordered in short lengths. One way to do this is to reinforce the walls in 4-foot increments with rebar that is 4 feet long plus the length of the lap.

Bars running in the horizontal direction are placed in a bond beam. The bond beam consists of a block with a channel in the top of the block, similar to the one shown in Figure 6-1. The rebar is placed in the channel, and the cells of the blocks are grouted solid. A typical bond beam consists of two #4 bars. The bond beams are commonly spaced 4 feet on center. Additional bond beams are located at the top of the walls and over openings. Like rebar in concrete, the rebar in the bond beam must be lapped. The combination of the lapping and the short lengths of rebar used to reinforce the wall vertically greatly increases the amount of rebar needed to complete the job. The estimator should read the specifications carefully and fully understand the reinforcing requirements for the project when preparing a bid.

Brick veneer is brick applied to the surface of a wood or block wall as an architectural finish material. The components of brick veneer include the brick, weather barrier (when installed over wood walls), lintels, and ties. Wood walls should be covered with a weather barrier, such as asphalt-impregnated felt or Tyvek, prior to installation of brick veneer. The weather barrier may be installed by the framers or the mason. The quantity takeoff of weather barriers is covered in Chapter 9. Steel lintels must be used to support the brick wherever the brick is not supported by a concrete wall. The lintels consist of steel angles that are fastened to a structural wall or imbedded in the brick veneer. The brick veneer must be attached to the surface of the wood or block wall. This is often done with masonry ties. The masonry ties are secured to wood walls with nails and are secured to block and brick by placing the ties in the mortar joints.

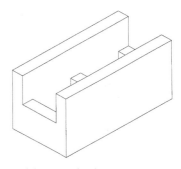

FIGURE 6-1 Bond-beam Block

In addition, masonry ties should be included at the corners to strengthen the corners.

Let's look at estimating block and structural brick walls.

BLOCK AND STRUCTURAL BRICK WALLS

The work package for block and structural brick walls includes the block or brick, reinforcement, mortar, grout, insulation (such as perlite) used to fill the ungrouted block cavities, and imbeds and anchor bolts used to connect the building components to the wall. The block and brick walls are treated as a sheet good. When calculating the number of blocks, the estimator must separate standard blocks from bond beam blocks and other specialty shapes. The vertical rebar is treated as a counted item; the horizontal rebar is treated as a linear good. The mortar and grout are treated as quantity-from-quantity goods. The best way to determine the quantities of mortar and grout is based on past experience. Finally, anchor bolts and imbeds are treated as counted items. Steel imbeds and anchor bolts through structural steel are often provided by the structural steel supplier and need to be delivered before the structural steel so they can be installed by the mason. Anchor bolts not through the structural steel also need to be provided so they can be installed by the mason. The quantity takeoff for a block or structural brick wall is shown in the following example.

EXAMPLE 6-1

Determine the number of blocks, the rebar, mortar, and grout needed to construct a 100-foot-long by 10-foot-high block wall constructed of 7 5/8-inch-high by 7 5/8-inch-wide by 15 5/8-long blocks. The mortar joint is 3/8 inch thick. The wall is horizontally reinforced by bond beams at 4 feet on center containing two #4 bars. The wall is vertically reinforced with #5 rebar at 32 inches on center. All rebar is to be lapped 30 bar diameters. The rebar for the bond beam is to be delivered to the site in 20-foot lengths, and the vertical rebar is to be delivered cut to the required lengths. Allow 5 cubic feet of mortar and 17 cubic feet of grout per 100 square feet of block wall. Anchor bolts are placed in the top of the wall at 32 inches on center. How many of the blocks need to be bond beam blocks?

Solution: Using Eq. (4-14) to find the number of rows, we get the following:

$$\text{Number}_{Rows} = \frac{(10 \text{ ft})(12 \text{ in/ft})}{(7\ 5/8 \text{ in} + 3/8 \text{ in})} = 15 \text{ rows}$$

Using Eq. (4-17) to get the number of columns, we get the following:

$$\text{Number}_{Columns} = \frac{(100 \text{ ft})(12 \text{ in/ft})}{(15\ 5/8 \text{ in} + 3/8 \text{ in})} = 75 \text{ columns}$$

Using Eq. (4-18), we calculate the number of blocks as follows:

$$\text{Number} = (15 \text{ rows})(75 \text{ columns}) = 1,125 \text{ blocks}$$

Three bond beams are required: one at 4 feet, one at 8 feet, and one at 10 feet. The number of bond beam blocks is calculated using Eq. (4-18) as follows:

$$\text{Number} = (3 \text{ rows})(75 \text{ columns}) = 225 \text{ blocks}$$

The project requires a total of 1,125 blocks of which 225 are bond beam blocks. For the bond beams, six 100-foot-long #4 bars are required for a total length of 600 feet. The lap for the #4 rebar is 30 bar diameters, or 15 inches (30 × 0.5 in). The number of 20-foot bars is calculated using Eq. (4-6) as follows:

$$\text{Number} = \frac{600 \text{ ft}}{(20 \text{ ft} - 1.25 \text{ ft})} = 32 \text{ each}$$

The number of vertical bars is calculated using Eq. (4-1) as follows:

$$\text{Number} = \frac{(100 \text{ ft})(12 \text{ in/ft})}{32 \text{ in}} + 1 = 39 \text{ each}$$

The required lap on the vertical bar is 18.75 inches (30 × 0.625 in). Each of these vertical bars will be ordered in the following increments: 5 feet 7 inches long for the first 4 feet of the wall and 5 feet 11 inches long for the remaining 6 feet of the wall. If we ordered 5-foot 7-inch-long bars for the second four feet of the wall and 1-foot 11-inch-long bars for the last two feet of the wall, the last bar will lap the second to last bar 19 of its 23-inch length. It is more economical and is just as easy to replace the last two bars with a single bar. The wall will require 39 #5 rebar 5 feet 7 inches long and 39 #5 rebar 5 feet 11 inches long. The area of the wall is calculated as follows:

$$\text{Area} = (100 \text{ ft})(10 \text{ ft}) = 1,000 \text{ ft}^2$$

The number of cubic yards of mortar needed is calculated as follows:

$$\text{Volume} = (1,000 \text{ ft}^2)\left(\frac{5 \text{ ft}^3}{100 \text{ ft}^2}\right)\left(\frac{1 \text{ yd}^3}{27 \text{ ft}^3}\right) = 1.9 \text{ yd}^3$$

The number of cubic yards of grout needed is calculated as follows:

$$\text{Volume} = (1,000 \text{ ft}^2)\left(\frac{17 \text{ ft}^3}{100 \text{ ft}^2}\right)\left(\frac{1 \text{ yd}^3}{27 \text{ ft}^3}\right) = 6.3 \text{ yd}^3$$

The number of anchor bolts is calculated using Eq. (4-1) as follows:

$$\text{Number} = \frac{(100 \text{ ft})(12 \text{ in/ft})}{32 \text{ in}} + 1 = 39 \text{ each}$$

The estimator must take openings into account when estimating block walls. Openings reduce the number of blocks needed and may require the use of special blocks around the openings. Additional rebar is often placed around the opening as shown in Figure 6-2. The blocks above the opening are often supported by multiple bond beams. To allow the bond beams to provide a finished surface, a lintel block with a flat bottom, as shown in Figure 6-3, may be used for the bottom bond beam.

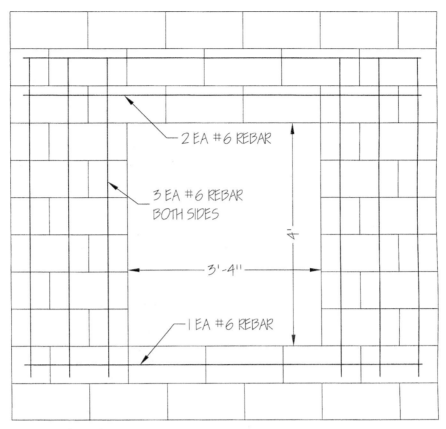

FIGURE 6-2 Reinforcing around an Opening

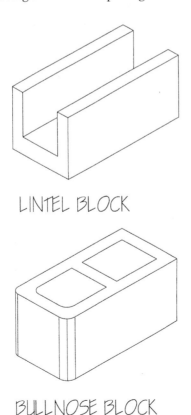

FIGURE 6-3 Specialty Blocks

Other specialty blocks, such as the bullnose block shown in Figure 6-3, may be used on the sides of the opening.

When estimating block, the estimator must determine the additional reinforcing for the opening, the specialty blocks needed, and the quantity of standard block that can be deducted for the opening. Block waste around the opening is greater than for other areas of the wall; therefore, the estimator cannot determine the number of standard blocks to be deducted by simply deducting the area of the opening and the specialty blocks. Many estimators do not deduct any blocks for small openings and deduct only half of the area for medium-sized openings. In addition to accounting for the opening, support for the bond beam needs to be provided until the grout in the beam has cured. The following example looks at how to take openings into account when estimating block and structural brick walls.

EXAMPLE 6-2

Determine the number of specialty blocks needed, additional rebar required, and the number of standard blocks that can be deducted when building a wall with the opening shown in Figure 6-2. The wall is constructed of 7 5/8-inch-high by 7 5/8-inch-wide by 15 5/8-inch-long blocks with a 3/8-inch mortar joint. The blocks at the top and bottom of the opening are to be the lintel block shown in Figure 6-3, with the blocks at the bottom placed upside down to provide a smooth surface.

Bullnose blocks are to be provided at the sides of the opening. A 2 × 8 wood frame with a center support will be provided to support the bond beam until the grout has cured.

Solution: Lintel blocks are needed at the top and bottom of the opening. The number of lintel blocks needed for the top of the opening is calculated by dividing the width of the opening by the length of the block including one mortar joint as follows:

$$\text{Blocks} = \frac{40 \text{ in}}{(15\ 5/8 \text{ in} + 3/8 \text{ in})} = 2.5 \text{ blocks}$$

The number of blocks will be rounded up to three blocks. Three lintel blocks are needed at the top and three at the bottom for a total of six.

Bullnose blocks are needed at both sides. The number of bullnose blocks needed for one of the sides of the opening is calculated by dividing the height of the opening by the height of the block including one mortar joint as follows:

$$\text{Blocks} = \frac{(4 \text{ ft})(12 \text{ in/ft})}{(7\ 5/8 \text{ in} + 3/8 \text{ in})} = 6 \text{ blocks}$$

Twelve bullnose blocks are needed.

The horizontal #6 rebar above and below the opening extends about a block and a half (about 24 in) beyond the opening in both directions; therefore, three #6 bars 7 foot 4 inches (2 ft + 3 ft 4 in + 2 ft) long are needed. Bond beam blocks are needed for each of these bars. The number of bond beam blocks that are needed for one bar is calculated by dividing the length of the bar by the length of the block including one mortar joint as follows:

$$\text{Blocks} = \frac{88 \text{ in}}{(15\ 5/8 \text{ in} + 3/8 \text{ in})} = 5.5 \text{ blocks}$$

Six bond beam blocks are needed for each bar for a total of 18 blocks. Six of these blocks are replaced with lintel blocks; therefore, 12 bond beam blocks are needed.

The vertical rebar to the sides of the opening extends about two blocks above and one block below the opening; therefore, six #6 bars 6 feet (16 in + 4 ft + 8 in) long are needed.

The number of standard blocks that can be deducted for the opening is determined by multiplying the number of full blocks that can fit in the width of the opening by the number of full blocks that can fit in the height of the opening. In this case two full blocks can fit in the width of the opening and six full blocks can fit in the height of the opening. The number of standard blocks that can be deducted is 12 (6 × 2). The bond beam blocks, bullnose blocks, and lintel blocks will replace standard blocks. A total of 42 (12 + 12 + 12 + 6) standard blocks may be deducted from the estimate for the opening.

Two 3-foot 4-inch-long 2 × 8s are needed for the top and bottom of the opening. Three 3-foot 9-inch-long (4 ft − 1 1/2 in − 1 1/2 in) 2 × 8s are needed for the sides and center of the opening. Order three 8-foot-long 2 × 8s. ∎

EXCEL QUICK TIP 6-1
Block Wall

The numbers of standard blocks, bond beam blocks, bars of horizontal rebar, and bars of vertical rebar needed for a block wall are set up in a spreadsheet by entering the data and formatting the cells as follows:

	A	B	C
1	Wall Length	100	ft
2		-	in
3	Wall Height	10	ft
4		-	in
5	Block Length	15.625	in
6	Block Height	7.625	in
7	Joint Thickness	0.375	in
8	Bond Beam Spacing	48	in
9	Number of Horizontal Bars per Bond Beam	2	ea
10	Splice Length for Bond Beam Rebar	15.00	in
11	Order Length of Horizontal Rebar	20.00	ft
12	Vertical Rebar Spacing	32	in
13	Splice Length for Vertical Rebar	18.75	in
14	Mortar	5	cft/100 sft
15	Grout	17	cft/100 sft
16			
17	Rows	15	ea
18	Columns	75	ea
19	Bond Beams	3	ea
20	Standard Blocks	900	ea
21	Bond Beam Blocks	225	ea
22	Horizontal Rebar	32	ea
23	Length of Vertical Rebar (1)	67	in
24	Vertical Rebar (1)	39	ea
25	Length of Vertical Rebar (2)	71	in
26	Vertical Rebar (2)	39	ea
27	Mortar	1.9	cyd
28	Grout	6.3	cyd

In this spreadsheet, if the distance between the second-to-last bond beam and the top of the wall is 24 inches or less, the last vertical bar runs from the third-to-last bond beam to 1 inch below the top of the wall. The Length of Vertical Rebar (1) and the Vertical Rebar (1) are used to calculate the vertical rebar for the wall from the bottom of the wall to the second-to-last or third-to-last bond beam. The Length of Vertical Rebar (2) and the Vertical Rebar (2) are used to calculate the last vertical rebar for the wall, which runs from the second-to-last or third-to-last bond beam to 1 inch below the top of the wall. The following formulas need to be entered into the associated cells:

Cell	Formula
B17	=ROUNDUP((B3*12+B4)/(B6+B7),0)
B18	=ROUNDUP((B1*12+B2)/(B5+B7),0)
B19	=ROUNDUP((B3*12+B4)/B8,0)
B20	=B17*B18-B21
B21	=B18*B19
B22	=ROUNDUP((B1+B2/12)*B19*B9/(B11-B10/12),0)
B23	=ROUNDUP(B8+B13,0)
B24	=IF((B3*12+B4-(B19-1)*B8)>24,(B19-1)*B26,(B19-2)*B26)
B25	=IF((B3*12+B4-(B19-1)*B8)>24,B3*12+B4-(B19-1)*B8-1,B3*12+B4-(B19-2)*B8-1)
B26	=ROUNDUP((B1*12+B2)/B12,0)+1
B27	=(B1+B2/12)*(B3+B4/12)*B14/(100*27)
B28	=(B1+B2/12)*(B3+B4/12)*B15/(100*27)

The data for the block wall is entered in Cells B1 through B15. Cell B20 determines the total number of blocks and subtracts the number of bond beam blocks. Cell 23 determines the length of the vertical rebar for all of the wall except the wall above the second-to-last or third-to-last bond beam by adding the lap on the vertical rebar to the spacing between the bond beams. Cell C25 determines the length of the last vertical rebar for the wall by determining the distance between the second-to-last or third-to-last bond beam and the top of the wall and subtracting 1 inch. An IF function is used in Cells B24 and B25 to determine if the last vertical bar runs from the second-to-last or third-to-last bond beam based upon the distance between the second-to-last bond beam and the top of the wall. The data shown in the foregoing figure is from Example 6-1 and is formatted using the comma style, which replaces zeros with dashes.

BRICK VENEER

The work package for brick veneer includes brick, weather barrier, lintels, mortar, and ties. The brick and ties are treated as sheet goods and the mortar is treated as a quantity-from-quantity good. The best way to determine the quantities of mortar is based on past experience. The brick veneer must bear on a footing, foundation wall, or a steel ledger created by bolting steel angle iron to the wall. Steel angle iron is used as lintels over window and door openings and must be bolted to the wall or bear on the brick at the sides of the opening. A common bearing distance for steel lintels is 4 inches. The quantity takeoff for brick walls is shown in the following example.

EXAMPLE 6-3

Determine the number of bricks, corrugated masonry ties, and mortar needed to face a wall 50 feet long by 10 feet high. The brick is 2 1/2 inches high by 3 1/2 inches wide by 11 1/2 inches long. Allow for a 1/2-inch mortar joint. The spacing between the ties is to be 18 inches measured vertically and 36 inches measured horizontally. Historically, 10.2 cubic feet of mortar is required per 100 square feet of brick.

Solution: Using Eq. (4-14) to find the number of rows of brick, we get the following:

$$\text{Number}_{\text{Rows}} = \frac{(10 \text{ ft})(12 \text{ in/ft})}{(2\ 1/2 \text{ in} + 1/2 \text{ in})} = 40 \text{ rows}$$

Using Eq. (4-17) to get the number of columns, we get the following:

$$\text{Number}_{\text{Columns}} = \frac{(50 \text{ ft})(12 \text{ in/ft})}{(11\ 1/2 \text{ in} + 1/2 \text{ in})} = 50 \text{ columns}$$

Using Eq. (4-18), we calculate the number of blocks needed as follows:

$$\text{Number} = (40 \text{ rows})(50 \text{ columns}) = 2,000 \text{ blocks}$$

The ties are laid out in rows and columns along the wall. A column will occur every 3 feet along the wall. The columns will consist of seven rows of ties, which are located at 1 foot 6 inches, 3 feet, 4 feet 6 inches, 6 feet, 7 feet 6 inches, 9 feet, and 9 feet 9 inches above the base of the wall. A tie is not needed at the base of the wall because it will bear on a footing, foundation, or steel lintel. Using Eq. (4-1) to determine the number of vertical columns of ties, we get the following:

$$\text{Number} = \frac{50 \text{ ft}}{3 \text{ ft}} + 1 = 18 \text{ columns}$$

The number of ties is 126 (7 rows × 18 columns). The area of the wall is calculated as follows:

$$\text{Area} = (50 \text{ ft})(10 \text{ ft}) = 500 \text{ ft}^2$$

The number of cubic yards of mortar needed is calculated as follows:

$$\text{Volume} = (500 \text{ ft}^2)\left(\frac{10.2 \text{ ft}^3}{100 \text{ ft}^2}\right)\left(\frac{1 \text{ yd}^3}{27 \text{ ft}^3}\right) = 1.9 \text{ yd}^3$$

EXCEL QUICK TIP 6-2
Brick Wall

The numbers of bricks and brick ties needed for a brick wall are set up in a spreadsheet by entering the data and formatting the cells as follows:

	A	B	C
1	Wall Length	50	ft
2	-	-	in
3	Wall Height	10	ft
4	-	-	in
5	Brick Length	11.500	in
6	Brick Height	2.500	in
7	Joint Thickness	0.500	in
8	Vertical Tie Spacing	18	in
9	Horizontal Tie Spacing	36	in
10	Mortar	10.2	cft/100 sft
11			
12	Rows of Brick	40	ea
13	Columns of Brick	50	ea
14	Bricks	2,000	ea
15	Rows of Ties	7	ea
16	Columns of Ties	18	ea
17	Ties	126	ea
18	Mortar	1.9	cyd

The following formulas need to be entered into the associated cells:

Cell	Formula
B12	=ROUNDUP((B3*12+B4)/(B6+B7),0)
B13	=ROUNDUP((B1*12+B2)/(B5+B7),0)
B14	=B12*B13
B15	=ROUNDUP((B3*12+B4)/B8,0)
B16	=ROUNDUP((B1*12+B2)/B9,0)+1
B17	=B15*B16
B18	=(B1+B2/12)*(B3+B4/12)*B10/(100*27)

The data for the brick wall is entered in Cells B1 through B10. The data shown in the foregoing figure is from Example 6-6 and is formatted using the comma style, which replaces zeros with dashes.

EXAMPLE 6-4

A 4-foot wide opening is placed in a brick wall. The top of the opening is supported by an L3 × 3 × 3/8 steel lintel. Determine the minimum length of the lintel.

Solution: Allow 4 inches of bearing on both sides of the lintel. The lintel needs to be 4 feet 8 inches (4 in + 4 ft + 4 in) long.

CONCLUSION

Masonry consists of block, structural brick, and brick veneer. The work package for a block or structural brick wall includes the block or brick, reinforcement, mortar, grout, insulation, and imbeds and anchor bolts used to connect the building components to the wall. The block and brick are treated as a sheet good. The vertical rebar is treated as a counted item; the horizontal rebar is treated as a linear good. The mortar and grout are treated as quantity-from-quantity goods. The work package for brick veneer includes brick, weather barrier, lintels, and ties. The brick and ties are treated as a sheet good and the mortar is treated as a quantity-from-quantity good.

PROBLEMS

1. Determine the number of 7 5/8-inch-high by 5 5/8-inch-wide by 15 5/8-inch-long concrete blocks required to complete a wall 80 feet long by 12 feet high. Allow for a 3/8-inch mortar joint.

2. How many of the blocks in Problem 1 need to be bond beam blocks if the bond beams are 4 feet on center?

3. Determine the number of 20-foot-long #4 bars needed for the wall in Problem 1 if the wall is horizontally reinforced by bond beams at 4 feet on center. Each bond beam has two each #4 bars. The rebar is to be lapped 18 inches.

4. Determine the quantity and length of vertical rebar needed for the wall in Problem 1 if the wall is vertically reinforced with #5 rebar at 32 inches on center. The rebar is to be lapped 18 inches.

5. Determine the quantity of mortar needed for the wall in Problem 1 if 4.5 cubic feet of mortar is required per 100 square feet of block.

6. Determine the quantity of grout needed for the wall in Problem 1 if 18 cubic feet of grout is required per 100 square feet of block.

7. Determine the number of 7 5/8-inch-high by 5 5/8-inch-wide by 15 5/8-inch-long concrete blocks required to complete a wall 160 feet long by 8 feet high. Allow for a 3/8-inch mortar joint.

8. How many of the blocks in Problem 7 need to be bond beam blocks if the bond beams are 4 feet on center?

9. Determine the number of 20-foot-long #4 bars needed for the wall in Problem 7 if the wall is horizontally reinforced by bond beams at 4 feet on center. Each bond beam has two #4 bars. The rebar is to be lapped 18 inches.

10. Determine the quantity and length of vertical rebar needed for the wall in Problem 7 if the wall is vertically reinforced with #5 rebar at 32 inches on center. The rebar is to be lapped 18 inches.

11. Determine the quantity of mortar needed for the wall in Problem 7 if 4.3 cubic feet of mortar is required per 100 square feet of block.

12. Determine the quantity of grout needed for the wall in Problem 7 if 16 cubic feet of grout is required per 100 square feet of block.

13. Determine the number of specialty blocks needed, additional rebar required, and number of standard blocks that can be deducted when building a wall with the opening shown in Figure 6-4. The wall is constructed of 7 5/8-inch-high by 7 5/8-inch-wide by 15 5/8-inch-long blocks with a 3/8-inch mortar joint. The blocks at the top and bottom of the opening are to be the lintel block shown in Figure 6-3, with the blocks at the bottom placed upside down to provide a smooth surface. Bullnose blocks are to be provided at the sides of the opening. A 2 × 8 wood frame with a center support will be provided to support the bond beam until the grout has cured.

14. Determine the number of specialty blocks needed, additional rebar required, and number of standard blocks that can be deducted when building a wall

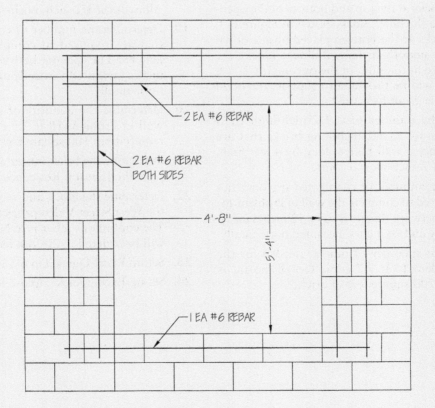

FIGURE 6-4 Opening

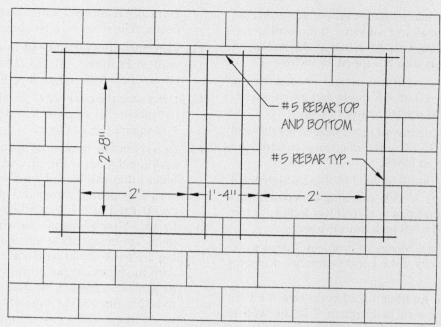

FIGURE 6-5 Openings

with the openings shown in Figure 6-5. The wall is constructed of 7 5/8-inch-high by 7 5/8-inch-wide by 15 5/8-inch-long blocks with a 3/8-inch mortar joint. The blocks at the top and bottom of the openings are to be the lintel block shown in Figure 6-3, with the blocks at the bottom placed upside down to provide a smooth surface. Bullnose blocks are to be provided at the sides of the opening. A 2 × 8 wood frame will be provided to support the bond beam until the grout has cured.

15. Determine the number of 2 1/2-inch-high by 3 1/2-inch-wide by 11 1/2-inch-long bricks that are required to face a wall 80 feet long by 4 feet high. Allow for a 1/2-inch mortar joint.

16. Determine the number of corrugated masonry ties that are required to complete the wall in Problem 15. The spacing between the ties is to be 12 inches measured vertically and 24 inches measured horizontally.

17. Determine the quantity of mortar needed for the wall in Problem 15 if 9.7 cubic feet of mortar is required per 100 square feet of brick.

18. Determine the number of 3 1/2-inch-high by 3 1/2-inch-wide by 11 1/2-inch-long bricks that are required to face a wall 40 feet long by 8 feet high. Allow for a 1/2-inch mortar joint.

19. Determine the number of corrugated masonry ties that are required to complete the wall in Problem 18. The spacing between the ties is to be 24 inches measured vertically and 24 inches measured horizontally.

20. Determine the quantity of mortar needed for the wall in Problem 18 if 7.5 cubic feet of mortar is required per 100 square feet of brick.

21. Determine the brick veneer and ties needed to complete the Johnson Residence given in Appendix G.

22. Determine the block and rebar needed to complete the West Street Video project given in Appendix G. The continuous rebar is to be lapped 18 inches and will be ordered in 20-foot lengths.

23. Set up Excel Quick Tip 6-1 in Excel.

24. Set up Excel Quick Tip 6-2 in Excel.

CHAPTER SEVEN

METALS

In this chapter you will learn how to apply the principles in Chapter 4 to steel beams and columns, joists, metal deck, trusses, stairs and handrails, and miscellaneous steel. This chapter includes sample spreadsheets that may be used in the quantity takeoff.

The steel for a building is fabricated and primed at a steel fabrication shop and shipped to the site for erection, where it is bolted or field welded together. Because it is fabricated off-site and assembled on-site, careful dimensioning and fabrication of each piece must occur for the steel to fit together when it is erected at the jobsite. Many structural engineers leave the actual design of the connections of the steel up to the fabricator, allowing fabricators to use the connection type that is most economical for them. When the design of the connectors is left up to the fabricator, the fabricator is required to submit shop drawings to the structural engineer for approval. The engineer will check to make sure the connections are adequate and approve the design or require changes to the shop drawings.

TYPES OF STRUCTURAL STEEL

Buildings can be constructed of different types of steel. The most common type of steel is A36 carbon steel. Other types of steel include A529 carbon steel; A440 high-strength steel; A441 high-strength, low-alloy steel; A572 high-strength, low-alloy steel; A242 corrosion-resistant, high-strength, low-alloy steel; and A588 corrosion-resistant, high-strength, low-alloy steel. The estimator needs to make sure he or she knows what type of structural steel the engineer is specifying.

COMMON SHAPES FOR STRUCTURAL STEEL

There are a number of different shapes of structural steel used in the construction of buildings. Common shapes are shown in Figure 7-1.

The properties, dimensions, and weight per foot for steel shapes can be found in the *Manual of Steel Construction* published by the American Institute of

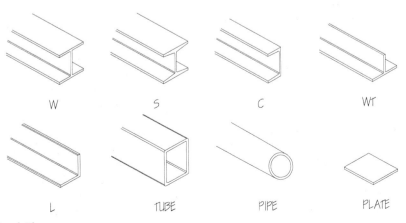

FIGURE 7-1 Common Steel Shapes

Steel Construction. Commonly used shapes include the following:

Wide-Flange Beam: Wide-flange beams are used as columns, girders, and beams. A wide-flange beam is designated by a "W" followed by its nominal depth and weight per foot. For example, a W12 × 40 beam has a nominal depth of 12 inches and a weight of 40 pounds per foot. Its actual depth is 11.94 inches.

American Standard Beams: American standard beams are sometimes used as girders and beams. An American Standard beam is designated by an "S" followed by its nominal depth and weight per foot. For example, an S18 × 70 beam has a nominal depth of 18 inches and a weight of 70 pounds per foot. The actual depth of this beam is also 18 inches.

American Standard Channels: American standard channels are used as beams and stair stringers. An American Standard channel is designated by a "C" followed by its nominal depth and weight per foot. For example, a C12 × 30 channel has a nominal depth of 12 inches and a weight of 30 pounds per foot. The actual depth of this channel is also 12 inches.

Structural Tees: Structural tees are used in steel trusses. Structural tees are created by cutting a beam along the length of the beam's web to create a "T" shape. A structural tee cut from a wide-flange beam is designated by "WT" followed by its nominal depth and weight per foot. A structural tee cut from an American standard beam is designated by "ST" followed by its nominal depth and weight per foot. For example, a WT4 × 29 structural tee has a nominal depth of 4 inches and a weight of 29 pounds per foot. Its actual depth is 4.375 inches.

Angles: Angles are used as cross-bracing, lintels, ledgers, and connectors and in the construction of steel trusses. Angles are designated by an "L" followed by the length of its longest leg, the length of the other leg, and the thickness of the angle. For example, an L5 × 3 × 1/2 would have one leg with a length of 5 inches, one leg with a length of 3 inches, and a thickness of 1/2 inch. The weight of this angle is 12.8 pounds per foot.

Structural Tubing: Structural tubing is used as columns and cross-bracing and in the construction of steel trusses. Structural tubing is designated by the length of its longest cross-sectional axis, the length of the other axis, and the thickness of the wall of the tubing. For example, a 4 × 4 × 1/2 tube steel would have a cross section of 4 inches by 4 inches and would have a wall thickness of 1/2 inch. Its weight is 21.63 pounds per foot.

Pipe: Pipe is used for columns. Pipe is designated by "Pipe" followed by the nominal diameter of the pipe and the type of pipe. For example, Pipe 4 Std. has a nominal diameter of 4 inches and is the standard weight of pipe. This pipe has a wall thickness of 0.237 inches, a weight of 10.79 pounds per foot, an inside diameter of 4.026 inches, and an outside diameter of 4.500 inches.

Plate: Plate steel is used as base plates and connectors. Plate steel is designated by the thickness. The weight of plate steel is about 490 pounds per cubic foot.

EXAMPLE 7-1

Determine the weight in pounds for a 50-foot-long W30 × 90.

Solution: The W30 × 90 weighs 90 pounds per foot. The beam weighs 4,500 pounds (50 ft × 90 lb/ft).

BEAMS, GIRDERS, AND COLUMNS

When estimating beams, girders, and columns, it is important to estimate not only the number of structural steel members needed, but also the weight of each member so the crane can be properly sized. General contractors often estimate beams, girders, and columns based on the weight of the steel members and adds a percentage to the weight to allow for connections and bolts because the design of the connections are often left up to the steel fabricator. Estimates by steel fabricators must include the connectors and bolts. Let's look at estimating the structural steel members.

The length of a structural steel member depends on how it is connected to the surrounding steel members. Common connections are shown in Figure 7-2, along with how the connection is drawn on the structural steel plan view for all but the splice connections. The first type of connection is a splice and is used to connect two shorter beams or columns into a longer beam or column. The next type of connection is a beam-to-beam (or beam-to-girder) connection, in which the end of one beam is joined to the web of another beam. In this case the beam being connected must run to the web of the beam that it is being connected to. The next three types of connections are connections used to connect a beam to a column made from a wide flange beam. In one type of connection the beam is connected to the web of the column. In this case the beam must run to the web of the column. In another type of connection the beam is connected to the flange of the column. In this case the beam must run to the flange of the column. In the third type of connection the beam rests on a plate attached to the top of the column. In this case the beam runs over the top of the column to the edge of the building or the next column. Making a connection to a tubular-steel column is similar to connecting the beam to the flange of the column, with the beam running to the exterior of the column or connecting the beam to a plate attached to the top of the column.

The following example shows how the lengths of individual structural steel members are calculated.

Metals 117

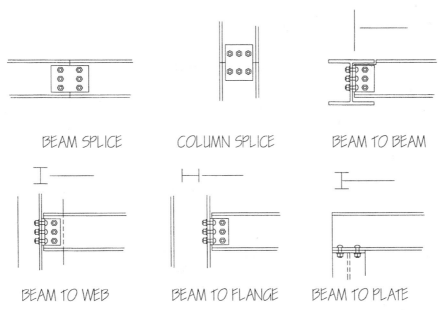

FIGURE 7-2 Typical Connections

EXAMPLE 7-2

Prepare a structural steel materials list for the roof-framing plan shown in Figure 7-3. The columns are 15 feet high. The flange width for the W18 × 86 is 11 1/8 inches. The properties of the columns are given in Table 7-1. How many pounds of steel need to be purchased for the roof? What is the weight of the heaviest structural steel member needed to construct the roof?

Solution: One C12 × 30 is needed to the right of the opening. The channel will run from the webs of the W12 × 45 beams and has a length of approximately 10 feet. The weight of the channel is 300 pounds (10 ft × 30 lb/ft).

Four W12 × 45 beams are needed between the C1 and C2 columns. These beams will run from the flange of column

TABLE 7-1 Column Properties

Column	Size	Flange Width (in)	Depth (in)
C1	W12 × 40	8.005	11.94
C2	W12 × 53	9.995	12.06

C1 to the flange of column C2. The length of the beam is 30 feet less the depth of column C1 less half the depth of column C2, or 28.5 feet (30 ft − 12 in − 12 in/2). The weight of one of these beams is 1,283 pounds (28.5 ft × 45 lb/ft).

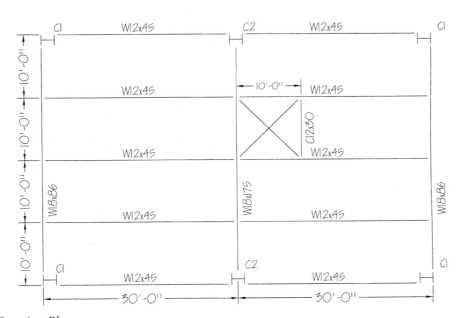

FIGURE 7-3 Steel Framing Plan

Six W12 × 45 beams are needed between the W18 × 86 girders and the W18 × 175 girder. These beams will run from the web of the two girders. The web of the W8 × 86 will be set back 5 9/16 inches (11.125 in/2) so that the flange of the W8 × 86 is flush with the flange of the W12 × 40 column. The length of the beam is approximately 30 feet less 5 inches, or 29 feet 7 inches. The weight of one of these beams is 1,331 pounds (29.583 ft × 45 lb/ft).

Two W18 × 86 girders are needed to go between the C1 columns. These girders will run from the webs of the two columns. The length of the girder is approximately 40 feet less the flange width of the column, or 39 feet 4 inches (40 ft − 8 in). The weight of one of these girders is 3,383 pounds (39.33 ft × 86 lb/ft).

One W18 × 175 girder is needed to go between the C2 columns. This girder will run from the webs of the two columns. The length of the girder is approximately 40 feet less the flange width of the column, or 39 feet 2 inches (40 ft − 10 in). The weight of one of these girders is 6,854 pounds (39.167 ft × 175 lb/ft).

Four 15-foot-long C1 columns are needed and will weigh 600 pounds (15 ft × 40 lb/ft) each. Two 15-foot-long C2 columns are needed and will weigh 795 pounds (15 ft × 53 lb/ft) each. The total weight of the steel is calculated as follows:

$$\text{Weight} = 1(300 \text{ lb}) + 4(1{,}283 \text{ lb}) + 6(1{,}331 \text{ lb}) + 2(3{,}383 \text{ lb}) + 1(6{,}854 \text{ lb}) + 4(600 \text{ lb}) + 2(795 \text{ lb})$$

$$\text{Weight} = (31{,}028 \text{ lb})\left(\frac{1 \text{ ton}}{2{,}000 \text{ lb}}\right) = 15.514 \text{ tons}$$

The heaviest member is the W18 × 175 girder at 6,854 pounds.

EXCEL QUICK TIP 7-1
Structural Steel

Calculating the weight of structural steel is set up in a spreadsheet by entering the data and formatting the cells as follows:

	A	B	C	D	E	F	G
1			Length		Weight		
2	Item	Quantity	Feet	Inches	lb/ft	Each	Total
3	C12x30	1	10	-	30.0	300	300
4	W12x45	4	28	6	45.0	1,283	5,130
5	W12x45	6	29	7	45.0	1,331	7,988
6	W18x86	2	39	4	86.0	3,383	6,765
7	W18x175	1	39	2	175.0	6,854	6,854
8	W12x40	4	15	-	40.0	600	2,400
9	W12X53	2	15	-	53.0	795	1,590
10						Weight (lb)	31,027
11						Weight (tons)	15.51

The following formulas need to be entered into the associated cells:

Cell	Formula
F3	=(C3+D3/12)*E3
G3	=F3*B3
G10	=SUM(G3:G9)
G11	=G10/2000

The formula in Cell F3 will need to be copied to Cells F4 through F9, and the formula in Cell G3 will need to be copied to Cells G4 through G9. The data for the steel is entered in Cells A3 through E9. The data shown in the foregoing figure is from Example 7-2 and is formatted using the comma style, which replaces zeros with dashes.

When estimating steel connections, the estimator must account for the angles, plate steel, bolts, and welds that make up most connectors. Estimating connectors is shown in the following example.

EXAMPLE 7-3

Determine the items needed for the connection shown in Figure 7-4.

Solution: The connection will require 9 inches of L3 × 3 × 1/4 angle, 11 inches of 1/4 fillet weld, and three 1/2-inch-diameter A325 bolts.

JOISTS AND JOIST GIRDERS

Floor-framing systems are often made up of joists and joist girders. The requirements for joists and joist girders are set by the Steel Joist Institute. The most common joist in use is the K-series joist, which is designated as shown in Figure 7-5. The joist in Figure 7-5 is a 24-inch-deep K-series joist with a section number of three. The larger the section number, the stronger is the joist.

Joist girders are joists that are designed to support joists just as girders support beams. Joist girders are designated as shown in Figure 7-6. The number of spaces is

Metals 119

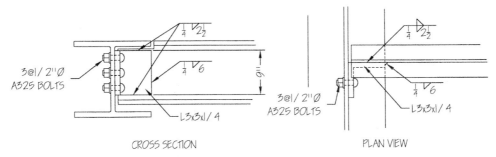

FIGURE 7-4 Connection Details

FIGURE 7-5 Joist Designation

FIGURE 7-6 Joist Girder Designation

one more than the number of joists that are supported by the joist girder. The load on each panel point is the load applied by each joist in kips, where 1 kip is 1,000 pounds. If the joist girder supports two joists at each point, one on each side of the joist girder, the load on each panel point is the combined load applied by the pair of joists. The joist girder in Figure 7-6 is a 24-inch-deep joist girder with five spaces (supporting four joists), and each panel point supports 8 kips, or 8,000 pounds.

When estimating joists and joist girders, the estimator needs to determine the number of joist and joist girders, their lengths, and their weights. The approximate weights for joists and joist girders are given in the *Catalogue of Standard Specifications and Load Tables for Steel Joists and Joist Girders* published by the Steel Joist Institute. Estimating joists and joist girders is shown in the following example.

EXAMPLE 7-4

Prepare a steel joist materials list for the roof-framing plan shown in Figure 7-7. The approximate weights of the joists are given in Table 7-2. What is the total number of pounds of steel joists that need to be purchased for the roof? What is the weight of the heaviest steel joist needed to construct the roof?

TABLE 7-2 Joist Weights

Joist	Approximate Weight (lb/ft)
24K12	16.0
24G8N8K	47.0
36G8N16K	73.0

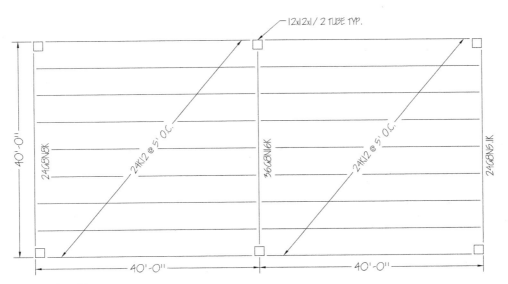

FIGURE 7-7 Joist Framing Plan

Solution: Four 24K12 joists are needed to run between the columns. The length of these joists is approximately 38 feet 6 inches (40 ft − 12 in − 6 in). They weigh 616 pounds (38.5 ft × 16.0 lb/ft) each. The number of joists in one bay is calculated using Eq. (4-3) as follows:

$$\text{Number} = \frac{40 \text{ ft}}{5 \text{ ft}} - 1 = 7 \text{ each}$$

Fourteen (2 bays × 7 joists/bay) 24K12 joists are needed to run between the joist girders. The length of these joists is approximately 40 feet, and each joist weighs 640 pounds (40 ft × 16.0 lb/ft). Two 24G8N8K joist girders are needed. The length of each of these joist girders is 38 feet (40 ft − 12 in − 12 in). These girders weigh 1,786 pounds (38 ft × 47.0 lb/ft) each. One 36G8N16K is needed. The length of this joist girder is 38 feet (40 ft − 12 in − 12 in). This girder weighs 2,774 pounds (38 ft × 73.0 lb/ft). The total weight of the steel is calculated as follows:

$$\text{Weight} = 4(616 \text{ lb}) + 14(640 \text{ lb}) + 2(1,786 \text{ lb}) + 1(2,774 \text{ lb})$$

$$= (17,770 \text{ lb})\left(\frac{1 \text{ ton}}{2,000 \text{ lb}}\right) = 8.885 \text{ tons}$$

The heaviest member is the 36G8N16K joist girder at 2,774 pounds.

METAL DECK

Metal deck is often placed on top of the steel framing and steel joists. The deck comes in sheets and is treated as a sheet good. Estimating metal deck is shown in the following example.

EXAMPLE 7-5

Determine the amount of metal deck that needs to be purchased to construct the floor shown in Figure 7-7. The deck comes in panels 4 feet wide by 40 feet long. The panels must be lapped 6 inches.

Solution: Each panel will run the entire width of the building; therefore, the number of rows is one. Determine the number of columns of metal deck that are needed using Eq. (4-15) as follows:

$$\text{Number}_{\text{Columns}} = \frac{80 \text{ ft}}{(4 \text{ ft} - 0.5 \text{ ft})} = 23 \text{ columns}$$

The number of sheets of metal deck needed is 23 (1 rows × 23 columns).

EXCEL QUICK TIP 7-2
Metal Deck Floor or Roof

The number of sheets of metal deck needed for a floor or roof is set up in a spreadsheet by entering the data and formatting the cells as follows:

	A	B	C
1	Length of Building	40	ft
2	Width of Building	80	ft
3	Length of Sheet	40	ft
4	Width of Sheet	4	ft
5	Lap at Ends of Sheet	-	in
6	Lap at Sides of Sheet	6	in
7			
8	Number of Rows	1.00	ea
9	Number of Columns	23	ea
10	Number of Sheets	23	ea

The following formulas need to be entered into the associated cells:

Cell	Formula
B8	=B1/(B3-B5/12)
B9	=ROUNDUP(B2/(B4-B6/12),0)
B10	=ROUNDUP(B8*B9,0)

The data for the metal deck is entered in Cells B1 through B6. This spreadsheet assumes that the waste from one column can be used on the next column and that the lengths of the sheets are run parallel to the length of the building. The data shown in the foregoing figure is from Example 7-5 and is formatted using the comma style, which replaces zeros with dashes.

STEEL TRUSSES

Like all other items constructed of steel, steel trusses are estimated by listing the steel components, including connectors, needed to construct the steel truss. Estimating the components needed to fabricate a steel truss is shown in the following example.

EXAMPLE 7-6

Determine the items, including welds, needed to fabricate the steel truss shown in Figure 7-8. The right side of the truss is the same as the left side of the truss. What is the estimated weight of the truss if an L3 × 3 × 1/2 weighs 9.4 pounds per foot?

Solution: Two 13-foot 5-inch-long WT4 × 20 (a tee cut from a wide-flange beam) are needed for the top cord and one 24-foot-long WT4 × 20 is needed for the bottom cord. Four 2-foot 9-inch-long L3 × 3 × 1/2 (one pair on each side of the truss), four 6-foot 2-inch-long L3 × 3 × 1/2 (one pair on each side of the truss), and two 5-foot 9-inch-long L3 × 3 × 1/2 (one pair at the center of the truss) are needed for the truss. Two 20-inch-long V-welds (one on each side of the truss) and one 3-inch-long V-weld are needed, for a total of 43 inches of V-weld. Each L3 × 3 × 1/2 will need 7 inches (2 in + 3 in + 2 in) of fillet weld on each end. There are 10 L3 × 3 × 1/2s; therefore, 140 inches (10 × 7 in/end × 2 ends) of fillet weld are needed.

Metals **121**

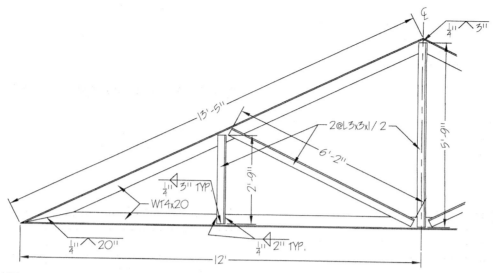

FIGURE 7-8 Steel Truss

The estimated weight of the truss is found by multiplying the length of the steel members by their weights as follows:

Weight = (2 × 13 ft 5 in + 24 ft)(40 lb/ft) + (4 × 2 ft 9 in + 4 × 6 ft 2 in + 2 × 5 ft 9 in)(9.4 lb/ft)

Weight = $(2{,}477 \text{ lb})\left(\dfrac{1 \text{ ton}}{2{,}000 \text{ lb}}\right)$ = 1.24 tons

STAIRS AND HANDRAIL

Like all other items constructed of steel, steel stairs and handrail are estimated by listing the steel components, including connectors, needed to construct the steel stairs and handrail. Estimating the components needed to fabricate a steel stairs and handrail is shown in the following example.

EXAMPLE 7-7

Determine the steel materials needed to fabricate the steel stairs and handrail shown in Figure 7-9. The treads are 1.5 inches deep, and the stairs are 4 feet wide. The depth of the flange on the channel is 3 inches, and the thickness of the web is 5/16 inch. The handrail is on one side of the stairs only.

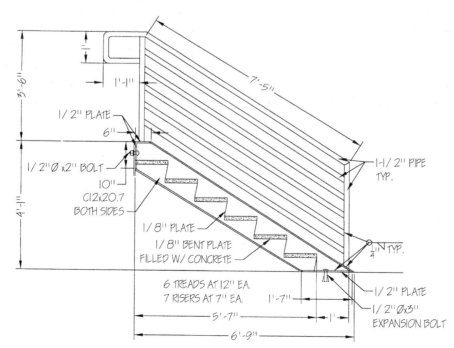

FIGURE 7-9 Steel Stairs

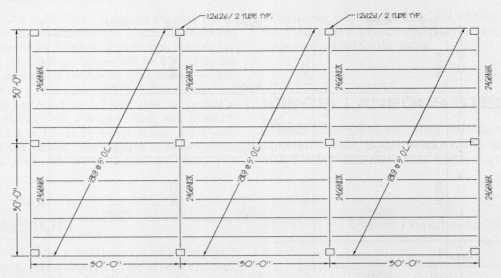

FIGURE 7-13 Steel Joist Plan

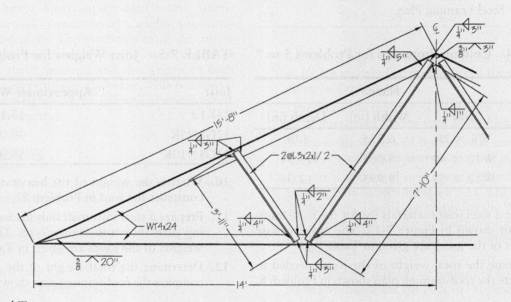

FIGURE 7-14 Steel Truss

TABLE 7-6 Joist Weights for Problems 11 to 13

Joist	Approximate Weight (lb/ft)
18K9	10.2
24G6N6K	23.0
24G6N12K	45.0

13. What is the weight of the heaviest joist needed to construct the roof in Problem 11?

14. Determine the amount of metal deck that needs to be purchased to construct the floor shown in Figure 7-12. The deck comes in panels 4 feet wide by 40 feet long. The panels must be lapped 6 inches.

15. Determine the amount of metal deck that needs to be purchased to construct the floor shown in Figure 7-13. The deck comes in panels 4 feet wide by 40 feet long. The panels must be lapped 6 inches at the sides and 12 inches at the ends.

16. Determine the items, including welds, needed to fabricate the steel truss shown in Figure 7-14. The right side of the truss is the same as the left side of the truss.

17. What is the estimated weight of the truss in Figure 7-14 if an L3 × 2 × 1/2 weighs 7.7 pounds per foot?

18. Determine the items, including connectors, needed to fabricate the steel truss shown in Figure 7-15. The right side of the truss is the same as the left side of the truss.

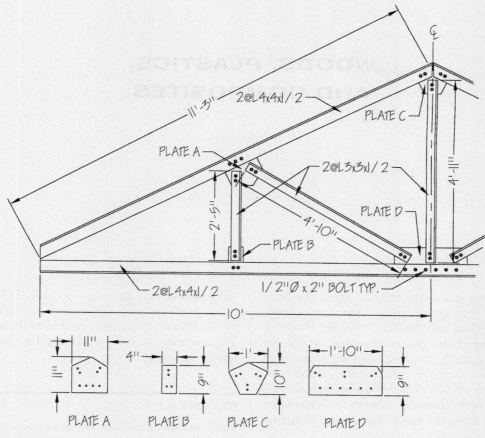

FIGURE 7-15 Steel Truss

19. What is the estimated weight of the truss in Figure 7-15 if an L3 × 3 × 1/2 weighs 9.4 pounds per foot, an L4 × 4 × 1/2 weighs 12.8 pounds per foot, and the plate weighs 20.4 pounds per square foot?
20. Determine the structural steel, steel joist, and metal deck needed to complete the West Street Video project given in Appendix G.
21. Set up Excel Quick Tip 7-1 in Excel.
22. Set up Excel Quick Tip 7-2 in Excel.
23. Modify Excel Quick Tip to allow the length of building, width of building, length of sheet, and width of sheet to be entered in feet and inches.

CHAPTER EIGHT

WOODS, PLASTICS, AND COMPOSITES

In this chapter you will learn how to apply the principles in Chapter 4 to wood floor systems (wood sills, columns, beams, joists, joist headers, trimmer, blocking, joist hangers, tongue and groove sheeting), walls (top and bottom plates, studs, headers, blocking, sheathing), roof-framing (trusses, rafters, fascia, soffit), wood trim, and cabinetry. This chapter includes sample spreadsheets that may be used in the quantity takeoff. It also includes example takeoffs from the residential garage drawings given in Appendix G.

Rough carpentry consists of framing the floors, walls, and roof of the structure. A typical wood structure is shown in Figure 8-1. Finish carpentry consists of installing finish material including wood trim, shelving, and handrails. Cabinetry includes the installation of cabinets and countertops. We begin by looking at framing the building's floor system.

FLOOR SYSTEMS

The floor system consists of joists covered with OSB or plywood sheathing and bears on sills, walls, or girders supported by posts. Sometimes 2× or 3× material is substituted for the sheathing as is the case with wood decks. Let's look at each of the components of the floor system beginning with the sill.

Sills

When the floor system is supported by concrete walls, the floor system must be placed on a naturally durable wood (for example, redwood) or pressure-treated wood sill. A foam sealing material is often required to be placed below the sill to prevent air infiltration between the concrete wall and the wood sill. This sealing material is provided with the lumber package and installed by the framer. Wood sills are bid as linear components. The quantity takeoff for wood sill is shown in the following example.

EXAMPLE 8-1

The perimeter of the floor system in Figure 8-2 sits on a redwood sill. Determine the number of 16-foot-long pieces of redwood that are needed to complete the floor system.

Solution: The perimeter of the building is 132 feet (40 ft + 26 ft + 40 ft + 26 ft). The number of pieces is calculated using Eq. (4–7) as follows:

$$\text{Number} = \frac{132 \text{ ft}}{16 \text{ ft}} = 9 \text{ each}$$

Large quantities of sill plate are often purchased in random lengths. When lumber is purchased in random lengths, the pricing is quoted in board feet. The quantity takeoff of random-length wood sill is shown in the following example.

EXAMPLE 8-2

How many board feet of sill are needed to complete the floor system in Figure 8-2 if the sill is constructed of random lengths of 2 × 4 redwood?

Solution: From Example 8-1 the perimeter of the building is 132 feet. The number of board feet (bft) is calculated as follows:

$$\text{Board Feet} = \frac{(\text{Depth})(\text{Width})(\text{Length})}{(12 \text{ in/ft})}$$

$$= \frac{(2 \text{ in})(4 \text{ in})(132 \text{ ft})}{(12 \text{ in/ft})} = 88 \text{ bft}$$

Woods, Plastics, and Composites **127**

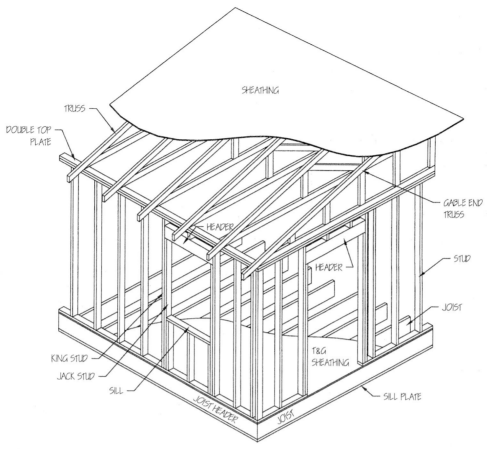

FIGURE 8-1 Typical Wood Structure

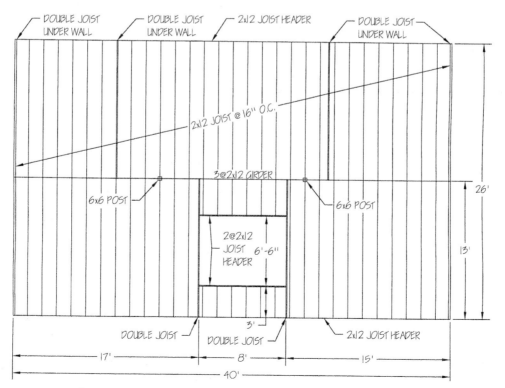

FIGURE 8-2 Floor-Framing Plan

EXCEL QUICK TIP 8-1
Board Foot Worksheet

The conversion of lineal feet to board feet is set up in a spreadsheet by entering the data and formatting the cells as follows:

	A	B	C
1	Board Depth	2	in
2	Board Width	4	in
3	Board Length	132	ft
4			
5	Board Feet	88.0	bft

The following formula needs to be entered into Cell B5:

=B1*B2*B3/12

The data for lumber is entered in Cells B1 through B3. The data shown in the foregoing figure is from Example 8-2.

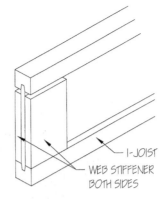

FIGURE 8-3 Web Stiffener

Posts and Girders

Posts and girders may be used to support a floor system and open up the space below. The girder may consist of a glue-laminated beam (GLB) or two or three $2 \times s$ nailed together. The posts should be secured to footing using a framing anchor, such as a Simpson post base (PB), and should be secured to the girder using a framing anchor, such as a Simpson post cap (BC). Posts and framing hardware are bid as counted items and girders are bid as linear components or as counted items. The quantity takeoff for posts and girders is shown in the following example.

EXAMPLE 8-3

Determine the materials needed to complete the posts and girder supporting the floor joists in Figure 8-2. The posts are connected to footings using a Simpson PB66 post base and are connected to the girder using a Simpson BC6 post cap. The distance between the top of the footing and the bottom of the girder is 7 feet. The posts are equally spaced along the girder.

Solution: The 6×6 posts come in 8-foot lengths. Two 8-foot-long 6×6 posts, two Simpson PB66 post bases, and two Simpson BC6 post caps are needed. The posts are 13 feet 4 inches (40 ft/3) on center; therefore, 14-foot-long 2×12s will be used for the girder. Nine (3 spaces × 3) 14-foot-long 2×12s are needed for the girder.

Floor Joists

Floors are constructed of floor joists that span from bearing wall or girder to bearing wall or girder. The floor joists may be constructed of 2×6, 2×8, 2×10, or 2×12 lumber or of engineered wood I-joists.

Engineered I-joists are gaining in popularity because they can produce a superior-quality floor system at prices similar to that of standard wood joists. Engineered I-joists are available in lengths of up to 60 feet, which allows a single joist to span across many of the bearing walls or girders in a building, which decreases the deflection of the floor joists. Engineered I-joists may require blocking and web stiffeners (a piece of wood on either side of the flange in the space between the webs). A web stiffener is shown in Figure 8-3. Short pieces of I-joists are often used as blocking. The manufacturer's data should be consulted to determine the need for blocking and web stiffeners.

Joist headers are required at the ends of the joists. Joist headers around the perimeter of the floor are often referred to as rim joist or rim board. When using $2 \times$ lumber the header is constructed of $2 \times$ lumber with the same depth as the joists. Manufacturers of engineered I-joists produce engineered wood products for use as joist headers.

Joist headers and trimmers are used when openings are required in the floor system. The framing for a typical opening is shown in Figure 8-4. The joist header is constructed of the same material used for the joists or the joist header at the edge of the building. The joist header and joist trimmers are required to be doubled if the opening is greater than 4 feet wide. When the opening is greater than 6 feet wide the joist header is to be connected to the joist trimmers with framing anchors.[1] The location of the joist trimmers may or may not line up with the standard joist spacing. If the trimmer joist does not line up with the standard joist spacing, as is the case in Figure 8-4, more joists will be needed than if the trimmer joist lined up with the standard joist spacing.

The joists are doubled under bearing walls running parallel to the joist including perimeter walls and may be doubled under nonbearing walls running parallel to the joists. The estimator must take into account the doubled joist when performing the takeoff. Joists are taken off as a counted item.

The joists must be blocked at specified intervals to prevent them from rolling onto their sides. The joists must also be blocked when the joists pass over walls to

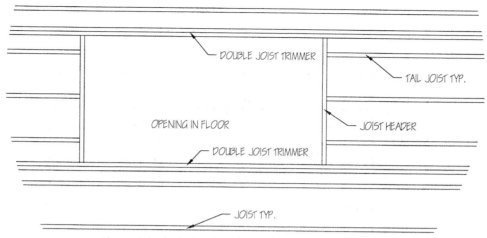

FIGURE 8-4 Framing for Opening in Floor

prevent the spread of fire. The blocking is usually constructed of the same material used for the joists or joist headers. Joist blocking is taken off as a linear component. The takeoff of the joists for a floor system is shown in the following example.

EXAMPLE 8-4

Determine the materials needed to construct the joists for the floor system shown in Figure 8-2. The joists are blocked over the girder and at the midpoint, between the girder and the exterior walls on joists over 8 feet long. The joist headers at the opening are attached to the trimmer joists with Simpson U210-2 hangers.

Solution: Beginning with the 13-foot-long joists, the number of joists needed to meet the spacing of 16 inches on center for the top half of the drawing is calculated using Eq. (4–1) as follows:

$$\text{Number} = \frac{(40 \text{ ft})(12 \text{ in/ft})}{16 \text{ in}} + 1 = 31$$

The number of joists needed to meet the spacing of 16 inches on center for the bottom half of the drawing located to the left of the opening is calculated using Eq. (4–1) as follows:

$$\text{Number} = \frac{(17 \text{ ft})(12 \text{ in/ft})}{16 \text{ in}} + 1 = 14$$

The number of joists needed to meet the spacing of 16 inches on center for the bottom half of the drawing located to the right of the opening is calculated using Eq. (4–1) as follows:

$$\text{Number} = \frac{(15 \text{ ft})(12 \text{ in/ft})}{16 \text{ in}} + 1 = 13$$

A total of 58 (31 + 14 + 13) joists are needed to meet the spacing requirements. Seven additional joists are needed to double up joists already included in this number (4 for the ends of the building, 1 at each side of the stairs, and 1 for the double joist under the wall on the left of the drawing). Two additional joists are needed for the double joist under the wall on the right because the wall is not located above the joists used to meet the spacing requirements. This brings the total number of joists to 67 (58 + 7 + 2). Fourteen-foot-long 2 × 12s will be ordered for these joists.

The number of joists needed on each side of the building for the joist header is calculated using Eq. (4–7) as follows:

$$\text{Number} = \frac{40 \text{ ft}}{14 \text{ ft}} = 3$$

Three additional 14-foot-long joists are needed as joist headers on each side of the building. Four 8-foot-long joists are needed as joist headers around the opening.

The joists above the opening are 3 feet 6 inches long (13 ft − 3 ft − 6 ft 6 in). The number of joists is calculated using Eq. (4–3) as follows:

$$\text{Number} = \frac{(8 \text{ ft})(12 \text{ in/ft})}{16 \text{ in}} = 6$$

Two 3-foot 6-inch-long joists can be cut from one 8-foot-long 2 × 12; therefore, three more 8-foot-long 2 × 12s are needed. Six 3-foot-long joists are needed below the opening. Three of these joists can be cut from a 10-foot-long 2 × 12; therefore, two more 10-foot-long 2 × 12s are needed.

Three rows of blocking are needed. The top row is located in the top half of the drawing and is 40 feet long. The second row is located over the girder and is 40 feet long. The third row is located in the bottom half of the drawing and is not needed in the opening. The length of blocking needed for this row is 32 feet (17 ft + 15 ft). A total of 112 feet (40 ft + 40 ft + 32 ft) of blocking is needed. Order twelve 10-foot-long 2 × 12s for the blocking.

Four Simpson U210-2 hangers are needed to attach the joist headers to the trimmer joists.

Order seventy-three (67 + 2 × 3) 14-foot-long 2 × 12s, fourteen 10-foot-long 2 × 12s, and seven 8-foot-long 2 × 12s for the floor and four Simpson U210-2 hangers.

> ### EXCEL QUICK TIP 8-2
> #### Joists
>
> The number of joists needed for a floor is set up in a spreadsheet by entering the data and formatting the cells as follows:
>
> | | A | B | C | D | E | F | G |
> |----|--------------------|--------|-----|--------|-----|--------|-----|
> | 1 | | Area 1 | | Area 2 | | Area 3 | |
> | 2 | Building Length | 40 | ft | 17 | ft | 15 | ft |
> | 3 | Joist Spacing | 16 | in | 16 | in | 16 | in |
> | 4 | End Condition | 1 | | 1 | | 1 | |
> | 5 | | | | | | | |
> | 6 | Joists | 31 | ea | 14 | ea | 13 | ea |
> | 7 | | | | | | | |
> | 8 | Additional Joists | 9 | | | | | |
> | 9 | | | | | | | |
> | 10 | Total # of Joists | 67 | | | | | |
>
> The following formulas need to be entered into the associated cells:
>
> | Cell | Formula |
> |------|---------|
> | B6 | =IF(B3="",0,ROUNDUP(B2*12/B3,0)+B4) |
> | B10 | =B6+D6+F6+B8 |
>
> After entering the formulas, Cell B6 must be copied to Cells D6 and F6. This spreadsheet is designed for three separate building areas. The data for the floor is entered in Cells B2 through B4, D2 through D4, and F2 through F4. If an area is not to be used, the joist spacing should be left blank. If a joist is required at both ends of the floor area, a "1" is entered as the end condition. If a joist is required at only one end of the floor area, a "0" is entered as the end condition. If joists are not required at either ends of the floor area, a "−1" is entered as the end condition. The number of additional joists needed for joist trimmers, doubled joists, and joists that do not fall on the standard spacing are entered in Cell B8. The IF statement in Cell B6 sets the quantity of joists to zero when the joist spacing is left blank. The data shown in the foregoing figure is from Example 8-4 for the 14-foot-long joists.

Floor Sheathing

Commonly 23/32-inch T&G (tongue and groove) OSB (oriented-strand board) or plywood is used as floor sheathing. The use of T&G sheathing provides for a stiffer flooring system with less squeaking than sheathing without tongues and grooves. When using T&G sheathing, the estimator must make sure that the sheathing can be cut in such a manner that there is a tongue and a groove for each joint that runs perpendicular to the joists. This often limits the use of waste material even if it is of sufficient size, because it lacks the proper tongue or groove. Sheathing should be placed with the longest direction perpendicular to the joists, and the sheets should span at least three joists. Floor sheathing is commonly glued and nailed with a slip-resistant nail, such as a ring-shank nail, to minimize squeaking. The takeoff of the sheathing for a floor system is shown in the following example.

EXAMPLE 8-5

Determine the number of 4-foot by 8-foot sheets of 23/32-inch T&G OSB sheathing needed to construct the floor system shown in Figure 8-2.

Solution: The number of rows is calculated using Eq. (4–13) as follows:

$$\text{Number}_{\text{Rows}} = \frac{26 \text{ ft}}{4 \text{ ft}} = 6.5$$

Because the sheathing is tongue and groove, the leftover half-sheet cannot be used; therefore, seven rows are needed. The number of columns is calculated using Eq. (4–16) as follows:

$$\text{Number}_{\text{Columns}} = \frac{40 \text{ ft}}{8 \text{ ft}} = 5 \text{ columns}$$

The number of sheets is calculated using Eq. (4–18) as follows:

$$\text{Number} = (7 \text{ rows})(5 \text{ columns}) = 35 \text{ sheets}$$

> ### EXCEL QUICK TIP 8-3
> #### Floor Sheathing
>
> The number of sheets of T&G floor sheathing for a rectangular floor is set up in a spreadsheet by entering the data and formatting the cells as follows:
>
> | | A | B | C |
> |---|-----------------|-------|----|
> | 1 | Building Length | 40 | ft |
> | 2 | Building Width | 26 | ft |
> | 3 | | | |
> | 4 | No. of Rows | 7.00 | ea |
> | 5 | No. of Columns | 5.00 | ea |
> | 6 | | | |
> | 7 | Sheathing | 35.00 | ea |

The following formulas need to be entered into the associated cells:

Cell	Formula
B4	=ROUNDUP(B2/4,0)
B5	=CEILING(B1/8,1/3)
B7	=ROUNDUP(B4*B5,0)

The data for the floor is entered in Cells B1 and B2. The spreadsheet assumes the sheets are 4 feet by 8 feet and the long direction of the sheet is run parallel to the length of the building. The number of rows is rounded up to the next whole sheet, and the number of columns is rounded up to the next one-third of a sheet using the CEILING function, which equals the distance covered by three joists when the joists are spaced at 16 inches on center. The data shown in the foregoing figure is from Example 8-5.

Wood Decks

In heavy timber construction and on outdoor decks, 2 × and 3 × lumber laid flat is often used as a decking material in lieu of OSB or plywood. The lumber may be laid with a space between the boards or laid right next to each other. Wood decks are treated as a sheet good. When purchasing random lengths of decking, the length of the decking is assumed to run the entire length of the deck. The takeoff of wood decking is shown in the following example.

EXAMPLE 8-6

How many board feet of 3 × 6 redwood decking are needed to construct a 20-foot by 20-foot deck? The boards are to be placed without any space between them.

Solution: The width of a 3 × 6 is 5 1/2 inches; therefore, the boards are spaced 5 1/2 inches apart. Begin by finding the number of lineal feet of 3 × 6 redwood needed for the deck, using the row and column method. Let the rows be made up of one 20-foot-long board. The number of columns is calculated using Eq. (4–16) as follows:

$$\text{Number}_{\text{Columns}} = \frac{(20 \text{ ft})(12 \text{ in/ft})}{5\ 1/2 \text{ in}} = 44 \text{ columns}$$

The number of lineal feet of 3 × 6 boards is calculated by multiplying the number of columns by their length as follows:

$$\text{Lineal Feet} = (20 \text{ ft/column})(44 \text{ columns}) = 880 \text{ ft}$$

The number of board feet is calculated as follows:

$$\text{Board Feet} = \frac{(\text{Thickness})(\text{Width})(\text{Lineal Feet})}{12 \text{ in/ft}}$$
$$= \frac{(3 \text{ in})(6 \text{ in})(880 \text{ ft})}{12 \text{ in/ft}} = 1{,}320 \text{ bft}$$

> ### EXCEL QUICK TIP 8-4
> #### Decking
>
> The number of board feet needed for a deck is set up in a spreadsheet by entering the data and formatting the cells as follows:
>
> | | A | B | C |
> |---|---|---|---|
> | 1 | Deck Length | 20 | ft |
> | 2 | Deck Width | 20 | ft |
> | 3 | Nominal Width | 6 | in |
> | 4 | Actual Width | 5.50 | in |
> | 5 | Nominal Thickness | 3 | in |
> | 6 | Spacing between Boards | - | in |
> | 7 | | | |
> | 8 | Lineal Feet | 880 | ft |
> | 9 | Board Feet | 1,320 | bft |
>
> The following formulas need to be entered into the associated cells:
>
> | Cell | Formula |
> |---|---|
> | B8 | =ROUNDUP(B2*12/(B4+B6),0)*B1 |
> | B9 | =B3*B5*B8/12 |
>
> The data for deck and lumber is entered in Cells B1 through B6. The data shown in the foregoing figure is from Example 8-6 and is formatted using the comma style, which replaces zeros with dashes.

WALLS

Walls consist of top and bottom plates, studs, headers, hold-downs, blocking, and sheathing. Let's look at each of these components.

Top and Bottom Plates

When the wall is placed on concrete or masonry, the bottom plate must be a naturally durable wood (for example, redwood) or pressure-treated wood. A foam sealing material is often placed below the bottom plate of exterior walls when the walls bear on concrete or masonry. When the wall is placed on a wood floor, the bottom plate is of the same lumber as the studs. For the bottom plate, the door openings are not deducted from the length because the bottom plate is run through the opening and is cut out after the framing is complete. This helps ensure that the walls are framed straight.

Bearing walls require two top plates. Nonbearing walls may be built with one top plate, but they are often built with two top plates to allow the same length of studs to be used for both the bearing and nonbearing walls. Top and bottom plates are taken off in the same way that sills are taken off.

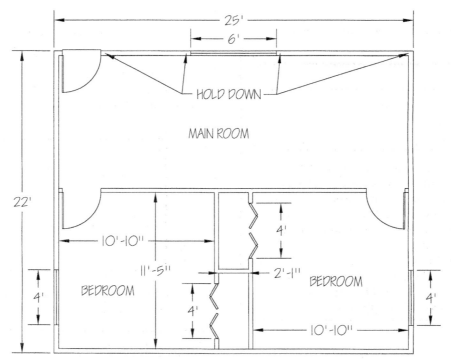

FIGURE 8-5 Wall Layout

EXAMPLE 8-7

How many feet of treated plate and untreated plate are needed to construct the walls shown in Figure 8-5? The walls are constructed of 2 × 4s and the exterior walls are sheathed with 7/16-inch OSB. The wall bears on a concrete floor.

Solution: The thickness of the exterior walls is 4 inches, and the thickness of the interior walls is 3 1/2 inches. The total length of the walls is calculated as follows:

Length = 2(22 ft) + 3(24 ft 4 in) + 2(11 ft 5 in) + 2 ft 1 in
= 141 ft 11 inches

Because the walls bear on a concrete floor, the bottom plate needs to be constructed of treated plate; therefore, 142 feet of treated plate are needed. The top requires a double top plate of untreated lumber; therefore, 284 feet (2 × 142 ft) of untreated plate are needed.

Studs

Precut studs are available for 8- and 9-foot ceilings. Precut studs for an 8-foot ceiling are 92 5/8 inches long, and when used with a single bottom plate and double top plate, they provide a rough ceiling height of 97 1/8 inches. The extra 1 1/8 inches allows for floor and ceiling finishes. Precut studs for a 9-foot ceiling are 104 5/8 inches long, and when used with a single bottom plate and double top plate, they provide a rough ceiling height of 109 1/8 inches.

The number of studs in a wall is determined by the stud spacing and the number of ends, corners, intersections, openings, and hold downs. The number of studs needed to meet the spacing requirements is determined by Eq. (4–2).

When a wall ends without forming a corner, an additional stud is needed, as was shown in Chapter 4. This takes the end conditions into account. When the wall forms a corner, the end condition will be accounted for with the corner.

Two extra studs are needed at corners. One of these studs provides for the end condition, and the other stud provides backing for the interior finish. Two methods of framing a corner are shown in Figure 8-6. The difference in these two corners is how the backing stud is placed. In this book, two additional studs are added for each corner.

When walls intersect, backing is needed for the interior finish and to secure the intersecting walls together. Two methods of framing an intersection are shown in Figure 8-7. The top method requires two additional studs to be provided for each intersection. The bottom method requires a wider stud (a 2 × 6 for a 2 × 4 wall or a 2 × 8 for a 2 × 6 wall) be provided for each intersection. In this book, two additional studs are added for each intersection.

Each opening requires additional studs. A typical door opening is shown in Figure 8-8. Narrow openings require a jack stud and a king stud on both sides of the opening. The jack studs are used to support the header. Wider openings may need two or three jack studs on each side of the opening to support the header, depending on the load supported by the header and the width of the opening. The number of jack studs should be determined

Woods, Plastics, and Composites

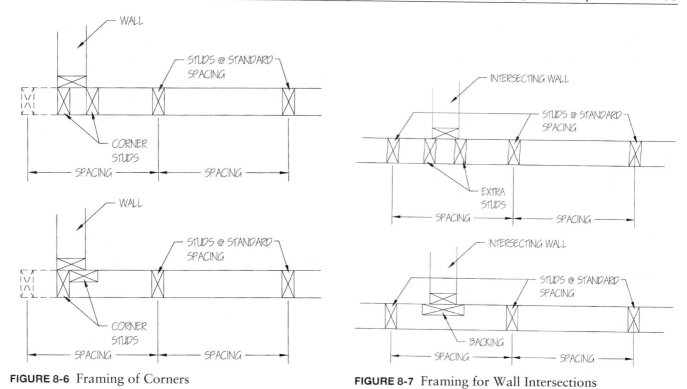

FIGURE 8-6 Framing of Corners

FIGURE 8-7 Framing for Wall Intersections

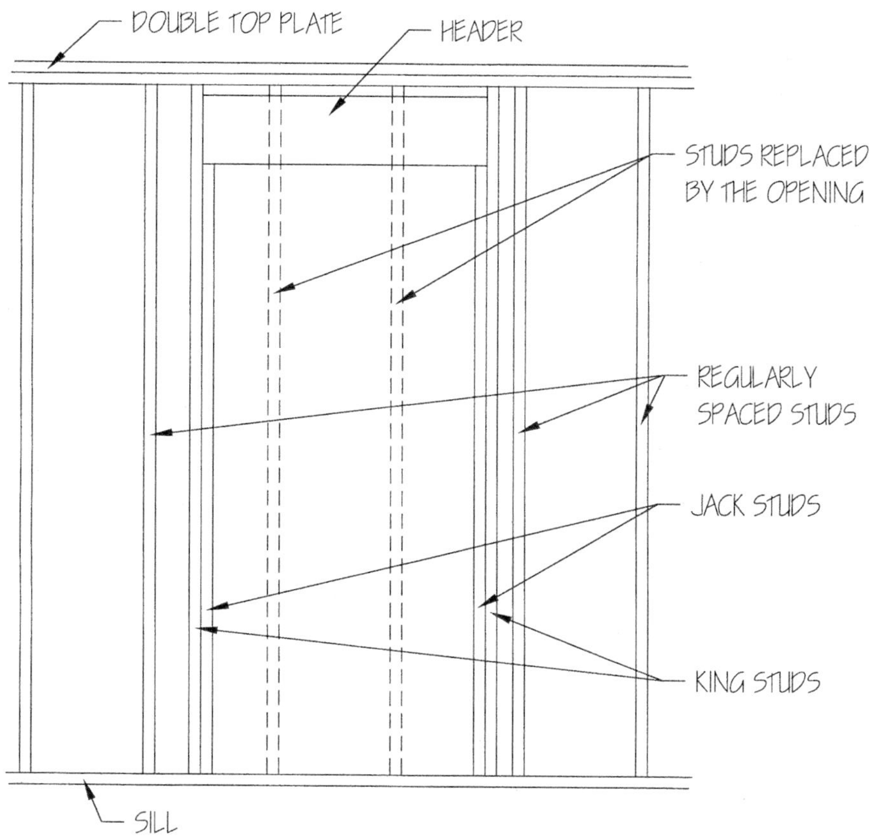

FIGURE 8-8 Framing of a Door Opening

by the design professional or the building code when it has not been specified by the design professional. For the opening shown in Figure 8-8, two additional studs are needed because two of the studs needed to meet the spacing requirements have been deleted and two king studs and two jack studs have been added. The actual number of additional studs will vary based on the location of the door. If the opening in Figure 8-8 were moved a few inches to the right, the right king stud would meet the stud spacing requirement, and only one additional stud would be needed. For wider openings a greater number of studs will be replaced by the openings; however, this is often offset by the need for a greater number of jack studs. In this book, two additional studs are added for each doorway.

Window openings are similar to door openings except that there is framing below the opening that reduces the deletion of the regularly spaced studs. The framing of a window opening is shown in Figure 8-9. Cripple studs and a sill are used to frame the wall below the opening. A cripple stud is located at each end of the sill, and cripple studs are placed under the sill to maintain the regular spacing of the studs. The number of additional studs that are needed will change depending on the location of the opening, the height of the opening, and the width of the opening. If the opening in Figure 8-9 were moved a few inches to the right, the right king stud would meet the stud spacing requirement and reduce the number of additional studs by one. Less material is required for the cripple studs in taller openings than in shorter openings. For example, if the top of the sill in Figure 8-9 was below 49 inches, two cripple studs could be cut from one 92 5/8-inch stud and two studs would be required for the cripple studs. If the top of the sill were above 49 inches, four studs would be required for the cripple studs. Wider openings may need additional jack studs. For an opening up to 8 feet wide with single jack studs, up to seven additional studs may be needed to frame the opening. If the opening required double jack studs, up to nine additional studs may be needed to frame the opening. In this book six additional studs are added for each window opening that requires only two jack studs; eight additional studs are added for each window opening that requires a pair of jack studs on each side.

Posts or doubled or tripled studs are needed wherever a hold down occurs. If the hold down occurs at the edge of a door or window, the hold down may be attached to the jack and king studs. When a hold down occurs at a corner, an additional stud is needed to form a solid blocked corner as shown in Figure 8-10. When a hold down occurs at other places, up to three additional

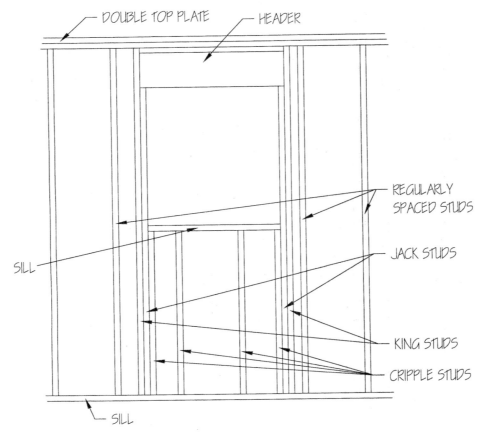

FIGURE 8-9 Framing of a Window Opening

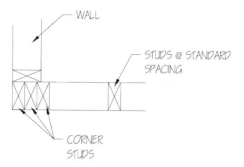

FIGURE 8-10 Framing of Corners with Hold Downs

studs are needed. In this book no studs are added when a hold down occurs at the edge of a door or window opening, one additional stud is added for each hold down located at a corner, and three additional studs are added for all other hold downs.

The quantity takeoff for studs is shown in the following example.

EXAMPLE 8-8

How many studs are needed to frame the wall in Figure 8-5? The stud spacing is 16 inches on center. Allow for one additional stud for each wall that does not have a corner; two additional studs for each corner, intersection, or doorway; six additional studs for each window less than 6 feet in width; eight additional studs for each window 6 feet or more in width; and one additional stud for each hold down located at a corner. Do not add additional studs for hold downs located at the edge of windows or doors.

Solution: From Example 8-7 the length of the wall is 142 feet. The number of studs needed for the standard spacing is calculated using Eq. (4–2) as follows:

$$\text{Number} = \frac{(142 \text{ ft})(12 \text{ in/ft})}{16 \text{ in}} = 107 \text{ each}$$

All four interior walls lack corners at both ends; therefore, four additional studs are needed for these walls. There are 4 corners, 8 intersections, and 5 doorways that will require 2 additional studs each, for a total of 34 additional studs. There are 2 windows that will require 6 additional studs each and 1 window that will require 8 additional studs for a total of 20 additional studs. There are 2 hold downs located at corners that will require an additional stud per hold down for 2 additional studs. The total number of studs is 167 (107 + 4 + 34 + 20 + 2). ■

EXCEL QUICK TIP 8-5
Studs

The number of studs needed for walls is set up in a spreadsheet by entering the data and formatting the cells as follows:

	A	B	C
1	Wall Length	142	ft
2	Stud Spacing	16	in
3	Walls without Corners	4	ea
4	Corners	4	ea
5	Intersections	8	ea
6	Doorways	5	ea
7	Windows, Small	2	ea
8	Windows, Large	1	ea
9	Hold Downs at Corners	2	ea
10	Other Hold Downs	-	ea
11			
12	Studs	167	ea

The following formula needs to be entered into Cell B12:

```
=ROUNDUP(B1*12/B2,0)+B3+B4*2+B5*2+B6*2+B
7*6+B8*8+B9+B10*3
```

The data for walls is entered in Cells B1 through B10. The data shown in the foregoing figure is from Example 8-8 and is formatted using the comma style, which replaces zeros with dashes.

Headers

Headers are constructed of solid wood, laminated lumber, or 2 × materials. Because of cost, most headers are constructed of 2 × material and require plywood or OSB spacers such that the width of the header matches the width of the wall. Headers in a 2 × 4 wall can be made from two pieces of 2 × material and a 1/2-inch plywood or 7/16-inch OSB spacer. Headers in a 2 × 6 wall can be made from three pieces of 2 × material and two 1/2-inch or 7/16-inch spacers. The headers must extend over all of the jack studs, requiring the header to be 3 inches wider than the opening for openings with single jack studs and 6 inches wider than the opening for openings with double jack studs. Openings for doors with jambs are typically framed 2 inches wider than the door size to allow for the doorjamb and shimming. A 36-inch-wide door with single jack studs would require a 41-inch-long header (36 in + 3 in + 2 in). Window openings are typically framed 1/2 to 1 inch wider than the opening to allow for placement of the window.

EXAMPLE 8-9

Prepare a materials list for the headers required to complete the walls in Figure 8-5. All headers are made from 2 × 12s, and the wall is made from 2 × 4s. The doors are 32 inches wide. Openings 6 feet and wider require double jack studs.

Solution: Each header will consist of two 2 × 12s with a 7/16-inch OSB spacer. The 32-inch-wide doors will require a 34-inch-wide opening and a 37-inch (34 in + 3 in) header. Three 37-inch headers are needed and can be cut from two

10-foot-long 2 × 12s. Approximately 10 square feet of spacer are needed for these headers.

The 4-foot windows and closet doors will require 49-inch openings and 52-inch (49 in + 3 in) headers. Because the bifold closet doors do not have jambs, the opening only needs to be framed 1 inch wider than the doors to allow for two layers of half-inch drywall. Four 52-inch headers are needed and can be cut from four 10-foot-long 2 × 12s. Approximately 20 square feet of spacer are needed for these headers.

The 6-foot window will require a 73-inch-wide opening and a 79-inch (73 in + 6 in) header. This header can be cut from two 8-foot-long 2 × 12s. Approximately 7 square feet of spacer are needed for this header. The total amount of spacer required is approximately 37 square feet (10 ft^2 + 20 ft^2 + 7 ft^2) and can be cut from two sheets of OSB.

Order six 10-foot-long 2 × 12s, two 8-foot-long 2 × 12s, and two sheets of 7/16-inch OSB.

Hold Downs

Shear panels are used to stiffen the building against lateral forces generated by wind and earthquakes. Shear panels are created by covering the wall framing with a rigid material such as OSB or plywood. Wide shear panels may be tied to the foundation with anchor bolts. Narrow shear panels require hold downs. Hold downs are specialty framing anchors that tie the shear panel to the foundation. Hold downs come in a variety of types and sizes.

Blocking

Blocking is used to strengthen load-bearing walls, prevent the spread of fire, and provide backing for wall-mounted items, such as grab bars. Blocking used to strengthen load-bearing walls is placed between the studs near the midpoint of the studs as shown in Figure 8-11. Blocking used to prevent the spread of fire is placed at points where walls intersect with mechanical drops and parallel to stair stringers. Blocking in walls is estimated using the same procedures as estimating blocking in floors.

EXAMPLE 8-10

How many lineal feet of blocking are needed for the walls in Figure 8-5 if the studs in the exterior walls are blocked at their midpoints, as shown in Figure 8-11?

Solution: Only the perimeter wall will need to be blocked. Blocking is not needed at openings for doors and windows. The length of the wall that needs to be blocked is calculated as follows:

Length = 2(22 ft) + 2(24 ft 4 in) − 2(4 ft) − 6 ft − 32 in
= 76 ft

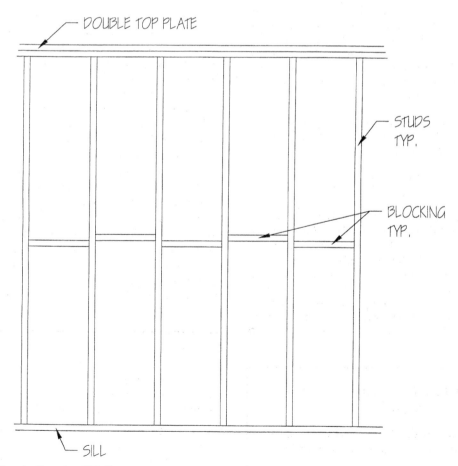

FIGURE 8-11 Blocking in Bearing Walls

Sheathing

Wood-framed exterior walls are commonly sheathed with 7/16-inch OSB or 1/2-inch plywood. Interior walls may also be sheeted with OSB or plywood to create a shear panel. The wood sheathing is treated as a sheet good. When calculating the number of rows, one must round the rows to a common fraction of a sheet such as 1/8, 1/6, 1/5 1/4, 1/3, or 1/2. Openings for doors and windows in the wall are commonly ignored because the material cut from these openings usually cannot be used as wall sheathing, but it may be used for headers built after the wall is sheathed. The quantity takeoff of wall sheathing is shown in the following example.

EXAMPLE 8-11

How many 4-foot by 8-foot sheets of 7/16-inch OSB are needed to sheet the outside of the exterior walls shown in Figure 8-5 if the wall height is 9 feet?

Solution: The number of rows is calculated using Eq. (4–13) as follows:

$$\text{Number}_{\text{Rows}} = \frac{9 \text{ ft}}{8 \text{ ft}} = 1.125 \text{ rows}$$

The number of rows is rounded up to the nearest 1/8 of a sheet. The length of the outside surface of the exterior wall is calculated as follows:

$$\text{Length} = 2(22 \text{ ft}) + 2(25 \text{ ft}) = 94 \text{ ft}$$

The number of columns is calculated using Eq. (4–16) as follows:

$$\text{Number}_{\text{Columns}} = \frac{94 \text{ ft}}{4 \text{ ft}} = 24 \text{ columns}$$

The number of sheets is calculated using Eq. (4–18) as follows:

$$\text{Number} = (1.125 \text{ rows})(24 \text{ columns}) = 27 \text{ sheets}$$

When sheathing gable ends, the waste from the sheets used near the peak of the gable ends can often be used for the sheets near the eaves of the gable ends as shown in the following example.

EXAMPLE 8-12

Determine the number of sheets of 7/16-inch OSB required for the end of the building shown in Figure 8-12.

Solution: The rise for the gable end is calculated as follows:

$$\text{Rise} = \left(\frac{32 \text{ ft}}{2}\right)\left(\frac{4}{12}\right) = 5.33 \text{ ft} = 5 \text{ ft } 4 \text{ in}$$

The height of the wall at the peak is 14 foot 4 inches. The number of rows of sheathing is calculated using Eq. (4–13) as follows:

$$\text{Number}_{\text{Rows}} = \frac{14.33 \text{ ft}}{8 \text{ ft}} = 2 \text{ rows}$$

The number of columns of sheathing is calculated using Eq. (4–16) as follows:

$$\text{Number}_{\text{Columns}} = \frac{32 \text{ ft}}{4 \text{ ft}} = 8 \text{ columns}$$

The first row of sheathing will run from the bottom of the wall to a height of 8 feet and will consist of eight sheets of sheathing. The second row of sheathing will run from a height of 8 feet to the top of the wall, which is 1 foot at the eaves and 6 foot 4 inches (1 ft + 5 ft 4 in) at the peak. Two pieces of sheathing can be cut from one sheet on the top row; therefore, only four sheets are required. The layout of the sheathing is shown in Figure 8-13. The number on each piece of sheathing represents the sheet from which the sheathing was cut.

Because the height at the eaves (1 ft) plus the height at the peak (6 ft 4 in) is less than 8 feet, the scrap from the short side of the gable end can be used at the peak as shown in Figure 8-14. A total of 12 sheets of sheathing are needed.

ROOF SYSTEMS

Roof systems may be stick framed using rafters, ceiling joists, and collar ties or may be constructed of manufactured trusses. Complex roofs that use trusses often have stick-framed overbuilds. Let's look at using rafters to construct a roof:

Rafters

A typical rafter is shown in Figure 8-15.

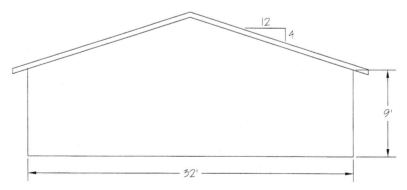

FIGURE 8-12 Building End Elevation

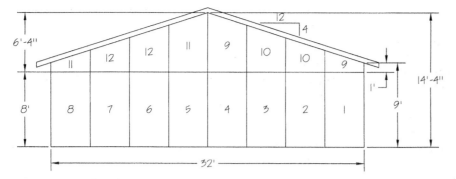

FIGURE 8-13 Sheathing Layout

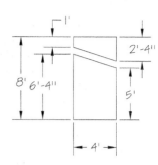

FIGURE 8-14 Layout for Sheet 9

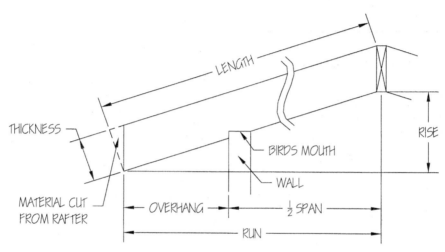

FIGURE 8-15 Rafter

The length of the rafter must be measured parallel to the rafter from the top corner of the rafter near the ridge to the bottom corner of the rafter at the eaves. The length of the rafter along the top or bottom edge may be determined from the Pythagorean theorem using the rise and the run of the rafter. To this length, we must add the length needed to allow the ends of the rafter to be cut at an angle, which is equal to the slope of the rafter times the width of the board used for the rafter. The length of the rafter is calculated using the following equation:

$$\text{Length} = (\text{Run}^2 + \text{Rise}^2)^{0.5} + (\text{Thickness})\left(\frac{\text{Rise}}{\text{Run}}\right) \quad (8\text{–}1)$$

Letting the slope equal the rise over the run and substituting the slope into Eq. (8–1), we get the following equation:

$$\text{Length} = (\text{Run})(1 + \text{Slope}^2)^{0.5} + (\text{Thickness})(\text{Slope}) \quad (8\text{–}2)$$

The slope may also be expressed as the relationship between the rise and the run such as a 4:12 slope or a 6:12 slope. Determining the length of a rafter is shown in the following example.

EXAMPLE 8-13

What length of rafters needs to be ordered for the building shown in Figure 8-16? The rafters are 2 × 8s with a slope of 4:12.

Solution: The thickness of a 2 × 8 is 7.25 inches. The run of the rafter is calculated as follows:

$$\text{Run} = \frac{\text{Span}}{2} + \text{Overhang} = \frac{20 \text{ ft}}{2} + 1 \text{ ft} = 11 \text{ ft}$$

The length of the rafter may be calculated using Eq. (8–2) as follows:

$$\text{Length} = (11 \text{ ft})\left[1 + \left(\frac{4}{12}\right)^2\right]^{0.5} + \frac{(7.25 \text{ in})}{(12 \text{ in/ft})}\left(\frac{4}{12}\right) = 11.8 \text{ ft}$$

Alternatively, it may be calculated by finding the rise as follows:

$$\text{Rise} = (\text{Run})(\text{Slope}) = (11 \text{ ft})\left(\frac{4}{12}\right) = 3.67 \text{ ft}$$

and then calculating the length of the rafter using Eq. (8–1) as follows:

$$\text{Length} = \left[(11 \text{ ft})^2 + (3.67 \text{ ft})^2\right]^{0.5} + \frac{(7.25 \text{ in})}{(12 \text{ in/ft})}\left(\frac{3.67 \text{ ft}}{11 \text{ ft}}\right)$$
$$= 11.8 \text{ ft}$$

Order 12-foot rafters.

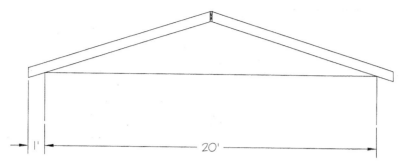

FIGURE 8-16 Rafter Section

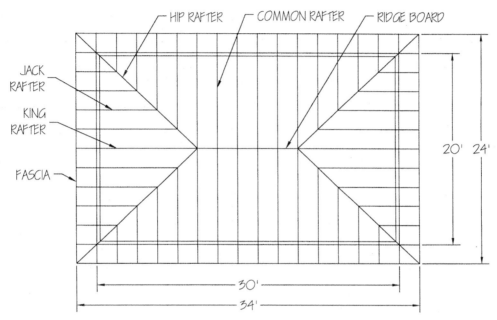

FIGURE 8-17 Roofing Plan for Hip Roof

A hip roof is shown in Figure 8-17. The length of a hip or valley rafter for a roof with the same slope on all surfaces may be calculated by finding the run of the hip or valley rafter using the Pythagorean theorem and using Eq. (8–1) or Eq. (8–2) to find the length. Alternatively, the length of the hip rafter may be determined from the run of the common rafter using the following equation:

$$\text{Length} = (\text{Run}_{\text{Common}})(2 + \text{Slope}^2_{\text{Common}})^{0.5} + (\text{Thickness})(\text{Slope}_{\text{Hip}}) \quad (8\text{--}3)$$

The hip and valley rafters have a flatter slope than the common rafters because their run is 41% longer than the run of the common rafters even though they have the same rise. The slope of a hip and valley rafter for a roof of a constant slope can be found by replacing the 12 in the slope of the roof with 17. A roof with a 4:12 slope will require hip and valley rafters with a 4:17 slope. Similarly, a roof with a 6:12 slope will require hip and valley rafters with a 6:17 slope. The length of hip and valley rafters is calculated in the same manner as the length of common rafters.

For the roof in Figure 8-17, the king rafter is the same length as the common rafters. The jack rafters will be shorter. When ordering lumber for the jack rafters in a hip roof, it is common practice to order lumber the same length as the common rafters and cut two jack rafters from each piece of lumber.

To prevent the wall from spreading apart due to the forces on the rafters, collar ties spaced at no more than 4 feet on center or ceiling joists connecting the walls must be used. Collar ties are commonly 1 × 4s that connect rafters and are located at least one-third of the rise below the peak of the rafters.

Ceiling joist may be used to prevent the spread of the rafters in lieu of collar ties. Ceiling joists are taken off in the same manner as floor joists.

The quantity takeoff for a hip roof is shown in the following example.

EXAMPLE 8-14

The hip roof shown in Figure 8-17 is constructed of 2 × 8 rafters spaced 24 inches on center. The roof has a slope of 4:12. Determine the quantities and lengths of 2 × 8s needed for the roof. Ceiling joists will be used to prevent the walls from spreading. Do not include the ceiling joists in your estimate.

Solution: The roof is 10 feet longer than it is wide; therefore, the common rafters must cover 10 feet on each side of the roof. The number of common rafters on each side of the roof is calculated from Eq. (4–1) as follows:

$$\text{Number} = \frac{10 \text{ ft}}{2 \text{ ft}} + 1 = 6 \text{ each}$$

There are 6 common rafters on each side of the roof for a total of 12. There are also two king rafters, one on each end, that are the same length as the common rafters. The run for the common and king rafters is 12 feet (24 ft/2). The length of the common and king rafters is calculated using Eq. (8–2) as follows:

$$\text{Length} = (12 \text{ ft})\left[1 + \left(\frac{4}{12}\right)^2\right]^{0.5} + \frac{(7.25 \text{ in})}{(12 \text{ in/ft})}\left(\frac{4}{12}\right) = 12.9 \text{ ft}$$

Fourteen-foot-long 2 × 8s will be used for the common and king rafters. There are four hip rafters. The length of the hip rafters is calculated using Eq. (8–3) as follows:

$$\text{Length} = (12 \text{ ft})\left[2 + \left(\frac{4}{12}\right)^2\right]^{0.5} + \frac{(7.25 \text{ in})}{(12 \text{ in/ft})}\left(\frac{4}{17}\right) = 17.6 \text{ ft}$$

Eighteen-foot-long 2 × 8s will be used for the hip rafters. Each of the four corners consists of two jack rafters with the following runs: 2, 4, 6, 8, and 10 feet. The length of these rafters is calculated using Eq. (8–2) as follows:

$$\text{Length}_{2 \text{ ft}} = (2 \text{ ft})\left[1 + \left(\frac{4}{12}\right)^2\right]^{0.5} + \frac{(7.25 \text{ in})}{(12 \text{ in/ft})}\left(\frac{4}{12}\right) = 2.3 \text{ ft}$$

$$\text{Length}_{4 \text{ ft}} = (4 \text{ ft})\left[1 + \left(\frac{4}{12}\right)^2\right]^{0.5} + \frac{(7.25 \text{ in})}{(12 \text{ in/ft})}\left(\frac{4}{12}\right) = 4.4 \text{ ft}$$

$$\text{Length}_{6 \text{ ft}} = (6 \text{ ft})\left[1 + \left(\frac{4}{12}\right)^2\right]^{0.5} + \frac{(7.25 \text{ in})}{(12 \text{ in/ft})}\left(\frac{4}{12}\right) = 6.5 \text{ ft}$$

$$\text{Length}_{8 \text{ ft}} = (8 \text{ ft})\left[1 + \left(\frac{4}{12}\right)^2\right]^{0.5} + \frac{(7.25 \text{ in})}{(12 \text{ in/ft})}\left(\frac{4}{12}\right) = 8.6 \text{ ft}$$

$$\text{Length}_{10 \text{ ft}} = (10 \text{ ft})\left[1 + \left(\frac{4}{12}\right)^2\right]^{0.5} + \frac{(7.25 \text{ in})}{(12 \text{ in/ft})}\left(\frac{4}{12}\right) = 10.7 \text{ ft}$$

From one 14-foot-long 2 × 8, one can cut one 2.3-foot-long and one 10.7-foot-long rafters, or one 4.4-foot-long and one 8.6-foot-long rafter, or two 6.5-foot-long rafters. Five 14-foot-long 2 × 8s will produce two 2.3-foot-long rafters, two 4.4-foot-long rafters, two 6.5-foot-long rafters, two 8.6-foot-long rafters, and two 10.7-foot-long rafters; this is enough rafters to complete one corner. Twenty 14-foot-long 2 × 8s are needed for the corners. A total of 34 (12ea + 2ea + 20 ea) 14-foot-long 2 × 8s are needed. One 10-foot-long 2 × 8 is needed for the ridge board.

> ### EXCEL QUICK TIP 8-6
> ### Length of a Rafter
>
> The calculation of the length of a rafter is set up in a spreadsheet by entering the data and formatting the cells as follows:
>
> | | A | B | C |
> |---|---|---|---|
> | 1 | Span | 20 | ft |
> | 2 | Overhang | 24 | in |
> | 3 | Slope | 4 | :12 |
> | 4 | Thickness | 7.25 | in |
> | 5 | | | |
> | 6 | Rise | 4.00 | ft |
> | 7 | Common Rafter | 12.9 | ft |
> | 8 | Hip Rafter | 17.6 | ft |
>
> The following formulas need to be entered into the associated cells:
>
> | Cell | Formula |
> |------|---------|
> | B6 | =(B1/2+B2/12)*B3/12 |
> | B7 | =(B1/2+B2/12)*(1+(B3/12)^2)^0.5+B4*B3/(12*12) |
> | B8 | =(B1/2+B2/12)*(2+(B3/12)^2)^0.5+B4*B3/(17*12)- |
>
> The data for rafter is entered in Cells B1 through B4. The data shown in the foregoing figure is from Example 8-14.

The materials needed are as follows: one 10-foot-long, thirty-four 14-foot-long, and four 18-foot-long 2 × 8s.

Trusses

Trusses are used to span larger distances using less lumber. Trusses consist of a top cord, a bottom cord, and truss members that are used to transfer the loads. Trusses are specified based on the slope of the top and bottom cords, the span of the truss, and the length of the overhang (often referred to as the length of the tails). A special kind of truss is a gable-end truss, which is used at the ends of a building and anywhere the walls extend up beyond the ceiling line, such as in the case of firewalls. A gable-end truss is shown in Figure 8-18. The purpose of the gable-end truss is to provide an easy-to-frame-and-sheet end that matches the slope and dimension of the trusses. As such, a gable-end truss must be supported along its full length by a bearing wall. The gable-end truss provides an alternative to stick framing the gable ends.

The quantity takeoff for a roof using trusses is shown in the following example.

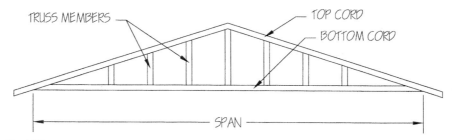

FIGURE 8-18 Gable-End Truss

EXAMPLE 8-15

What size and number of trusses need to be ordered for the roof shown in Figure 8-19? The gable ends of the building are to be constructed as part of the walls. The slope of the roof is 4:12. The trusses are spaced 24 inches on center and are made of 2 × 4s. How many 8-foot-long 2 × 4s need to be ordered if the lookouts are constructed of 2 × 4s spaced 24 inches on center?

Solution: The trusses need to cover 26 feet (30 ft − 2 ft − 2 ft) of roof. The number of trusses needed is calculated using Eq. (4–1) as follows:

$$\text{Number} = \frac{26 \text{ ft}}{2 \text{ ft}} + 1 = 14 \text{ each}$$

The trusses are to have a 4:12 slope, a 20-foot span, and 2-foot-long tails. The lookouts are on the sloped end of the roof; therefore, the sloped length of the roof from the eave of the roof to the peak of the roof must be calculated using the Pythagorean theorem as follows:

$$\text{Rise} = \left(\frac{24 \text{ ft}}{2}\right)\left(\frac{4}{12}\right) = 4 \text{ ft}$$

$$\text{Length} = \left[\left(\frac{24 \text{ ft}}{2}\right)^2 + (4 \text{ ft})^2\right]^{0.5} = 12.7 \text{ ft}$$

Excluding the lookout at the peak and the lookout at the eave (the lookout at the eave of the roof is part of the fascia), the number of lookouts needed for the roof from the eave to the peak is calculated using Eq. (4–3) as follows:

$$\text{Number} = \frac{12.7 \text{ ft}}{2 \text{ ft}} - 1 = 6 \text{ ea}$$

Adding 1 for the peak, the number of lookouts needed on each end of the building is 13 (6 + 1 + 6) for a total of 26 4-foot-long lookouts. Two lookouts can be cut from one 8-foot-long 2 × 4; therefore, 13 (26/2) 8-foot-long 2 × 4s are needed.

Sheathing

The quantity takeoff for roof sheathing is done in the same manner as for floor sheathing. The sheathing for each surface of the roof should be done separately because changes in the surface of the roof increase the waste. Unlike floor sheathing, roof sheathing is seldom T&G,

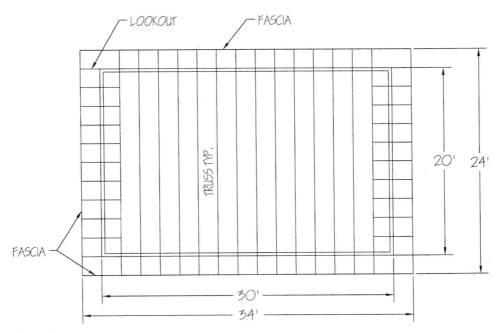

FIGURE 8-19 Roofing Plan

which allows more of the waste to be used. The takeoff for roof sheathing is shown in the following example.

EXAMPLE 8-16

What is the area of the roof in Figure 8-19? How many sheets of 7/16-inch OSB are needed for the roof? What is the waste factor for this roof?

Solution: The sloped area of one side of the roof is calculated using the Pythagorean theorem to determine the slope length of the roof, which is multiplied by the width of the roof. This area is multiplied by two to get the area of the roof. In Example 8-15, the sloped length of the roof in Figure 8-19 is found to be 12.7 feet. The area of the roof is calculated as follows:

$$\text{Area} = 2(12.7\text{ ft})(34\text{ ft}) = 864\text{ ft}^2$$

The number of rows of sheathing needed for one side of the roof is calculated using Eq. (4–13) as follows:

$$\text{Number}_{\text{Rows}} = \frac{12.7\text{ ft}}{4\text{ ft}} = 3.2\text{ rows}$$

The number of rows is rounded to one-fifth of a sheet. The number of columns is calculated using Eq. (4–16) as follows:

$$\text{Number}_{\text{Columns}} = \frac{34\text{ ft}}{8\text{ ft}} = 4.5\text{ columns}$$

The number of columns is rounded to half of a sheet because the sheet must span at least two trusses, or 4 feet. The number of sheets needed for the roof is calculated using Eq. (4–18) as follows:

$$\text{Number} = (2\text{ sides})(3.2\text{ rows/side})(4.5\text{ columns}) = 29\text{ sheets}$$

The area of the sheets is 928 square feet (29 sheets × 32 ft²/sheet). The waste for the roof is 7.4% (928/864 − 1).

Fascia

When fascia is being installed on the gable ends of a roof, the length of the fascia is calculated in the same manner as the length of the rafters is calculated using Eq. (8–1) or Eq. (8–2). When fascia is installed on the eaves, the height of the fascia is greater than the width of the framing member to which it is being installed, as shown in Figure 8–20. For example, the fascia being attached to a truss constructed of 2 × 4s will need to be constructed from a 2 × 6.

The size of the lumber needed to construct the fascia is calculated using the following equation:

$$\text{Width}_{\text{Fascia}} = (\text{Width}_{\text{Rafter}})(1 + \text{Slope}^2)^{0.5} + (\text{Thickness}_{\text{Fascia}})(\text{Slope}) \quad (8\text{–}4)$$

The quantity takeoff for fascia is shown in the following example.

EXAMPLE 8-17

How many feet of 2 × fascia are needed for the roof in Figure 8-19? What size of fascia is needed?

Solution: The width of the fascia is calculated using Eq. (8–4) as follows:

$$\text{Width}_{\text{Fascia}} = (3.5\text{ in})\left[1 + \left(\frac{4}{12}\right)^2\right]^{0.5} + (1.5\text{ in})\left(\frac{4}{12}\right) = 4.2\text{ in}$$

Use 2 × 6s for the fascia. The length of the fascia on the ends of the roof is calculated using Eq. (8–2) as follows:

$$\text{Length} = (24\text{ ft}/2)\left[1 + \left(\frac{4}{12}\right)^2\right]^{0.5} + \frac{(5.5\text{ in})}{(12\text{ in/ft})}\left(\frac{4}{12}\right) = 12.8\text{ ft}$$

The 2 × 6 fascia needed to frame the perimeter of the roof (including the lookouts at the eaves) is four 14-foot-long 2 × 6s for the sloped ends of the building (whose length is 12.8 ft) and six 12-foot-long 2 × 6s for the side of the building (whose length is 34 ft).

Soffit

Sometimes sheathing is used for the soffit. The quantity takeoff for soffit is done using the row and column method, where the number of rows of soffit is a fraction (1/2, 1/3, 1/4, 1/5, 1/6, 1/7, or 1/8) of a sheet. The quantity takeoff for soffit is shown in the following example.

EXAMPLE 8-18

Determine the number of sheets of 4-foot by 8-foot by 7/16-inch OSB that are needed for the soffit of the roof in Figure 8-19. The soffit is installed parallel to the surface of the roof. The soffit is held back the width of the fascia (1 1/2 inches) from the edge of the roof.

Solution: The soffit will be divided into six areas: two 30-foot-long areas on the sides of the building and four 12.7-foot-long by 2-foot-wide areas at the ends of the building. First, we will look at the two 30-foot-long areas. The width of these areas is determined by the Pythagorean theorem as follows:

$$\text{Rise} = (1\text{ ft }10\ 1/2\text{ in})\left(\frac{4}{12}\right) = 0.62\text{ ft}$$

$$\text{Length} = [(1\text{ ft }10\ 1/2\text{ in})^2 + (0.62\text{ ft})^2]^{0.5} = 2.0\text{ ft}$$

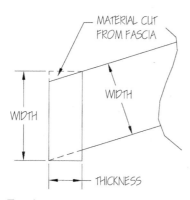

FIGURE 8-20 Fascia

The number of rows for the sides is calculated using Eq. (4–13) as follows:

$$\text{Number}_{\text{Rows}} = \frac{2.0 \text{ ft}}{8 \text{ ft}} = 0.25 \text{ rows}$$

The number of columns for the sides is calculated using Eq. (4–16) as follows:

$$\text{Number}_{\text{Columns}} = \frac{30 \text{ ft}}{4 \text{ ft}} = 7.5 \text{ columns}$$

The number of sheets needed for the sides of the roof is calculated using Eq. (4–18) as follows:

Number = (2 sides)(0.25 rows/side)(7.5 columns) = 4 sheets

The number of rows for the ends is calculated using Eq. (4–13) as follows:

$$\text{Number}_{\text{Rows}} = \frac{(1 \text{ ft} 10 \text{ 1/2 in})}{4 \text{ ft}} = 0.5 \text{ rows}$$

The number of columns for the ends is calculated using Eq. (4–16) as follows:

$$\text{Number}_{\text{Columns}} = \frac{12.7 \text{ ft}}{8 \text{ ft}} = 1.75 \text{ columns}$$

The number of columns is rounded to the quarter-sheet increment, which is 2 feet. The number of sheets needed for the side of the roof is calculated using Eq. (4–18) as follows:

Number = (4 end)(0.5 rows/end)(1.75 columns) = 3.5 sheets

The number of sheets needed is 8 (4 + 3.5).

FINISH CARPENTRY

Finish carpentry often includes the installation of trim, wood shelving, and wood paneling. Trim and shelving is bid by the lineal foot and paneling is bid by the sheet in the same manner as sheathing is bid. The quantity takeoff of wood trim is shown in the following example.

EXAMPLE 8-19

Determine the wood base and wood casing needed to complete the rooms shown in Figure 8-5. The casing is only located around the single-hung doors. The base comes in 8-, 12-, and 16-foot lengths and the casing comes in 7-foot lengths. Minimize the number of joints.

Solution: Six 12-foot-long pieces of base are needed for the main room. Three 12-foot-long pieces and one 8-foot-long piece of base are needed for each of the bedrooms. One 8-foot-long piece of base along with scraps left over from the door opening to the bedroom is needed for each closet. The following base needs to be ordered: twelve 12-foot-long and four 8-foot-long pieces of base.

Each side of a door requires 2 1/2 pieces of casing. Only the inside of the exterior door needs casing; therefore, 13 (5 × 2.5) pieces of door casing are needed.

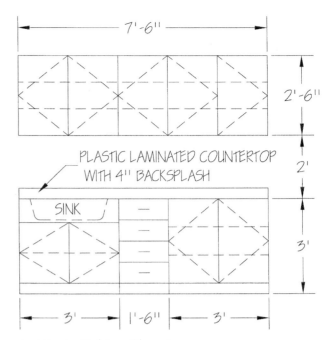

FIGURE 8-21 Cabinet Elevation

CABINETRY AND COUNTERTOPS

The type and size of each cabinet need to be quantified. Countertops are bid by the lineal foot. For corners, the length of the countertop is measured at the longest side of the countertop. The quantity takeoff of cabinets and countertops is shown in the following example:

EXAMPLE 8-20

Determine the cabinets and lineal feet of countertop needed for the cabinet layout shown in Figure 8-21.

Solution: Starting with the left, the following base cabinets are needed: a 36-inch-wide by 36-inch-high sink base, an 18-inch-wide by 36-inch-high four-drawer base, and a 36-inch-wide by 36-inch-high standard base. Starting with the left, the following wall cabinets are needed: two 36-inch-wide by 30-inch-high wall cabinets and one 18-inch-wide by 30-inch-high wall cabinet. A plastic laminated countertop with a 4-inch backsplash 7 feet 6 inches long is needed.

SAMPLE TAKEOFF FOR THE RESIDENTIAL GARAGE

A sample takeoff for rough carpentry, lumber, trusses, finish carpentry, and wood trim from a set of plans is shown in the following example.

EXAMPLE 8-21

Determine the rough carpentry, lumber, trusses, finish carpentry, and wood trim needed to complete the residential garage found in Appendix G.

Solution: The thickness of the exterior wall is 4 inches (3 1/2-inch-thick stud plus 1/2-inch-thick exterior sheathing). Find the length of the wall as follows:

Length = 4 ft + 25 ft 4 in + 24 ft + 25 ft 4 in + 4 ft = 82 ft 8 in

Eighty-three feet of wall with one opening for a single-hung door and one opening for an overhead door need to be framed. An 8-foot-long redwood plate is ordered for the west wall, a 12-foot-long and a 14-foot-long redwood plate are ordered for the north wall, two 12-foot-long redwood plates are ordered for the east wall, and a 12-foot-long and a 14-foot-long redwood plate are ordered for the south wall. Order one 8-foot-long, four 12-foot-long, and two 14-foot-long redwood 2 × 4 plates.

Two 50-foot-long rolls of foam seal are needed under the bottom of wall. Nine 10-foot-long Z-flashings are needed for the exterior walls.

The number of studs needed for a 16-inch on-center spacing is calculated using Eq. (4–2) as follows:

$$\text{Number} = \frac{(82 \text{ ft } 8 \text{ in})(12 \text{ in/ft})}{16 \text{ in}} = 62 \text{ each}$$

Two additional studs are needed for each of the four corners, 2 for the doorway, 6 for the overhead door, and 1 for each of the two hold downs located at the corners, for a total of 80 studs (62 + 2 × 4 + 2 + 6 + 2 × 1).

The door header will require one 8-foot-long 2 × 6 and a sheet of 7/16 OSB for a spacer. The garage header will require a 16-foot 6-inch-long 3 1/2-inch by 12-inch GLB.

The top plates for the north, east, and south walls need twice the quantities needed for the bottom plate. For the west wall, four 12-foot-long plates are needed. Order twelve 12-foot-long and four 14-foot-long 2 × 4 plates.

The number of standard trusses needed is calculated using Eq. (4–3) as follows:

$$\text{Number} = \frac{26 \text{ ft}}{2 \text{ ft}} - 1 = 12 \text{ each}$$

Twelve 24-foot trusses with a 4:12 slope and 18-inch tails are needed along with two 24-foot gable-end trusses. Twenty-four hurricane ties are needed to secure the standard trusses to the roof.

Next, the 2 × 4 framing for the soffit will be taken off beginning with the north side. A 29-foot-long ledger is needed at the wall and a 29-foot fascia board is needed at the edge of the roof. Sixteen (one at each truss and one at both ends of the roof) 15-inch-long pieces of blocking are needed between the ledger and the fascia board. The same is needed for the south side of the roof.

Next, the 2 × 4 framing on the west end of the building will be taken off. A 2 × 4 ledger is needed along the truss from the north side of the building to the peak of the roof. The length is determined by Eq. (8–2) as follows:

$$\text{Length} = \left(\frac{24 \text{ ft}}{2}\right)\left[1 + \left(\frac{4}{12}\right)^2\right]^{0.5} + \frac{(3.5 \text{ in})}{(12 \text{ in/ft})}\left(\frac{4}{12}\right) = 12.8 \text{ ft}$$

The fascia board will run parallel to the truss from the edge of the roof to the peak of the building. The length is determined by Eq. (8–2) as follows:

$$\text{Length} = \left(\frac{27 \text{ ft}}{2}\right)\left[1 + \left(\frac{4}{12}\right)^2\right]^{0.5} + \frac{(3.5 \text{ in})}{(12 \text{ in/ft})}\left(\frac{4}{12}\right) = 14.3 \text{ ft}$$

Blocking is located at 24 inches on center on the sloped end of the roof; therefore, the sloped length of the roof from the eave of the roof to the peak of the roof must be calculated using the Pythagorean theorem as follows:

$$\text{Rise} = \left(\frac{27 \text{ ft}}{2}\right)\left(\frac{4}{12}\right) = 4.5 \text{ ft}$$

$$\text{Length} = \left[\left(\frac{27 \text{ ft}}{2}\right)^2 + (4.5 \text{ ft})^2\right]^{0.5} = 14.3 \text{ ft}$$

Excluding the blocking at the peak and the blocking at the eave (the blocking at the eave of the roof is part of the fascia), the number of pieces of blocking needed for the roof from the eave to the peak is calculated using Eq. (4–3) as follows:

$$\text{Number} = \frac{14.3 \text{ ft}}{2 \text{ ft}} - 1 = 7 \text{ ea}$$

Adding 1 for the peak, we find that the number of pieces of blocking needed on the west end of the building is 15 (7 + 1 + 7). Each of these is 15 inches long. The same quantities are needed for the east side.

The total length of ledger needed is 110 feet (29 + 12.8 + 12.8 + 29 + 12.8 + 12.8). Order fourteen 8-foot-long 2 × 4s for the ledger. The total length of fascia is 116 feet (29 + 14.3 + 14.3 + 29 + 14.3 + 14.3). Order fifteen 8-foot-long 2 × 4s for the fascia. The number of 15-inch-long blockings is 62 (16 + 15 + 16 + 15). Six pieces 15 inches long can be cut from one 8-foot-long 2 × 4. Order eleven 8-foot-long 2 × 4s for the blocking. A total of 40 (14 + 15 + 11) 8-foot-long 2 × 4s are needed for the soffit framing.

Two 24-foot-long pieces of blocking are needed for the ceiling (see detail K5 on Sheet 7 in Appendix G). Order six 8-foot-long 2 × 4s for the ceiling blocking. Order a total of forty-six 8-foot-long 2 × 4s, to frame the soffit and block the ceiling.

Next, the roof sheathing will be taken off. The width of one side of the roof is 14.3 feet. The area of the roof that needs to be sheathed is calculated as follows:

$$\text{Area} = 2(14.3 \text{ ft})(29 \text{ ft}) = 829 \text{ ft}^2$$

The number of rows of sheathing needed for one side of the roof is calculated using Eq. (4–13) as follows:

$$\text{Number}_{\text{Rows}} = \frac{14.3 \text{ ft}}{4 \text{ ft}} = 4 \text{ rows}$$

The number of columns is calculated using Eq. (4–16) as follows:

$$\text{Number}_{\text{Columns}} = \frac{29 \text{ ft}}{8 \text{ ft}} = 4 \text{ columns}$$

The number of sheets needed for the roof is calculated using Eq. (4–18) as follows:

Number = (2 sides)(4 rows/side)(4 columns) = 32 sheets

Three rows of plywood clips are needed on each side of the roof with a plywood clip located between each pair of trusses; therefore, 13 (1 less than the number of trusses) columns of clips are needed. The number of clips is calculated using Eq. (4–18) as follows:

Number = (2 sides)(3 rows/side)(13 columns) = 78 each

The finish carpentry includes the installation of the trim. The trim consists of the fascia and the casing on the inside of the door. The T1-11 siding, the soffit, and 1 × 4 cedar trim associated with the siding are covered under siding in Chapter 9. The total length of fascia is 116 feet (29 + 14.3 + 14.3 + 29 + 14.3 + 14.3). The height of the fascia is calculated using Eq. (8–4) as follows:

$$\text{Width}_{\text{Fascia}} = (3.5\,\text{in})\left[1 + \left(\frac{4}{12}\right)^2\right]^{0.5} + (1.5\,\text{in})\left(\frac{4}{12}\right) = 4.2\,\text{in}$$

The fascia is cut from 1 × 6s. Six 10-foot pieces of 1 × 6 trim are needed for the north and south sides of the building. Two 16-foot pieces of trim are needed for each gable end. Order six 10-foot-long 1 × 6 and four 16-foot-long 1 × 6 pieces of cedar for the fascia.

Three 7-foot pieces of casing are needed for the inside of the single-hung door.

The quantities needed for the garage, grouped by the cost codes in Appendix B, are shown in Table 8-1.

TABLE 8-1 **Quantities for Residential Garage**

06-100 Rough Carpentry		
2 × 4 wall	83	ft
Single-hung door opening	1	ea
Overhead door opening	1	ea
Install trusses	14	ea
Install ledger	110	ft
Install fascia board	116	ft
Install 15" blocking	62	ea
Install ceiling blocking	48	ft
Install roof sheathing	829	sft
06-110 Lumber		
Walls		
2 × 4-8' redwood	1	ea
2 × 4-12' redwood	4	ea
2 × 4-14' redwood	2	ea
Foam sill 50'	2	ea
Z-flashing 10' long	9	ea
2 × 4-92 5/8" stud	80	ea
2 × 6-8'	1	ea
7/16" × 4' × 8' OSB	1	ea
3-1/2" × 12" GLB	16.5	ft
2 × 4-12'	12	ea
2 × 4-14'	4	ea
Roof		
Hurricane ties	24	ea
2 × 4-8'	46	ea
7/16" × 4' × 8' OSB	32	ea
Plywood clips	78	ea
06-120 Trusses		
24' 4:12 standard trusses w/18" tails	12	ea
24' 4:12 gable trusses w/18" tails	2	ea
06-200 Finish Carpentry		
Install shaped fascia	116	ft
Install door casing	1	set
06-210 Wood Trim		
1 × 6-10' cedar	6	ea
1 × 6-16' cedar	4	ea
7' casing	3	ea

CONCLUSION

Woods and plastics consist of rough carpentry, finish carpentry, and cabinets and countertops. Rough carpentry consists of framing the floors, walls, and roof of the structure. Finish carpentry consists of installing finish material including wood trim, shelving, and handrails. Cabinetry includes the installation of cabinets and countertops.

PROBLEMS

1. The floor system in Figure 8-22 sits on a redwood sill on a concrete wall. Determine the number of 16-foot-long pieces of redwood that are needed to complete the floor system.

2. Determine the materials needed to complete the posts and girder supporting the floor joists in Figure 8-22. The posts are connected to footings using a Simpson PB44 post base and are connected

to the girder using a Simpson BC4 post cap. The distance between the top of the footing and the bottom of the girder is 7 feet. The posts are spaced 12 feet apart.

3. Determine the materials needed to construct the joists and joist headers for the floor system shown in Figure 8-22.

4. For the floor system shown in Figure 8-22, the joists are blocked over the girder, at the concrete wall at the overhang, and at the midpoint between the girder and the exterior wall on the joists over 8 feet long. Determine the materials needed for the blocking.

5. Determine the number of 4-foot by 8-foot sheets of 23/32-inch T&G OSB sheathing that are needed to construct the floor system shown in Figure 8-22.

6. The perimeter of the floor system in Figure 8-23 sits on a redwood sill. Determine the number of

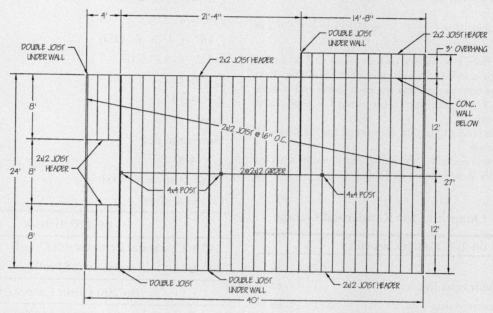

FIGURE 8-22 Floor-Framing Plan

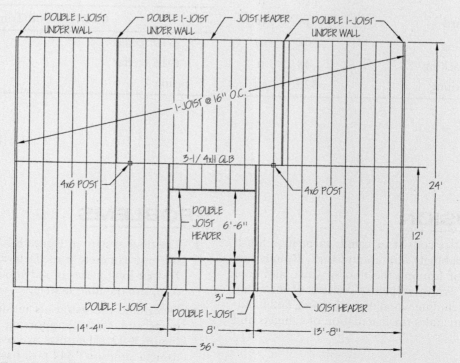

FIGURE 8-23 Floor-Framing Plan

16-foot-long pieces of redwood that are needed to complete the floor system.

7. Determine the materials needed to complete the posts and girder supporting the floor joists in Figure 8-23. The posts are connected to footings using a Simpson PB46 post base and are connected to the girder using a Simpson BC46 post cap. The distance between the top of the footing and the bottom of the girder is 7 feet. The posts are equally spaced along the gluelam beam.

8. Determine the materials needed to construct the joists and joist headers for the floor system shown in Figure 8-23. An engineered rim board is used for the joist headers.

9. For the floor system shown in Figure 8-23, the joists are blocked over the girder and at the midpoint between the girder and the exterior wall on the joists over 8 feet long. I-joists are to be used for the blocking. How many feet of I-joists are needed for the blocking?

10. Determine the number of 4-foot by 8-foot sheets of 23/32-inch T&G OSB sheathing that are needed to construct the floor system shown in Figure 8-23.

11. How many board feet of 2 × 6 redwood decking are needed to construct a 20-foot by 30-foot deck? The boards are to be placed without any space between them.

12. How many board feet of 2 × 4 redwood decking are needed to construct a 10-foot by 20-foot deck? The boards are to be placed with a 1/2-inch space between them.

13. How many board feet of 2 × 4 treated and untreated plate are needed to construct 100 feet of 2 × 4 wall? The wall bears on a concrete floor.

14. How many board feet of 2 × 4 treated and untreated plate are needed to construct 75 feet of 2 × 4 wall? The wall bears on a wood floor.

15. How many studs are needed to construct 100 feet of 2 × 4 wall? The wall has one 8-foot-wide window, four 4-foot-wide windows, two 36-inch-wide doors, one 30-inch-wide door, four corners, two intersections, and eight hold downs located at corners. The stud spacing is 16 inches on center. Allow for two additional studs for each corner, intersection, or doorway; six additional studs for each window less than 6 feet in width; eight additional studs for each window 6 feet or more in width; and one additional stud for each hold down located at a corner.

16. Prepare a materials list for the headers required to complete the wall in Problem 15. All headers are made from 2 × 12s and the wall is made from 2 × 4s. Openings 6 feet or more in width require double jack studs.

17. How many studs are needed to construct 180 feet of 2 × 6 wall? The wall has two 6-foot-wide windows, four 4-foot-wide windows, three 3-foot-wide windows, one 60-inch-wide door, one 36-inch-wide door, one 30-inch-wide door, six corners, eight intersections, and four hold downs located at corners. The stud spacing is 24 inches on center. Allow for two additional studs for each corner, intersection, or doorway; six additional studs for each window less than 6 feet in width; eight additional studs for each window 6 feet or more in width; and one additional stud for each hold down located at a corner.

18. Prepare a materials list for the headers required to complete the wall in Problem 17. All headers are made from 2 × 12s and the wall is made from 2 × 6s. Openings 6 feet or more in width require double jack studs.

19. How many sheets of 7/16-inch OSB are needed to construct 100 feet of wall 8 feet high?

20. How many sheets of 7/16-inch OSB are needed to construct 100 feet of wall 9 feet 3 inches high?

21. How many feet of treated and untreated plate are needed to construct the walls shown in Figure 8-24? The walls are constructed of 2 × 4s with the exterior walls being sheathed with 7/16-inch OSB. The walls bear on a concrete floor.

22. How many studs are needed to frame the wall in Figure 8-24? The stud spacing is 16 inches on center. Allow for one additional stud for each wall that does not have a corner; two additional studs for each corner, intersection, or doorway; six additional studs for each window less than 6 feet in width; and eight additional studs for each window 6 feet or more in width. The closet doors are 6 feet wide and the single-hung doors are 30 inches wide.

23. Prepare a materials list for the headers required to complete the walls in Figure 8-24. All headers are made from 2 × 12s and the wall is made from 2 × 4s. The single-hung doors are 32 inches wide and the closet doors are 6 feet wide. Openings 6 feet or more in width require double jack studs.

24. How many lineal feet of blocking are needed for the walls in Figure 8-24 if the studs in the exterior walls are blocked at their midpoints as shown in Figure 8-11?

25. How many 4-foot by 8-foot sheets of 7/16-inch OSB are needed to sheet the outside of the exterior walls shown in Figure 8-24 if the wall height is 9 feet?

26. How many feet of treated and untreated plate are needed to construct the walls shown in Figure 8-25? The walls are constructed of 2 × 4s with the

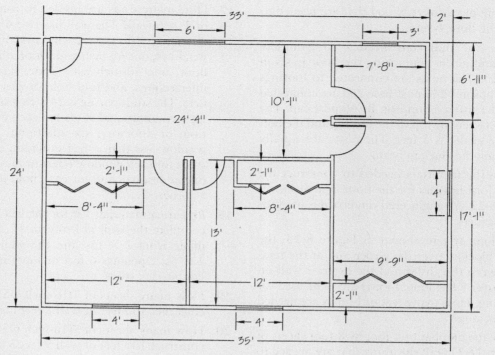

FIGURE 8-24 Wall Layout

exterior walls being sheathed with 7/16-inch OSB. The walls bear on a concrete floor.

27. How many studs are needed to frame the wall in Figure 8-25? The stud spacing is 16 inches on center. Allow for one additional stud for each wall that does not have a corner; two additional studs for each corner, intersection, or doorway; six additional studs for each window less than 6 feet in width; and eight additional studs for each window 6 feet or more in width.

28. Prepare a materials list for the headers required to complete the walls in Figure 8-25. All headers are made from 2 × 12s and the wall is made from 2 × 4s. The exterior doors are 36 inches wide and the interior single-hung doors are 32 inches wide. Openings 6 feet or more in width require double jack studs.

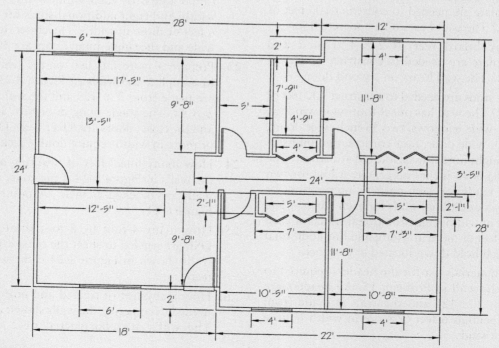

FIGURE 8-25 Wall Layout

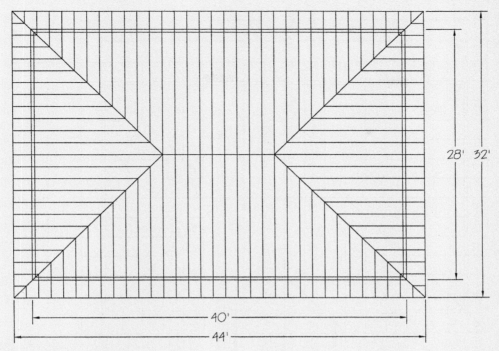

FIGURE 8-26 Roofing Plan for Hip Roof

29. How many lineal feet of blocking are needed for the walls in Figure 8-25 if the studs in the exterior walls are blocked at their midpoints as shown in Figure 8-11?

30. How many 4-foot by 8-foot sheets of 7/16-inch OSB are needed to sheet the outside of the exterior walls shown in Figure 8-25 if the wall height is 10 feet 4 inches?

31. A roof framed with 2 × 8 rafters has a span of 30 feet, an overhang of 18 inches, and a slope of 6:12. What length of common rafters needs to be ordered for the roof?

32. A roof framed with 2 × 6 rafters has a span of 18 feet, an overhang of 12 inches, and a slope of 4:12. What length of common rafters needs to be ordered for the roof?

33. The hip roof shown in Figure 8-26 is constructed of 2 × 8 rafters spaced 16 inches on center. The hip rafters are 1 1/2-inch-wide by 12-inch-high GLBs. The roof has a slope of 6:12. Prepare a list of lengths and quantities for the 2 × 8s and GLBs. Ceiling joists will be used to prevent the walls from spreading. Do not include the ceiling joists in your estimate.

34. The hip roof shown in Figure 8-27 is constructed of 2 × 6 rafters spaced 24 inches on center. The hip and valley rafters are also constructed of 2 × 6s. The roof has a slope of 4:12. Prepare a list of lengths and quantities for the 2 × 6s. Ceiling joists will be used to prevent the walls from spreading. Do not include the ceiling joists in your estimate.

Hint: The lengths of the valley rafters are calculated in the same way as those for the hip rafters.

35. What size of trusses needs to be ordered for the roof shown in Figure 8-28? The gable ends of the building are to be constructed as part of the walls. The slope of the roof is 4:12. The trusses are spaced 24 inches on center and are made of 2 × 4s. How many 8-foot-long 2 × 4s need to be ordered if the lookouts are constructed of 2 × 4s spaced 24 inches on center?

36. What size of trusses needs to be ordered for the roof shown in Figure 8-29? The slope of the roof is 4:12. The trusses are spaced 24 inches on center and are made of 2 × 4s.

37. What is the area of the roof in Figure 8-28? How many sheets of 7/16-inch OSB are needed for the roof? What is the waste factor for this roof?

38. What is the area of the roof in Figure 8-29? How many sheets of 7/16-inch OSB are needed for the roof? What is the waste factor for this roof?

39. How many feet of 2 × fascia are needed for the roof in Figure 8-28? What size of fascia is needed?

40. How many feet of 2 × fascia are needed for the perimeter of the roof in Figure 8-29? What size of fascia is needed?

41. Determine the number of sheets of 4-foot by 8-foot by 7/16-inch OSB that are needed for the soffit of the roof in Figure 8-28. The soffit is installed parallel to the roof.

42. Determine the number of sheets of 4-foot by 8-foot by 7/16-inch OSB that are needed for the soffit

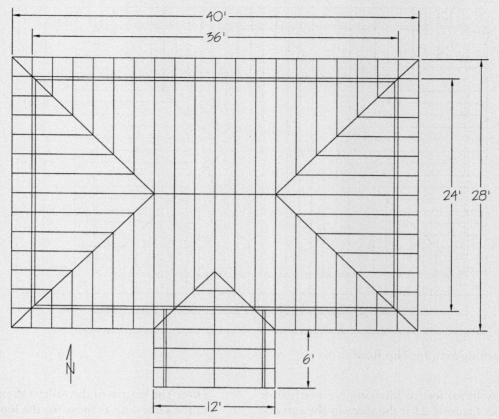

FIGURE 8-27 Roofing Plan for Hip Roof

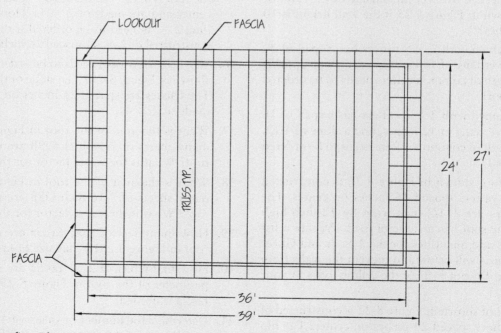

FIGURE 8-28 Roofing Plan

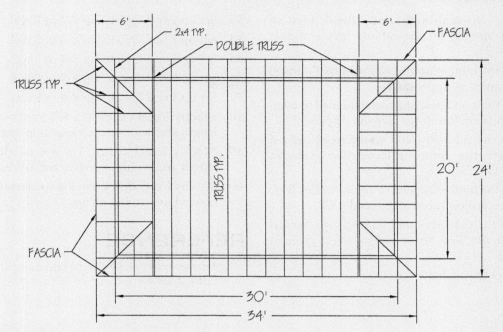

FIGURE 8-29 Roofing Plan

of the roof in Figure 8-29. The soffit is installed horizontally.

43. Determine the wood base and wood casing needed to complete the rooms shown in Figure 8-24. The casing is only located around the single-hung doors. The base comes in 8-, 12-, and 16-foot lengths and the casing comes in 7-foot lengths.

44. Determine the wood base and wood casing needed to complete the rooms shown in Figure 8-25. The casing is only located around the single-hung doors. The base comes in 8-, 12-, and 16-foot lengths and the casing comes in 7-foot lengths.

45. Determine the cabinets and lineal feet of countertop needed for the cabinet layout shown in Figure 8-30.

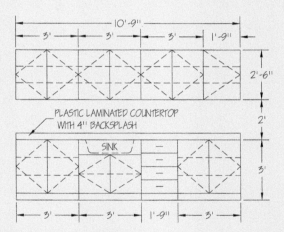

FIGURE 8-30 Cabinet Elevation

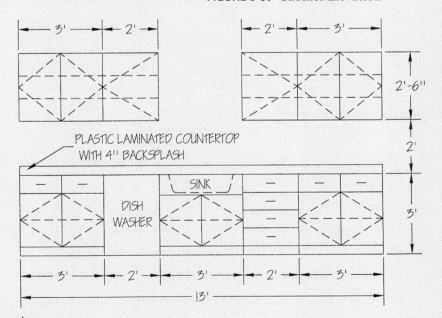

FIGURE 8-31 Cabinet Elevation

46. Determine the cabinets and lineal feet of countertop needed for the cabinet layout shown in Figure 8-31.
47. Determine the lumber needed to frame the Johnson Residence given in Appendix G.
48. Determine the wood base and casing needed to complete the Johnson Residence given in Appendix G.
49. Determine the cabinetry and countertops needed to complete the Johnson Residence given in Appendix G.
50. Determine the lumber needed to complete the West Street Video project given in Appendix G.
51. Determine the cabinetry and countertops needed to complete the West Street Video project given in Appendix G.
52. Set up Excel Quick Tip 8-1 in Excel.
53. Set up Excel Quick Tip 8-2 in Excel.
54. Set up Excel Quick Tip 8-3 in Excel.
55. Set up Excel Quick Tip 8-4 in Excel.
56. Set up Excel Quick Tip 8-5 in Excel.
57. Set up Excel Quick Tip 8-6 in Excel.
58. Modify Excel Quick Tip 8-3 to allow the building length and width to be entered in feet and inches.
59. Modify Excel Quick Tip 8-4 to allow the deck length and width to be entered in feet and inches.
60. Modify Excel Quick Tip 8-6 to allow the span to be entered in feet and inches.

REFERENCE

1. *International Building Code*, International Code Council, 2014, 2308.4.4.

CHAPTER NINE

THERMAL AND MOISTURE PROTECTION

In this chapter you will learn how to apply the principles in Chapter 4 to waterproofing and dampproofing, building paper and vapor barriers, insulation, exterior insulation finish system (EIFS), shingles, siding, and membrane roofing. This chapter includes sample spreadsheets that may be used in the quantity takeoff. It also includes example takeoffs from the residential garage drawings in Appendix G.

This chapter deals with the insulation and exterior finishes that protect the building against heat loss and the infiltration of water. Let's begin by looking at waterproofing and dampproofing.

WATERPROOFING AND DAMPPROOFING

Waterproofing and dampproofing consist of a water-resistant coat that is sprayed or brushed onto the exterior of the foundation. Sometimes a waterproof membrane or layers of building paper are included in the waterproofing system. To increase the effectiveness of the waterproofing, a drain board may be placed on the outside of the waterproof membrane. The drain board allows water to travel freely in the vertical direction, thus minimizing the chance that it will penetrate the waterproof membrane. For a drain board to be effective, there must be a drainage system at the bottom of the drain board to prevent the accumulation of water.

Sprayed-on waterproofing is bid by the square foot or as a quantity-from-quantity good. Waterproof membranes and building paper is bid as a sheet or roll good. The quantity takeoff for sprayed-on waterproofing is shown in the following example.

EXAMPLE 9-1

Figure 9-1 shows the foundation plan and a wall section for a building. The entire exterior perimeter of the building is to be waterproofed with a sprayed-on membrane starting 6 inches from the top of the wall, continuing down the wall, and covering the top of the footing. How many square feet of waterproofing are needed for the building? If one gallon of waterproofing material covers 80 square feet, how many gallons of waterproofing material are needed?

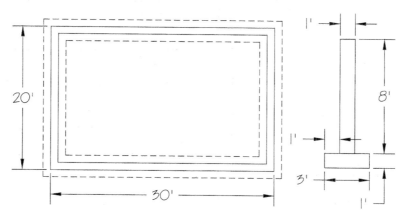

FIGURE 9-1 Foundation Plan and Section

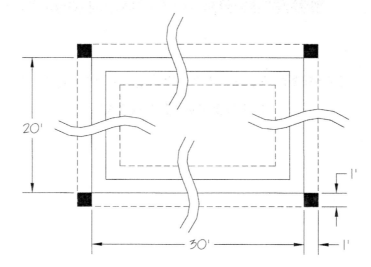

FIGURE 9-2 Corners

Solution: The perimeter of the building is 100 feet (20 ft + 30 ft + 20 ft + 30 ft). The height of the waterproofing is the height of the wall less 6 inches plus the width of the top of the footing on one side of the wall, or 8.5 feet (8 ft − 0.5 ft + 1 ft). The area of the waterproofing is 850 square feet (100 ft × 8.5 ft). To this, one needs to add the four 1-foot by 1-foot corners (shown in Figure 9-2) on top of the footing. The total area is 854 square feet. The number of gallons of waterproofing is determined using Eq. (4-20) as follows:

$$\text{Quantity} = \frac{854 \text{ sft}}{80 \text{ sft/gallon}} = 10.7 \text{ gallons}$$

EXCEL QUICK TIP 9-1
Waterproofing

The calculation of the area of the waterproofing is set up in a spreadsheet by entering the data and formatting the cells as follows:

	A	B	C
1	Wall Length 1	30	ft
2	Wall Length 2	20	ft
3	Wall Length 3	30	ft
4	Wall Length 4	20	ft
5	Wall Length 5	-	ft
6	Wall Length 6	-	ft
7	Wall Height	8	ft
8	Start __ Below Top of Wall	6	in
9	Width of Top of Footing	12	in
10	Extra Corners	4	ea
11	Coverage	80	sft/gal
12			
13	Perimeter	100	ft
14	Area of Waterproofing	854	sft
15	Gallons of Waterproofing	10.6	gal

The following formulas need to be entered into the associated cells:

Cell	Formula
B13	=SUM(B1:B6)
B14	=B12*(B7-B8/12+B9/12)+B10*(B9/12)*(B9/12)
B15	=B14/B11

The data for the foundation is entered in Cells B1 through B10. The data shown in the foregoing figure is from Example 9-1 and is formatted using the comma style, which replaces zeros with dashes.

BUILDING PAPER AND VAPOR BARRIERS

Building paper and vapor barriers are used behind masonry or siding, under shingles, and between the insulation and the inside finish of the exterior walls. Building paper and vapor barriers are bid as a rolled good and must be lapped to be effective. The quantity takeoff for building paper behind masonry or siding is shown in the following example.

EXAMPLE 9-2

Determine the number of rolls of 30-pound asphalt-impregnated felt needed to cover a wall 90 feet long by 8 feet high. The felt comes in rolls 3 feet wide with 2 squares (200 ft^2) per roll. The felt must be lapped 12 inches on the ends and 6 inches on the sides.

Solution: The number of rows of felt is calculated using Eq. (4-12) as follows:

$$\text{Number}_{\text{Rows}} = \frac{8 \text{ ft}}{(3 \text{ ft} - 0.5 \text{ ft})} = 3.2 \text{ rows}$$

Thermal and Moisture Protection

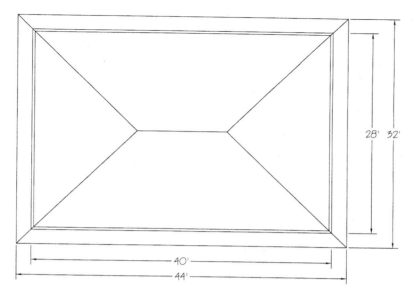

FIGURE 9-3 Roof Plan

Because the partial row is equal to the lap, the number of rows can be rounded down. Three rows of felt are needed. The first row will run from the bottom of the wall to 3 feet high, the second row will run from 2 feet 6 inches to 5 feet 6 inches, and the last row will run from 5 feet to 8 feet, the top of the wall. The length of a roll is 66 feet 8 inches (200 ft²/3 ft). The number of columns is calculated using Eq. (4-15) as follows:

$$\text{Number}_{\text{Columns}} = \frac{90 \text{ ft}}{(66.67 \text{ ft} - 1 \text{ ft})} = 1.4 \text{ columns}$$

The number of rolls needed for the wall is calculated using Eq. (4-18) as follows:

$$\text{Number} = (3 \text{ rows})(1.4 \text{ columns}) = 5 \text{ rolls}$$

When using asphalt-impregnated felt on roofs with a slope greater than or equal to 2:12 and less than 4:12, the felt must be lapped 19 inches. When using asphalt-impregnated felt on roofs with a slope greater than or equal to 4:12, the felt must be lapped 2 inches.[1] In areas of the country where there is the possibility of ice forming on the eaves, two layers of felt cemented together or self-adhering bitumen sheet (such as Ice and Water Shield) are required from the edge of the roof to a point 24 inches inside the exterior wall line of the building.[2] Local codes may increase this distance.

When taking off roofs, distances that are perpendicular to the slope of the roof can be measured directly off of the plans, whereas distances that are parallel to the slope of the roof must be converted from a plan view (horizontal) distance to a sloped distance. This is done using the following equation:

$$\text{Length}_{\text{Sloped}} = \text{Length}_{\text{Plan View}}(1 + \text{Slope}^2)^{0.5} \quad (9\text{-}1)$$

The plan view area is substituted into Eq. (9-1) to get the following equation, which is used to convert the plan view area into the actual surface area:

$$\text{Area}_{\text{Surface}} = \text{Area}_{\text{Plan View}}(1 + \text{Slope}^2)^{0.5} \quad (9\text{-}2)$$

Hips and valleys on roofs increase the waste and the amount of material needed to complete the roof. For the hip roof in Figure 9-3, the felt layout for the trapezoidal areas is shown in Figure 9-4. During installation, the dashed areas of the felt are folded over onto the triangular areas of the roof. A similar situation occurs on the triangular areas. This increases the amount of material needed to complete the roof.

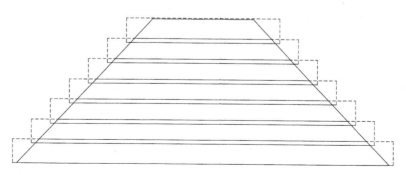

FIGURE 9-4 Felt Layout

To do a proper takeoff for a roof, each roof surface must be taken off individually. The felt folded over onto the adjoining roof surfaces is taken into account by increasing the width of the surface by the width of the roll of felt plus the required end lap. In the case of a 36-inch-wide roll of felt with a 12-inch end lap, the length of each row is increased by 4 feet (48 in). It is important to note that the increase in width is perpendicular to the length of the roll (parallel to the slope of the roof surface) rather than parallel to the hip or valley.

Rather than calculate the length of each of the rows, one can approximate the lengths by using the average length. Unless the top row of felt has sufficient material to be folded over the ridge and meet the lapping requirements, an additional piece of felt is needed for the ridge whose length equals the length of the ridge plus twice the required edge lap. The takeoff of felt is shown in the following example.

EXAMPLE 9-3

Determine the number of rolls of 15-pound asphalt-impregnated felt needed for the roof in Figure 9-3. The felt is to be lapped 6 inches on the side and 12 inches on the ends. The felt is available in rolls 3 feet wide with 4 squares (400 ft²) per roll. The slope on the roof is 6:12.

Solution: The sloped length of the roof from the eaves to the ridge is calculated using Eq. (9-1) as follows:

$$\text{Length}_{\text{Sloped}} = (16 \text{ ft})\left[1 + \left(\frac{6}{12}\right)^2\right]^{0.5} = 17.89 \text{ ft}$$

The number of rows is calculated using Eq. (4-12) as follows:

$$\text{Number}_{\text{Rows}} = \frac{17.89 \text{ ft}}{(3 \text{ ft} - 0.5 \text{ ft})} = 7.156 \text{ rows}$$

Because the width of the partial row is less than the width of the lap, the number of rows is rounded down to 7. The rows will run from the eave to 3 feet, from 2 feet 6 inches to 5 feet 6 inches, from 5 feet to 8 feet, from 7 feet 6 inches to 10 feet 6 inches, from 10 feet to 13 feet, from 12 feet 6 inches to 15 feet 6 inches, and from 15 feet to 18 feet, covering the side of the roof. The average length of a row for trapezoidal areas is the average of the length of the ridge and the eave plus the width of the felt plus the required end lap and is calculated as follows:

$$\text{Length}_{\text{Trapezoidal}} = \frac{(44 \text{ ft} + 12 \text{ ft})}{2} + 3 \text{ ft} + 1 \text{ ft} = 32 \text{ ft}$$

The area of felt needed for a trapezoidal area is equal to the average length multiplied by the number of rows multiplied by the width of the felt and is calculated as follows:

$$\text{Area}_{\text{Trapezoidal}} = (32 \text{ ft})(7 \text{ rows})(3 \text{ ft/row}) = 672 \text{ ft}^2$$

The average length of a row for the triangular areas is calculated as follows:

$$\text{Length}_{\text{Triangular}} = \frac{(32 \text{ ft} + 0 \text{ ft})}{2} + 3 \text{ ft} + 1 \text{ ft} = 20 \text{ ft}$$

The area of felt needed for a triangular area is calculated as follows:

$$\text{Area}_{\text{Triangular}} = (20 \text{ ft})(7 \text{ rows})(3 \text{ ft/row}) = 420 \text{ ft}^2$$

A 13-foot-long (12 ft + 2 × 0.5 ft) piece of felt is needed for the ridge. The area of this felt is 39 square feet (13 ft × 3 ft). The area of felt needed for the entire roof is calculated as follows:

$$\text{Area} = 2(\text{Area}_{\text{Trapezoidal}}) + 2(\text{Area}_{\text{Triangular}}) + \text{Area}_{\text{Ridge}}$$
$$= 2(672 \text{ ft}^2) + 2(420 \text{ ft}^2) + 39 \text{ ft}^2 = 2{,}223 \text{ ft}^2$$

The number of rolls of felt is calculated as follows:

$$\text{Rolls} = \frac{2{,}223 \text{ ft}^2}{(400 \text{ ft}^2/\text{roll})} = 6 \text{ rolls} \qquad ■$$

Alternatively, the quantity of felt is approximated by the following equation:

$$\text{Area}_{\text{Felt}} = \text{Area}_{\text{Roof}}(1 + \text{Slope}^2)^{0.5}\left[\frac{\text{Width}}{(\text{Width} - \text{Lap}_{\text{Side}})}\right]$$
$$+ \text{Length}_{\text{Hip/Valley}}\left[\frac{(1 + \text{Slope}^2)}{2}\right]^{0.5}$$
$$\times (\text{Width} + \text{Lap}_{\text{End}})\left[\frac{\text{Width}}{(\text{Width} - \text{Lap}_{\text{Side}})}\right]$$
$$+ [\text{Length}_{\text{Ridge}} + 2(\text{Ridges})(\text{Lap}_{\text{End}})]\,\text{Width}$$

(9-3)

where

$\text{Area}_{\text{Felt}}$ = Area of the Felt Needed for the Roof
$\text{Area}_{\text{Roof}}$ = Area of the Roof in Plan View
Slope = Slope of the Roof (Rise/Run)
Width = Width of the Felt
Lap_{Side} = Required Lap at the Sides of the Felt
$\text{Length}_{\text{Hip/Valley}}$ = Length of the Hips and Valleys in Plan View
Lap_{End} = Required Lap at the Ends of the Felt
$\text{Length}_{\text{Ridge}}$ = Length of Ridges
Ridges = Number of Ridges

The purposes of the various components of Eq. (9-3) are shown in Table 9-1.

EXAMPLE 9-4

Solve Example 9-3 using Eq. (9-3).

Solution: The lengths of the hips in plan view are calculated using the Pythagorean theorem as follows:

$$\text{Length} = [(16 \text{ ft})^2 + (16 \text{ ft})^2]^{0.5} = 22.62 \text{ ft}$$

The total length of the hips is 91 feet (4 × 22.62 ft). The quantity of felt needed is calculated as follows:

$$\text{Area}_{\text{Felt}} = (44 \text{ ft})(32 \text{ ft})[1 + (6/12)^2]^{0.5}\left[\frac{3 \text{ ft}}{(3 \text{ ft} - 0.5 \text{ ft})}\right]$$
$$+ (91 \text{ ft})\left[\frac{(1 + (6/12)^2)}{2}\right]^{0.5}(3 \text{ ft} + 1 \text{ ft})\left[\frac{3 \text{ ft}}{(3 \text{ ft} - 0.5 \text{ ft})}\right]$$
$$+ [12 \text{ ft} + 2(1)(1 \text{ ft})](3 \text{ ft})$$
$$= 2{,}276 \text{ ft}^2$$

The number of rolls of felt is calculated as follows:

$$\text{Rolls} = \frac{2{,}276 \text{ ft}^2}{(400 \text{ ft}^2/\text{roll})} = 6 \text{ rolls} \qquad ■$$

Thermal and Moisture Protection

TABLE 9-1 Components of Eq. (9-3)

Equation Component	Purpose
$(1 + \text{Slope}^2)^{0.5}$	Converts the area from the plan view to the actual area of the roof using Eq. (9-2)
$\left[\dfrac{\text{Width}}{(\text{Width} - \text{Lap}_{\text{Side}})}\right]$	Takes into account the lap on the side of the rolls
$\left[\dfrac{(1 + \text{Slope}^2)}{2}\right]^{0.5}$	Converts the length of hips and valleys in the plan view to the length along the roof surface as measured perpendicular to the felt
$(\text{Width} + \text{Lap}_{\text{End}})$	Takes into account the lap on the hips and valleys
$2(\text{Ridges})(\text{Lap}_{\text{End}})$	Takes into account the lap at the ends of the ridges

When Ice and Water Shield is used at the eaves, the number of rows of Ice and Water Shield required is determined. The Ice and Water Shield reduces the area covered by the felt. The felt must cover the area not covered by the Ice and Water Shield and must lap over the edge of the Ice and Water Shield. The felt needed to cover the remaining area is calculated using the procedures shown in Examples 9-3 and 9-4.

EXCEL QUICK TIP 9-2
Roofing Felt

The calculation of the number of rolls of roofing felt needed for a roof is set up in a spreadsheet by entering the data and formatting the cells as follows:

	A	B	C
1	Roof Length (Plan View)	44	ft
2	Roof Width (Plan View)	32	ft
3	Slope	6	:12
4	Underlayment Width	3	ft
5	Squares Per Roll	4	squ
6	Side Lap	6	in
7	End Lap	12	in
8	Total Length of Hips/Valleys	91	ft
9	Total Length of Ridges	12	ft
10	Number of Ridges	1	ea
11			
12	Area	2,276	sft
13	Rolls	6	ea

The following formulas need to be entered into the associated cells:

Cell	Formula
B12	`=B1*B2*(1+(B3/12)^2)^0.5*(B4/(B4-B6/12))` `+B8*((1+(B3/12)^2)/2)^0.5*(B4+B7/12)` `*(B4/(B4-B6/12))+(B9+2*B10*B7/12)*B4`
B13	`=ROUNDUP(B12/(B5*100),0)`

The data for roof is entered in Cells B1 through B10. The data shown in the foregoing figure is from Example 9-4.

INSULATION

The two most common types of insulation are fiberglass batt insulation and blown insulation. Batt insulation may be paperbacked or unfaced. Paperbacked insulation is installed with the paper on the interior side of the wall. Unfaced insulation requires a vapor barrier, such as a plastic sheet, to be placed on the interior side of exterior walls. Batt insulation is available in rolls or precut blankets packaged in bundles. Batt insulation comes in two common widths: 15 inches for use in $2 \times$ framing with a spacing of 16 inches on center and 23 inches for a spacing of 24 inches on center. The quantity of insulation needed for a wall or ceiling is bid as a sheet good with a 1-inch spacing between the sheets; a sample calculation is shown in the following example.

EXAMPLE 9-5

How many square feet of wall insulation are needed to insulate 90 feet of an 8-foot-high wall? The stud spacing is 16 inches on center. How many rolls of insulation are needed if there are 40 square feet per roll?

Solution: The number of rows is one row 8 feet high. The number of columns is calculated using Eq. (4-17) as follows:

$$\text{Number}_{\text{Columns}} = \frac{(90 \text{ ft})(12 \text{ in/ft})}{(15 \text{ in} + 1 \text{ in})} = 68 \text{ columns}$$

The number of sheets of 8-foot by 15-inch-wide insulation needed is 68. The area of insulation needed is equal to number of sheets multiplied by the area of the sheet, which is equal to the length of the sheet multiplied by the width of the sheet:

$$\text{Area} = (68)(8 \text{ ft})(15 \text{ in})\left(\frac{1 \text{ ft}}{12 \text{ in}}\right) = 680 \text{ ft}^2$$

The number of rolls is calculated as follows:

$$\text{Number} = \frac{680 \text{ ft}^2}{(40 \text{ ft}^2/\text{roll})} = 17 \text{ rolls}$$

Alternatively, the square footage of insulation for an insulated surface is approximated by the following equation:

$$\text{Area}_{\text{Insulation}} = (\text{Area}_{\text{Wall}})\left(\frac{\text{Width}}{\text{Spacing}}\right) \qquad (9\text{-}4)$$

where

Width = Width of the Insulation
Spacing = Spacing of the Framing Members

The width of the insulation should be 1 inch less than the stud spacing. The use of this equation is shown in the following example:

EXAMPLE 9-6

Solve Example 9-5 using Eq. (9-4).

Solution: The area of the wall is calculated as follows:

$$\text{Area}_\text{Walls} = (90 \text{ ft})(8 \text{ ft}) = 720 \text{ ft}^2$$

The area of the insulation is calculated using Eq. (9-4) as follows:

$$\text{Area}_\text{Insulation} = (720 \text{ ft}^2)\left(\frac{15 \text{ in}}{16 \text{ in}}\right) = 675 \text{ ft}^2$$

The number of rolls is calculated as follows:

$$\text{Number} = \frac{675 \text{ ft}^2}{(40 \text{ ft}^2/\text{roll})} = 17 \text{ rolls}$$

EXCEL QUICK TIP 9-3
Batt Insulation

The calculation of the number of rolls of insulation needed is set up in a spreadsheet by entering the data and formatting the cells as follows:

	A	B	C	D	E	F	G
1	Insulation Width	15	in				
2	Square Foot per Roll	40	sft				
3							
4			Rectangular Areas				
5	Number	1	ea		ea		ea
6	Width/Height	8	ft		ft		ft
7	Length	90	ft		ft		ft
8	Area	720	sft	-	sft	-	sft
9			Triangular Areas				
10	Number		ea		ea		ea
11	Base		ft		ft		ft
12	Height		ft		ft		ft
13	Area	-	sft	-	sft	-	sft
14							
15	Total Area	720	sft				
16	Area of Insulation	675	sft				
17	Rolls	17	ea				

The following formulas need to be entered into the associated cells:

Cell	Formula
B8	=B5*B6*B7
B13	=B10*B11*B12/2
B15	=B8+D8+F8+B13+D13+F13
B16	=B15*B1/(B1+1)
B17	=ROUNDUP(B16/B2,0)

Once these formulas have been entered, Cell B8 needs to be copied to Cells D8 and F8, and Cell B13 needs to be copied to Cells D13 and F13.

The data for insulation is entered in Cells B1 and B2, and the data for the insulated surface is entered in Cells B5 through B7, D5 through D7, F5 through F7, B10 through B12, D10 through D12, and F10 through F12. The data shown in the foregoing figure is from Example 9-6 and is formatted using the comma style, which replaces zeros with dashes.

Blown insulation may be made from fiberglass, shredded newspaper, or spun rock. Blown insulation is bid as a quantity-from-quantity good using historical data or data from the manufacturer on how many bags of insulation are required to cover 1,000 square feet. The quantity takeoff for blown insulation is shown in the following example.

EXAMPLE 9-7

How many bags of blown insulation are needed to insulate a 1,250-square foot ceiling with R-38 insulation? The manufacturer's recommended coverage rate is 58 bags per 1,000 square feet.

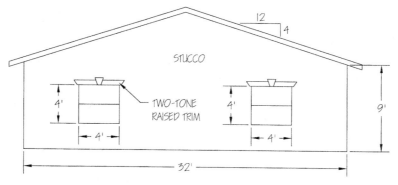

FIGURE 9-5 Building End

Solution: The quantity is calculated as follows:

$$\text{Bags} = (1{,}250 \text{ ft}^2)\left(\frac{58 \text{ bags}}{1{,}000 \text{ ft}^2}\right) = 73 \text{ bags}$$

EXTERIOR INSULATION FINISH SYSTEM

Exterior insulation finish system (EIFS) consists of a ridged insulation layer covered with a synthetic stucco system. The stucco system includes a base coat, a reinforcing layer, a brown coat, and a colored finish coat. The insulation is bid as a sheet good using the same principles as for bidding sheathing. The synthetic stucco system is bid as a quantity-from-quantity good using historical data or coverage rates from the manufacturer. Any specialty features, such as raised areas or areas of a different color, need to be accounted for. The quantity takeoff for EIFS is shown in the following example.

EXAMPLE 9-8

Determine the quantities of insulation, mesh reinforcement, and stucco needed to cover the end of the building shown in Figure 9-5. The insulation comes in 4-foot by 8-foot sheets. The reinforcement comes in rolls 100 feet long by 3 feet wide and needs to be lapped 3 inches. One cubic foot of base or brown coat will cover 75 square feet of wall, and one bucket of finish coat will cover 130 square feet of wall.

Solution: The rise for the gable end is calculated as follows:

$$\text{Rise} = \left(\frac{32 \text{ ft}}{2}\right)\left(\frac{4}{12}\right) = 5.33 \text{ ft}$$

The height of the wall at the peak is 14 feet 4 inches. The number of rows of insulation is calculated using Eq. (4-13) as follows:

$$\text{Number}_{\text{Rows}} = \frac{14.33 \text{ ft}}{8 \text{ ft}} = 2 \text{ rows}$$

The number of columns of insulation is calculated using Eq. (4-16) as follows:

$$\text{Number}_{\text{Columns}} = \frac{32 \text{ ft}}{4 \text{ ft}} = 8 \text{ columns}$$

The first row of insulation will run from the bottom of the wall to a height of 8 feet and will consist of eight sheets of insulation. The second row of insulation will run from a height of 8 feet to the top of the wall, which is 1 foot at the eaves and 6 feet 4 inches at the peak. Because the height at the eaves (1 ft) plus the height at the peak (6 ft 4 in) is less than 8 feet, the scrap from the short side of the gable end can be used at the peak, allowing two pieces of insulation to be cut from one sheet on the top row. Four sheets are required for the top row. The layout of the insulation is shown in Figure 9-6. The number on each piece of insulation represents the sheet from which the insulation was cut. A total of 12 sheets of insulation are needed.

The number of rows of reinforcement is calculated using Eq. (4-12) as follows:

$$\text{Number}_{\text{Rows}} = \frac{14.33 \text{ ft}}{(3 \text{ ft} - 0.25 \text{ ft})} = 6 \text{ rows}$$

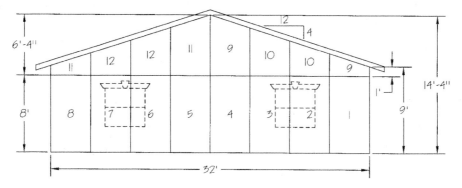

FIGURE 9-6 Insulation Layout

The first row will run from the bottom of the wall to 3 feet high, the second row will run from 2 feet 9 inches to 5 feet 9 inches, the third row will run from 5 feet 6 inches to 8 feet 6 inches, the fourth row will run from 8 feet 3 inches to 11 feet 3 inches, the fifth row will run from 11 feet to 14 feet, and the last row will run from 13 feet 9 inches to 14 feet 4 inches. The first four rows will need to be 32 feet wide. For the gable ends, the rows are 3 feet shorter on each end for every foot above the point where the gable end begins to slope. The fifth row, which is 2 feet above the point where the roof begins to slope, will need to be 20 feet wide (32 ft − 2 × 2 ft × 3 ft/ft). The last row can be cut from the scraps. The length of reinforcement needed is calculated as follows:

$$\text{Length} = 4(32 \text{ ft}) + 1(20 \text{ ft}) = 148 \text{ ft}$$

The number of rolls is calculated as follows:

$$\text{Rolls} = \frac{148 \text{ ft}}{(100 \text{ ft/roll})} = 2 \text{ rolls}$$

The area of the wall, exclusive of the windows, is calculated as follows:

$$\text{Area} = (32 \text{ ft})(9 \text{ ft}) + 0.5(32 \text{ ft})(5.33 \text{ ft}) - 2(4 \text{ ft})(4 \text{ ft}) = 341 \text{ ft}^2$$

The quantity of the base coat is calculated using Eq. (4-20) as follows:

$$\text{Quantity} = \frac{341 \text{ ft}^2}{(75 \text{ ft}^2/\text{ft}^3)} = 4.6 \text{ ft}^3$$

The quantity of the brown coat is calculated using Eq. (4-20) as follows:

$$\text{Quantity} = \frac{341 \text{ ft}^2}{(75 \text{ ft}^2/\text{ft}^3)} = 4.6 \text{ ft}^3$$

The quantity of the finish coat is calculated using Eq. (4-20) as follows:

$$\text{Quantity} = \frac{341 \text{ ft}^2}{(130 \text{ ft}^2/\text{bucket})} = 3 \text{ buckets}$$

In addition, there are two trim pieces, which will require an additional bucket of finish coat that is a different color. ■

SHINGLE ROOFS

Asphalt shingles are often specified by their expected life, with 20-, 25-, and 30-year shingles being the most common. Asphalt shingles may be three-tab or architectural-grade shingles. Flashings are required at all penetrations (such as vent pipes) and where the roof intersects a wall or chimney. Drip edge is required at all edges of the roof, including edges on a gable end.

Shingle roofs consist of underlayment (building paper and Ice and Water Shield), a starter row of shingles located at the eaves of the roof, standard shingles placed over most of the surface of the roof, and cap shingles placed on the hips and ridges. The standard exposure for shingles is 5 inches, or in other words, the shingles are lapped so that the rows of shingles are 5 inches apart. Shingles are bid by the square, which equals the number of shingles needed to cover 100 square feet of roofing at a specified exposure.

Starter rows of shingles consist of shingles with the area of the shingle that is normally exposed cut off, and as a result they are 5 inches narrower than a standard shingle and are entirely covered by the next row of shingles. For some shingle types, a special shingle is available for use as the starter row. Starter rows are required for all roof edges that are parallel to the rows of shingles. The gable ends of gable roofs do not require a starter row; however, all the edges of a hip roof require a starter row. The quantity of shingles required for the starter row is equal to the length of the edge multiplied by the exposure of the shingle. This result is divided by 100 square feet to convert the quantity into squares. The quantity of shingles for the starter row is calculated using the following equation:

$$\text{Squares}_{\text{Starter}} = \frac{(\text{Exposure})(\text{Length})}{(100 \text{ ft}^2/\text{square})} \quad (9\text{-}5)$$

The quantity of field shingles for a gable roof is equal to the area of the roof in squares (100 ft²). When determining the area of the roof, the estimator must take the slope of the roof into account as was done with sheathing. This is done with the following equation:

$$\text{Squares}_{\text{Field}} = \frac{(\text{Length})(\text{Width})(1 + \text{Slope}^2)^{0.5}}{(100 \text{ ft}^2/\text{square})} \quad (9\text{-}6)$$

where

Length = Length of the Roof in Plan View
Width = Width of the Roof in Plan View
Slope = Rise/Run (4/12 for a 4:12 slope)

Substituting the plan view area for length and width into Eq. (9-6), we get the following equation:

$$\text{Squares}_{\text{Field}} = \frac{(\text{Area})(1 + \text{Slope}^2)^{0.5}}{(100 \text{ ft}^2/\text{square})} \quad (9\text{-}7)$$

The area of a hip roof is calculated using Eq. (9-6) or Eq. (9-7) just as for the area of a gable roof. Although a hip roof has the same area as a gable roof with the same slope and dimensions, the presence of hips and valleys increases the amount of material needed to cover a roof because they are required to overlap at the hips and valleys. Therefore, a building covered with a hip roof will require more materials than the same building covered with a gabled roof with the same slope, even though the roof areas are the same. The added material needed for the hips and valleys in a hip roof is approximated by the following equation:

$$\text{Squares}_{\text{Hip/Valley}} = \frac{\text{Length}_{\text{Hip/Valley}}\left[\frac{(1 + \text{Slope}^2)}{2}\right]^{0.5}(\text{Exposure})}{(100 \text{ ft}^2/\text{square})} \quad (9\text{-}8)$$

where

$\text{Length}_{\text{Hip/Valley}}$ = Length of the Hips and Valleys in Plan View

Cap shingles are shingles 1 foot wide and are used to cover hips and ridges. The quantity of cap shingles is equal to the length of the hips and ridges times the width (1 ft)

divided by 100 square feet to convert the quantity into squares. Ridges are horizontal and can be measured directly from the plans. The quantity of ridge shingles needed for a ridge is calculated using the following equation:

$$\text{Squares}_{\text{Ridge}} = \frac{(\text{Length}_{\text{Ridge}})(1\text{ ft})}{(100\text{ ft}^2/\text{square})} \quad (9\text{-}9)$$

Because hips are sloped, the slope of the hip must be taken into account when calculating the quantity of shingles needed for the hip and is calculated using the following equation:

$$\text{Squares}_{\text{Hip}} = \frac{(\text{Length}_{\text{Hip}})(1 + \text{Slope}_{\text{Hip}}^2)^{0.5}(1\text{ ft})}{(100\text{ ft}^2/\text{square})} \quad (9\text{-}10)$$

where

$\text{Length}_{\text{Hip}}$ = Length of the Hip in Plan View
$\text{Slope}_{\text{Hip}}$ = Rise/Run (4/17 for a 4:12 slope)

Note that the slope is divided by 17 rather than 12 because the slope of the hip rafter is flatter than the slope of the roof as discussed in Chapter 8.

The amount of shingles needed is calculated by summing the shingles needed for the starter row, the field, the waste at the hips and valleys, and the cap shingles placed on the ridges and hips. The quantity takeoff for a hip roof is shown in the following example.

EXAMPLE 9-9

Determine the number of squares of shingles needed for the hip roof shown in Figure 9-3. The shingles have a 5-inch exposure and the slope of the roof is 6:12. The cap and starter strip will be cut from a standard shingle.

Solution: From Example 9-4, the length of the hips in plan view is 91 feet. The quantity of shingles needed for the starter row is calculated using Eq. (9-5) as follows:

$$\text{Squares}_{\text{Starter}} = \frac{(5\text{ in})(1\text{ ft}/12\text{ in})(32\text{ ft} + 44\text{ ft} + 32\text{ ft} + 44\text{ ft})}{(100\text{ ft}^2/\text{square})}$$
$$= 0.63\text{ square}$$

The quantity of shingles needed for the field is calculated using Eq. (9-6) as follows:

$$\text{Squares}_{\text{Field}} = \frac{(44\text{ ft})(32\text{ ft})(1 + (6/12)^2)^{0.5}}{(100\text{ ft}^2/\text{square})} = 15.74\text{ squares}$$

The quantity of shingles needed for the hips is calculated using Eq. (9-8) as follows:

$$\text{Squares}_{\text{Hip/Valley}} = \frac{(91\text{ ft})\left[\frac{(1 + (6/12)^2)}{2}\right]^{0.5}(5\text{ in})(1\text{ ft}/12\text{ in})}{(100\text{ ft}^2/\text{square})}$$
$$= 0.30\text{ square}$$

The quantity of cap shingles needed for the ridge is calculated using Eq. (9-9) as follows:

$$\text{Squares}_{\text{Ridge}} = \frac{(12)(1\text{ ft})}{(100\text{ ft}^2/\text{square})} = 0.12\text{ square}$$

The quantity of cap shingles needed for the hips is calculated using Eq. (9-10) as follows:

$$\text{Squares}_{\text{Hip}} = \frac{(91\text{ ft})[1 + (6/17)^2]^{0.5}(1\text{ ft})}{(100\text{ ft}^2/\text{square})} = 0.97\text{ square}$$

The total number of squares is calculated as follows:

$$\text{Squares} = 0.63\text{ square} + 15.74\text{ squares} + 0.30\text{ square}$$
$$+ 0.12\text{ square} + 0.97\text{ square}$$
$$= 17.76\text{ squares}$$

EXCEL QUICK TIP 9-4
Roofing

The calculation of the number of squares of shingles needed for a roof is set up in a spreadsheet by entering the data and formatting the cells as follows:

	A	B	C
1	Roof Length (Plan View)	44	ft
2	Roof Width (Plan View)	32	ft
3	Slope	6	:12
4	Shingle Exposure	5	in
5	Length of Starter Row	152	ft
6	Total Length of Hips/Valleys	91	ft
7	Total Length of Ridges	12	ft
8			
9	Starter Strip	0.63	squ
10	Field Shingles	15.74	squ
11	Hip/Valley Shingles	0.30	squ
12	Ridge Cap Shingles	0.12	squ
13	Hip/Valley Cap Shingles	0.97	squ
14	Total Shingles	17.76	squ

The following formulas need to be entered into the associated cells:

Cell	Formula
B9	=(B4/12)*B5/100
B10	=B1*B2*(1+(B3/12)^2)^0.5/100
B11	=B6*((1+(B3/12)^2)/2)^0.5*(B4/12)/100
B12	=B7/100
B13	=B6*(1+(B3/17)^2)^0.5/100
B14	=SUM(B9:B13)

The data for the roof is entered in Cells B1 and B7. The data shown in the foregoing figure is from Example 9-9.

The drip edge is bid as a linear component. When calculating the lineal feet of drip edge needed for the roof, the slope of the roof must be taken into account on all gable ends. This is done using Eq. (9-1). The calculation of the lineal feet of drip edge is shown in the following example.

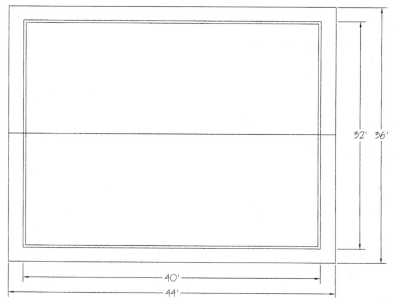

FIGURE 9-7 Roof Plan

EXAMPLE 9-10

Determine the number of lineal feet of drip edge needed for the roof in Figure 9-7. The slope of the roof is 4:12. The drip edge comes in 10-foot lengths and needs to be lapped 2 inches.

Solution: The length of the gable ends of the roof is calculated using Eq. (9-1) as follows:

$$\text{Length}_{\text{Sloped}} = \left(\frac{36 \text{ ft}}{2}\right)\left[1 + \left(\frac{4}{12}\right)^2\right]^{0.5} = 19 \text{ ft}$$

The total length is calculated as follows:

$$\text{Length} = 4(19 \text{ ft}) + 2(44 \text{ ft}) = 164 \text{ ft}$$

The number of pieces of drip edge is calculated using Eq. (4-6) as follows:

$$\text{Number} = \frac{164 \text{ ft}}{[10 \text{ ft} - (2 \text{ in})(1 \text{ ft}/12 \text{ in})]} = 17 \text{ each}$$

SIDING, SOFFIT, AND FASCIA

Siding includes vinyl and aluminum siding, wood board siding, and wood sheet siding (such as T1-11). Let's begin by looking at vinyl and aluminum siding.

Vinyl and aluminum siding consists of a starter strip located at the bottom of the wall; outside corner pieces on the outside corners; J-molding around windows and doors, on the inside corners, and at the top of the wall; and long, narrow siding panels. Aluminum soffit and fascia consist of an L-shaped piece of aluminum used for the fascia, a J-molding located on the wall side of the soffit, and narrow soffit panels, which may be vented or nonvented. The siding panels are designated by their width of coverage; for example, a 10-inch panel is wider than 10 inches because it includes the material that goes under the next panel, but it covers a 10-inch wide swath of wall. When performing the quantity takeoff, one measures the starter strip, outside corners, and the J-molding off the plans. The number of squares of siding needed for rectangular areas is the width times the height. For triangular and trapezoidal areas, additional siding needs to be added to the width of the area to account for waste at the ends. This is done by the following equation:

$$\text{Area} = \left[\left(\frac{\text{Length}_{\text{Top}} + \text{Length}_{\text{Bottom}}}{2}\right) + \frac{\text{Exposure}}{\text{Slope}}\right]\text{Height}$$
(9-11)

where

$\text{Length}_{\text{Top}}$ = Length at the Top of the Area
$\text{Length}_{\text{Bottom}}$ = Length at the Bottom of the Area
Exposure = Exposure of the Siding
Slope = Slope (Rise/Run) of Triangular or Trapezoidal Area (for example, 4/12)
Height = Height of the Area

The siding is taken off by the square foot using the same procedures as for taking off shingles. Soffit and fascia are taken off using the same procedures as for taking off sheathing for soffit and fascia (see Chapter 8).

When the area of the openings, such as windows, is small they are often ignored. To deduct an opening, the number of rows of siding is calculated, and two rows are subtracted for the top and bottom of the opening. The number of rows is then multiplied by the width of the opening. The takeoff of vinyl and aluminum siding is shown in the following example.

EXAMPLE 9-11

Determine the quantities of starter strip, outside corners, J-moldings, and 10-inch-wide siding needed to cover the end of the building shown in Figure 9-8. The starter strip, outside

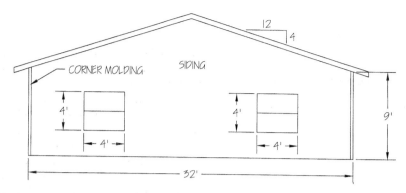

FIGURE 9-8 Building End

corners, and J-moldings are available in 10-foot lengths. The siding comes in packages of 200 square feet. When calculating the siding, do not deduct the area of the windows.

Solution: Four 10-foot lengths of starter strip and two 10-foot lengths of outside corner are needed. J-molding is required around the windows and at the top of the wall. The length of the J-molding at the top of the wall is calculated using Eq. (9-1) as follows:

$$\text{Length}_{\text{Sloped}} = \left(\frac{32 \text{ ft}}{2}\right)\left[1 + \left(\frac{4}{12}\right)^2\right]^{0.5} = 16.9 \text{ ft}$$

To minimize splicing, order two 10-foot lengths for each side of the roof and two 10-foot lengths for each window, for a total of eight 10-foot lengths. The rise for the gable end is calculated as follows:

$$\text{Rise} = \left(\frac{32 \text{ ft}}{2}\right)\left(\frac{4}{12}\right) = 5.33 \text{ ft}$$

Divide the wall into two areas: a 32-foot-wide by 9-foot-high rectangular area and a 32-foot-wide by 5.33-foot-high triangular area. The area of the rectangular area is calculated as follows:

$$\text{Area} = (32 \text{ ft})(9 \text{ ft}) = 288 \text{ ft}^2$$

The siding has an exposure of 10 inches, or 0.833 ft. The triangular area is calculated using Eq. (9-11) as follows:

$$\text{Area} = \left[\frac{(0 \text{ ft} + 32 \text{ ft})}{2} + \frac{(0.833 \text{ ft})}{(4/12)}\right](5.33 \text{ ft}) = 99 \text{ ft}^2$$

The total area is 387 square feet (288 ft² + 99 ft²). The number of packages of siding is calculated as follows:

$$\text{Number} = \frac{387 \text{ ft}^2}{(200 \text{ ft}^2/\text{package})} = 2 \text{ packages}$$

EXCEL QUICK TIP 9-5
Siding

The calculation of the number of square feet of siding needed for a wall is set up in a spreadsheet by entering the data and formatting the cells as follows:

	A	B	C	D	E	F	G
1	Exposure	10	in				
2	Slope	4	:12				
3							
4			Rectangular Areas				
5	Number	1	ea		ea		ea
6	Width/Height	9.00	ft		ft		ft
7	Length	32.00	ft		ft		ft
8	Area	288	sft	-	sft	-	sft
9			Triangular Areas				
10	Number	1	ea		ea		ea
11	Base	32.00	ft		ft		ft
12	Height	5.33	ft	-	ft	-	ft
13	Area	99	sft	-	sft	-	sft
14							
15	Total Area	387	sft				

The following formulas need to be entered into the associated cells:

Cell	Formula
B8	=B5*B6*B7
B12	=(B11/2)*$B$2/12
B13	=B10*(B11/2+$B$1/$B$2)*B12
B15	=B8+D8+F8+B13+D13+F13

Once these formulas have been entered, Cell B8 needs to be copied to Cells D8 and F8; Cell B12 needs to be copied to Cells D12 and F12; and Cell B13 needs to be copied to Cells D13 and F13.

The exposure for the siding is entered in Cell B1, the slope is entered into Cell B3, and the data for the wall is entered in Cells B5 through B7, D5 through D7, F5 through F7, B10, B11, D10, D11, F10, and F11. The data shown in the foregoing figure is from Example 9-11 and is formatted using the comma style, which replaces zeros with dashes.

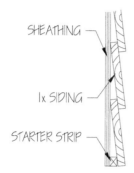

FIGURE 9-9 Board Siding Section

Board siding consists of wood boards (such as 1 × 8s) or engineered siding (such as Hardiplank). A section for wood board siding is shown in Figure 9-9.

Board siding requires a starter strip and wood trim on the corners, around the doors and windows, and at the tops of the wall. When estimating wood board siding, the ends are butt jointed and the sides of the board are lapped, greatly reducing the coverage of the board. For example, a 1 × 8 board siding that is lapped 1 inch has an exposure of 6.25 inches because the 1 × 8 is 7.25 inches wide and 1 inch is lost because of lapping. The nominal and actual sizes of wood siding are shown in Table 9-2.

When performing the quantity takeoff, one measures the starter strip and trim off of the plans, as discussed in Chapter 8. The siding is taken off as a sheet good. When dealing with triangular and trapezoidal areas waste must be accounted for by determining the length of the boards before their ends are cut to match the slope. The average length of the boards can be approximated by using the following equation:

$$\text{Length}_{Ave} = \frac{(\text{Length}_{Top} + \text{Length}_{Bottom})}{2} + \frac{\text{Exposure}}{\text{Slope}}$$

(9-12)

where

Length_{Top} = Length at the Top of the Area
Length_{Bottom} = Length at the Bottom of the Area
Exposure = Exposure of the Siding
Slope = Slope (Rise/Run) of Triangular or Trapezoidal Area (for example, 4/12)

Openings are handled in the same way as they were for vinyl and aluminum siding. Soffit and fascia are taken off using the same procedure as for taking off sheathing for soffit and fascia (see Chapter 8). The takeoff of wood board siding is shown in the following example.

EXAMPLE 9-12

Determine the quantities of starter strip, 1 × 4 trim, and 1 × 8 wood siding needed to cover the end of the building shown in Figure 9-10. The siding is lapped 1 inch on the sides and the ends are butt jointed. The starter strip and 1 × 4 trim are available in 8-, 10-, and 12-foot lengths. The siding comes in 12-foot lengths. When calculating the siding, do not deduct the area of the windows.

Solution: Four 8-foot lengths of starter strip are needed. The exposure of the siding is 6.25 inches. The rise for the gable end is calculated as follows:

$$\text{Rise} = \left(\frac{32 \text{ ft}}{2}\right)\left(\frac{4}{12}\right) = 5.33 \text{ ft}$$

The height at the peak of the gable end is 14.33 feet (9 ft + 5.33 ft). The number of 32-foot-long rows needed for the rectangular area is calculated using Eq. (4-12) as follows:

$$\text{Number}_{Rows} = \frac{(9 \text{ ft})(12 \text{ in/ft})}{(7.25 \text{ in} - 1 \text{ in})} = 18 \text{ rows}$$

The number of columns is calculated using Eq. (4-16) as follows:

$$\text{Number}_{Columns} = \frac{32 \text{ ft}}{12 \text{ ft}} = 2.67 \text{ columns}$$

TABLE 9-2 Wood Siding Sizes

Nominal Size	Actual Size
1″ × 4″	3/4″ × 3-1/2″
1″ × 5″	3/4″ × 4-1/2″
1″ × 6″	3/4″ × 5-1/2″
1″ × 7″	3/4″ × 6-1/4″
1″ × 8″	3/4″ × 7-1/4″
1″ × 10″	3/4″ × 9-1/4″
1″ × 12″	3/4″ × 11-1/4″

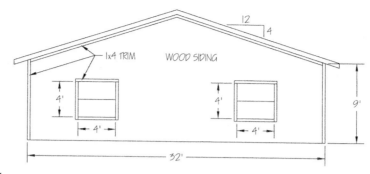

FIGURE 9-10 Building End

The number of pieces of siding needed for the rectangular area is calculated using Eq. (4-18) as follows:

$$\text{Number} = (18 \text{ rows})(2.67 \text{ columns}) = 48 \text{ each}$$

The average length of each of the rows in the triangular area is calculated using Eq. (9-12) as follows:

$$\text{Length}_{Ave} = \frac{(0 \text{ ft} + 32 \text{ ft})}{2} + \frac{(7.25 \text{ in} - 1 \text{ in})(1 \text{ ft}/12 \text{ in})}{(4/12)}$$
$$= 17.6 \text{ ft}$$

The number of rows needed for the entire end is calculated using Eq. (4-12) as follows:

$$\text{Number}_{Rows} = \frac{(14.33 \text{ ft})(12 \text{ in/ft})}{(7.25 \text{ in} - 1 \text{ in})} = 28 \text{ rows}$$

Therefore, 10 rows (28 − 18) of siding are needed for the gable end. The average number of columns is calculated using Eq. (4-16) as follows:

$$\text{Number}_{Columns} = \frac{17.6 \text{ ft}}{12 \text{ ft}} = 1.47 \text{ columns}$$

The number of pieces of siding needed for the triangular area is calculated using Eq. (4-18) as follows:

$$\text{Number} = (10 \text{ rows})(1.47 \text{ columns}) = 15 \text{ each}$$

Order sixty-three (48 + 15) 12-foot-long 1 × 8s. Two 10-foot lengths of 1 × 4s are needed for the corners. A 4-foot 7-inch piece of trim is needed for the top of the window, two 4-foot 3.5-inch pieces of trim are needed for the sides of the windows, and a 4-foot piece of trim is needed for the bottom of the window. Order three 10-foot and one 8-foot lengths of 1 × 4s for the windows. The length of the top of the wall is calculated using Eq. (9-1) as follows:

$$\text{Length}_{Sloped} = \left(\frac{32 \text{ ft}}{2}\right)\left[1 + \left(\frac{4}{12}\right)^2\right]^{0.5} = 16.9 \text{ ft}$$

Order an 8-foot length and a 10-foot length for each side of the roof. A total of seven 10-foot lengths and three 8-foot lengths of 1 × 4s are needed.

Sheets of wood siding require wood trim on the corners, around the windows and doors, and at the top of the wall. In performing the quantity takeoff, trim is measured off of the plans, the siding is taken off using the same procedure as for taking off sheets of sheathing, and the soffit and fascia are taken off using the same procedures as for taking off sheathing for soffit and fascia (see Chapter 8).

MEMBRANE ROOFING

Membrane roofing consists of a single-ply polyvinyl chloride (PVC), ethylene propylene diene monomer (EPDM), or other types of membrane that is used to cover a roof. Membrane roofing comes in sheets or rolls; for example, a PVC membrane is available in 75-foot-wide by 300-foot-long sheets folded up on a pallet. The sheets must be glued or welded together to form a single membrane. The membrane must be fastened at the edges of the roof using a batten strip and flashing to prevent water from getting between the membrane and the wall. The membrane must be fastened to the roof deck or held in place with rock ballast. Flashings are required on all penetrations. The membrane is taken off as a sheet or roll good, and the batten strip is taken off as a linear good. The takeoff of membrane roofing is shown in the following example.

EXAMPLE 9-13

Determine the number of 25-foot-wide by 100-foot-long sheets of membrane roofing needed for a roof 110 feet wide by 210 feet long. Allow 3 feet at the perimeter of the roof for the membrane to be attached to the parapet wall. The roofing must be lapped 3 inches at the sides and ends.

Solution: The membrane needs to cover an area 116 feet (110 ft + 2 × 3 ft) wide by 216 feet (210 ft + 2 × 3 ft) long. The number of rows needed for the roof is calculated using Eq. (4-12) as follows:

$$\text{Number}_{Rows} = \frac{116 \text{ ft}}{[25 \text{ ft} - 3 \text{ in}(1 \text{ ft}/12 \text{ in})]} = 5 \text{ rows}$$

The number of columns is calculated using Eq. (4-15) as follows:

$$\text{Number}_{Columns} = \frac{216 \text{ ft}}{[100 \text{ ft} - 3 \text{ in}(1 \text{ ft}/12 \text{ in})]} = 2.2 \text{ columns}$$

The number of sheets of roofing needed is calculated using Eq. (4-18) as follows:

$$\text{Number} = (5 \text{ rows})(2.2 \text{ columns}) = 11 \text{ sheets}$$

SAMPLE TAKEOFF FOR THE RESIDENTIAL GARAGE

A sample takeoff for insulation, wood siding, and roofing from a set of plans is shown in the following example.

EXAMPLE 9-14

Determine the insulation, wood siding (including trim), and roofing (including underlayment) needed to complete the residential garage given in Appendix G.

Solution: From Example 8-21 there is 83 feet of 8-foot-high wall. There is also 16 feet of 2-foot-high wall above the garage door, but only 1 foot can be insulated. For the 8-foot-high wall, there is one 8-foot-high row of insulation. The number of columns is calculated using Eq. (4-17) as follows:

$$\text{Number}_{\text{Columns}} = \frac{(83 \text{ ft})(12 \text{ in/ft})}{(15 \text{ in} + 1 \text{ in})} = 63 \text{ columns}$$

The number of 1-foot-high columns needed above the garage is calculated using Eq. (4-17) as follows:

$$\text{Number}_{\text{Columns}} = \frac{(16 \text{ ft})(12 \text{ in/ft})}{(15 \text{ in} + 1 \text{ in})} = 12 \text{ columns}$$

Two columns of insulation 7 feet high (the height of the door and header) can be deducted for the single-hung door. The area of insulation is calculated as follows:

$$\text{Area} = [63(8 \text{ ft}) + 12(1 \text{ ft}) - 2(7 \text{ ft})](15 \text{ in})(1 \text{ ft}/12 \text{ in}) = 628 \text{ ft}^2$$

The number of rolls is calculated as follows:

$$\text{Number} = \frac{628 \text{ ft}^2}{(40 \text{ ft}^2/\text{roll})} = 16 \text{ rolls}$$

A vapor barrier is needed on the unfaced insulation. Use 8-foot by 12-foot sheets of 3-mil-thick plastic. One 8-foot-high row is needed. The number of columns is calculated using Eq. (4-16) as follows:

$$\text{Number}_{\text{Columns}} = \frac{83 \text{ ft}}{12 \text{ ft}} = 7 \text{ columns}$$

Seven sheets are needed for the walls, and one additional sheet is needed above the garage door.

The area of the ceiling to be insulated is 26 feet wide by 24 feet long, or 624 square feet. The insulation comes in 32-foot-long rolls. The number of rows is calculated using Eq. (4-13) as follows:

$$\text{Number}_{\text{Rows}} = \frac{24 \text{ ft}}{32 \text{ ft}} = 0.75 \text{ row}$$

The number of columns is calculated using Eq. (4-17) as follows:

$$\text{Number}_{\text{Columns}} = \frac{(26 \text{ ft})(12 \text{ in/ft})}{(23 \text{ in} + 1 \text{ in})} = 13 \text{ columns}$$

The number of rolls of insulation needed for the ceiling is calculated using Eq. (4-18) as follows:

$$\text{Number} = (0.75 \text{ row})(13 \text{ columns}) = 10 \text{ rolls}$$

The area of insulation is calculated as follows:

$$\text{Area} = (13 \text{ columns})(23 \text{ in})(1 \text{ ft}/12 \text{ in})(24 \text{ ft}) = 598 \text{ ft}^2$$

Next, determine the number of sheets of T1-11 siding needed for the garage. The siding will be broken down into three areas: the 8-foot-high walls, the 2-foot-high wall above the garage door, and the gable ends. From Example 8-21 there is 83 feet of 8-foot-high wall. The rise for the gable end is calculated as follows:

$$\text{Rise} = \left(\frac{24 \text{ ft}}{2}\right)\left(\frac{4}{12}\right) = 4 \text{ ft}$$

The area of siding needed to be installed is as follows:

$$\text{Area} = \text{Area}_{\text{8-Foot Walls}} + \text{Area}_{\text{Above Garage Door}} + 2(\text{Area}_{\text{Gable End}})$$
$$= (8 \text{ ft})(83 \text{ ft}) + (2 \text{ ft})(16 \text{ ft}) + 2\left(\frac{(4 \text{ ft})(24 \text{ ft})}{2}\right) = 792 \text{ ft}^2$$

The number of rows of T1-11 needed for the 8-foot-high walls is one, and the number of columns is calculated using Eq. (4-16) as follows:

$$\text{Number}_{\text{Columns}} = \frac{83 \text{ ft}}{4 \text{ ft}} = 21 \text{ columns}$$

Twenty-one sheets are needed for the 8-foot-high wall. Four 2-foot-high by 4-foot-wide sheets are needed above the garage door and can be cut from one sheet of T1-11. The number of columns on a gable end is calculated using Eq. (4-16) as follows:

$$\text{Number}_{\text{Columns}} = \frac{24 \text{ ft}}{4 \text{ ft}} = 6 \text{ columns}$$

The three columns on each side of the peak can be cut from one sheet; thus, four sheets are required. A total of 26 sheets (21 + 1 + 4) of T1-11 are needed.

Next, the T1-11 for the soffit will be calculated. The soffit is 18 inches wide. Determine the length of the soffit, making sure that the corners are not double counted. The sloped length of the roof from the eave to the ridge is calculated using Eq. (9-1) as follows:

$$\text{Length}_{\text{Sloped}} = \left(\frac{27 \text{ ft}}{2}\right)\left[1 + \left(\frac{4}{12}\right)^2\right]^{0.5} = 14.2 \text{ ft}$$

The length of the soffit is calculated as follows:

$$\text{Length} = 4(14.2 \text{ ft}) + 2(26 \text{ ft}) = 109 \text{ ft}$$

Five 18-inch-wide by 4-foot-long pieces of soffit can be cut from one sheet of T1-11. The number of sheets is calculated as follows:

$$\text{Number} = \frac{109 \text{ ft}}{(5 \times 4 \text{ ft})} = 6 \text{ sheets}$$

The area of the soffit is calculated as follows:

$$\text{Area} = (1.5 \text{ ft})(109 \text{ ft}) = 164 \text{ ft}^2$$

In addition, 58 feet (2 × 29 ft) of 3-inch-wide screen and two 12-inch-wide by 14-inch-high vents are needed.

The amount of 1 × 4 cedar trim needed, along with its use, is shown in Table 9-3. The sloped trim will be cut in the field. The trim will be ordered in 8-, 10-, and 12-foot lengths. The total length of sloped trim that needs to be cut and installed is as follows:

$$\text{Length} = 4 \text{ ft} + 17 \text{ ft} + 2(24 \text{ ft}) + 2(2 \text{ ft}) + 2(1 \text{ ft}) = 75 \text{ ft}$$

The total length of unsloped trim that needs to be installed is as follows:

$$\text{Length} = 8(9 \text{ ft}) + 2(7 \text{ ft}) + 2(7 \text{ ft}) + 2(26 \text{ ft}) + 4(13 \text{ ft}) + 4(2 \text{ ft})$$
$$= 212 \text{ ft}$$

The perimeter of the roof is calculated as follows:

$$\text{Perimeter} = 4(14.2 \text{ ft}) + 2(29 \text{ ft}) = 115 \text{ ft}$$

The area of the roof is calculated as follows:

$$\text{Area} = 2(14.2 \text{ ft})(29 \text{ ft}) = 824 \text{ ft}^2$$

Determine the number of pieces of 10-foot-long drip edge, allowing 0.2 foot for lap, using Eq. (4-6) as follows:

$$\text{Number} = \frac{115 \text{ ft}}{(10.0 \text{ ft} - 0.2 \text{ ft})} = 12 \text{ ea}$$

TABLE 9-3 1 × 4 Cedar Trim

	Actual		Order		
Use	Quantity (ea)	Length (ft)	Quantity (ea)	Length (ft)	Sloped
Corners	8	9	8	10	No
Single-hung door (sides)	2	7	2	8	No
Single-hung door (top)	1	4	0.5	8	Yes
Overhead door (sides)	2	7	2	8	No
Overhead door (top)	1	17	1	8	Yes
			1	10	
Top of side walls	2	26	4	8	No
			2	10	
Top of end walls	4	13	4	8	No
			2	10	
Gable ends at seam	2	24	4	12	Yes
Vents (tops)	2	2	0.5	8	Yes
Vents (sides)	4	2	1	8	No
Vents (bottoms)	2	1	0.5	8	Yes

Determine the number of rolls of 15-pound felt. The number of rows per side is calculated using Eq. (4-12) as follows:

$$\text{Number}_{\text{Rows}} = \frac{14.2 \text{ ft}}{(3 \text{ ft} - 0.5 \text{ ft})} = 5.7 \text{ rows}$$

Round to 6 rows, and use the extra to lap at the ridge, thus eliminating the need for a row at the ridge. The area of felt needed for the roof is calculated as follows:

$$\text{Area} = (29 \text{ ft})(2 \text{ sides})(6 \text{ rows/side})(3 \text{ ft/row}) = 1{,}044 \text{ ft}^2$$

The number of rolls of felt is calculated as follows:

$$\text{Rolls} = \frac{1{,}044 \text{ ft}^2}{(400 \text{ ft}^2/\text{roll})} = 3 \text{ rolls}$$

The quantity of shingles needed for the starter row is calculated using Eq. (9-5) as follows:

$$\text{Squares}_{\text{Starter}} = \frac{(5 \text{ in})(1 \text{ ft}/12 \text{ in})(29 \text{ ft} + 29 \text{ ft})}{(100 \text{ ft}^2/\text{square})} = 0.24 \text{ square}$$

The quantity of shingles needed for the field is equal to the area of the roof, which is 824 square feet, or 8.24 squares. The quantity of cap shingles needed for the ridge is calculated using Eq. (9-9) as follows:

$$\text{Squares}_{\text{Ridge}} = \frac{(29)(1 \text{ ft})}{(100 \text{ ft}^2/\text{square})} = 0.29 \text{ square}$$

The total number of squares is calculated as follows:

$$\text{Squares} = 0.24 \text{ square} + 8.24 \text{ squares} + 0.29 \text{ square}$$
$$= 8.77 \text{ squares}$$

Each bundle of shingles is 1/3 square. Round to 9 squares. Twenty-seven (3 × 9) bundles of shingles are needed.

The quantities needed for the garage, grouped by the cost codes in Appendix B, are shown in Table 9-4.

TABLE 9-4 Quantities for Residential Garage

07-200 Insulation		
R-13 15″ × 32′ unfaced	16	ea
3-mil plastic 8′ × 12′	8	ea
Install R-13 insulation	628	sft
R-25 23″ × 32′ unfaced	10	ea
Install R-25 insulation	598	sft
07-400 Siding		
T1-11	26	ea
Install siding	792	sft
T1-11 soffit	6	ea
Install soffit	164	sft
3″-wide screen	58	ft
12″ × 14″ vent	2	ea
1 × 4 cedar 8′	16	ea
1 × 4 cedar 10′	13	ea
1 × 4 cedar 12′	4	ea
Cut and install trim	75	ft
Install trim	212	ft
07-500 Roofing		
Drip edge 10′	12	ea
15-lb felt, 4-square roll	3	ea
20-yr three-tab shingles	9	sq
Install roofing	824	sft

CONCLUSION

Thermal and moisture protection includes exterior finishes (such as siding, stucco, and roofing) that protect the building against moisture infiltration, and insulation that protects the building against heat loss. Many of the components used for thermal and moisture protection are bid as a sheet or roll good.

PROBLEMS

1. How much waterproofing is needed to cover the foundation walls of a 24-foot by 50-foot basement? The wall is 8 feet high. The wall sits on a footing that is 12 inches high and extends 16 inches beyond the wall on the exterior side. Assume that the waterproofing starts 12 inches below the top of the foundation wall and covers the top surface of the footings. If one gallon of waterproofing material covers 75 square feet, how many gallons of waterproofing material are needed?

2. How much waterproofing is needed to cover the foundation walls of a 32-foot by 48-foot basement? The wall is 8 feet high. The wall sits on a footing that is 12 inches high and extends 12 inches beyond the wall on the exterior side. Assume that the waterproofing starts 24 inches below the top of the foundation wall and covers the top surface of the footings. If one gallon of waterproofing material covers 90 square feet, how many gallons of waterproofing material are needed?

3. Determine the number of rolls of 15-pound asphalt-impregnated felt needed for the roof in Figure 9-11. The felt is to be lapped 6 inches on the side and 12 inches on the ends. The felt is available in rolls 3 feet wide with 4 squares (400 square feet) per roll. The slope on the roof is 6:12.

4. Determine the number of rolls of 30-pound asphalt-impregnated felt needed for the roof in Figure 9-12. The felt is to be lapped 6 inches on the side and 12 inches on the ends. The felt is available in rolls 3 feet wide with 2 squares (200 square feet) per roll. The slope on the roof is 4:12.

5. How many square feet of wall insulation are needed to insulate 25 feet of an 8-foot-high wall? The stud spacing is 16 inches on center.

6. How many rolls of insulation are needed for Problem 5 if the insulation comes in rolls 15 inches wide and 32 feet long? Do not include a waste factor.

7. How many square feet of wall insulation are needed to insulate 175 feet of a 10-foot-high wall? The stud spacing is 24 inches on center.

8. How many rolls of insulation are needed for Problem 7 if the insulation comes in rolls 23 inches wide and 32 feet long? Do not include a waste factor.

9. How many bags of blown insulation are needed to insulate a 1,835-square-foot ceiling with R-30 insulation? The manufacturer's recommended coverage rate is 41 bags per 1,000 square feet.

10. How many bags of blown insulation are needed to insulate a 1,120-square-foot ceiling with R-25 insulation? The manufacturer's recommended coverage rate is 34 bags per 1,000 square feet.

11. Determine the quantities of insulation, mesh reinforcement, and stucco needed to cover the end of the building shown in Figure 9-13. The insulation comes in 4-foot by 8-foot sheets. The reinforcement comes in

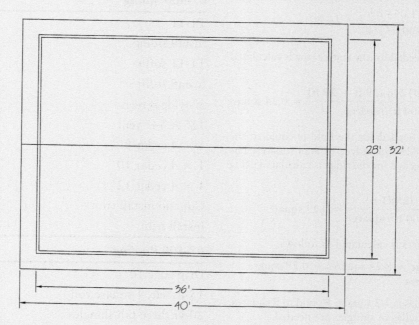

FIGURE 9-11 Roof Plan

Thermal and Moisture Protection

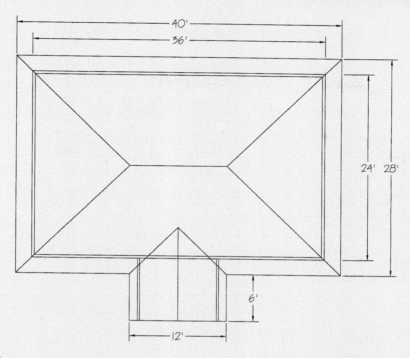

FIGURE 9-12 Roof Plan

rolls 100 feet long by 3 feet wide and needs to be lapped 3 inches. One cubic foot of base or brown coat will cover 75 square feet of wall, and one bucket of finish coat will cover 130 square feet of wall.

12. Determine the quantities of insulation, mesh reinforcement, and stucco needed to cover the end of the building shown in Figure 9-14. The insulation comes in 4-foot by 8-foot sheets. The reinforcement comes in rolls 100 feet long by 3 feet wide and needs to be lapped 3 inches. One cubic foot of base or brown coat will cover 60 square feet of wall, and one bucket of finish coat will cover 120 square feet of wall.

13. How many squares of roofing are required to cover the building in Figure 9-11? The shingles have a 5-inch exposure, and the slope on the roof is 6:12. The cap and starter strip will be cut from a standard shingle.

14. Drip edge is to be installed around the entire perimeter of the roof in Figure 9-11. How many feet of drip edge are needed for the roof?

15. How many pieces of drip edge are needed for Problem 14 if the drip edge comes in 10-foot lengths and needs to be lapped 2 inches?

16. How many squares of roofing are required to cover the building in Figure 9-12? The shingles have a 5-inch exposure, and the slope on the roof is 4:12. The cap and starter strip will be cut from a standard shingle.

17. Drip edge is to be installed around the entire perimeter of the roof in Figure 9-12. How many feet of drip edge are needed for the roof?

18. How many pieces of drip edge are needed for Problem 17 if the drip edge comes in 10-foot lengths and needs to be lapped 2 inches?

19. Determine the quantities of starter strip, outside corners, J-moldings, and 10-inch-wide vinyl siding needed to cover the end of the building shown in Figure 9-15. The starter strip, outside corners, and J-moldings are available in 10-foot lengths. The siding comes in packages of 200 square feet. When calculating the siding, do not deduct the area of the windows.

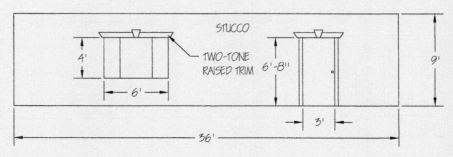

FIGURE 9-13 Building End

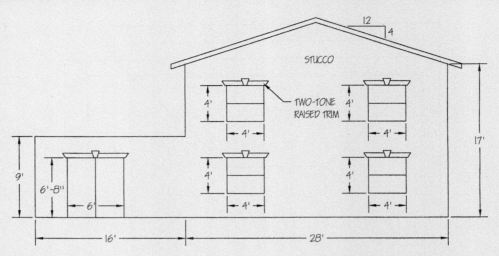

FIGURE 9-14 Building End

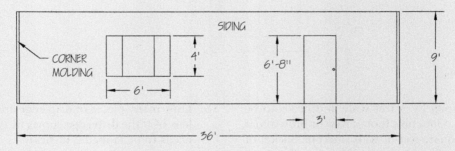

FIGURE 9-15 Building End

20. Determine the quantities of starter strip, outside corners, J-moldings, and 10-inch-wide vinyl siding needed to cover the end of the building shown in Figure 9-16. The starter strip, outside corners, and J-moldings are available in 10-foot lengths. The siding comes in packages of 200 square feet. When calculating the siding, do not deduct the area of the windows and doors.

21. Determine the quantities of starter strip, 1 × 4 trim, and 1 × 8 wood siding needed to cover the end of the building shown in Figure 9-17. The siding is lapped 1 inch on the sides and butt jointed on the ends. The starter strip and 1 × 4 trim are available in 8-, 10-, and 12-foot lengths. The siding comes in 12-foot lengths. When calculating the siding, do not deduct the area of the windows.

22. Determine the quantities of starter strip, 1 × 4 trim, and 1 × 8 wood siding needed to cover the end of the building shown in Figure 9-18. The siding is

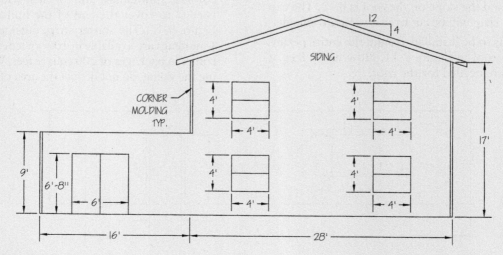

FIGURE 9-16 Building End

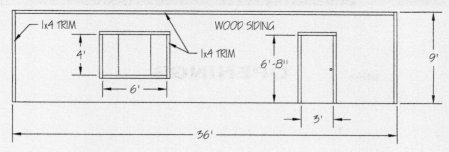

FIGURE 9-17 Building End

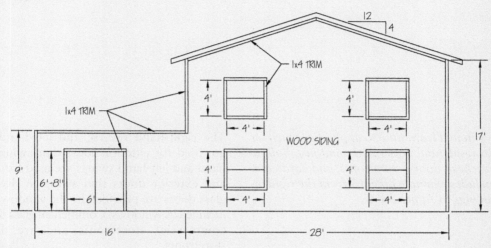

FIGURE 9-18 Building End

lapped 1 inch on the sides and butt jointed on the ends. The starter strip and 1 × 4 trim are available in 8-, 10-, and 12-foot lengths. The siding comes in 12-foot lengths. When calculating the siding, do not deduct the area of the windows and door.

23. Determine the number of 15-foot-wide by 200-foot-long sheets of membrane roofing needed for a roof 80 feet wide by 90 feet long. Allow 3 feet per side at the perimeter of the roof for the membrane to be attached to the parapet wall. The roofing must be lapped 3 inches at the sides and ends.
24. Determine the number of 25-foot-wide by 75-foot-long sheets of membrane roofing needed for a roof 300 feet wide by 400 feet long. Allow 4 feet per side at the perimeter of the roof for the membrane to be attached to the parapet wall. The roofing must be lapped 6 inches at the sides and ends.
25. Determine the insulation needed to complete the Johnson Residence given in Appendix G.
26. Determine the vinyl siding, soffit, and fascia needed to complete the Johnson Residence given in Appendix G.
27. Determine the stucco needed to complete the Johnson Residence given in Appendix G.
28. Determine the roofing, including underlayment, needed to complete the Johnson Residence given in Appendix G.
29. Determine the insulation needed to complete the West Street Video project given in Appendix G.
30. Determine the membrane roofing and sheet metal needed to complete the West Street Video project given in Appendix G.
31. Set up Excel Quick Tip 9-1 in Excel.
32. Set up Excel Quick Tip 9-2 in Excel.
33. Set up Excel Quick Tip 9-3 in Excel.
34. Set up Excel Quick Tip 9-4 in Excel.
35. Set up Excel Quick Tip 9-5 in Excel.
36. Modify Excel Quick Tip 9-1 to allow the wall lengths and wall height to be entered in feet and inches.
37. Modify Excel Quick Tip 9-2 to allow the roof length and width to be entered in feet and inches.
38. Modify Excel Quick Tip 9-4 to allow the roof length and width to be entered in feet and inches.

REFERENCES

1. *International Building Code,* International Code Counsel, 2014, Section 1507.2.8.
2. *International Building Code,* International Code Counsel, 2014, Section 1507.2.8.2.

CHAPTER TEN

OPENINGS

In this chapter you will learn how to apply the principles in Chapter 4 to residential doors and windows, hollow metal doors, overhead doors, storefronts, and hardware. This chapter includes example takeoffs from the residential garage drawings in Appendix G.

Openings consist of the doors, frames, glazing, and hardware to cover openings in both the interior and exterior walls of buildings. Let's begin by looking at doors.

DOORS

Doors include residential prehung doors and commercial wood and hollow metal doors. Most often doors are marked with a numbered symbol that relates to a door schedule found in the plans. In this book, the doors are marked with numbers inside a hexagon (a six-sided polygon). Each door may have its own unique identification, or each type of door may have its own unique identification. Doors come in four standard swings, which are shown in Figure 10-1.

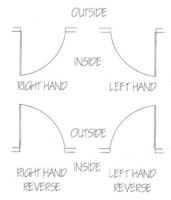

FIGURE 10-1 Standard Door Swings

The right-hand reverse and left-hand reverse swings are used for exterior doors that swing out. The right-hand and left-hand swings are used for all interior doors and exterior doors that swing in. Residential sliding glass doors are provided by the window supplier and are included as windows. Commercial glass doors are a part of commercial storefronts and are discussed under storefronts.

Doors are bid as a counted item. When bidding doors the estimator locates the door on the floor plan and uses the symbol next to the door along with the door schedule to determine the size and type of door needed for the opening. In addition, the estimator should identify any holes that need to be predrilled or mortised in the door for the hardware and the size of the door-jambs. The size of the door jambs is determined in part by the thickness of the wall in which the door is located. The estimating of doors is shown in the following example.

EXAMPLE 10-1

Using the door schedule in Figure 10-3, determine the number, size, and type of doors needed to complete the building shown in Figure 10-2.

Solution: The following prehung doors are needed: one right-hand 6-foot 8-inch-high by 3-foot-wide by 1 3/8-inch-thick solid-core wood exterior door, one right-hand 6-foot 8-inch-high by 2-foot 8-inch-wide by 1 3/8-inch-thick hollow-core birch interior door, one left-hand 6-foot 8-inch-high by 2-foot 8-inch-wide by 1 3/8-inch-thick hollow-core birch interior door, and two pairs of 6-foot 8-inch-high by 2-foot wide by 1 3/8-inch-thick hollow-core birch interior bifold doors. The prehung doors have 4 5/8-inch jambs. The holes that need to be drilled in the doors are discussed in Example 10-6 where hardware is discussed.

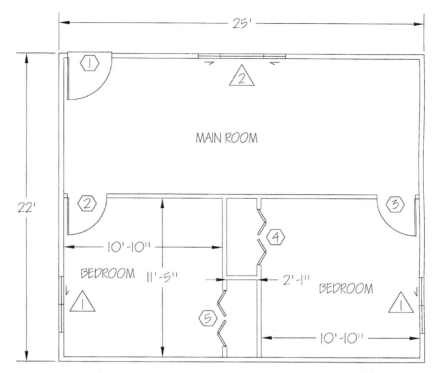

FIGURE 10-2 Floor Plan

MARK	HEIGHT	WIDTH	THICK.	TYPE	HWD
1	6'8"	3'0"	1-3/8"	SC WOOD	1
2	6'8"	2'8"	1-3/8"	HC BIRCH	2
3	6'8"	2'8"	1-3/8"	HC BIRCH	2
4	6'8"	4'0"	1-3/8"	PAIR HC BIRCH BIFOLD	3
5	6'8"	4'0"	1-3/8"	PAIR HC BIRCH BIFOLD	3

FIGURE 10-3 Door Schedule

WINDOWS

Windows include windows and sliding glass doors. Most often windows are marked with a numbered or lettered symbol that relates to a window schedule or a set of window drawings found in the plans. In this book, the windows are marked with a number inside a triangle. Each window may have its own unique identification, or each type of window may have its own unique identification. When bidding sliding windows the estimator must note which panels are fixed and which panels are operable. The estimator must also identify any windows that require safety glazing (tempered glass). With glass placed in hollow metal window frames, the frames are bid with the hollow metal doors and the glass is included as part of glazing.

Windows are bid as a counted item. When bidding windows, the estimator locates the window on the floor plan and uses the symbol next to the window along with the window schedule or drawings to determine the size and type of window needed for the opening. The estimating of windows is shown in the following example.

EXAMPLE 10-2

Using the window schedule in Figure 10-4, determine the number, size, and type of windows needed to complete the building shown in Figure 10-2.

Solution: The following windows are needed: one 4-foot-high by 4-foot-wide sliding window with the left panel fixed and the right panel operable, one 4-foot-high by 4-foot-wide sliding window with the right panel fixed and the left panel

MARK	HEIGHT	WIDTH	TYPE
1	4'0"	4'0"	SLIDER
2	4'0"	6'0"	SLIDER W/ 2 SLIDING PANES

FIGURE 10-4 Window Schedule

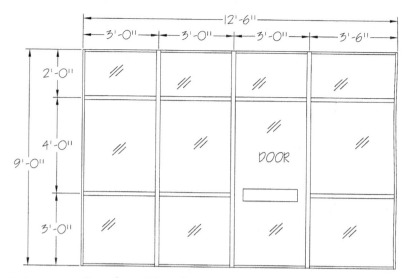

FIGURE 10-5 Storefront Elevation

operable, and one 4-foot-high by 6-foot-wide sliding window with the center panel fixed and the right and left panels operable.

COMMERCIAL STOREFRONT

Commercial storefront is a glass wall that is made up of glass together with metal tubing (commonly aluminum) and may include doors. It is commonly used at the front of retail stores, hence the name commercial storefront. Commercial storefront is provided by the glazing contractor, who provides the glass, metal tubing, and hardware (except for the cylinder) for the doors. A cylinder to match the rest of the hardware is provided by the hardware supplier for the storefront doors.

Often, commercial storefront is bid based on the square footage of the storefront and the hardware is bid as a counted item. The estimating of storefront using the square-footage method is shown in the following example.

GROUP 1:
1-1/2 PAIR HINGES
1 EA SINGLE-KEYED LOCK
1 EA PANIC HARDWARE
1 SET WEATHER STRIPPING
1 EA THRESHOLD
1 EA FLOOR MOUNTED DOORSTOP

FIGURE 10-6 Hardware Schedule

EXAMPLE 10-3

Determine the square footage of storefront and hardware needed to construct the storefront in Figure 10-5 using the hardware schedule shown in Figure 10-6.

Solution: The area of the storefront is 112.5 square feet (9 ft × 12.5 ft). The hardware needed is 1 1/2 pairs (3 each) of hinges, a single-keyed lock with the cylinder being provided by the hardware supplier, panic hardware, one set (17 ft) of weather stripping, one 3-foot-wide threshold, and one floor-mounted doorstop.

Glazing contractors should bid the storefront based on the necessary components required to fabricate the storefront. When bidding storefront based on the necessary components required to fabricate the storefront, the estimator often uses nominal dimensions for the storefront and glazing and then verifies the required sizes by field measuring the opening. The different components of a storefront are shown in Figure 10-7.

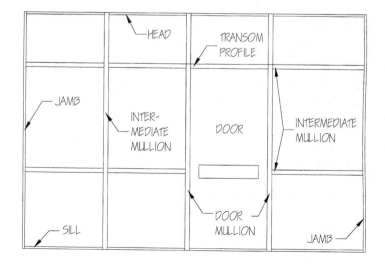

FIGURE 10-7 Storefront

TABLE 10-1 Metal Tubing Required

Quantity	Item	Notes
18 ft	Jamb	Left and right ends
12 ft 6 in	Head	Top
9 ft 6 in	Sill	Bottom, except at door
14 ft	Door mullion	Left and right sides of door
3 ft	Transom profile	Above door
32 ft	Intermediate mullion	All other locations

TABLE 10-2 Glazing Required

Quantity	Item
3 ea	2-0 × 3-0 glass
1 ea	4-0 × 3-0 glass
1 ea	4-0 × 3-0 tempered glass
1 ea	3-0 × 3-0 glass
1 ea	3-0 × 3-0 tempered glass
1 ea	7-0 × 3-0 tempered glass door
1 ea	2-0 × 3-6 glass
1 ea	4-0 × 3-6 tempered glass
1 ea	3-0 × 3-6 tempered glass

The estimator must also identify any glazing that requires safety glazing (tempered glass). The rules for safety glass are found in Section 2406 of the *International Building Code*.[1]

Estimating commercial storefront by the components needed to fabricate the storefront is shown in the following example.

EXAMPLE 10-4

Determine the components and hardware needed to fabricate the storefront in Figure 10-5 using the hardware schedule shown in Figure 10-6. The glass in the door and the two panels on either side of the door needs to be tempered glass.

Solution: The metal tubing needed for the door is shown in Table 10-1, and the glazing needed is shown in Table 10-2.

The hardware needed is 1 1/2 pairs (3 each) of hinges, a single-keyed lock with the cylinder being provided by the hardware supplier, panic hardware, one set (17 ft) of weather stripping, one 3-foot-wide threshold, and one floor-mounted doorstop.

GLAZING

Glazing includes installing glass in wood and hollow metal frames. Glazing is bid as a counted item. The estimator often uses nominal dimensions for the glazing when bidding and then verifies the required sizes with the frame manufacturer or by field measuring the frame. The estimator must also identify any windows that require safety glazing (tempered glass). Estimating glazing is shown in the following example.

EXAMPLE 10-5

Determine the glazing needed for the transom and sidelight shown in Figure 10-8. The two panels on the side of the door need to be tempered glass. Use nominal dimensions.

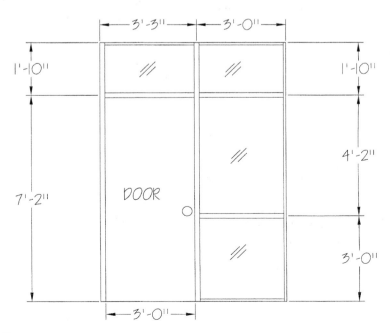

FIGURE 10-8 Door Elevation

```
GROUP 1:                          GROUP 2:
1-1/2 PAIR HINGES                 1-1/2 PAIR HINGES
1 EA SINGLE-KEYED DEADBOLT        1 EA PRIVACY LOCKSET
1 EA KEYED LOCKSET                1 EA WALL MOUNTED DOORSTOP
1 SET WEATHER STRIPPING
1 EA THRESHOLD                    GROUP 3:
1 EA WALL MOUNTED DOORSTOP        1 PAIR BRASS PULLS
```

FIGURE 10-9 Hardware Schedule

Solution: The following panels of glass are needed: one 1 foot 10 inches high by 3 feet 3 inches wide, one 1 foot 10 inches high by 3 feet wide, one 4 feet 2 inches high by 3 feet wide tempered, and one 3 feet high by 3 feet wide tempered.

HARDWARE

Hardware is taken off as a counted item and includes hinges (butts), locksets, cylinders, silencers, doorstops, thresholds, weather stripping, closers, and panic hardware. The hardware group for each door is shown on the door schedule, and the hardware items included in each hardware group is shown on the plans by the door schedule or in the specifications. Frequently, some of the hardware comes with the doors. For example, a prehung exterior residential door includes the frame, the hinges, the weather stripping, and the threshold. Hardware that comes with the door is not included in the hardware takeoff because it has already been included in the quantity takeoff with the door. Estimating hardware is shown in the following example.

EXAMPLE 10-6

Using the hardware schedule shown in Figure 10-9, determine the hardware needed for the doors in Figures 10-2 and 10-3.

Solution: The hinges, weather stripping, and threshold come with the prehung doors. One single-keyed deadbolt, one key lockset, and one wall-mounted doorstop are needed for Door 1. Door 1 should be ordered predrilled for the deadbolt and lockset; otherwise, they will have to be drilled in the field. Two privacy locksets and two wall-mounted doorstops are needed for Doors 2 and 3. These doors should be predrilled for a lockset. Four brass pulls are needed for Doors 4 and 5.

SAMPLE TAKEOFF FOR THE RESIDENTIAL GARAGE

A sample takeoff from a set of plans is shown in the following example:

EXAMPLE 10-7

Determine the doors, overhead doors, and hardware needed to complete the residential garage given in Appendix G.

Solution: The garage needs one 3-foot by 6-foot 8-inch, six-panel metal exterior prehung door with 4 5/8-inch wood jambs and 1 1/2 pairs of hinges. The door should be drilled for the deadbolt and lockset. A prehung exterior door includes the hinges, weather stripping, and threshold.

To complete the hardware for the door, one single-keyed deadbolt, one keyed lockset, and one floor-mounted doorstop are needed.

The garage needs one 16-foot by 7-foot prefinished, insulated, sectional overhead door with one screw-drive opener, two remotes, and one keyless entry.

The quantities needed for the garage, grouped by the cost codes in Appendix B, are shown in Table 10-3.

TABLE 10-3 Quantities for Residential Garage

08-100 Metal Doors		
3-0 × 6-8, six-panel metal prehung door with 4 5/8 wood jambs, 1 1/2 pairs of hinges, weather stripping, and threshold, drilled for deadbolt and lockset	1	ea
08-300 Overhead Doors		
16-ft × 7-ft prefinished, insulated, sectional overhead door	1	ea
1/2-hp screw-drive opener	1	ea
Keyless entry	1	ea
Remotes	2	ea
08-700 Hardware		
Single-keyed deadbolt	1	ea
Keyed lockset	1	ea
Floor-mounted doorstop	1	ea

Openings 177

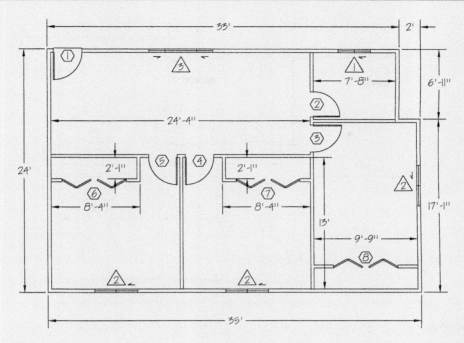

FIGURE 10-10 Floor Plan

CONCLUSION

Doors and windows consist of the doors, frames, glazing, and hardware to cover the opening in both the interior and exterior walls of buildings. These items are bid as counted items, with the exception of storefront, which is bid by the square foot or lineal foot for the tubing.

PROBLEMS

1. Using the door schedule in Figure 10-11, determine the number, size, and type of doors needed to complete the building shown in Figure 10-10.
2. Using the door schedule in Figure 10-13, determine the number, size, and type of doors needed to complete the building shown in Figure 10-12.
3. Using the window schedule in Figure 10-14, determine the number, size, and type of windows needed to complete the building shown in Figure 10-10.
4. Using the window schedule in Figure 10-15, determine the number, size, and type of windows needed to complete the building shown in Figure 10-12.
5. Determine the square footage of storefront needed to construct the storefront in Figure 10-16.
6. Determine the square footage of storefront needed to construct the storefront in Figure 10-17.

MARK	HEIGHT	WIDTH	THICK.	TYPE	HWD
1	6'8"	3'0"	1-3/8"	SC WOOD	1
2	6'8"	2'6"	1-3/8"	HC BIRCH	2
3	6'8"	2'8"	1-3/8"	HC BIRCH	2
4	6'8"	2'8"	1-3/8"	HC BIRCH	2
5	6'8"	2'8"	1-3/8"	HC BIRCH	2
6	6'8"	6'0"	1-3/8"	PAIR HC BIRCH BIFOLD	3
7	6'8"	6'0"	1-3/8"	PAIR HC BIRCH BIFOLD	3
8	6'8"	6'0"	1-3/8"	PAIR HC BIRCH BIFOLD	3

FIGURE 10-11 Door Schedule

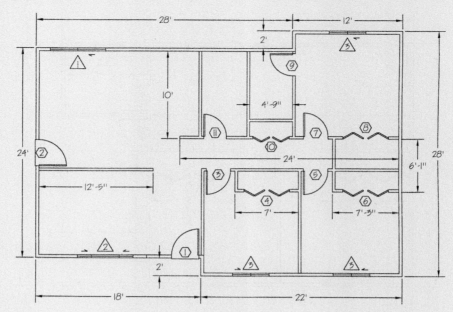

FIGURE 10-12 Floor Plan

MARK	HEIGHT	WIDTH	THICK.	TYPE	HWD
1	6'8"	3'0"	1-3/8"	SC WOOD	1
2	6'8"	3'0"	1-3/8"	SC WOOD	1
3	6'8"	2'8"	1-3/8"	HC BIRCH	2
4	6'8"	5'0"	1-3/8"	PAIR HC BIRCH BIFOLD	3
5	6'8"	2'8"	1-3/8"	HC BIRCH	2
6	6'8"	5'0"	1-3/8"	PAIR HC BIRCH BIFOLD	3
7	6'8"	2'8"	1-3/8"	HC BIRCH	2
8	6'8"	5'0"	1-3/8"	PAIR HC BIRCH BIFOLD	3
9	6'8"	2'6"	1-3/8"	HC BIRCH	2
10	6'8"	4'0"	1-3/8"	PAIR HC BIRCH BIFOLD	3
11	6'8"	2'6"	1-3/8"	HC BIRCH	2

FIGURE 10-13 Door Schedule

MARK	HEIGHT	WIDTH	TYPE
1	1'8"	3'0"	SLIDER
2	4'0"	4'0"	SLIDER
3	4'0"	6'0"	SLIDER W/ 2 SLIDING PANES

FIGURE 10-14 Window Schedule

MARK	HEIGHT	WIDTH	TYPE
1	6'8"	6'0"	SLIDING GLASS DOOR
2	4'0"	6'0"	SLIDER W/ 2 SLIDING PANES
3	4'0"	4'0"	SLIDER

FIGURE 10-15 Window Schedule

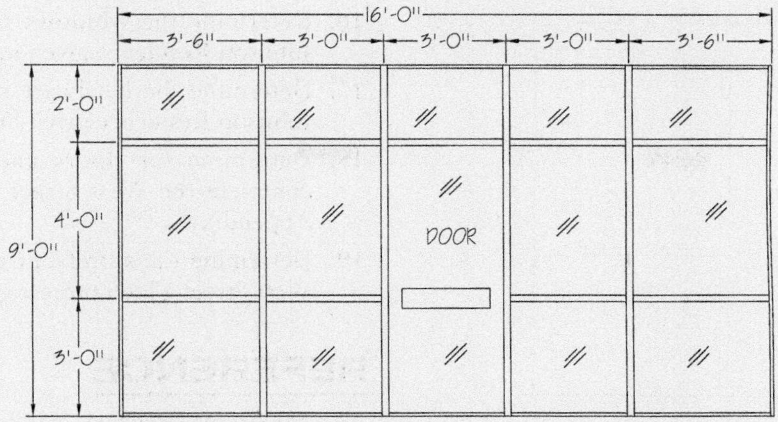

FIGURE 10-16 Storefront Elevation

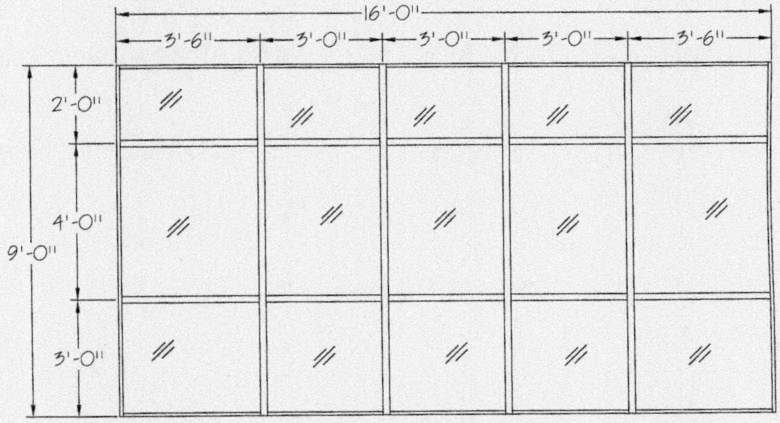

FIGURE 10-17 Storefront Elevation

7. Determine the metal tubing needed to fabricate the storefront in Figure 10-16.
8. Determine the glazing needed to fabricate the storefront in Figure 10-16. The glass in the door and the two panels on either side of the door needs to be tempered glass. Use nominal dimensions.
9. Determine the metal tubing needed to fabricate the storefront in Figure 10-17.
10. Determine the glazing needed to fabricate the storefront in Figure 10-17. No tempered glass is needed. Use nominal dimensions.
11. Determine the glazing needed for the transom and sidelight shown in Figure 10-18. The glass in the two panels to the side of the door needs to be tempered glass. Use nominal dimensions.
12. Determine the glazing needed for the transom and sidelight shown in Figure 10-19. The glass needs to be tempered glass. Use nominal dimensions.
13. Using the hardware schedule shown in Figure 10-9, determine the hardware needed for the doors in Figures 10-10 and 10-11.
14. Using the hardware schedule shown in Figure 10-9, determine the hardware needed for the doors in Figures 10-12 and 10-13.
15. Determine the doors needed to complete the Johnson Residence given in Appendix G.

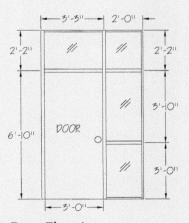

FIGURE 10-18 Door Elevation

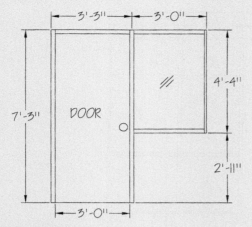

FIGURE 10-19 Door Elevation

16. Determine the windows needed to complete the Johnson Residence given in Appendix G.
17. Determine the hardware needed to complete the Johnson Residence given in Appendix G.
18. Determine the doors and hardware needed to complete the West Street Video project given in Appendix G.
19. Determine the storefront needed to complete the West Street Video project given in Appendix G.

REFERENCE

1. *International Building Code*, International Code Counsel, 2014, Section 2406.

CHAPTER ELEVEN

FINISHES

In this chapter you will learn how to apply the principles in Chapter 4 to metal studs, gypsum board, acoustical ceiling tile, carpet and pad, ceramic tile, sheet vinyl, vinyl composition tile (VCT), rubber base, and paint. This chapter includes sample spreadsheets that may be used in the quantity takeoff. It also includes example takeoffs from the residential garage drawings given in Appendix G.

Finishes include interior partition (nonbearing) walls and the finishes applied to floors, walls, and ceilings. We begin by looking at metal stud partitions.

METAL STUD PARTITIONS

Nonbearing metal stud partitions consist of light-gage steel studs set in a C-shaped runner at the top and bottom of the partition. The studs are fastened to the runner with screws. The bottom runner is fastened to the floor. The top runner may be located at the bottom of the floor or roof deck above, in which case it is fastened to the floor or roof deck, or it may stop at the ceiling line, in which case the wall must be braced to the floor or roof deck above. This bracing is done with metal studs. Unlike wood-framed walls, where the wood stud must be long enough to run from the top of the bottom plate to the bottom of the lower top plate, the metal studs must run from the floor to the top of the wall.

The number of studs needed to construct a metal stud partition is calculated in the same manner as determining the number of studs needed to construct a wood wall (see Chapter 8). Because the partitions are nonbearing, the openings are framed differently. Instead of a king stud and a jack stud on either side of an opening, two king studs are placed on either side of the opening and a piece of runner is screwed between the king studs to act as the header. Each door opening requires a piece of runner 6 inches longer than the width of the opening. When installing wood-framed doors in metal stud partitions, one of the king studs is replaced with a wood stud or fire-treated wood stud to provide backing for the installation of the door frame. The quantity takeoff for a metal stud partition is shown in the following example.

EXAMPLE 11-1

Determine the number of metal studs and feet of runner needed to construct 125 feet of interior wall with four corners, six intersections, six walls without corners, and six 3-foot-wide doorways. Stud spacing is 16 inches on center. Allow two extra studs for each corner, doorway, or intersection and one extra stud for each wall without corners. How many pieces of runner are needed if the runner comes in 10-foot lengths and is lapped 2 inches?

Solution: The number of studs needed for the standard spacing is calculated using Eq. (4-2) as follows:

$$\text{Number} = \frac{\text{Distance}}{\text{Spacing}}$$

$$\text{Number} = \frac{(125 \text{ ft})(12 \text{ in/ft})}{16 \text{ in}} = 94 \text{ each}$$

Six interior walls lack corners at both ends; therefore, six additional studs are needed for these walls. There are four corners, six doorways, and six intersections that will require two additional studs each, for 32 additional studs. The total number of studs is 132 (94 + 6 + 32).

Determine the number of pieces of runner using Eq. (4-6) as follows:

$$\text{Number} = \frac{\text{Length}}{\text{Length}_{\text{Piece}} - \text{Lap}}$$

$$\text{Number} = \frac{(125 \text{ ft} + 125 \text{ ft})}{(10 \text{ ft} - 2 \text{ in})} = 26$$

182 CHAPTER ELEVEN

For the door openings, six 3-foot 6-inch-long pieces of runner are needed above the doors. Only 2 of these pieces can be cut from one piece of runner; therefore, 3 additional pieces of runner are needed, for a total of 29 pieces.

GYPSUM BOARD

Gypsum board (drywall) consists of the gypsum board, metal or plastic trim, fasteners, tape, and drywall compound (mud). Gypsum board comes in different thicknesses with 1/2 and 5/8 inches being the most common. The width of the sheets is 4 feet, and the length of the sheets varies from 8 to 16 feet. Gypsum board comes in three common types: standard, Type-X (fire rated), and water resistant. Type-X is used to create fire-rated partitions and ceilings, and their location is identified on the plans. Water-resistant gypsum board, which can be identified by its green color, is required in wet locations. It is not uncommon for water-resistant gypsum board to be used in entire bathrooms, laundry rooms, and other rooms subject to continuous moisture. Gypsum board is bid as a sheet good using the same procedures as when bidding sheathing.

Trim includes corner molding on the outside corners and J-molding on exposed ends of gypsum board. Trim is bid by the linear foot or piece. Fasteners, tape, and drywall compound are quantity-from-quantity goods and are bid based on historical data or data from the manufacturer. The quantity takeoff for gypsum board is shown in the following example.

EXAMPLE 11-2

Determine the number of sheets of gypsum board, the number of pieces of trim, the pounds of screws, boxes of joint compound, and rolls of tape needed for the rooms in Figure 11-1. The ceiling height is 8 feet, and the ceiling joists run north to south. The gypsum board on the walls is to be run with the long direction of the board running horizontal. The gypsum board may be ordered in 8-, 10-, 12-, 14-, or 16-foot-long sheets. The trim is available in 10-foot lengths. Historical data indicates that 1.1 pounds of screws, 0.4 box of joint compound, and 0.13 roll of tape are needed for 100 square feet of drywall. The doors are 30 inches wide.

Solution: The gypsum board on the ceiling of the main room will be run with the longest direction running east and west. Use one 14-foot-long and one 12-foot-long sheet for the columns for a total length of 26 feet. The number of rows is determined by Eq. (4-13) as follows:

$$\text{Number}_{\text{Rows}} = \frac{\text{Height}}{\text{Height}_{\text{Piece}}}$$

$$\text{Number}_{\text{Rows}} = \frac{9 \text{ ft } 8 \text{ in}}{4 \text{ ft}} = 2.5 \text{ rows}$$

For the partial row, a 14-foot-long sheet can be cut in half. Order two 12-foot-long sheets and three 14-foot-long sheets for the ceiling. The gypsum board on the ceiling of the bedrooms will be run with the long direction running east and west, perpendicular to the ceiling joists. Use one 12-foot-long sheet for the columns in the bedrooms. The number of rows is determined by Eq. (4-13) as follows:

$$\text{Number}_{\text{Rows}} = \frac{(11 \text{ ft } 5 \text{ in})}{4 \text{ ft}} = 3 \text{ rows}$$

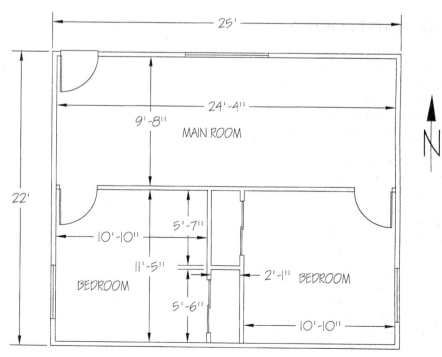

FIGURE 11-1 Wall Layout

Order three 12-foot-long sheets for each bedroom. In addition, one 8-foot-long sheet is needed for each closet.

The drywall on the walls will need to be two rows high. For the top row of drywall for the north and south walls of the main room, a 14-foot-long sheet and a 12-foot-long sheet are needed for each wall. The bottom row of the north wall does not need to run through the door; therefore, one 10-foot-long sheet and one 12-foot-long sheet are needed. For the bottom row of the south wall, two 10-foot-long sheets are needed. Two 10-foot-long sheets are needed for the east wall and two for the west wall. The walls of each bedroom will require six 12-foot-long sheets and two 8-foot-long sheets. The 8-foot-long sheets will be used for the bottom row of the two walls with doors. Two 8-foot-long sheets will cover the back and one side wall of a closet. One-and-a-half 8-foot-long sheets will cover the inside front and the other side wall of a closet. A half sheet is required on the bottom row because drywall will not be needed at the door opening. Each closet will require 3.5 sheets of 8-foot-long drywall. Order the following sheets of gypsum board: thirteen 8-foot-long, seven 10-foot-long, twenty-three 12-foot-long, and five 14-foot-long sheets.

A 4-foot piece of J-molding is needed above each closet door. Four 7-foot-long pieces (sides) and one 4-foot-long piece (above) of corner are needed for each closet door opening. Order one 10-foot-long piece of J-molding and nine 10-foot-long pieces of corner molding.

To calculate the pounds of nails, boxes of joint compound, and rolls of tape, one needs the area of gypsum board needed for the ceiling and walls. The area of the ceiling is the length times the width and is calculated as follows:

$$\begin{aligned} \text{Area} &= (\text{Length})(\text{Width}) \\ &= (24 \text{ ft } 4 \text{ in})(9 \text{ ft } 8 \text{ in}) + (11 \text{ ft } 5 \text{ in})(10 \text{ ft } 10 \text{ in}) \\ &\quad + (11 \text{ ft } 5 \text{ in})(10 \text{ ft } 10 \text{ in}) + (5 \text{ ft } 7 \text{ in})(2 \text{ ft } 1 \text{ in}) \\ &\quad + (5 \text{ ft } 6 \text{ in})(2 \text{ ft } 1 \text{ in}) \\ &= 506 \text{ ft}^2 \end{aligned}$$

The area of the walls is their perimeter times their height and is calculated as follows:

$$\begin{aligned} \text{Area} &= [2(24 \text{ ft } 4 \text{ in}) + 2(9 \text{ ft } 8 \text{ in}) + 2(11 \text{ ft } 5 \text{ in}) \\ &\quad + 2(10 \text{ ft } 10 \text{ in}) + 2(11 \text{ ft } 5 \text{ in}) + 2(10 \text{ ft } 10 \text{ in}) \\ &\quad + 2(5 \text{ ft } 7 \text{ in}) + 2(2 \text{ ft } 1 \text{ in}) + 2(5 \text{ ft } 6 \text{ in}) \\ &\quad + 2(2 \text{ ft } 1 \text{ in})](8 \text{ ft}) \\ &= 1{,}500 \text{ ft}^2 \end{aligned}$$

The total area is 2,006 square feet ($506 \text{ ft}^2 + 1{,}500 \text{ ft}^2$). The number of pounds of screws needed is calculated using Eq. (4-21) as follows:

$$\text{Quantity} = (\text{Quantity}_{\text{Base}})(\text{Average Quantity Required})$$

$$\text{Number} = (2{,}006 \text{ ft}^2)\left(\frac{1.1 \text{ lb}}{100 \text{ ft}^2}\right) = 22.1 \text{ lb}$$

EXCEL QUICK TIP 11-1
Drywall

The calculation of the square footage of drywall needed for a rectangular room is set up in a spreadsheet by entering the data and formatting the cells as follows:

	A	B	C	D	E	F	G	H	I	J	K
1		Room 1		Room 2		Room 3		Room 4		Room 5	
2	Room Length	24	ft	11	ft	11	ft	5	ft	5	ft
3		4	in	5	in	5	in	7	in	6	in
4	Room Width	9	ft	10	ft	10	ft	2	ft	2	ft
5		8	in	10	in	10	in	1	in	1	in
6	Room Height	8	ft	8	ft	8	ft	8	ft	8	ft
7		-	in	-	in	-	in	-	in	-	in
8											
9	Ceiling Area	235	sft	124	sft	124	sft	12	sft	11	sft
10	Wall Area	544	sft	356	sft	356	sft	123	sft	121	sft
11											
12	**Totals**										
13	Ceiling Area	506	sft								
14	Wall Area	1,500	sft								

The following formulas need to be entered into the associated cells:

Cell	Formula
B9	=(B2+B3/12)*(B4+B5/12)
B10	= (2*(B2+B3/12)+2*(B4+B5/12))* (B6+B7/12)
B13	=B9+D9+F9+H9+J9
B14	=B10+D10+F10+H10+J10

The formulas in Cells B9 and B10 need to be copied to Cells D9 and D10, F9 and F10, H9 and H10, and J9 and J10. The data for the room is entered in Cells B2 through B7, D2 through D7, F2 through F7, H2 through H7, and J2 through J7. The data shown in the foregoing figure is from Example 11-1 and is formatted using the comma style, which replaces zeros with dashes.

The number of boxes of joint compound needed is calculated using Eq. (4-21) as follows:

$$\text{Number} = (2{,}006 \text{ ft}^2)\left(\frac{0.4 \text{ box}}{100 \text{ ft}^2}\right) = 8 \text{ boxes}$$

The number of rolls of tape needed is calculated using Eq. (4-21) as follows:

$$\text{Number} = (2{,}006 \text{ ft}^2)\left(\frac{0.13 \text{ roll}}{100 \text{ ft}^2}\right) = 3 \text{ rolls}$$

TILE

Tile comes in a variety of sizes. The number of tiles needed is estimated using the row and column method. If backing board, such as Hardibacker, is required it is estimated using the same procedures used to estimate sheathing (see Chapter 8). Thickset mortar, thinset glue, and grout are quantity-from-quantity goods and are bid based on historical data or the estimated coverage rate provided by the manufacturer. When estimating the materials needed for tile, the estimator must take into account any layout requirements set by the architect, such as a minimum size of tile. Tile estimating is shown in the following example.

EXAMPLE 11-3

Determine the number of 7 3/4-inch by 7 3/4-inch tiles needed to tile a 30-inch by 48-inch entry. The tile is to have 1/4-inch grout joints. Where possible the tiles should be centered in the space with at least one-half of a tile at the edge of the space.

Solution: The number of rows is calculated using Eq. (4-14) as follows:

$$\text{Number}_{\text{Rows}} = \frac{\text{Height}}{\text{Height}_{\text{Piece}} + \text{Space}}$$

$$\text{Number} = \frac{30 \text{ in}}{(7.75 \text{ in} + 0.25 \text{ in})} = 4 \text{ rows}$$

The number of columns is calculated using Eq. (4-17) as follows:

$$\text{Number}_{\text{Columns}} = \frac{\text{Length}}{\text{Length}_{\text{Piece}} + \text{Space}}$$

$$\text{Number} = \frac{48 \text{ in}}{(7.75 \text{ in} + 0.25 \text{ in})} = 6 \text{ columns}$$

The total number of tiles is calculated using Eq. (4-18) as follows:

$$\text{Number} = (\text{Number}_{\text{Rows}})(\text{Number}_{\text{Columns}})$$

$$\text{Number} = (4 \text{ rows})(6 \text{ columns}) = 24 \text{ each}$$

The entry is laid out as shown in Figure 11-2 with the minimum size of tile being about 7 inches wide.

EXCEL QUICK TIP 11-2
Ceramic Tile

The calculation of the number of ceramic tiles needed for a rectangular room is set up in a spreadsheet by entering the data and formatting the cells as follows:

	A	B	C
1	Length	-	ft
2		48	in
3	Width	-	ft
4		30	in
5	Tile Size	7.75	in
6	Joint Width	0.25	in
7			
8	Tiles		24	ea

The following formula needs to be entered into Cell C8.

```
=ROUNDUP((B3*12+B4)/(B5+B6),0)*
ROUNDUP((B1*12+B2)/(B5+B6),0)
```

The data for the room and tile is entered in Cells B1 through B6. Both the row and the column are rounded up to the next full tile to ensure that at least one half of tile is located around the perimeter of the room. The data shown in the foregoing figure is from Example 11-3 and is formatted using the comma style, which replaces zeros with dashes.

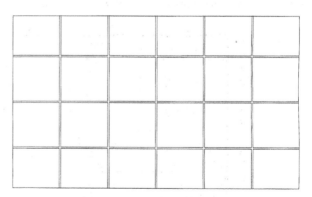

FIGURE 11-2 Tile Layout

SUSPENDED ACOUSTICAL CEILINGS

Suspended acoustical ceilings consist of ceiling tiles suspended in a metal grid. The metal grid consists of wall moldings, main tees, 4-foot cross tees, and 2-foot cross tees. The ceiling grid is suspended from the structure by wire. Wall molding is required around the entire perimeter of the room, and the main tees are run at 4 feet on center throughout the room. The main tees have a connection on both ends that allow them to be connected end to end to form continuous tees long enough to span the room. Four-foot cross tees are run between the main tees and between the main tees and the walls. The 4-foot cross tees are run perpendicular to the main runners such that a 4-foot by 2-foot grid is created. To achieve a 2-foot by 2-foot grid, 2-foot cross tees are run between the 4-foot cross tees. Two-foot cross tees can also be used between the main runners and the walls. Each end of a cross tee has a connection that allows it to be connected to a main or cross tee. When using leftover pieces of tees, one must make sure the piece has the connection needed to connect to the main and cross tees. At the edge of the ceiling, the main and cross tees are cut to length and riveted to the wall molding.

Suspended acoustical ceilings often are bid by the square foot; however, this does not provide the necessary information to order the correct materials for the ceiling. The tiles in the acoustical ceiling are bid based on the row and column method deducting any tiles that are being replaced by lighting fixtures, diffusers, registers, and grilles. The wall moldings and main tees are bid as linear components. The cross tees are bid as a counted item. When estimating the materials needed for an acoustical ceiling, the estimator must take into account any layout requirements set by the architect, such as a minimum size of tile. When dealing with odd-shaped rooms, often the easiest way to determine the components needed for the grid is to prepare a layout for the room. The estimate of materials needed for an acoustical ceiling is shown in the following example.

EXAMPLE 11-4

A 13-foot-wide by 15-foot-long ceiling is to receive a 2-foot by 2-foot acoustical ceiling. The ceiling includes four 2-foot by 2-foot fluorescent light fixtures and one 2-foot by 2-foot mechanical register. Determine the number of 12-foot-long wall moldings, 12-foot-long main runners, 4-foot cross tees, 2-foot cross tees, and 2-foot by 2-foot tiles needed for the ceiling. The tiles should be centered in the room, and, where possible, there should be at least half a tile at the edge of the room.

Solution: Use the row and column method to calculate the number of tiles. The number of rows is calculated using Eq. (4-13) as follows:

$$\text{Number} = \frac{13 \text{ ft}}{2 \text{ ft}} = 7 \text{ rows}$$

To center the rows, the following widths of rows are needed (from bottom to top): one at 18 inches, five at 24 inches, and one at 18 inches. The number of columns is calculated using Eq. (4-16) as follows:

$$\text{Number}_{\text{Columns}} = \frac{\text{Length}}{\text{Length}_{\text{Piece}}}$$

$$\text{Number} = \frac{15 \text{ ft}}{2 \text{ ft}} = 8 \text{ columns}$$

To center the columns, the following widths of columns are needed (from left to right): one at 18 inches, six at 24 inches, and one at 18 inches. The total number of squares in the grid is calculated using Eq. (4-18) as follows:

$$\text{Number} = (7 \text{ rows})(8 \text{ columns}) = 56 \text{ each}$$

Five of these squares are filled with light fixtures and a mechanical register; therefore, 51 (56 − 5) tiles are needed.

The number of wall moldings can be calculated using Eq. (4-7) as follows:

$$\text{Number} = \frac{\text{Length}}{\text{Length}_{\text{Piece}}}$$

$$\text{Number} = \frac{(13 \text{ ft} + 15 \text{ ft} + 13 \text{ ft} + 15 \text{ ft})}{12 \text{ ft}} = 5 \text{ each}$$

The main tees will be run in the long direction. A 15-foot-long main tee is needed between every two rows of tiles. The main tees are located between the second and third rows, the fourth and fifth rows, and the sixth and seventh rows. Three main tees are required and are 15 feet long. The number of 12-foot-long main tees needed is calculated as follows:

$$\text{Number} = \frac{(3)(15 \text{ ft})}{12 \text{ ft}} = 4 \text{ each}$$

Ordering four main tees will require the first main tee to be a 12-foot piece and a 3-foot piece, the second main tee to be the leftover 9-foot piece and a 6-foot piece, and the last main tee to be the leftover 6-foot piece and a 9-foot piece because only two pieces can be cut from a 12-foot-long main tee while maintaining the required end connections. If the ceiling was constructed by starting each of the main tees with a full-length 12-foot-long piece, 1 1/2 full-length pieces will be needed for each main tee, for a total of five 12-foot-long pieces.

The main runners divide the ceiling into four 15-foot-long rectangles with the following widths: 3.5, 4, 4, and 1.5 feet. These rectangles are subdivided by cross tees running the short direction of the room. For each of the first three rectangles, seven (one less than the number of tiles required for the length of the rectangle) 4-foot cross tees are needed, for a total of twenty-one 4-foot cross tees. Each of these rectangles is further subdivided into 2-foot by 2-foot or smaller squares using 2-foot cross tees. Eight (the number of tiles required for the length of the rectangle) 2-foot cross tees are needed for each of these rectangles, for a total of 24. For the 1.5-foot-wide rectangle, seven 2-foot cross tees are needed, bringing the total number of 2-foot cross tees to 31. The layout for the ceiling is shown in Figure 11-3. ■

WOOD AND LAMINATE FLOORS

Laminate floors, such as Pergo, and many wood floors are not fastened to the underlying floor but float over it. A gap is left at all edges of the floor. The gap and the floating nature of the floor allow the flooring to move, expand, and contract. To protect the flooring against moisture, a vapor barrier may be required in certain applications. The flooring comes in planks and is bid using the row and column method. Because the flooring is connected using a form of tongue and groove, waste from one row may only be used when it has the necessary

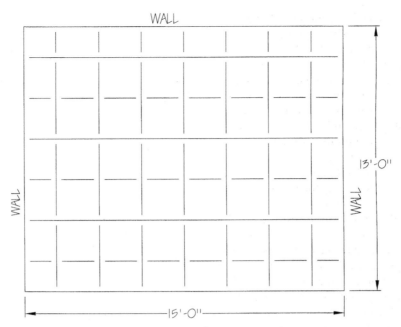

FIGURE 11-3 Ceiling Layout

tongue or groove. In addition, the pieces must be staggered, which may also increase the waste.

A sound-deadening underlayment is often used directly below the flooring. There are different types of underlayment that provide differing levels of cushioning and sound attenuation. The underlayment is bid as a rolled good.

Trim must be provided at all edges of the flooring to cover the gap between the floor and the wall. Trim pieces may include stair nosing for use on stairs; base and shoe molding to terminate the flooring at walls; end molding to terminate the flooring where base or shoe molding will not work, such as around a raised fireplace hearth; T-molding as a transition between floors of the same height; and reducer strips as a transition between floors of different heights. Trim pieces are bid as a counted item or a linear good depending on where it is used.

EXAMPLE 11-5

Determine the number of 7.3-inch by 50.6-inch pieces of laminate flooring and rolls of underlayment that are needed for a 10-foot by 10-foot floor. The flooring must be staggered at least 24 inches and comes eight pieces per package. The underlayment comes in rolls 3 feet wide by 50 feet long.

Solution: Determine the number of rolls of underlayment. The number of rows is calculated using Eq. (4-13) as follows:

$$\text{Number} = \frac{10 \text{ ft}}{3 \text{ ft}} = 3.5 \text{ rows}$$

The number of columns is calculated using Eq. (4-16) as follows:

$$\text{Number} = \frac{10 \text{ ft}}{50 \text{ ft}} = 0.20 \text{ columns}$$

The total number of rolls of underlayment is calculated using Eq. (4-18) as follows:

$$\text{Number} = (3.5 \text{ rows})(0.2 \text{ columns}) = 1 \text{ each}$$

Determine the required number of pieces of flooring. The number of rows is calculated using Eq. (4-13) as follows:

$$\text{Number} = \frac{(10 \text{ ft})(12 \text{ in/ft})}{50.6 \text{ in}} = 2.5 \text{ rows}$$

The number of rows is rounded to one half of a piece of flooring to ensure proper staggering. The number of columns is calculated using Eq. (4-16) as follows:

$$\text{Number} = \frac{(10 \text{ ft})(12 \text{ in/ft})}{7.3 \text{ in}} = 17 \text{ columns}$$

The total number of pieces of flooring is calculated using Eq. (4-18) as follows:

$$\text{Number} = (2.5 \text{ rows})(17 \text{ columns}) = 43 \text{ each}$$

The number of packages of flooring is calculated as follows:

$$\text{Number} = \frac{43 \text{ each}}{(8 \text{ each/package})} = 6 \text{ packages}$$

SHEET VINYL

Sheet vinyl is available in 12-foot widths. Vinyl must be installed in one direction, and the pattern must be matched. When the room is wider than 12 feet, the vinyl must be seamed. This may be done by installing one large piece and a number of smaller pieces of vinyl as shown in Figure 11-4 or by installing two large pieces as shown in Figure 11-5. The advantage of installing the vinyl as shown in Figure 11-4 is that less material is required.

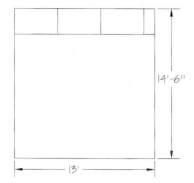

FIGURE 11-4 Vinyl Seaming Plan

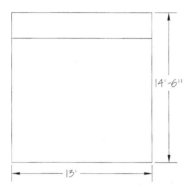

FIGURE 11-5 Vinyl Seaming Plan

The advantage of installing the vinyl as shown in Figure 11-5 is there are fewer seams, which improves the aesthetics of the installation and the wear of the vinyl because it often wears out at the seams faster than elsewhere.

To calculate the quantity of vinyl using the seaming plan in Figure 11-4, one calculates the number of rows of vinyl using Eq. (4-13). The number of rows is rounded up to the twelfth, sixth, quarter, third, or half of a row. The number in the denominator of the fraction of a row represents the number of smaller pieces of vinyl that are used. For example, if the fraction of a row is 1/12, then 12 smaller pieces will be used. The length of the room is multiplied by the fraction of a row (or divided by the number of pieces) to determine the length of vinyl needed for the small pieces. The length of vinyl needed for the large pieces equals the number of large pieces times the length of the room. To ensure that the pattern matches, the length of the vinyl needs to be rounded up to whole increments of the pattern. For example, if the pattern in the vinyl repeats every 16 inches, the length would be rounded up to increments of 16 inches. In Excel, the CEILING function would be used to round to increments of 16 inches. The calculation of vinyl using this method is shown in the following example.

EXAMPLE 11-6

How many yards of vinyl need to be ordered for a 13-foot-long by 14-foot 6-inch-wide room? The pattern repeats every 12 inches.

Solution: The number of rows is calculated using Eq. (4-13) as follows:

$$\text{Number}_{\text{Rows}} = \frac{14.5 \text{ ft}}{12 \text{ ft}} = 1.21$$

Round the number of rows up to 1.25. The additional length needed for the small pieces is 4 feet (13 ft × 1/4). The length is rounded up to the whole foot because the pattern repeats every foot. The length of vinyl needed is 17 feet (13 ft + 4 ft). The number of yards of vinyl is calculated as follows:

$$\text{Area} = (17 \text{ ft})(12 \text{ ft})\left(\frac{1 \text{ yd}^2}{9 \text{ ft}^2}\right) = 22.7 \text{ yd}^2$$

To calculate the quantity of vinyl using the seaming plan in Figure 11-5, one calculates the number of rows of vinyl using Eq. (4-13) and rounds up to the whole number of rows. The length of vinyl needed equals the number of rows times the length of the room. The calculation of vinyl is shown in the following example.

EXAMPLE 11-7

How many yards of vinyl need to be ordered for a 13-foot-long by 14-foot 6-inch-wide room? The pattern repeats every 12 inches. Lay the vinyl out to minimize the seams.

Solution: The number of rows needed is calculated using Eq. (4-13) as follows:

$$\text{Number}_{\text{Rows}} = \frac{(14.5 \text{ ft})}{(12 \text{ ft})} = 1.21$$

Round the number of rows up to 2. The length of vinyl needed is 26 feet (2 × 13 ft). The number of yards of vinyl is calculated as follows:

$$\text{Area} = (26 \text{ ft})(12 \text{ ft})\left(\frac{1 \text{ yd}^2}{9 \text{ ft}^2}\right) = 34.7 \text{ yd}^2$$

VINYL COMPOSITION TILE

Vinyl composition tile (VCT) is available in 12-inch by 12-inch squares. The number of tiles is estimated in the same manner as for ceramic tile, with one exception: there is no space between the tiles, which eliminates the need for grout. The tiles are glued down and the glue is bid as a quantity-from-quantity good. The coverage for a gallon of glue can be obtained from the manufacturer or from historical data.

RUBBER BASE

Rubber base comes in rolls or pieces and is bid as a linear good. The rubber base may be bent around corners, or the architect may require that manufactured corners be used. Base is bid as a linear good, and manufactured corners are bid as a counted item.

CARPET AND PAD

Carpet is handled in the same manner as sheet vinyl. In residential applications, the carpet is laid over a pad. Tackless fastening strips are required around the perimeter of the carpet, and trim is required anytime the carpet transitions to vinyl. The pad is bid as a sheet good using the same procedure as in estimating underlayment (pad) for wood and laminate floors. The tackless fastening strips and trim are bid as linear goods.

In commercial applications the carpet is glued down. The glue is bid as a quantity-from-quantity good. The amount of glue needed for the carpet is based on historical or manufacturer's data.

PAINT

Paint is bid based on the square footage for walls, ceilings, and floors. Trim is bid based on the lineal foot when the trim is a separate color or a different type of paint than the surrounding area. When the trim is painted with the same paint as the surrounding area, it is bid as part of the surrounding area. Doors are bid as counted items. The amount of paint needed depends on the type of paint being used and the surface that is being painted. The paint usage is bid as a quantity-from-quantity good based on historical data or coverage rates from the manufacturer. Sometimes residential painters bid painting based on the square footage of floor space. The bidding of paint is shown in the following example.

EXAMPLE 11-8

How much paint is needed to paint 100 lineal feet of an 8-foot-high block wall with one coat of primer and two coats of latex paint? From historical data, 1 gallon of primer will cover 250 square feet of wall and 1 gallon of latex paint will cover 400 square feet of wall with a single coat.

Solution: The area of the wall is calculated as follows:

$$\text{Area} = (100 \text{ ft})(8 \text{ ft}) = 800 \text{ ft}^2$$

The amount of primer and paint needed is calculated using Eq. (4-20) as follows:

$$\text{Quantity} = \frac{\text{Quantity}_{\text{Base}}}{\text{Coverage}}$$

$$\text{Volume}_{\text{Primer}} = \frac{800 \text{ ft}^2}{(250 \text{ ft}^2/\text{gal})} = 4 \text{ gal}$$

$$\text{Volume}_{\text{Paint}} = \frac{(2 \text{ coats})(800 \text{ ft}^2)}{(400 \text{ ft}^2/\text{gal})} = 4 \text{ gal}$$

SAMPLE TAKEOFF FOR THE RESIDENTIAL GARAGE

A sample takeoff for gypsum board and paint from a set of plans is shown in the following example.

EXAMPLE 11-9

Determine the number of sheets of gypsum board, the number of pieces of trim, the pounds of screws, boxes of joint compound, rolls of tape, gallons of primer, and gallons of paint needed for the residential garage given in Appendix G. The gypsum board on the walls is to be run with the long direction of the board running horizontal. The gypsum board may be ordered in 8-, 10-, 12-, 14-, or 16-foot-long sheets. Historical data indicates that 1.1 pounds of screws, 0.4 box of joint compound, and 0.13 roll of tape are needed for 100 square feet of drywall. From historical data, 1 gallon of PVA primer will cover 300 square feet of drywall, 1 gallon of interior latex paint will cover 400 square feet of drywall with a single coat, 1 gallon of oil base primer will cover 250 square feet of T1-11 siding, and 1 gallon of exterior latex paint will cover 375 square feet of siding with a single coat or four single-hung doors with two coats.

Solution: The gypsum board on the ceiling will be run east-west, perpendicular to the trusses. Use one 14-foot-long sheet and one 12-foot-long sheet for the columns, for a total length of 26 feet. The number of rows is determined by Eq. (4-13) as follows:

$$\text{Number}_{\text{Rows}} = \frac{(23 \text{ ft 4 in})}{4 \text{ ft}} = 6 \text{ rows}$$

The drywall on the walls will need to be two rows high. The number of sheets needed for the north wall is two 14-foot-long sheets and two 12-foot-long sheets. The number of sheets needed for the south wall is one 14-foot-long sheet and three 12-foot-long sheets because the single-hung door will allow us to use two 12-foot-long sheets on the bottom row. The number of sheets needed for the east wall is two rows and two columns of 12-foot-long sheets, for a total of four sheets. For the west wall two 12-foot-long sheets are needed for the top row, and one 8-foot-long sheet is needed for the bottom row. Order one 8-foot-long, seventeen 12-foot-long, and nine 14-foot-long sheets.

The area of gypsum board is needed to calculate the pounds of nails, boxes of joint compound, and rolls of tape. The area of the ceiling is the length times the width and is calculated as follows:

$$\text{Area} = (25 \text{ ft 4 in})(23 \text{ ft 4 in}) = 591 \text{ ft}^2$$

The area of the walls is their perimeter times their height less the openings, and is calculated as follows:

$$\text{Area} = [2(25 \text{ ft 4 in}) + 2(23 \text{ ft 4 in})]8$$
$$- (16 \text{ ft})(7 \text{ ft}) - (3 \text{ ft})(6 \text{ ft 8 in})$$
$$= 647 \text{ ft}^2$$

The total area is 1,238 square feet (591 ft² + 647 ft²). The number of pounds of screws needed is calculated using Eq. (4-21) as follows:

$$\text{Number} = (1{,}238 \text{ ft}^2)\left(\frac{1.1 \text{ lb}}{100 \text{ ft}^2}\right) = 14 \text{ lb}$$

The number of boxes of joint compound needed is calculated using Eq. (4-21) as follows:

$$\text{Number} = (1{,}238 \text{ ft}^2)\left(\frac{0.4 \text{ box}}{100 \text{ ft}^2}\right) = 5 \text{ boxes}$$

The number of rolls of tape needed is calculated using Eq. (4-21) as follows:

$$\text{Number} = (1{,}238 \text{ ft}^2)\left(\frac{0.13 \text{ roll}}{100 \text{ ft}^2}\right) = 2 \text{ rolls}$$

Now we will look at the paint. The area of drywall to be painted is 1,238 square feet. The amount of primer and paint needed is calculated using Eq. (4-20) as follows:

$$\text{Volume}_{\text{Primer}} = \frac{1{,}238 \text{ ft}^2}{(300 \text{ ft}^2/\text{gal})} = 5 \text{ gal}$$

$$\text{Volume}_{\text{Paint}} = \frac{(2 \text{ coats})(1{,}238 \text{ ft}^2)}{(400 \text{ ft}^2/\text{gal})} = 7 \text{ gal}$$

From Example 9-14, the area of T1-11 to be painted is 956 square feet (792 ft² + 164 ft²). The amount of primer and paint needed is calculated using Eq. (4-20) as follows:

$$\text{Volume}_{\text{Primer}} = \frac{956 \text{ ft}^2}{(250 \text{ ft}^2/\text{gal})} = 4 \text{ gal}$$

$$\text{Volume}_{\text{Paint}} = \frac{(2 \text{ coats})(956 \text{ ft}^2)}{(375 \text{ ft}^2/\text{gal})} = 6 \text{ gal}$$

One quart is needed for the single-hung door.

The quantities needed for the garage, grouped by the cost codes in Appendix B, are shown in Table 11-1.

TABLE 11-1 Quantities for Residential Garage

09-200 Drywall			09-900 Paint		
Hang and finish drywall	1238	sft	Interior primer	5	gal
4′ × 8′ × 1/2″ gypsum board	1	ea	Interior latex	7	gal
4′ × 12′ × 1/2″ gypsum board	17	ea	Paint drywall with 3 coats	1238	sft
4′ × 14′ × 1/2″ gypsum board	9	ea	Exterior primer	4	gal
1 5/8″ screws	14	lbs	Exterior latex	6	gal
Joint compound	5	box	Paint T1-11 with 3 coats	956	sft
Joint tape	2	rolls	Exterior latex	0.25	gal
			Paint single-hung door	1	ea

CONCLUSION

Finishes include interior partition (nonbearing) walls and the finishes applied to the floors, walls, and ceilings. Metal stud walls are bid by using the same procedures as for wood stud walls. Drywall, tile, acoustical ceiling tile, wood and laminate floors, vinyl, carpet, pad, and vinyl composition tile are all bid using the row and column method. Painting is bid by the square foot or lineal foot. Fasteners, drywall tape and mud, adhesives, grout, primer, and paint are quantity-from-quantity goods and are bid based on historical data or data from the manufacturer.

PROBLEMS

1. Determine the number of metal studs needed to construct 83 feet of interior wall with six corners, eight intersections, and four 3-foot-wide doorways. Stud spacing is 16 inches on center. Allow two extra studs for each corner, intersection, or doorway.

2. How many pieces of runner are needed for Problem 1 if the runner comes in 10-foot-long lengths and is lapped 2 inches?

3. How many 4-foot-wide by 8-foot-high sheets of drywall are needed for the wall in Problem 1 if the wall is 8 feet high and drywall is placed on both sides of the wall?

4. Determine the pounds of screws, boxes of joint compound, and rolls of tape needed for the wall in Problem 1. Historical data indicates that 1.1 pounds of screws, 0.4 box of joint compound, and 0.13 roll of tape are needed for 100 square feet of drywall.

5. Determine the number of metal studs needed to construct the walls for the tenant finish in Figure 11-6. The studs for the perimeter walls are already in place, but drywall has not been installed on the walls. Stud spacing is 16 inches on center. Allow two extra studs for each corner or intersection and four extra studs for each doorway.

6. How many pieces of runner are needed for the walls of the tenant finish in Figure 11-6 if the

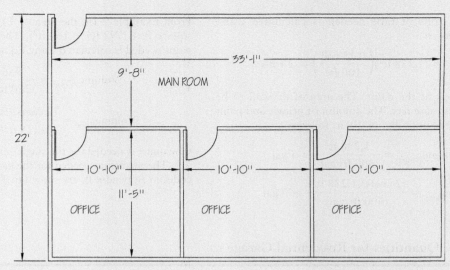

FIGURE 11-6 Wall Layout

runner comes in 10-foot-long lengths and is lapped 2 inches? The perimeter walls are already in place.

7. How many 4-foot-wide by 9-foot-high sheets of drywall are needed for the walls of the tenant finish in Figure 11-6 if the walls are 9 feet high? The perimeter walls need to be drywalled on the tenant's side only.

8. Determine the number of pieces of trim, the pounds of screws, boxes of joint compound, and rolls of tape needed for the walls of the tenant finish in Figure 11-6. The ceiling height is 9 feet. The trim is available in 10-foot lengths. Historical data indicates that 1.2 pounds of screws, 0.45 box of joint compound, and 0.15 roll of tape are needed for 100 square feet of drywall.

9. Determine the number of 11 3/4-inch by 11 3/4-inch tiles needed to tile a 10-foot 4-inch by 12-foot floor. The tile is to have 1/4-inch grout joints. Where possible, the tiles should be centered in the room with at least one-half of a tile at the edge of the room.

10. Determine the number of 7 3/4-inch by 7 3/4-inch tiles needed for the floor in Figure 11-7. The tile is to have 1/4-inch grout joints. Where possible, the tiles should be centered in the room with at least one-half of a tile at the edge of the room.

11. A 21-foot-wide by 23-foot-long ceiling is to receive a 2-foot by 2-foot acoustical ceiling. The ceiling includes twelve 4-foot by 2-foot fluorescent light fixtures and five 2-foot by 2-foot mechanical diffusers. Determine the number of 12-foot-long wall moldings, 12-foot-long main runners, 4-foot cross tees, 2-foot cross tees, and 2-foot by 2-foot tiles needed for the ceiling. Where possible, the tiles should be centered in the room with at least one-half of a tile at the edge of the room.

12. The room in Figure 11-8 is to receive a 2-foot by 2-foot acoustical ceiling. The ceiling includes five 2-foot by 2-foot fluorescent light fixtures and two 2-foot by 2-foot mechanical diffusers. Determine the number of 12-foot-long wall moldings,

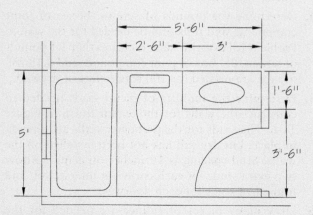

FIGURE 11-7 Bathroom Floor Plan

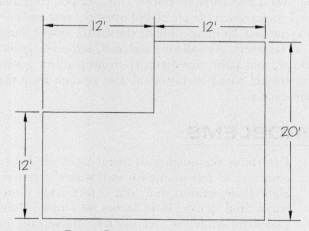

FIGURE 11-8 Room Layout

12-foot-long main runners, 4-foot cross tees, 2-foot cross tees, and 2-foot by 2-foot tiles needed for the ceiling. Where possible, the tiles should be centered in the room with at least one-half of a tile at the edge of the room.

13. Determine the number of 7.3-inch by 50.6-inch pieces of laminate flooring and rolls of underlayment needed for a 22-foot by 24-foot floor. The flooring must be staggered at least 24 inches and comes eight pieces per package. The underlayment comes in rolls 3 feet wide by 50 feet long. Run the long direction of the flooring parallel to the short direction of the room.

14. Determine the number of 7.3-inch by 50.6-inch pieces of laminate flooring and rolls of underlayment needed for the room in Figure 11-8. The flooring must be staggered at least 24 inches and comes eight pieces per package. The underlayment comes in rolls 3 feet wide by 50 feet long. Run the long direction of the flooring parallel to the short direction of the room.

15. How many yards of vinyl need to be ordered for a 17-foot-long by 22-foot-wide room?

16. How many yards of vinyl need to be ordered for the room in Figure 11-8?

17. How many 12-inch by 12-inch vinyl composition tiles (VCT) need to be ordered for a 17-foot-long by 22-foot-wide room?

18. How many 12-inch by 12-inch vinyl composition tiles (VCT) need to be ordered for the room in Figure 11-8?

19. How many feet of rubber base need to be ordered for a 17-foot-long by 22-foot 6-inch-wide room?

20. How many feet of rubber base need to be ordered for the room in Figure 11-8?

21. How many yards of carpet need to be ordered for a 17-foot-long by 22-foot-wide room? The carpet is 12 feet wide.

22. How many yards of carpet need to be ordered for the room in Figure 11-8? The carpet is 12 feet wide.

23. How many rolls of pad need to be ordered for a 17-foot-long by 22-foot-wide room? The pad comes in rolls 6 feet wide by 50 feet long.

24. How many rolls of pad need to be ordered for the room in Figure 11-8? The pad comes in rolls 6 feet wide by 50 feet long.

25. How much paint is needed to paint 75 lineal feet of a 12-foot-high block wall with one coat of primer and two coats of latex paint? From historical data, 1 gallon of primer will cover 250 square feet of wall and 1 gallon of latex paint will cover 400 square feet of wall with a single coat. Only one side of the wall is painted.

26. How much paint is needed to paint the walls in Figure 11-8 with one coat of primer and two coats of latex paint? The ceiling is 8 feet 6 inches high. From historical data, 1 gallon of primer will cover 300 square feet of wall and 1 gallon of latex paint will cover 400 square feet of wall with a single coat. Only one side of the wall is painted.

27. Determine the drywall needed to complete the Johnson Residence given in Appendix G.

28. Determine the ceramic tile needed to complete the Johnson Residence given in Appendix G.

29. Determine the laminate flooring and underlayment (pad) needed to complete the Johnson Residence given in Appendix G.

30. Determine the carpet and pad needed to complete the Johnson Residence given in Appendix G.

31. Determine the paint needed to complete the Johnson Residence given in Appendix G.

32. Determine the metal stud and track needed to complete the West Street Video project given in Appendix G.

33. Determine the drywall needed to complete the West Street Video project given in Appendix G.

34. Determine the acoustical ceiling components needed to complete the West Street Video project given in Appendix G.

35. Determine the ceramic tile needed to complete the West Street Video project given in Appendix G.

36. Determine the carpet needed to complete the West Street Video project given in Appendix G.

37. Determine the paint needed to complete the West Street Video project given in Appendix G.

38. Set up Excel Quick Tip 11-1 in Excel.

39. Set up Excel Quick Tip 11-2 in Excel.

CHAPTER TWELVE

FIRE SUPPRESSION

In this chapter you will learn how to apply the principles in Chapter 4 to fire sprinkler systems.

The design drawings and specifications may provide a complete fire sprinkler design or they may identify the requirements that the systems must meet. In the latter case, the fire sprinkler system is built under a design-build subcontract, which precludes the general contractor from identifying and quantifying the components, leaving the estimator to rely on average square foot pricing or bids from the subcontractor. When the fire sprinkler system has been designed, the estimator can identify the components and provide an accurate bid for the sprinkler system.

Fire sprinkler systems consist of the underground service, the valve assembly, fire department connections, the riser pipe, main lines, branch lines, and the sprinkler heads. The underground service provides water from the utility's pipeline to the building. The portion of the service provided by the fire sprinkler subcontractor usually stops 5 feet outside the building, with the remaining portion of the underground service being provided by the site utility contractor. A valve assembly (which includes check valves, drain valves, and a flow alarm) is located where the fire sprinkler system enters the building. A typical valve assembly for a wet pipe system is shown in Figure 12-1. The valve assembly allows the system to be serviced and sets off an alarm when the water flows.

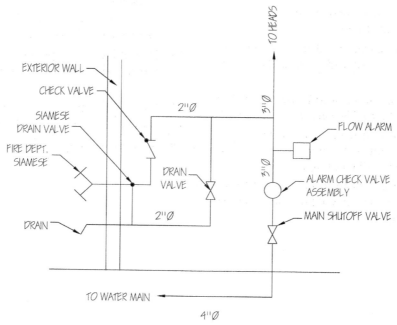

FIGURE 12-1 Valve Assembly

Fire Suppression 193

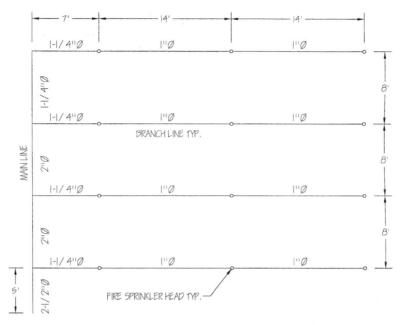

FIGURE 12-2 Main and Branch Lines

The fire department connection allows the fire department to connect their hoses to the fire sprinkler system. In Figure 12-1, a fire department Siamese connection is provided as part of the valve assembly. The riser pipe runs vertically through the building delivering water to the different levels of the building. In addition, valves and drains may be provided along the riser pipe at each floor, allowing the individual floors to be isolated. The main line runs horizontally through the building supplying water to the branch lines. The branch lines run horizontally providing water to the individual sprinkler heads. A layout of a main line and branch lines is shown in Figure 12-2.

The main and branch lines are supported from the roof or floor above by pipe hangers. Typically, steel pipe is supported at 12 feet on center, and pipes 4 inches in diameter and larger require sway bracing anytime they have a change in direction greater than 45 degrees. In addition, pipe hangers are required within 1 foot of all changes in direction. A typical pipe hanger is shown in Figure 12-3.

The main and branch lines are hung so they slope to a drain; therefore, the pipe hangers will vary in length.

The final components of the fire sprinkler system are the sprinkler head and the piece of pipe, known as a drop, that connects the head to the branch line. Like the pipe hanger, the drop will vary in length to allow the lines to be sloped so the system can be drained. A typical sprinkler head is shown in Figure 12-4. Flexible drops are used when the sprinkler heads are to be centered in a ceiling tile.

Fire sprinkler systems, where there is sufficient design, are bid as counted items. Long runs of pipe are bid as a linear component. The bid of a fire sprinkler is shown in the following example.

EXAMPLE 12-1

Determine the fire sprinkler components needed to complete the fire sprinkler system shown in Figure 12-2. The drops range from 1 to 2 feet in length.

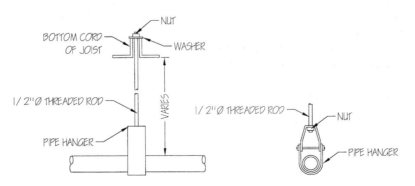

FIGURE 12-3 Pipe Hanger

CHAPTER TWELVE

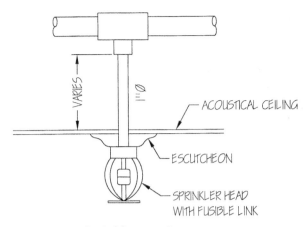

FIGURE 12-4 Sprinkler Head

TABLE 12-1 Sprinkler System Components

Quantity	Item
12 ea	Heads with fusible link
12 ea	1" dia. × 1-to 2-ft drop
12 ea	Escutcheon
4 ea	1" elbow
4 ea	1" × 1" × 1" tee
4 ea	1 1/4" × 1" × 1" tee
1 ea	1 1/4" elbow
1 ea	2" × 1 1/4" × 1 1/4" tee
1 ea	2" × 2" × 1 1/4" tee
1 ea	2 1/2" × 2" × 1 1/4" tee
8 ea	1" dia. × 14-ft pipe
4 ea	1" dia. × 7-ft pipe
1 ea	1 1/4" dia. × 8-ft pipe
2 ea	2" dia. × 8-ft pipe
1 ea	2 1/2" dia. × 5-ft pipe
20 ea	Pipe hangers

Solution: There are 12 sprinkler heads, each of which requires a sprinkler head with a fusible link, an escutcheon, and a 1-inch-diameter pipe 1 to 2 feet long for the drop. There are four branch lines. The components for each branch line, from right to left, are as follows: a 1" elbow, a 14-foot-long 1-inch-diameter pipe, a 1" × 1" × 1" tee, a 14-foot-long 1-inch-diameter pipe, a 1 1/4" × 1" × 1" tee, and a 7-foot-long 1 1/4-inch-diameter pipe. The components for a main line, from top to bottom, are as follows: a 1 1/4" elbow, an 8-foot-long 1 1/4-inch-diameter pipe, a 2" × 1 1/4" × 1 1/4" tee, an 8-foot-long 2-inch-diameter pipe, a 2" × 2" × 1 1/4" tee, an 8-foot-long 2-inch-diameter pipe, a 2 1/2" × 2" × 1 1/4" tee, and a 5-foot-long 2 1/2-inch-diameter pipe. The actual lengths of pipe are slightly shorter to account for the length of the fitting.

Pipe hangers are required every 12 feet. The number of pipe hangers required for a branch line is calculated using Eq. (4-1) as follows:

$$\text{Number} = \frac{\text{Distance}}{\text{Spacing}} + 1 \quad (12\text{-}1)$$

$$\text{Number} = \frac{(7 \text{ ft} + 14 \text{ ft} + 14 \text{ ft})}{12 \text{ ft}} + 1 = 4$$

The number of pipe hangers required for the main line is calculated using Eq. (4-1) as follows:

$$\text{Number} = \frac{(8 \text{ ft} + 8 \text{ ft} + 8 \text{ ft} + 5 \text{ ft})}{12 \text{ ft}} + 1 = 4$$

A total of 20 hangers (4 ea/branch × 4 branches + 4 ea for the main line) are required. The components for the sprinkler system are shown in Table 12-1.

CONCLUSION

A typical fire sprinkler system consists of the underground service, the valve assembly, the fire department connections, the riser pipe, main lines, branch lines, and the sprinkler heads. The pipes are supported by pipe hangers. Fire sprinkler systems are bid as a counted item. Long runs of pipe are bid as a linear component.

PROBLEMS

1. Determine the components needed to complete the fire sprinkler system shown in Figure 12-5. The drops range from 1 to 2 feet in length.

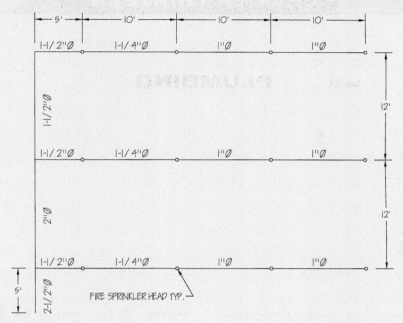

FIGURE 12-5 Fire Sprinkler Layout

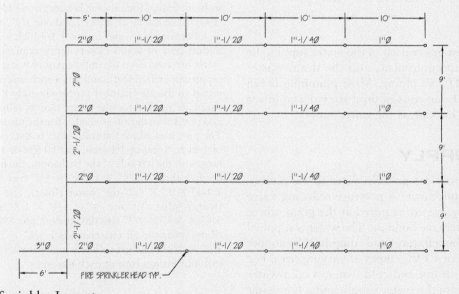

FIGURE 12-6 Fire Sprinkler Layout

2. Determine the components needed to complete the fire sprinkler system shown in Figure 12-6. The drops range from 2 to 3 feet in length.

3. Determine the components needed to complete the fire sprinkler system for the West Street Video project given in Appendix G.

CHAPTER THIRTEEN

PLUMBING

In this chapter you will learn how to apply the principles in Chapter 4 to plumbing systems. Whole books have been written on estimating plumbing; therefore, this chapter will limit its scope to the basics of estimating plumbing.

Plumbing systems consist of the water supply, the fixtures and equipment, and the drain-waste-and-vent (DWV) piping. Most plumbing is bid as a counted item. Long runs of pipe are bid as a linear good. We begin by looking at the water supply.

WATER SUPPLY

The water supply provides potable water to the plumbing fixtures and equipment. A pressure-reducing valve (PRV) and shutoff valve are required at the point where the water supply enters the building. The water is supplied to the various plumbing fixtures and equipment by copper or plastic (e.g., Pex or CPVC) pipes of varying sizes. For fixtures that use both hot and cold water, a cold water pipe must be run from the water supply and a hot water pipe must be run from the water heater. Copper pipe is connected with copper fittings that are sweated (soldered) to the pipe. Pex is coupled with compression fittings, and CPVC is coupled with CPVC fittings solvent welded to the pipe. Water lines are bid using the same procedure as for fire sprinkler systems. New water systems must be capped at the fixture to allow the system to be pressure tested. Stub-out fittings for fixtures come precapped. The bidding of water supply is shown in the following example.

EXAMPLE 13-1

Determine the water line components needed to complete the copper water supply for the bathroom shown in Figures 13-1 and 13-2.

Solution: We begin with the hot water line. Approximately 19 feet of 3/4-inch pipe are needed. Beginning with the left side of the bathroom, the following fittings are needed: four 3/4" × 3/4" × 1/2" tees (one at each lavatory), four 1/2" × 6" stub-out fittings (one at each lavatory), one 3/4" × 1/2" 90-degree elbow (near the air chamber), and one 1/2" × 12" air chamber.

For the cold water line, a 1 1/4-inch water line runs to the middle pair of water closets (approximately 8 feet), a 1-inch water line runs from the middle pair of water closets to the urinal (approximately 3 feet), and a 3/4-inch water line runs from the urinal to the air chamber (approximately 8 feet). Five 1/2-inch risers are needed for the water closets (allow 18 inches each), and one 1/2-inch riser is needed for the urinal (allow 30 inches). The pipe needed for the cold water is 8 feet of 1 1/4-inch, 3 feet of 1-inch, 8 feet of 3/4-inch, and 10 feet of 1/2-inch pipe. Beginning with the left side of the bathroom, the following fittings are needed: three 1 1/4" × 1 1/4" × 1/2" tees (water closets), one 1 1/4" × 1" × 1/2" tee (water closet), one 1" × 1" × 1/2" tee (water closet), one 1" × 3/4" × 1/2" tee (urinal), four 3/4" × 3/4" × 1/2" tees (lavatories), one 3/4" × 1/2" 90-degree elbow (near the air chamber), one 1/2" × 12" air chamber, six 1/2" × 1/2" 90-degree elbows (one per riser), and ten 1/2" × 6" stub-out fittings (one at each fixture).

DRAIN-WASTE-AND-VENT SYSTEM

The drain-waste-and-vent (DWV) is used to carry soiled water from the building and consists of three types of components: drain, waste, and vent. The drain piping and waste piping are used to carry waste water from the building. The distinction between drain piping and waste piping is that waste piping includes water with solid waste from water closets (toilets) and drain piping does not. P-traps are required at each fixture (except water closets and urinals, which have a built-in trap) to prevent sewer gas from leaving the DWV system and entering the room. This is accomplished by creating a water barrier in the fixture or DWV system near the point where the

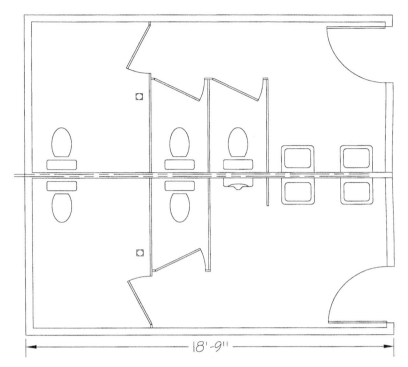

FIGURE 13-1 Water Supply Plan

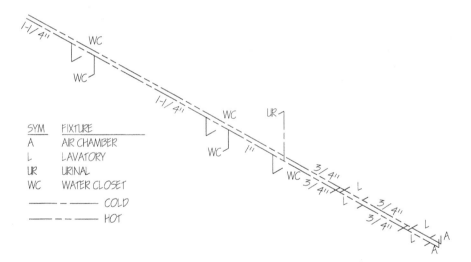

FIGURE 13-2 Water Supply Isometric

fixture connects to the DWV system. For the p-trap to work properly, the DWV system must be vented near it. This venting must occur before the top of the DWV pipe falls below the elevation of the flow line of the pipe as it leaves the p-trap. If this does not occur, the water will be siphoned from the p-trap and allow gasses to enter the building. Figure 13-3 shows the design of the p-trap.

The vent piping must be vented outside the building and a flashing must be provided for the vent pipe where it leaves the building. Cleanouts must be provided to allow blockages in the drain and waste piping to be cleaned out. The quantity takeoff for the DWV system is shown in the following example.

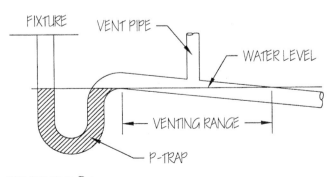

FIGURE 13-3 P-trap

CHAPTER THIRTEEN

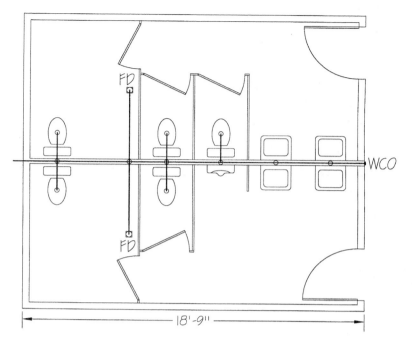

FIGURE 13-4 DWV Plan

EXAMPLE 13-2

Determine the DWV components needed to complete the DWV system for the bathroom shown in Figures 13-4 and 13-5. The horizontal waste piping is 2 feet below the finished floor, the horizontal vent piping is 9 feet above the finished floor, and the vent terminates 14 feet above the finished floor.

Solution: The DWV piping will be taken off in three steps: the main (horizontal) waste, the vertical waste servicing the fixtures, and the vent. Beginning with the cleanout, the lengths of pipe needed for the main waste line are as follows: 4 feet of 2-inch pipe run vertically, 8 feet of 2-inch pipe run horizontally, and 12 feet of 4-inch pipe. The following fittings are needed: one 2″ cleanout plug, one 2″ cleanout adapter, two 2″

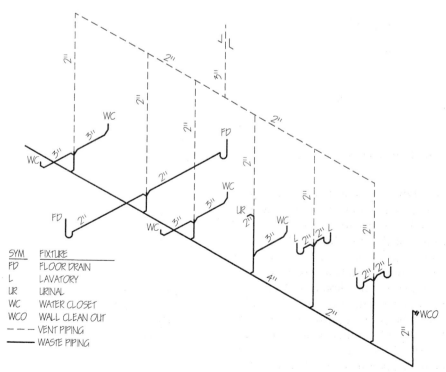

FIGURE 13-5 DWV Isometric

90-degree elbows, one 2″ sanitary tee, one 4″ × 2″ bushing, one 4″ × 4″ × 2″ reducing sanitary tee, two 4″ × 4″ × 3″ reducing sanitary tees, one 4″ × 4″ × 2″ reducing sanitary tee, and one 4″ × 4″ × 3″ reducing sanitary tee.

Each of the pair of lavatory risers will require 4 feet of 2-inch pipe, one double 2″ sanitary tee, two 6-inch-long 2-inch pieces of pipe, and two male adapters. The risers for the urinal and water closet require 1 foot of 3-inch pipe, one 3″ sanitary tee, 2 feet of 3-inch pipe, a 3″ 90-degree elbow, and a 3″ toilet flange servicing the water closet, one 3″ × 2″ bushing, 3 feet of 2-inch pipe, one 2″ sanitary tee, and one 6-inch-long 2-inch piece of pipe servicing the urinal. Each of the pairs of water closets will require 1 foot of 3-inch pipe, one 3″ double sanitary tee, one 3″ × 2″ bushing, two 2-foot-long 3-inch pipes, two 3″ 90-degree elbows, and two 3″ toilet flanges. The pair of floor drains will require 1 foot of 2-inch pipe, one 2″ double sanitary tee, two 4-foot-long 2-inch pipes, two 2″ p-traps, and two 2″ floor drains.

For the vertical venting pipe, beginning at the lavatories, three 7-foot-long 2-inch pipes, one 5-foot-long 3-inch pipe, and three 10-foot-long 2-inch pipes are needed. Fifteen feet of 2-inch pipe are needed for the horizontal vent pipe. The following fittings are needed for the vent pipe: two 2″ 90-degree elbows, four 2″ tees, one 3″ tee, and two 3″ × 2″ bushings.

In addition, seven 2″ test caps, five 3″ test caps, one 3″ thermoplastic roof flashing, and one decorative cleanout cover are needed.

FIXTURES AND EQUIPMENT

Fixtures are bid as a counted item. When bidding fixtures, the estimator must include shut-off valves for each fixture at the water supply, supply lines from the shut-off valve to the fixture, and a waste line (including the p-trap, if needed) from the fixtures to the DWV system. A shut-off valve is required for both the hot and cold water pipes at each fixture. All of these components are installed as part of the finish plumbing package. The quantity takeoff for plumbing fixtures is shown in the following example.

EXAMPLE 13-3

Determine the fixtures needed for the bathroom in Figures 13-1, 13-2, 13-4, and 13-5.

Solution: The bathroom requires five water closets (toilets), one urinal, and four lavatories (sinks). Each water closet will require a shut-off valve and a supply line to connect the water closet to the shut-off valve. The p-trap is built into the water closet. The urinal will require a flush valve assembly. Each lavatory will require two shut-off valves, two supply lines, a faucet with a drain assembly, and a 1 1/2″ p-trap.

CONCLUSION

Plumbing consists of the water supply lines, the drain-waste-vent (DWV) lines, and the fixtures. Plumbing components, pipe, fittings, and fixtures, are bid as counted items. Long runs of pipe are bid as a linear good.

PROBLEMS

1. Determine the water line components needed to complete the water supply for the bathroom shown in Figures 13-6 and 13-7.
2. Determine the DWV components needed to complete the DWV system for the bathroom shown in Figures 13-8 and 13-9. The horizontal waste piping is 2 feet below the finished floor, the horizontal vent piping is 9 feet above the finished floor, and the vent terminates 14 feet above the finished floor.
3. Determine the fixtures needed for the bathroom in Figures 13-6, 13-7, 13-8, and 13-9.
4. Determine the plumbing components needed to complete the Johnson Residence given in Appendix G.
5. Determine the plumbing components needed to complete the West Street Video project given in Appendix G.

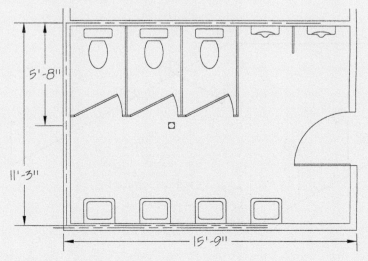

FIGURE 13-6 Water Supply Plan

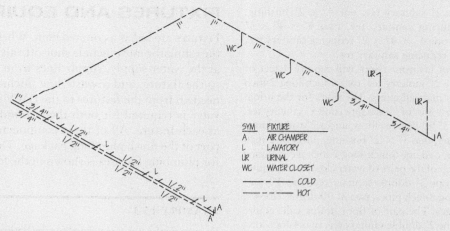

FIGURE 13-7 Water Supply Isometric

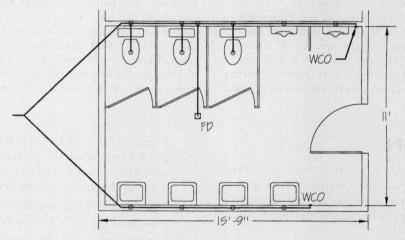

FIGURE 13-8 DWV Plan

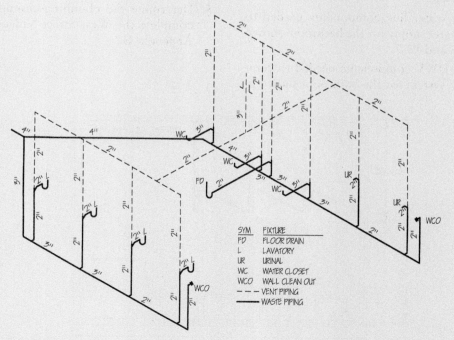

FIGURE 13-9 DWV Isometric

CHAPTER FOURTEEN

HEATING, VENTILATION, AND AIR-CONDITIONING (HVAC)

In this chapter you will learn how to apply the principles in Chapter 4 to HVAC systems. Whole books have been written on estimating HVAC systems; therefore, this chapter will limit its scope to the basics of estimating HVAC.

Heating, ventilation, and air-conditioning (HVAC) systems consist of equipment, ductwork, and piping used to heat and ventilate buildings and to condition (cool and dry) the air. Many components are bid as a counted item, including short runs of duct and pipe. Longer runs of duct and pipe are bid as a linear good. Residential and commercial HVAC systems are discussed separately. Let's begin with two common residential systems.

RESIDENTIAL HVAC SYSTEMS

Central heating and air-conditioning, the most common residential HVAC system, consists of a forced air furnace supplying heated air through metal ductwork located under the floor or above the ceiling. The supply duct, the duct that provides heated air, may consist of a main trunk line of rectangular duct running almost the entire length of the building with individual supply lines running from the trunk line to the registers (vents) in the individual rooms. Figures 14-1 and 14-2 show a heating system with a main trunk line. When the supply duct is run in the attic or crawl space, the trunk line may be eliminated by running the supply lines from the furnace to the individual rooms. For bathrooms and kitchens, the supply line may be run into the bottom of a base cabinet, and a register is placed in the base cabinet to supply the kitchen or bathroom with air, thereby using the bottom of the cabinet as part of the supply line as shown in Figure 14-3. The return air duct, which returns the air from the rooms to the furnace, usually draws air from the hallways and uses the natural space between framing members (joists and studs) as well as metal duct to transport the air as shown in Figure 14-4. Gas- and oil-fired furnaces must be vented to the outside to allow combustion gases to escape. The furnace may be equipped with a split air-conditioning unit consisting of a coil in the ductwork leaving the furnace and a condenser located outside the building. The coil is connected to the condenser using two flexible copper pipes known as a line set. Central heating may also include humidifying or dehumidifying equipment and air filtration equipment.

The takeoff of a residential central HVAC system is shown in the following example.

EXAMPLE 14-1

Determine the HVAC components needed for the residence in Figures 14-1 and 14-2. The furnace includes an air conditioner. A 6″ × 18″ grille is needed at the beginning of each return air. The distance from the top of the furnace to the roof is 20 feet. The vent pipe is 4 inches in diameter.

Solution: The equipment needed is a furnace, an air-conditioning coil, and an air-conditioning compressor. A plenum is needed between the furnace and the supply trunk line. For the supply trunk line, 16 feet of 8″ × 12″ rectangular duct, two 8″ × 12″ to 8″ × 8″ transitions, 17 feet of 8″ × 8″ square duct, and two 8″ × 8″ end caps are needed. For the individual supply lines, the following items are needed: nine 6-inch 90-degree elbows to connect to the trunk line, 90 feet (17 ft + 17 ft + 2 ft + 19 ft + 19 ft + 4 ft + 3 ft + 3 ft + 6 ft) of 6-inch-diameter duct, three 6-inch 90-degree elbows to connect to the base cabinets, three 2″ × 12″ registers, six 4″ × 12″ boots, and six 4″ × 12″ registers.

The following items are needed for the return air: two 6″ × 18″ grilles, 15 feet (1 ft + 4 ft + 1 ft + 1 ft + 7 ft + 1 ft) of 18-inch-wide sheet metal to create return air chases between the joists, two 8″ × 12″ end caps, 12 feet of 8″ × 12″ duct, and 8 feet of 12″ × 18″ duct to run from the trunk line to the bottom of the furnace.

The venting needs to run from the furnace to a point at least 2 feet above the roof, and 4 feet of pipe is allowed to enter the chase. The pipe in the chase will need to be

201

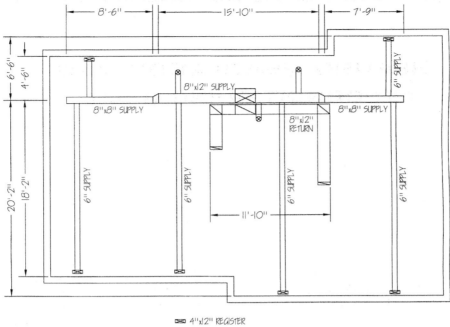

FIGURE 14-1 Plan View of Residential System

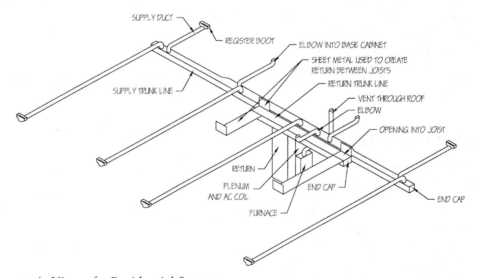

FIGURE 14-2 Isometric View of a Residential System

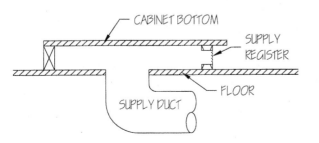

FIGURE 14-3 Supply Air in a Base Cabinet

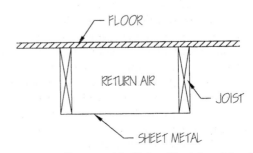

FIGURE 14-4 Return Air Using Framing Members

double walled. For the venting, the following items are needed: two 4-inch 90-degree single-wall elbows, 4 feet of 4-inch-diameter single-wall pipe, 22 feet of 4-inch-diameter double-wall pipe, one 4-inch roof flashing, and one 4-inch termination cap.

Other items needed for the furnace include one box of line set to connect the condenser to the coil, one thermostat, and one roll of low-voltage wiring. In addition, the electrician will need to provide power and a switch at the furnace, and a gas line will need to be provided from the gas source.

Another type of residential heating is radiant heating. Radiant heating uses hot water circulating in flexible plastic tubing located in the floor to warm the floor, which in turn radiates heat into the living space to warm the space. A typical tubing layout for a room is shown in Figure 14-5. This tubing is often embedded in concrete or gypcrete. In snowy climates, radiant heat located in driveways and sidewalks may be used to melt snow on these surfaces. The water used by radiant heat is heated by a boiler and is recirculated by a pump. Like furnaces, gas- and oil-fired boilers must be vented. To prevent damage to the system, an expansion tank and an air ejector are located on the supply side of the boiler. A supply manifold with valves and return manifold allow the boiler to provide water for multiple heating circuits. The temperature of each circuit may be individually controlled. The equipment for a typical radiant heating system is shown in Figure 14-6. Radiant heating systems cannot be used to cool the space;

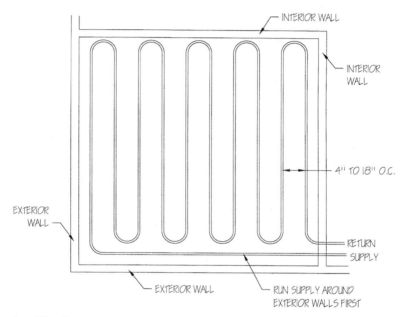

FIGURE 14-5 Radiant Heating Pipe Layout

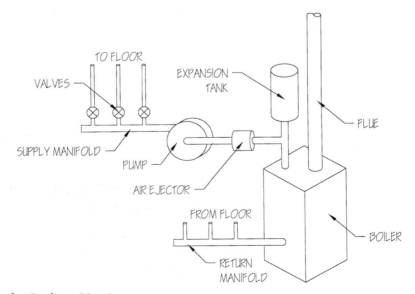

FIGURE 14-6 Equipment for Radiant Heating

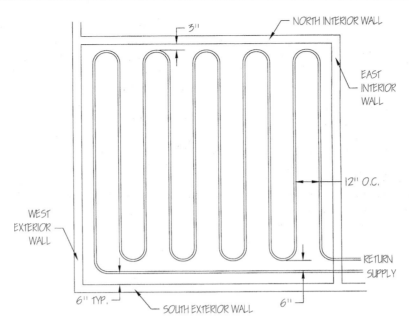

FIGURE 14-7 Pipe Spacing for Example 14-2

therefore, a separate cooling system must be included if cooling is required.

The takeoff of a radiant heating system uses the same principles as that for plumbing and is shown in the following examples.

EXAMPLE 14-2

Determine the piping needed for a radiant heating system for a 10-foot by 10-foot room. The pipe spacing is shown in Figure 14-7. Allow 40 feet of pipe to connect the pipe in the room to the boiler.

Solution: The radius on the bends will be one-half of the pipe spacing, or 6 (12/2) inches. The pipe along the south wall will begin its turn 1 foot from the wall, which equals the distance of the west pipe from the west wall (6 inches) plus the radius of the turn (6 inches). Nine feet of pipe will be needed for this pipe.

The pipe along the west wall will begin 1 foot from the south wall and will begin its turn 9 inches from the north wall. Eight feet 3 inches (8.25 ft) of pipe will be needed for this pipe.

The remaining pipes running north-south will begin their turn 18 inches from the south wall and will begin their turn 9 inches from the north wall. Each of these pipes will require 7 feet 9 inches (7.75 ft) of pipe. The distance from the pipe along the west wall to the pipe along the east wall is 9 feet (10 feet − 6 inches − 6 inches). Because we have already accounted for the pipe at the west end, we will use Eq. (4-2) to determine the number of pipes as follows:

$$\text{Number} = \frac{\text{Distance}}{\text{Spacing}}$$
$$\text{Number} = \frac{9 \text{ ft}}{1 \text{ ft}} = 9$$

The number of pipes running north-south must be an even number for the supply and return to enter the room at the same point. We have 10 pipes (9 + 1).

The length of pipe needed for a 180-degree turn is 1.6 feet ($\pi \times 0.5$ ft). Nine 180-degree turns are needed. The length of pipe needed for a 90-degree turn is 0.8 feet ($\pi \times 0.5$ ft/2). Two 90-degree turns are needed. The total length of pipe (including the 40 feet needed to connect to the boiler) is calculated as follows:

Length = 9 ft + 8.25 ft + 9(7.75 ft) + 9(1.6 ft) + 2(0.8 ft) + 40 ft
Length = 143 ft

EXAMPLE 14-3

Determine the HVAC equipment needed for a radiant heat system for a residence with six heating circuits.

Solution: The following equipment is needed: one boiler, one expansion tank, one air ejector, one pump, two manifolds with six outlets, and six valves.

Other types of heating and cooling used in residences include through-the-wall heat pumps, heat pumps that include ductwork, electric baseboard heaters, and swamp coolers.

In addition to heating and cooling equipment, residences may include gas fireplaces, gas piping, exhaust fans, and temperature controls. Gas piping is bid using the procedures used to bid pipe in Chapters 12 and 13.

COMMERCIAL HVAC SYSTEMS

Small commercial buildings (such as buildings in strip malls) and churches often use HVAC systems similar to that of the residential system shown in Figures 14-1 and 14-2, using multiple units as the building gets larger. Larger single-story buildings often use a roof-top unit (sometimes referred to as a package unit) that heats or cools the air and then distributes it through the building using ductwork. A package unit is similar to a central heating and air-conditioning system, except that it combines the heating and air-conditioning equipment into a single unit. As buildings get taller it becomes impractical to use ductwork for all of the air requirements because the duct would take up an enormous amount of space. In these cases, the heating and cooling is accomplished by heating or cooling water at a central point, distributing the water to the point where heating and cooling is needed, transferring heat from the water to air (or air to the water when cooling is needed) by use of a heat exchanger, and, finally, distributing the air to the needed areas, often by blowing the air through ductwork using a fan. An air handler combines a heat exchanger and fan to perform this function. In each of these cases, the heating and cooling equipment and ductwork are bid using the same procedure as for a residential system, and the water piping is bid using the procedures in Chapter 13. Often a commercial HVAC system uses the plenum, the space between a dropped ceiling and the roof or floor above, as the return air pathway. When this is done, return-air grilles are provided in the ceiling. Commercial systems may include a wide variety of equipment including chillers, boilers, dampers, fans, and variable-air-volume (VAV) boxes. The takeoff of a commercial HVAC system is shown in the following example.

EXAMPLE 14-4

Determine the HVAC components needed for the commercial tenant finish shown in Figure 14-8. Two feet of 6-inch-diameter flexible duct is used to connect the diffuser to the ductwork. The plenum is used for return air. The existing supply and return lines are made of copper.

Solution: The HVAC system requires one air handler and two VAV boxes. Beginning with the air handler, the following duct is needed: 6 feet of 12-inch-diameter duct with one 45-degree, 8-inch-diameter boot attached; one 45-degree, 8-inch-diameter elbow; 9 feet of 8-inch-diameter duct with one 6-inch-diameter boot attached; one 8-inch-diameter to 6-inch-diameter reducer; 11 feet of 6-inch-diameter duct; one 90-degree, 6-inch-diameter elbow; one 12-inch-diameter to 10-inch-diameter reducer; 6 feet of 10-inch-diameter duct; one 90-degree, 10-inch-diameter elbow; 6 feet of 10-inch-diameter duct with a 6-inch-diameter boot attached; one 10-inch-diameter to 8-inch-diameter reducer; 10 feet of 8-inch-diameter duct with a 6-inch-diameter boot attached; one 8-inch-diameter to 6-inch-diameter reducer; 6 feet of 6-inch-diameter duct; one 90-degree, 6-inch-diameter elbow; five 2-foot-long 6-inch-diameter flexible duct with two clamps each; and five diffusers. In addition, two thermostats with wiring are needed. The calculation of wiring is covered in Chapter 15. To connect the air handler to the supply and return lines two 2″ × 2″ × 1″ copper tees and two 4-foot-long pieces of copper pipe are needed.

Many commercial buildings use fin-tube convectors along the exterior walls of the building below the windows to assist in heating the building. The fin-tube convector uses a fin-tube to transfer heat from a pipe carrying hot water to the surrounding air. Convection, the rising of hot air and the falling of cool air, is used to move air past the fin-tube convector. A fin-tube convector is shown in Figure 14-9.

The takeoff of fin-tube convectors uses the same principles as those for plumbing and is shown in the following example.

EXAMPLE 14-5

Determine the HVAC components needed for the fin-tube convectors shown in Figure 14-10. The supply lines are run into the bottom of the fin-tube convectors from the floor below.

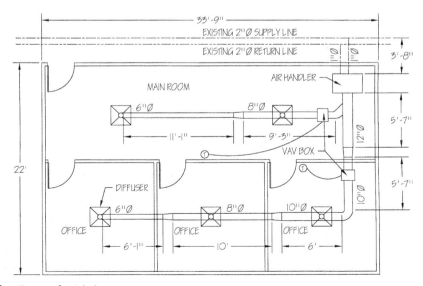

FIGURE 14-8 HVAC for Example 14-4

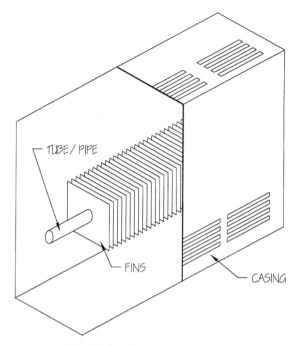

Solution: The supply line (beginning at the exiting line) requires one 2" × 2" × 1" copper tee, one piece of 1-inch-diameter pipe approximately 26 feet long to connect (from supply to interior wall), one 1" 90-degree elbow, one piece of 1-inch-diameter pipe 1 foot long, one 1" × 1" × 1" copper tee, one piece of 1-inch-diameter pipe 11 feet long, one 1" × 1" × 1" copper tee, one piece of 1-inch-diameter pipe 11 feet long, one 90-degree elbow, and three pieces of 1-inch-diameter pipe 2 feet long to connect the fin-tube convectors to the tees and elbow.

The return line (beginning at the exiting line) requires one 2" × 2" × 1" copper tee, one piece of 1-inch-diameter pipe approximately 26 feet long to connect (from supply to interior wall), one 1" 90-degree elbow, one piece of 1-inch-diameter pipe 11 feet long, one 1" × 1" × 1" copper tee, one piece of 1-inch-diameter pipe 11 feet long, one 1" × 1" × 1" copper tee, one piece of 1" pipe 11 feet long, one 90-degree elbow, and three pieces of 1-inch-diameter pipe 2 feet long to connect the fin-tube convectors to the tees and elbow.

Three 10-foot 10-inch-long fin-tube convectors will be needed.

FIGURE 14-9 Fin-Tube Convector

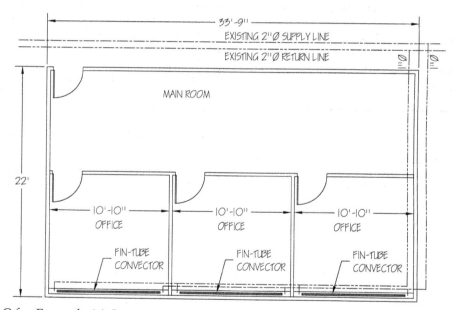

FIGURE 14-10 HVAC for Example 14-5

CONCLUSION

Heating, ventilation, and air-conditioning (HVAC) systems consist of equipment, ductwork, and piping used to heat and ventilate buildings and to condition (cool and dry) the air. Many components are bid as a counted item, including short runs of duct and pipe. Longer runs of duct and pipe are bid as a linear good.

PROBLEMS

1. Determine the HVAC components needed for the residence in Figure 14-11. The furnace includes an air conditioner. A 6" × 18" grille is needed at the beginning of each return air. The distance from the top of the furnace to the roof is 22 feet. The vent pipe is 4 inches in diameter.

Heating, Ventilation, and Air-Conditioning (HVAC) **207**

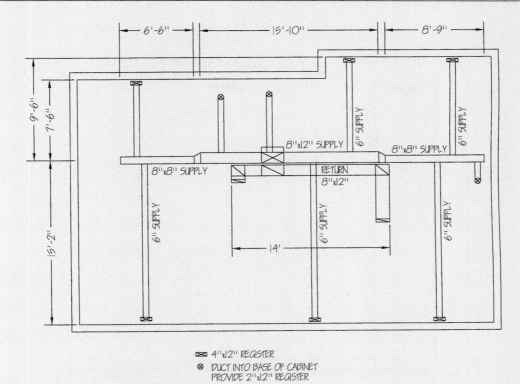

FIGURE 14-11 HVAC for Problem 1

2. Determine the piping needed for a radiant heating system for an 11-foot by 14-foot room. The pipe spacing is shown in Figure 14-7. Allow 65 feet of pipe to connect the pipe in the room to the boiler.

3. Determine the HVAC components needed for the commercial tenant finish shown in Figure 14-12. Two feet of 6-inch-diameter flexible duct is used to connect the diffuser to the ductwork. The plenum is

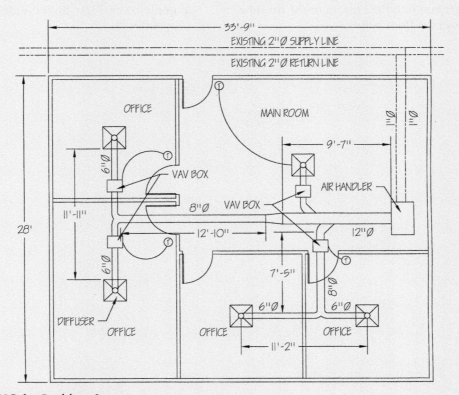

FIGURE 14-12 HAVC for Problem 3

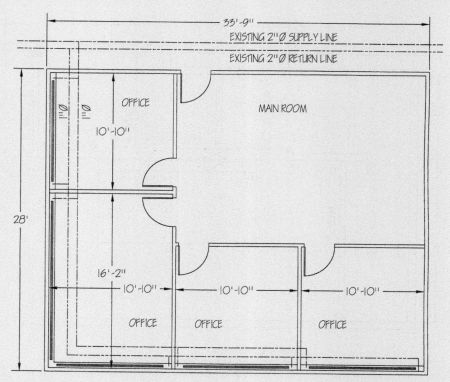

FIGURE 14-13 HVAC for Problem 4

used for return air. The existing supply and return lines are made of copper.

4. Determine the HVAC components needed for the fin-tube convectors shown in Figure 14-13. The supply lines are run into the bottom of the fin-tube convectors from the floor below.

5. Determine the HVAC components needed to complete the Johnson Residence given in Appendix G. The furnace includes an air conditioner.

6. Determine the HVAC components needed to complete the West Street Video project given in Appendix G.

CHAPTER FIFTEEN

ELECTRICAL

In this chapter you will learn how to apply the principles in Chapter 4 to electrical systems. Whole books have been written on estimating electrical systems; therefore, this chapter will limit its scope to the basics of estimating electrical systems. This chapter includes example takeoffs from the residential garage drawings.

The electrical system can be divided into two parts: (1) electrical devices such as light fixtures, outlets, panels, transformers, starters, motors, motor controls, and so forth and (2) the wiring that connects the devices. Electrical devices are typically bid as counted items; therefore, estimating the electrical devices is simply a matter of locating them on the plans and counting them up. Estimating the wiring is more difficult. For the wiring the estimator must determine the length of wiring and, if included, the conduit carrying the wire. To simplify the preparation of an estimate, sometimes the wiring is bid as a quantity-from-quantity good. Let's look how to estimate the wiring.

RESIDENTIAL WIRING

The most common wiring used in residential wiring is Romex, in which two or three conductors and a ground wire are sheathed in a plastic sheathing. Wiring for duplex outlets and lights consists of two #12 or #14 wires and a ground, with three #12 or #14 wires and a ground being run between three-way switches. Most dryers, ranges, cook tops, air conditioners, and electrical heaters run on 220 volts and require three conductors and a ground wire. Typically a range or an oven requires a 40-ampere breaker with #8 copper wires, and air conditioners and dryers require a 30-ampere breaker with #10 copper wires. All electrical equipment must be sized by the required volt-amp rating for the appliance and per code requirements.

Residential wiring is run horizontally or vertically in the walls and diagonally through ceiling spaces. When calculating the needed length of wiring, the estimator must not only take into account the horizontal distance between electrical devices, but must take into account the length of wire needed to run the wire vertically within the walls and make connections to the electrical devices. Many residential estimators include a standard length of wire for each duplex outlet, light fixture, and switch rather than determine the needed length of wire. When this is done, the wiring is being bid as a quantity-from-quantity good. Estimating the wire length is shown in the following example.

EXAMPLE 15-1

Determine the number of devices and wiring needed to complete the lighting shown in Figure 15-1. The light fixtures are to be recessed-can fixtures. The outlet is existing and will be used to supply the power to the lighting. Use Romex with #12 wires. The ceiling height is 9 feet.

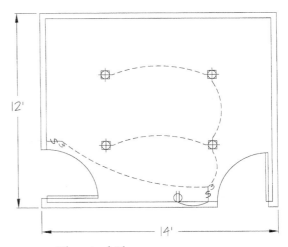

FIGURE 15-1 Electrical Plan

209

Solution: The lighting will require four can lights with trim and light bulbs and two three-way switches with single-gang boxes and covers. The outlet is already there. The distances between the electrical devices are shown in Figure 15-2. When preparing an estimate, these distances would be taken off using a scale, a plan measurer, or software takeoff package.

The wiring between the right switch and the lights and outlet is to be two #12 copper wires with a ground. When figuring this wiring we add 1 foot to each end of the wire to allow the wiring to be secured to the framing and connected to the electrical device. The wiring between lights can be run in the ceiling spaces. The length of wiring between the light fixtures is calculated as follows:

$$\text{Length} = (6.33 \text{ ft} + 2 \text{ ft}) + (4.33 \text{ ft} + 2 \text{ ft}) + (6.33 \text{ ft} + 2 \text{ ft}) = 23 \text{ ft}$$

The wiring between the switch and the light will need to run 6 vertical feet in the wall to go from the switch to the ceiling space. The length of wiring between the switch and the light fixture is 12 feet (3.8 ft + 6 ft + 2 ft). The wiring from the outlet to the switch will need to run 3 feet vertically and 2 feet horizontally. The length of wiring between the outlet and the switch is 7 ft (2 ft + 3 ft + 2 ft). The lighting will require 42 feet (23 ft + 12 ft + 7 ft) of wire with two #12 wires and a ground. The wiring between switches will require three #12 copper wires with a ground and must be run through the wall rising over the door frame or must be run in the ceiling space. If the wire is run through the ceiling space, it will need to be run 6 vertical feet in the wall to go from the switch to the ceiling space at both ends. The lighting will require 25 feet (6 ft + 10.25 ft + 6 ft + 2 ft) of wire with three #12 copper wires and a ground. In addition, wire nuts are needed to connect the wires, and staples are needed to secure the wires to the framing.

COMMERCIAL WIRING

From an estimator's point of view, commercial wiring differs from residential wiring in a number of key ways, including the following: (1) all commercial wiring must be run in conduit or raceways (except for low-voltage wiring), (2) the wiring is usually run parallel to the side of the building or structure within the ceiling spaces, increasing the length of the wire needed, (3) the design of the wiring is better defined, (4) commercial wiring is more extensive and services more equipment, (5) commercial wiring includes multiple electrical panels and (6) commercial wiring is wired with individual strands of wire in conduit. When estimating commercial wiring, the estimator must include the conduit, boxes, connector between the conduit and boxes, electrical devices, and wiring. The conduit may be ridged or flexible. As is the case with residential wiring, the estimator must take into account not only the horizontal distance between electrical devices, but also the length of wire needed to run the wire vertically within the walls and make connections to the electrical devices.

Figure 15-3 shows the typical wiring for a fluorescent light fixture in a dropped acoustical ceiling. To complete this light fixture, not only do the conduit and wiring need to be run to the fixture, but at each fixture the following items are needed: a junction box with a cover, two set-screw/threaded connectors to connect the

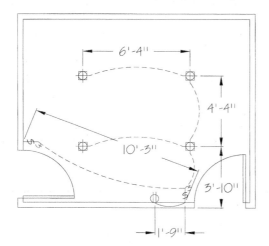

FIGURE 15-2 Distances between Electrical Devices

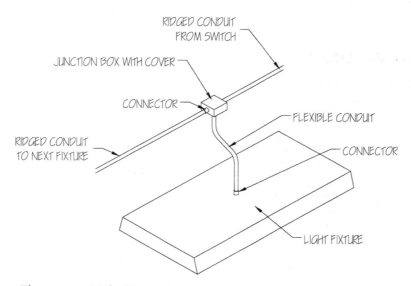

FIGURE 15-3 Wiring for a Fluorescent Light Fixture

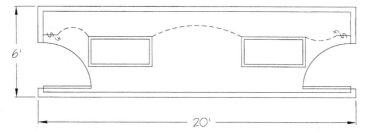

FIGURE 15-4 Electrical Plan

ridged conduit to the box, two set-screw/threaded connectors to connect the flexible conduit, and a length of flexible conduit whose length will depend on the difference in elevation between the ridged conduit and the dropped ceiling. Estimating commercial electrical components and wiring is shown in the following examples.

EXAMPLE 15-2

Determine the number of electrical devices, conduit, and wiring needed to complete the lighting shown in Figure 15-4. The light fixtures are to be 120-volt, 4-foot by 2-foot fluorescent light fixtures with three bulbs. The power will be provided to the light switch on the left. Use #12 copper wires in a 1/2-inch conduit. The ceiling height is 9 feet, and the conduit will be run at a height of 12 feet.

Solution: The lighting will require two 120-volt, 4-foot by 2-foot fluorescent light fixtures, six fluorescent light bulbs, two junction boxes with covers, and two three-way switches with single-gang boxes and covers. Two hot wires will need to be run between the switches, a hot wire will need to be run between the right light switch and the two light fixtures, a common wire will need to be run between the two light fixtures and the left light switch, and a ground wire will need to be run to all the switches and fixtures. The wiring diagram for the lights is shown in Figure 15-5.

Ridged conduit will be run from the switch at one end of the hall to the switch at the other end of the hall, passing through two junction boxes. The conduit can be bent in a radius to change directions. Nine feet of conduit is required at both switches to go from the switch to the height of 12 feet. Thirty-eight feet of conduit are needed (9 ft + 20 ft + 9 ft) as well as 6 set-screw/threaded connectors, which are used to connect the ridged conduit to the junction boxes and single-gang boxes. Two 5-foot lengths of flexible conduit are needed to run from the junction boxes to the light fixtures and are used to center the fixtures in the hall. Four set-screw/threaded connectors are needed to connect the flexible conduit to the junction boxes and the light fixtures. In addition, fasteners to secure the conduit in place are needed. The material the conduit is being fastened to will determine the type of fastener used.

Allow for an extra foot of wire on both ends of each wire. Forty feet (38 ft + 2 ft) of red wire is needed to connect the switches. Similarly, 40 feet of black wire are needed to connect the switches. Sixteen feet (9 ft + 5 ft + 2 ft) of black wire is needed to connect the right switch to the junction box above the right light fixture. Twelve feet (10 ft + 2 ft) of black wire is needed to connect the junction boxes together, and 7 feet (5 ft + 2 ft) of black wire is needed at each light fixture to connect the junction boxes to the light fixtures. A total of 82 feet (40 ft + 16 ft + 12 ft + 7 ft + 7 ft) of black wire is needed. For the white wire, 16 feet is needed to connect the left switch to the junction box above the left light fixture, 12 feet is needed to connect the junction boxes, and 7 feet is needed at each light fixture to connect the junction boxes to the light fixtures, for a total of 42 feet (16 ft + 12 ft + 7 ft + 7 ft). For the green ground wire, 16 feet is needed at each end to connect the switches to the junction boxes, 12 feet is needed to connect the junction boxes, and 7 feet is needed at each light fixture to connect the junction boxes to the light fixtures, for a total of 58 feet (16 ft + 16 ft + 12 ft + 7 ft + 7 ft).

EXAMPLE 15-3

Determine the electrical components needed to complete the electrical distribution system (excluding wire from the panelboards) for the schematic shown in Figure 15-6. The distance to the meter is 60 feet through a 3 1/2-inch conduit that includes three 90-degree elbows. The wireway is 5 feet long, and the three panels are mounted 1 foot above the wireway. The panelboard schedules are shown in Figures 15-7 through 15-9.

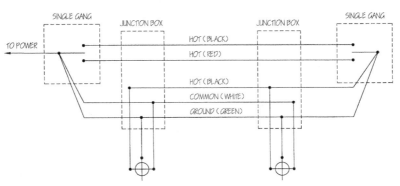

FIGURE 15-5 Wiring Diagram

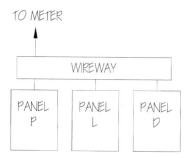

FIGURE 15-6 Distribution Schematic

Solution: Beginning at the meter, 60 feet of 3 1/2-inch conduit, three 3 1/2-inch 90-degree elbows, two 3 1/2-inch set-screw/threaded connectors (to connect the conduit to the meter and wireway), and one 5-foot wireway are needed to house the wiring from the meter to the panelboards.

The information for the panelboards, wiring size servicing the panelboards, and conduit sizes are found at the top of each panel schedule. For Panel P, a 120/208v three-phase 225-ampere panelboard with 30 spaces is needed. The panel is serviced by three #4/0 (0000) wires and one #12 wire. Each of these wires will need to be approximately 70 feet long to allow 2 feet for connection in the panelboard and meter, wire in the conduits, and wireway. The panelboard is connected to the wireway by 1 foot of 2-inch conduit with two 2-inch set-screw/threaded connectors. The rated amps and number of poles for the circuit breakers are specified in the "CB A-P" column of the panel schedules. The following breakers are needed for Panel P: two 90-ampere three-pole, one 35-ampere three-pole, one 30-ampere three-pole, three 30-ampere single-pole, two 25-ampere single-pole, one 20-ampere single-pole, and three 30-ampere double-pole breakers.

For Panel L, a 120/208v three-phase 150-ampere panelboard with 30 spaces is needed. The panelboard is connected to the wireway by 1 foot of 1 1/2-inch conduit with two 1 1/2-inch set-screw/threaded connectors. The panel is serviced

PANEL P POWER PANEL
120/208 VOLTS 225 AMPS MCB or (MLO) 30 SPACES
FEEDERS 3 #4/0 AND 1 #12 IN 2" CONDUIT

KEY	DESCRIPTION	CB A-P	WIRE	LOAD (WATTS OR VOLT-AMPS)		
1	15-HP MOTOR	90/3	#6	5,571		
2					5,571	
3						5,571
4	15-HP MOTOR	90/3	#6	5,571		
5					5,571	
6						5,571
7	5-HP MOTOR	35/3	#10	2,018		
8					2,018	
9						2,018
10	3-HP MOTOR	30/3	#14	1,269		
11					1,269	
12						1,269
13	1-HP MOTOR	30/1	#12	16		
14	1-HP MOTOR	30/1	#12		16	
15	1-HP MOTOR	30/1	#12			16
16	3/4-HP MOTOR	25/1	#12	14		
17	3/4-HP MOTOR	25/1	#12		14	
18	1/2-HP MOTOR	20/1	#14			10
19	STRIP HEAT	30/2	#10	2,250		
20					2,250	
21	STRIP HEAT	30/2	#10			2,250
22					2,250	
23	STRIP HEAT	30/2	#10		2,250	
24						2,250
25						
26						
27						
28						
29						
30						
			UNBALANCED LOADS	30	30	26
			PHASE TOTALS	18,859	18,959	18,955
			PANEL TOTAL		56,873	
	CONNECTED LOAD AMP	158	MAX. NEUTRAL AMPS	1		

FIGURE 15-7 Panel P Schedule

PANEL L   LIGHTING PANEL

120/208 VOLTS 150 AMPS MCB or (MLO) 30 SPACES
FEEDERS 4 #1/0 AND __ # ____ IN 1-1/2" CONDUIT

KEY	DESCRIPTION	CB A-P	WIRE	LOAD (WATTS OR VOLT-AMPS)		
1	LIGHTING	20/1	#12	1,800		
2	LIGHTING	20/1	#12		1,800	
3	LIGHTING	20/1	#12			1,800
4	LIGHTING	20/1	#12	1,800		
5	LIGHTING	20/1	#12		1,800	
6	LIGHTING	20/1	#12			1,800
7	LIGHTING	20/1	#12	1,800		
8	LIGHTING	20/1	#12		1,800	
9	LIGHTING	20/1	#12			1,800
10	LIGHTING	20/1	#12	1,800		
11	LIGHTING	20/1	#12		1,800	
12	LIGHTING	20/1	#12			1,800
13	LIGHTING	20/1	#12	1,800		
14	LIGHTING	20/1	#12		1,800	
15	LIGHTING	20/1	#12			1,800
16	LIGHTING	20/1	#12	1,800		
17	LIGHTING	20/1	#12		1,800	
18	LIGHTING	20/1	#12			1,800
19	LIGHTING	20/1	#12	1,800		
20	LIGHTING	20/1	#12		1,800	
21						
22						
23						
24						
25						
26						
27						
28						
29						
30						
			UNBALANCED LOADS	12,600	12,600	10,800
			PHASE TOTALS	12,600	12,600	10,800
			PANEL TOTAL	36,000		
	CONNECTED LOAD AMP	100	MAX. NEUTRAL AMPS	100		

FIGURE 15-8 Panel L Schedule

by four #1/0 (0) wires approximately 70 feet long. The panel requires twenty 20-ampere single-pole circuit breakers.

For Panel D, a 120/208v three-phase 150-ampere panelboard with 30 spaces is needed. The panelboard is connected to the wireway by 1 foot of 1 1/2-inch conduit with two 1 1/2-inch set-screw/threaded connectors. The panel is serviced by four #1/0 (0) wires approximately 70 feet long. The panel requires twenty-three 20-ampere single-pole circuit breakers.

SAMPLE TAKEOFF FOR THE RESIDENTIAL GARAGE

A sample takeoff from a set of plans is shown in the following example.

EXAMPLE 15-4

Prepare a quantity takeoff for the electrical system for the residential garage given in Appendix G.

Solution: We begin with the service from the house to the garage. To install this service we need to run a metal conduit down the side of the house, run PVC conduit from the house to the garage, and install the panel with breakers in the garage. From the site plan, the horizontal distance between the panel at the house and the panel at the garage is 33 feet. The wire will need to be buried 18 inches in the ground, and the boxes are mounted about 4 feet off of the ground, requiring at least 6 additional feet of wire at each end. To be safe we will order 50 feet of wire. The quantity takeoff for the service is as shown in Table 15-1. The conduit stops at the bottom of the garage wall and the wire will be run exposed through the wall to the panel in the garage. Next, we prepare the quantity takeoff for the electrical devices in the garage, exclusive of wiring. The items needed are shown in Table 15-2. Next, we prepare the quantity takeoff for the electrical wiring. The wire is to be Romex with two #12 copper wires and a ground. The wiring distances for the outlets are shown in Figure 15-10. There are two outlet circuits; therefore, the outlet for the overhead door and two wall outlets are connected on one circuit and the remaining two wall outlets are connected on the other circuit.

Panel D Schedule

PANEL D   DUPLEX OUTLETS
120/208 VOLTS   150 AMPS   MCB or (MLO)   30 SPACES
FEEDERS 4 #1/0 AND __ # ___ IN 1-1/2" CONDUIT

KEY	DESCRIPTION	CB A-P	WIRE	LOAD (WATTS OR VOLT-AMPS)		
1	OUTLETS	20/1	#12	1,440		
2	OUTLETS	20/1	#12		1,440	
3	OUTLETS	20/1	#12			1,440
4	OUTLETS	20/1	#12	1,440		
5	OUTLETS	20/1	#12		1,440	
6	OUTLETS	20/1	#12			1,440
7	OUTLETS	20/1	#12	1,440		
8	OUTLETS	20/1	#12		1,440	
9	OUTLETS	20/1	#12			1,440
10	OUTLETS	20/1	#12	1,440		
11	OUTLETS	20/1	#12		1,440	
12	OUTLETS	20/1	#12			1,440
13	OUTLETS	20/1	#12	1,440		
14	OUTLETS	20/1	#12		1,440	
15	OUTLETS	20/1	#12			1,440
16	OUTLETS	20/1	#12	1,440		
17	OUTLETS	20/1	#12		1,440	
18	OUTLETS	20/1	#12			1,440
19	OUTLETS	20/1	#12	1,440		
20	OUTLETS	20/1	#12		1,440	
21	OUTLETS	20/1	#12			1,440
22	OUTLETS	20/1	#12	1,440		
23	OUTLETS	20/1	#12		1,440	
24						
25						
26						
27						
28						
29						
30						
	UNBALANCED LOADS			11,520	11,520	10,800
	PHASE TOTALS			11,520	11,520	10,800
	PANEL TOTAL			33,120		

CONNECTED LOAD AMP __92__   MAX. NEUTRAL AMPS __92__

FIGURE 15-9 Panel D Schedule

TABLE 15-1   Quantity Takeoff for Servicer

Quantity	Item	Use
1 ea	Meter tap	Connect wire to existing meter
1 ea	1 1/4" set-screw/threaded connector	Connect conduit to existing meter
10 ft	1 1/4" conduit	Comes in 10-ft lengths
1 ea	1 1/4" compression/male threaded coupling	Connect ridged conduit to PVC conduit
1 ea	1 1/4" slip/female threaded PVC coupling	Connect ridged conduit to PVC conduit
2 ea	1 1/4" PVC 90-degree sweep	Turn from horizontal to vertical at both ends
40 ft	1 1/4" PVC conduit	Run wire underground
50 ft	3 ea #4 copper wire with ground	Wire from meter to panel
1 ea	70-ampere single-phase panel with main breaker slot	Panel for garage
1 ea	70-ampere main breaker	Main breaker for panel
1 ea	20-ampere breaker	Breaker for lighting
2 ea	20-ampere GFCI breaker	Breakers for outlets
1 ea	6' copper ground rod	Ground rod per plan

TABLE 15-2 Quantity Takeoff for Electrical Devices

Quantity	Item	Use
3 ea	Octagon box	Exterior lights
2 ea	Single-gang box	Fluorescent lights
5 ea	Single-gang box	Outlets
5 ea	Duplex outlets	Outlets
4 ea	Outlet covers	Interior outlets
1 ea	Weatherproof outlet cover	Exterior outlet
1 ea	Quad-gang box	Switches
1 ea	Quad-gang switch cover	Switches
4 ea	Single-pole switches	Switches
2 ea	Four-tube fluorescent fixtures	Interior light fixtures
8 ea	4′ fluorescent light bulbs	Interior light fixtures
3 ea	Brass coach lights	Exterior light fixtures
3 ea	9-w LED light bulbs	Exterior light fixtures

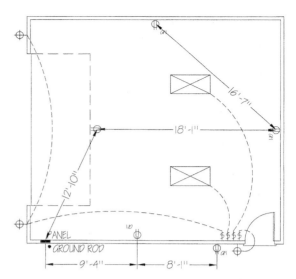

FIGURE 15-10 Horizontal Wiring Distances for Outlets

One foot of wire is allowed for each connection and 2 feet of wire is allowed at the panel. The wire from the panel to the overhead door outlet will need to be run 5 feet up the wall and 13 feet across the ceiling and will require 3 feet for connections at the panel and outlet, for a total of 21 feet. The wire from the overhead door outlet to the outlet on the east wall will need to be run 18 feet across the ceiling and 7 feet down the wall and will require 2 feet for connections at the outlets, for a total of 27 feet. The wire from the outlet on the east wall to the outlet on the north wall will need to be run 7 feet up the wall, 17 feet across the ceiling, and 7 feet down the wall, and it will require 2 feet for connections at the outlets, for a total of 33 feet. The wire from the panel to the outlet on the inside of the south wall will need to be run 3 feet down the wall, and 10 feet along the wall, and it will require 3 feet for connections at the panel and outlet, for a total of 16 feet. The wire from the outlet on the inside of the south wall to the outside outlet will need to be run 9 feet along the wall and 2 feet up the wall and it will require 2 feet for connections at the outlets, for a total of 13 feet. The total length of wire needed for the outlets is 110 feet (21 ft + 27 ft + 33 ft + 16 ft + 13 ft).

The wiring distances for the lights are shown in Figure 15-11. One foot of wire is allowed for each connection and 2 feet of wire is allowed at the panel. The wire from the panel to the switches is run 20 feet and will require 3 feet for

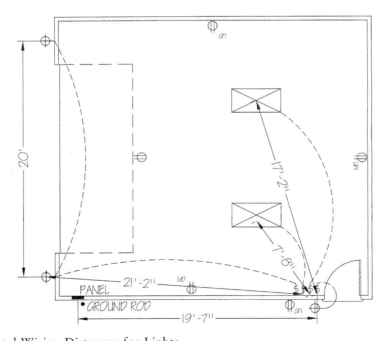

FIGURE 15-11 Horizontal Wiring Distances for Lights

TABLE 15-3 Quantities for Residential Garage

26-100 Electrical

Meter tap	1	ea	Duplex outlets	5	ea
1 1/4" set-screw/threaded connector	1	ea	Outlet covers	4	ea
1 1/4" conduit	10	ft	Weatherproof outlet cover	1	ea
1 1/4" compression/male threaded coupling	1	ea	Quad-gang box	1	ea
1 1/4" slip/female threaded PVC coupling	1	ea	Quad-gang switch cover	1	ea
1 1/4" PVC 90-degree sweep	2	ea	Single-pole switches	4	ea
1 1/4" PVC conduit	40	ft	Four-tube fluorescent fixtures	2	ea
Install conduit	40	ft	4' fluorescent light bulbs	8	ea
3 ea #4 copper wire with ground	50	ft	Brass coach lights	3	ea
70-ampere single-phase panel with main	1	ea	9-w LED light bulbs	3	ea
70-ampere main breaker	1	ea	250' roll 2 #12 with ground Romex	1	ea
20-ampere breaker	1	ea	NM/SE cable connectors	3	ea
20-ampere GFCI breaker	2	ea	Electrical staples	1	box
6' copper ground rod	1	ea	Red wire nuts	1	bag
Install panel	1	ea	Install light fixture	5	ea
Octagon box	3	ea	Install outlet	5	ea
Single-gang box	7	ea	Install switch	4	ea

connections at the panel and outlet, for a total of 23 feet. The wire from the switch to the north interior light will need to be run 5 feet up the wall and 17 feet across the ceiling and will require 2 feet for connections at ends, for a total of 24 feet. The wire from the switch to the south interior light will need to be run 5 feet up the wall and 8 feet across the ceiling and will require 2 feet for connections at ends, for a total of 15 feet. The wire from the switch to the light on the south side of the overhead door will need to be run 5 feet up the wall, 21 feet across the ceiling, and 2 feet down the wall and will require 2 feet for connections at ends, for a total of 30 feet. The wire from the light on the south side of the overhead door to the light on the north side of the overhead door will need to be run 2 feet up the wall, 20 feet across the ceiling, and 2 feet down the wall and will require 2 feet for connections at ends, for a total of 26 feet. The wire from the switch to the light on the south exterior wall of the garage will need to be run 4 feet up the wall and will require 2 feet for connections at ends, for a total of 6 feet. The total length for wire needed of the lighting is 124 feet (23 ft + 24 ft + 15 ft + 30 ft + 26 ft + 6 ft). Six feet of wire are needed to connect the panel to the ground rod.

A 250-foot roll of Romex is needed for the wiring. In addition, three NM/SE cable connectors to secure the Romex to the panel, a box of electrical staples to secure the Romex to the framing, and a bag of red wire nuts to connect the wires are needed.

The quantities needed for the garage, grouped by the cost codes in Appendix B, are shown in Table 15-3.

CONCLUSION

The electrical system consists of electrical devices such as light fixtures, outlets, panels, transformers, starters, motors, motor controls, and so forth and the wiring that connects the devices. Electrical devices are typically bid as counted items. For the wiring the estimator must determine the length of wiring and the conduit, including connectors, carrying the wire. The estimator must also include fasteners used to secure the wire or conduit and wire nuts used to connect the wires.

PROBLEMS

1. Determine the number of electrical devices needed to complete the electrical system in Figure 15-12. The light fixtures are to be 9-inch mushroom light fixtures.

2. Determine the wiring needed to complete the electrical system shown in Figure 15-12. Forty feet of wire is needed to run from the top-right corner of the room to the panel. All wiring in the room is to be run on the same circuit and is to be run with #12 Romex. The ceiling height is 8 feet.

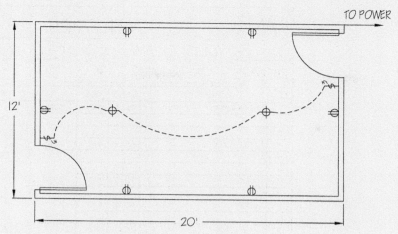

FIGURE 15-12 Electrical Plan

3. Determine the number of electrical devices needed to complete the lighting shown in Figure 15-13. The light fixtures are to be 120-volt, 4-foot by 2-foot fluorescent light fixtures with three bulbs.
4. Determine the conduit and wiring needed to complete the lighting shown in Figure 15-13. Assume that power will be provided to the light switches and #12 wires will be used. The ceiling height is 9 feet and the conduit will be run at a height of 12 feet.
5. Determine the electrical components needed to complete the electrical distribution system (excluding wire from the panelboards) for the schematic shown in Figure 15-6. The distance to the meter is 35 feet through a 2 1/2-inch conduit that includes three 90-degree elbows. The wireway is 5 feet long, and the three panels are mounted 2 feet above the wireway. The panelboard schedules are shown in Figures 15-14 through 15-16.
6. Determine electrical devices and wiring needed for the Johnson Residence given in Appendix G.
7. Determine the electrical devices, conduit, and wiring needed for the West Street Video project given in Appendix G.

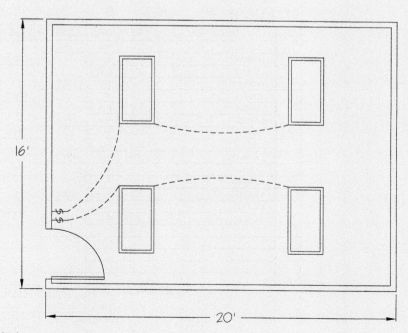

FIGURE 15-13 Electrical Plan

PANEL P   POWER PANEL

120/208 VOLTS 100 AMPS MCB or (MLO) 24 SPACES
FEEDERS 3 #3 AND 1 #12 IN 1-1/4" CONDUIT

KEY	DESCRIPTION	CB A-P	WIRE	LOAD (WATTS OR VOLT-AMPS)		
1	5-HP MOTOR	35/3	#10	2,018		
2					2,018	
3						2,018
4	5-HP MOTOR	35/3	#10	2,018		
5					2,018	
6						2,018
7	DRYER	30/2	#10	2,250		
8					2,250	
9	DRYER	30/2	#10			2,250
10				2,250		
11	1-HP MOTOR	30/1	#12		16	
12	1-HP MOTOR	30/1	#12			16
13	1-HP MOTOR	30/1	#12	16		
14	1-HP MOTOR	30/1	#12		16	
15	3/4-HP MOTOR	25/1	#12			14
16	3/4-HP MOTOR	25/1	#12	14		
17	3/4-HP MOTOR	25/1	#12		14	
18	1/2-HP MOTOR	20/1	#14			10
19	1/2-HP MOTOR	20/1	#14	10		
20	1/2-HP MOTOR	20/1	#14		10	
21						
22						
23						
24						
	UNBALANCED LOADS			40	56	40
	PHASE TOTALS			8,576	6,342	6,326
	PANEL TOTAL			21,244		

CONNECTED LOAD AMP ___59___   MAX. NEUTRAL AMPS ___1___

FIGURE 15-14 Panel P Schedule

PANEL L   LIGHTING PANEL

120/208 VOLTS 125 AMPS MCB or (MLO) 20 SPACES
FEEDERS 4 #1 AND __ #____ IN 1-1/2" CONDUIT

KEY	DESCRIPTION	CB A-P	WIRE	LOAD (WATTS OR VOLT-AMPS)		
1	LIGHTING	20/1	#12	1,800		
2	LIGHTING	20/1	#12		1,800	
3	LIGHTING	20/1	#12			1,800
4	LIGHTING	20/1	#12	1,800		
5	LIGHTING	20/1	#12		1,800	
6	LIGHTING	20/1	#12			1,800
7	LIGHTING	20/1	#12	1,800		
8	LIGHTING	20/1	#12		1,800	
9	LIGHTING	20/1	#12			1,800
10	LIGHTING	20/1	#12	1,800		
11	LIGHTING	20/1	#12		1,800	
12	LIGHTING	20/1	#12			1,800
13	LIGHTING	20/1	#12	1,800		
14	LIGHTING	20/1	#12		1,800	
15	LIGHTING	20/1	#12			1,800
16						
17						
18						
19						
20						
	UNBALANCED LOADS			9,000	9,000	9,000
	PHASE TOTALS			9,000	9,000	9,000
	PANEL TOTAL			27,000		

CONNECTED LOAD AMP ___75___   MAX. NEUTRAL AMPS ___75___

FIGURE 15-15 Panel L Schedule

PANEL D   DUPLEX OUTLETS

120/208 VOLTS 100 AMPS MCB or (MLO) 24 SPACES

FEEDERS 4 #3 AND __ # ____ IN 1-1/4" CONDUIT

KEY	DESCRIPTION	CB A-P	WIRE	LOAD (WATTS OR VOLT-AMPS)		
1	OUTLETS	20/1	#12	1,440		
2	OUTLETS	20/1	#12		1,440	
3	OUTLETS	20/1	#12			1,440
4	OUTLETS	20/1	#12	1,440		
5	OUTLETS	20/1	#12		1,440	
6	OUTLETS	20/1	#12			1,440
7	OUTLETS	20/1	#12	1,440		
8	OUTLETS	20/1	#12		1,440	
9	OUTLETS	20/1	#12			1,440
10	OUTLETS	20/1	#12	1,440		
11	OUTLETS	20/1	#12		1,440	
12	OUTLETS	20/1	#12			1,440
13	OUTLETS	20/1	#12	1,440		
14	OUTLETS	20/1	#12		1,440	
15						
16						
17						
18						
19						
20						
21						
22						
23						
24						
	UNBALANCED LOADS			7,200	7,200	5,760
	PHASE TOTALS			7,200	7,200	5,760
	PANEL TOTAL				20,160	

CONNECTED LOAD AMP ___56___   MAX. NEUTRAL AMPS ___56___

FIGURE 15-16 Panel D Schedule

CHAPTER SIXTEEN

EARTHWORK

In this chapter you will learn about the characteristics of soils as they relate to excavation and how to estimate excavation using the geometric method, average-width-length-depth method, average-end method, modified-average-end method, and cross-sectional method. It also includes example takeoffs from the residential garage drawings. This chapter includes a sample spreadsheets that may be used in the quantity takeoff.

Before we can learn about estimating excavation quantities, we must first understand the basic characteristics of soils.

CHARACTERISTICS OF SOILS

Soils are made up of three basic components: solids, water, and air. The composition of a soil may be described by four terms: wet unit weight (or wet density), dry unit weight (or dry density), unit weight of water, and water content.

The wet unit weight describes the density of the soil, both the solids and the water. The wet unit weight is calculated by dividing the total weight of the soil (solids and water) by the total volume of the soil and is expressed in pounds per cubic foot or pounds per cubic yard.

The dry unit weight describes the density of the solid portion of the soil. The dry unit weight is calculated by dividing the weight of the solids by the total volume of the soil and is expressed in pounds per cubic foot or pounds per cubic yard. The dry unit weight is used to measure how well the soil is compacted.

The unit weight of water is the weight of the water in the soil and is expressed in pounds per cubic foot. The unit weight of water equals the wet unit weight less the dry unit weight.

The water content describes the relationship between the solids and the water in the soils and is the weight of the water (not volume) expressed as a percentage of the weight of the solids.

The water content of the soils is important to the estimator because the soils must meet specific water content requirements for the soils to properly compact. Soils that are too wet will need to be dried out before compaction, whereas soils that are too dry will require the addition of moisture.

The dry unit weight is important to the estimator because it can be used to describe how the volume of the soil changes as it is excavated, transported, and compacted.

SWELL AND SHRINKAGE

During the excavation and placement process, the volume of the soils changes. The volume of the soils can be described in three conditions: bank (or in situ), loose, and compacted. These conditions are shown in Figure 16-1. The bank or in situ condition describes the conditions of the soil before it has been excavated.

When soils are excavated they swell, their volume increases, and their density decreases. Soils that have been excavated and have not been placed are described as being in a loose condition. Soils are transported in the loose condition.

The relationship between the bank and loose conditions is described as the *swell percentage*. The swell, or the amount the volume increases as the soils are excavated, is calculated using the following equation:

$$\text{Swell \%} = \left(\frac{D_B}{D_L} - 1\right)100 \qquad (16\text{-}1)$$

where

D_B = Bank Dry Density (Unit Weight)
D_L = Loose Dry Density (Unit Weight)

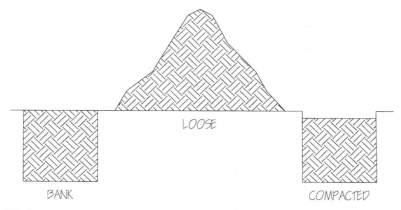

FIGURE 16-1 Volumes of Soils

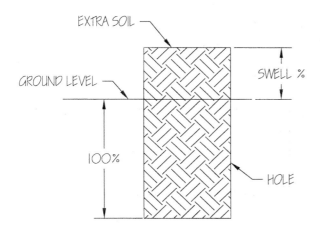

FIGURE 16-2 Swell Percentage

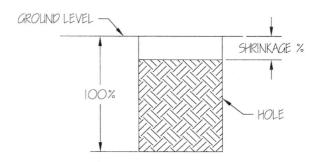

FIGURE 16-3 Shrinkage Percentage

Another way of describing the swell percentage is to imagine that we dig a hole and loosely place the excavated soil back in the hole, only to find that we have more soil than fits in the hole. This additional soil is due to swell. This example is shown in Figure 16-2.

When soils are placed and compacted, their volume decreases and their density increases. This decrease in volume is measured against the original volume of the soils in the bank condition. Soils that have been placed as backfill are described as being in the compacted condition.

The relationship between the bank and compacted conditions is described as the *shrinkage percentage*. The shrinkage percentage is calculated using the following equation:

$$\text{Shrinkage \%} = \left(1 - \frac{D_B}{D_C}\right)100 \qquad (16\text{-}2)$$

where

D_B = Bank Dry Density (Unit Weight)
D_C = Compact Dry Density (Unit Weight)

Another way of describing the shrinkage percentage is to imagine that we dig a hole and place the excavated soil back in the hole, compacting as we go. When all of the soil is placed in the hole we might find that the hole is not completely filled because we packed the soil at a higher level of compaction than when it was in its bank condition. The unfilled volume is due to shrinkage. This example is shown in Figure 16-3.

The calculation of the swell percentage and the shrinkage percentage is shown in the following example.

EXAMPLE 16-1

A soil has a bank dry density of 110 pounds per cubic foot, a loose dry density of 80 pounds per cubic foot, and a compacted dry density of 115 pounds per cubic foot. Determine the swell percentage and shrinkage percentage for the soil.

Solution: The swell percentage is calculated using Eq. (16-1) as follows:

$$\text{Swell \%} = \left(\frac{110 \text{ lb/ft}^3}{80 \text{ lb/ft}^3} - 1\right)100 = 37.5\%$$

The shrinkage percentage is calculated using Eq. (16-2) as follows:

$$\text{Shrinkage \%} = \left(1 - \frac{110 \text{ lb/ft}^3}{115 \text{ lb/ft}^3}\right)100 = 4.3\%$$

> **EXCEL QUICK TIP 16-1**
> **Swell and Shrinkage Percentages**
>
> The calculation of the swell and shrinkage percentages is set up in a spreadsheet by entering the data and formatting the cells as follows:
>
> | | A | B | C |
> |---|---|---|---|
> | 1 | Bank Dry Density | 110 | lb/cft |
> | 2 | Loose Dry Density | 80 | lb/cft |
> | 3 | Compacted Dry Density | 115 | lb/cft |
> | 4 | | | |
> | 5 | Swell Percentage | 37.5 | % |
> | 6 | Shrink Percentage | 4.3 | % |
>
> The following formulas need to be entered into the associated cells:
>
> | Cell | Formula |
> |------|---------|
> | B6 | =(B1/B2-1)*100 |
> | B7 | =(1-B1/B3)*100 |
>
> The data for the dry weights of the soil is entered in Cells B1 through B3. The data shown in the foregoing figure is from Example 16-1.

The relationships between the bank volume (V_B), the loose volume (V_L), and the compacted volume (V_C) are shown in Table 16-1. The relationships between the bank dry density (D_B), the loose dry density (D_L), and the compacted dry density (D_C) of a soil are also shown in Table 16-1.

The use of these equations is shown in the following example.

EXAMPLE 16-2

A construction project needs 1,000 cubic yards of compacted fill. Soil with a swell percentage of 37.5% and a shrinkage percentage of 4.3% is to be used for the fill. How many cubic yards of soil must be excavated? How many cubic yards of soil must be transported?

Solution: To determine the soil excavated, we need to convert the compacted volume to a bank volume. Compacted cubic yards (V_C) are converted to bank cubic yards (V_B) using Eq. (16-6) as follows:

$$V_B = \frac{(1{,}000 \text{ yd}^3)}{\left(1 - \frac{4.3}{100}\right)} = 1{,}045 \text{ yd}^3$$

To determine the volume of soil transported, we need to determine the loose volume. Compacted cubic yards (V_C) are converted to loose cubic yards (V_L) using Eq. (16-4) as follows:

$$V_L = (1{,}000 \text{ yd}^3)\frac{\left(1 + \frac{37.5}{100}\right)}{\left(1 - \frac{4.3}{100}\right)} = 1{,}437 \text{ yd}^3$$

Alternatively, the loose cubic yards (V_L) can be calculated from the previously determined bank cubic yards (V_B) using Eq. (16-3) as follows:

$$V_L = (1{,}045 \text{ yd}^3)\left(1 + \frac{37.5}{100}\right) = 1{,}437 \text{ yd}^3$$

> **EXCEL QUICK TIP 16-2**
> **Excavated and Transported Volumes**
>
> The calculation of the excavated and transported volumes of soil from the fill volume is set up in a spreadsheet by entering the data and formatting the cells as follows:
>
> | | A | B | C |
> |---|---|---|---|
> | 1 | Volume of Fill | 1,000 | cyds |
> | 2 | Swell Percentage | 37.5 | % |
> | 3 | Shrink Percentage | 4.3 | % |
> | 4 | | | |
> | 5 | Excavated (Bank) Volume | 1,045 | cyds |
> | 6 | Transported (Loose) Volume | 1,437 | cyds |
>
> The following formulas need to be entered into the associated cells:
>
> | Cell | Formula |
> |------|---------|
> | B5 | =B1/(1-B3/100) |
> | B6 | =B5*(1+B2/100) |
> | C5 | =C1 |
> | C6 | =C1 |
>
> The data for the soil is entered in Cells B1 and B3 and the units (e.g., cft, cyd) for the volume is entered in Cell C1. The data shown in the foregoing figure is from Example 16-2.

Determining the volume of excavation is only an approximation or estimate of the actual volume of soils that needs to be excavated because of variations in the surface of the area to be excavated and the inability of excavation equipment to quickly excavate exact geometric shapes. In spite of this, some methods of calculating excavation quantities are more accurate than others. Let's begin by looking at the most accurate of the methods, the geometric method.

Earthwork

TABLE 16-1 Volume and Density Conversion Equations

To Find	From	Use	
V_L	V_B	$V_L = V_B\left(1 + \dfrac{\text{Swell \%}}{100}\right)$	(16-3)
V_L	V_C	$V_L = V_C \dfrac{\left(1 + \dfrac{\text{Swell \%}}{100}\right)}{\left(1 - \dfrac{\text{Shrinkage \%}}{100}\right)}$	(16-4)
V_B	V_L	$V_B = \dfrac{V_L}{\left(1 + \dfrac{\text{Swell \%}}{100}\right)}$	(16-5)
V_B	V_C	$V_B = \dfrac{V_C}{\left(1 - \dfrac{\text{Shrinkage \%}}{100}\right)}$	(16-6)
V_C	V_L	$V_C = V_L \dfrac{\left(1 - \dfrac{\text{Shrinkage \%}}{100}\right)}{\left(1 + \dfrac{\text{Swell \%}}{100}\right)}$	(16-7)
V_C	V_B	$V_C = V_B\left(1 - \dfrac{\text{Shrinkage \%}}{100}\right)$	(16-8)
D_L	D_B	$D_L = \dfrac{D_B}{\left(1 + \dfrac{\text{Swell \%}}{100}\right)}$	(16-9)
D_L	D_C	$D_L = D_C \dfrac{\left(1 - \dfrac{\text{Shrinkage \%}}{100}\right)}{\left(1 + \dfrac{\text{Swell \%}}{100}\right)}$	(16-10)
D_B	D_L	$D_B = D_L\left(1 + \dfrac{\text{Swell \%}}{100}\right)$	(16-11)
D_B	D_C	$D_B = D_C\left(1 - \dfrac{\text{Shrinkage \%}}{100}\right)$	(16-12)
D_C	D_L	$D_C = D_L \dfrac{\left(1 + \dfrac{\text{Swell \%}}{100}\right)}{\left(1 - \dfrac{\text{Shrinkage \%}}{100}\right)}$	(16-13)
D_C	D_B	$D_C = \dfrac{D_B}{\left(1 - \dfrac{\text{Shrinkage \%}}{100}\right)}$	(16-14)

GEOMETRIC METHOD

The geometric method breaks the excavation into geometric shapes and calculates the volume for each of the shapes. The use of the geometric method is shown in the following example.

EXAMPLE 16-3

A contractor needs to install a 16-foot by 26-foot by 8-foot-high footing for a bridge. The bottom of the footing is to be 10 feet below grade. The sides of the excavation need to be sloped 1:1 (horizontal:vertical). A 2-foot space between the footing and the sides of the excavation must be provided to form the footing. Determine the volume of the excavation using the geometric method.

Solution: The width of the excavation at the bottom is 20 feet (2 ft + 16 ft + 2 ft), and the length of the excavation at the bottom is 30 feet (2 ft + 26 ft + 2 ft). The width of the excavation is 20 feet wider at the top than at the base. Cross sections of the excavation are shown in Figure 16-4. A three-dimensional representation of the excavation is shown in Figure 16-5. This excavation can be divided into a column, four prisms, and four pyramids as shown in Figure 16-6. The volume of the column is calculated as follows:

$$\text{Volume}_{\text{Column}} = (30 \text{ ft})(20 \text{ ft})(10 \text{ ft})\left(\dfrac{1 \text{ yd}^3}{27 \text{ ft}^3}\right) = 222 \text{ yd}^3$$

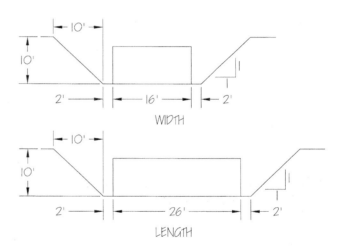

FIGURE 16-4 Excavation Cross Sections

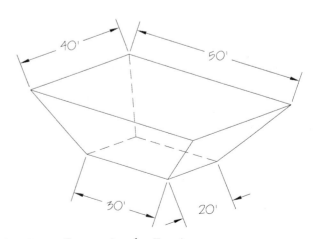

FIGURE 16-5 Excavation for Footing

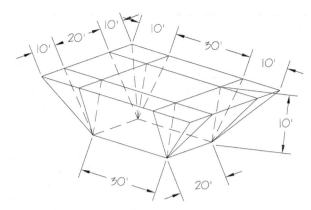

FIGURE 16-6 Excavation for Footing

There are two different sizes of prisms. Their volumes are calculated as follows:

$$\text{Volume}_{\text{Prism 1}} = \frac{(30 \text{ ft})(10 \text{ ft})(10 \text{ ft})}{2}\left(\frac{1 \text{ yd}^3}{27 \text{ ft}^3}\right) = 56 \text{ yd}^3$$

$$\text{Volume}_{\text{Prism 2}} = \frac{(20 \text{ ft})(10 \text{ ft})(10 \text{ ft})}{2}\left(\frac{1 \text{ yd}^3}{27 \text{ ft}^3}\right) = 37 \text{ yd}^3$$

The volume of the pyramids is calculated as follows:

$$\text{Volume}_{\text{Pyramid}} = \frac{(10 \text{ ft})(10 \text{ ft})(10 \text{ ft})}{3}\left(\frac{1 \text{ yd}^3}{27 \text{ ft}^3}\right) = 12 \text{ yd}^3$$

The total volume is calculated as follows:

$$\text{Volume} = 222 \text{ yd}^3 + 2(56 \text{ yd}^3) + 2(37 \text{ yd}^3) + 4(12 \text{ yd}^3)$$
$$= 456 \text{ yd}^3 \quad\blacksquare$$

AVERAGE-WIDTH-LENGTH-DEPTH METHOD

The average-width-length-depth method calculates the volume of excavation by multiplying the average width of the excavation by the average length of the excavation by the average depth of the excavation using the following equation:

$$\text{Volume} = (W_{\text{ave}})(L_{\text{ave}})(D_{\text{ave}}) \quad (16\text{-}15)$$

The average depth of the excavation is measured at the bottom of the excavation and does not include the side slopes. This method produces an approximate volume and is less accurate than the geometric method. The use of the average-width-length-depth method is shown in the following example.

EXAMPLE 16-4

Determine the excavation volume for Example 16-3 using the average-width-length-depth method.

Solution: The average depth is 10 feet. Using the dimensions from Figure 16-5, the average width and length are calculated as follows:

$$W_{\text{ave}} = \frac{(20 \text{ ft} + 40 \text{ ft})}{2} = 30 \text{ ft}$$

$$L_{\text{ave}} = \frac{(30 \text{ ft} + 50 \text{ ft})}{2} = 40 \text{ ft}$$

The volume is calculated using Eq. (16-15) as follows:

$$\text{Volume} = (30 \text{ ft})(40 \text{ ft})(10 \text{ ft})\left(\frac{1 \text{ yd}^3}{27 \text{ ft}^3}\right) = 444 \text{ yd}^3 \quad\blacksquare$$

AVERAGE-END METHOD

The average-end method calculates the volume by cutting cross sections through the excavation and determining the area of cut and fill for each of the cross sections. The volume of cut or fill between two cross sections is equal to the average of the two cross-sectional areas of the cut or fill multiplied by the perpendicular distance between the cross sections:

$$\text{Volume} = L\frac{(A_1 + A_2)}{2} \quad (16\text{-}16)$$

where

L = Perpendicular Distance between the Cross Sections
A_1 = Area of Cross Section 1
A_2 = Area of Cross Section 2

When more than two cross sections are taken and the distances between them are the same, the following equation may be used to calculate the volume:

$$\text{Volume} = L(0.5A_1 + A_2 + A_3 + \cdots + A_{n-1} + 0.5A_n)$$
$$(16\text{-}17)$$

where

L = Perpendicular Distance between the Cross Sections
A_1 = Area of Cross Section 1
A_2 = Area of Cross Section 2
A_3 = Area of Cross Section 3
A_n = Area of Cross Section n

The average-end method produces an approximate volume and is less accurate than the geometric method. The accuracy is improved by increasing the number of cross sections and locating cross sections at major changes in grades. Historically, this method has been used for road construction by cutting the cross sections perpendicular to the centerline of the road as shown in the following example.

TABLE 16-2 Area of Cuts

Station	Cut (ft²)
5+00	0
5+50	75
6+00	110
6+20	0

EXAMPLE 16-5

Using the average-end method, determine the required excavation for a new road giving the stations and their associated cuts shown in Table 16-2.

Solution: The distance between Stations 5+00 and 5+50 is 50 feet. The distance between Stations 5+50 and 6+00 is 50 feet. The distance between Stations 6+00 and 6+20 is 20 feet. The volumes of the cuts are calculated using Eq. (16-16) as follows:

$$\text{Volume}_{5+00 \text{ to } 5+50} = (50 \text{ ft})\frac{(0 \text{ ft}^2 + 75 \text{ ft}^2)}{2}\left(\frac{1 \text{ yd}^3}{27 \text{ ft}^3}\right) = 69 \text{ yd}^3$$

$$\text{Volume}_{5+50 \text{ to } 6+00} = (50 \text{ ft})\frac{(75 \text{ ft}^2 + 110 \text{ ft}^2)}{2}\left(\frac{1 \text{ yd}^3}{27 \text{ ft}^3}\right) = 171 \text{ yd}^3$$

$$\text{Volume}_{6+00 \text{ to } 6+20} = (20 \text{ ft})\frac{(110 \text{ ft}^2 + 0 \text{ ft}^2)}{2}\left(\frac{1 \text{ yd}^3}{27 \text{ ft}^3}\right) = 41 \text{ yd}^3$$

The total volume is calculated as follows:

$$\text{Volume} = 69 \text{ yd}^3 + 171 \text{ yd}^3 + 41 \text{ yd}^3 = 281 \text{ yd}^3$$

The average-end method may also be used to calculate the volume of a basement excavation. When using this method to calculate excavation volumes, a cross section should be cut at each horizontal step on the excavation. The use of the average-end method to calculate basement excavation is shown in the following example.

EXAMPLE 16-6

Determine the excavation volume for Example 16-3 using the average-end method.

Solution: The cross sections will be cut at the top and bottom of the excavation. Using the dimensions from Figure 16-5, the areas of the top and bottom of the excavation are calculated as follows:

$$\text{Area}_{\text{Top}} = (40 \text{ ft})(50 \text{ ft}) = 2{,}000 \text{ ft}^2$$
$$\text{Area}_{\text{Bottom}} = (20 \text{ ft})(30 \text{ ft}) = 600 \text{ ft}^2$$

The volume is calculated using Eq. (16-16) as follows:

$$\text{Volume} = (10 \text{ ft})\frac{(2{,}000 \text{ ft}^2 + 600 \text{ ft}^2)}{2}\left(\frac{1 \text{ yd}^3}{27 \text{ ft}^3}\right) = 481 \text{ yd}^3$$

MODIFIED-AVERAGE-END METHOD

Equation (16-16) may be modified as follows:

$$\text{Volume} = L\left(\frac{A_1 + A_2 + \sqrt{A_1 A_2}}{3}\right) \quad (16\text{-}18)$$

where

L = Perpendicular Distance between the Cross Sections
A_1 = Area of Cross Section 1
A_2 = Area of Cross Section 2

Equation (16-18) gives an answer closer to the answer provided by the geometric method, although not mathematically identical. The use of the modified-average-end method to calculate basement excavation is shown in the following example.

EXAMPLE 16-7

Determine the excavation volume for Example 16-3 using the modified-average-end method.

Solution: From Example 16-6, the area of the top is 2,000 ft² and the area of the bottom is 600 ft². The volume is calculated using Eq. (16-16) as follows:

$$\text{Volume} =$$
$$10 \text{ ft}\left(\frac{600 \text{ ft}^2 + 2{,}000 \text{ ft}^2 + \sqrt{(600 \text{ ft}^2)(2{,}000 \text{ ft}^2)}}{3}\right)\left(\frac{1 \text{ yd}^3}{27 \text{ ft}^3}\right)$$

$$\text{Volume} = 456 \text{ yd}^2$$

EXCEL QUICK TIP 16-3
Volume of a Rectangular Excavation

The calculation of the volume for a rectangular excavation is set up in a spreadsheet by entering the data and formatting the cells as follows:

	A	B	C
1	Width of Structure	16	ft
2	Length of Structure	26	ft
3	Depth of Excavation	10	ft
4	Slope of Sides	1	?H:1V
5	Toe of Slope to Structure	2	ft
6			
7	Width at Toe of Excavation	20	ft
8	Length at Toe of Excavation	30	ft
9	Width at Top of Excavation	40	ft
10	Length at Top of Excavation	50	ft
11			
12	Method	Volume	
13	Geometric	457	cyd
14	Average-Width-Length-Depth	444	cyd
15	Average-End	481	cyd
16	Modified-Average-End	456	cyd

The following formulas need to be entered into the associated cells:

Cell	Formula
B7	=B1+2*B5
B8	=B2+2*B5
B9	=B7+2*B3*B4
B10	=B8+2*B3*B4
B13	=(B7*B8*B3+B7*B3*(B3*B4)+B8*B3*(B3*B4)+4*B3*(B3*B4)*(B3*B4)/(3))/27
B14	=(B7+B9)/2*(B8+B10)/2*B3/27
B15	=B3*(B7*B8+B9*B10)/(2*27)
B16	=B3*(B7*B8+B9*B10+((B7*B8)*(B9*B10))^0.5)/(3*27)

The data for the excavation is entered in Cells B1 through B4. The Toe of Slope to Structure should be the distance between the toe of slope and the edge of the structure measured in Cells B1 and B2. The data shown in the foregoing figure is from Examples 16-3, 16-4, 16-6, and 16-7.

CROSS-SECTIONAL METHOD

The cross-sectional method is most commonly used for calculation of site excavation on sites where the grade around the building changes dramatically. Like the average-width-length-depth method, the average-end method, and the modified-average-end method, the cross-sectional method produces an approximate volume and is less accurate than the geometric method. The following steps need to be completed to calculate the volume using the cross-sectional method:

Step 1: Divide the site into grids.
Step 2: Calculate the cut or fill at each grid intersection.
Step 3: Separate the cuts and fills with a zero line.
Step 4: Calculate the volume of the cuts and fills.

Let's look at each of these steps.

Step 1: Divide the Site into Grids

The first step is to divide the proposed area or site into grids. The grids may be evenly spaced or aligned with site features. Figure 16-7 shows a site plan for the construction of a building, with the existing grades to be changed drawn as dashed lines and the proposed final grades drawn as solid lines. The site is divided into twelve 50-foot by 50-foot squares as shown in Figure 16-8.

Step 2: Calculate Cut or Fill at Each Grid Intersection

The next step is to calculate the cut or fill at each grid intersection where the grades are going to change. Often the grades around the perimeter of the site are not going to change; therefore, there is neither a cut nor a fill at these intersections. The cut or fill is calculated by subtracting the existing grade from the proposed grade using the following equation:

$$\text{Cut/Fill} = \text{Grade}_{\text{Proposed}} - \text{Grade}_{\text{Existing}} \quad (16\text{-}19)$$

If the result of Eq. (16-19) is positive, a fill (the adding of soil) is required; if the result is negative, a cut (the removal of soil) is required. The existing and proposed

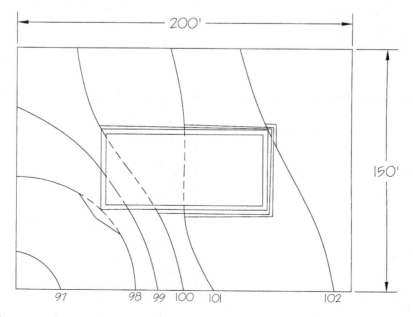

FIGURE 16-7 Site Plan

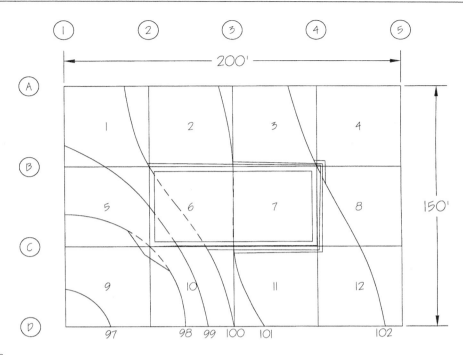

FIGURE 16-8 Grids

grades are determined by estimating the existing or proposed elevation from the contours or other information (for example, slab elevation) on the drawings. When an intersection falls between two grade lines, the grade at the intersection of the grid lines is estimated by interpolating between two adjacent contour lines using the following equation:

$$\text{Grade}_p = (\text{Grade}_2 - \text{Grade}_1)\frac{(\text{Dist}_{p-1})}{(\text{Dist}_{2-1})} + \text{Grade}_1 \quad (16\text{-}20)$$

where

Grade_p = Grade at Intersection
Grade_1 = Grade of First Contour Line
Grade_2 = Grade of Second Contour Line
Dist_{p-1} = Distance between the Intersection and the First Contour Line
Dist_{2-1} = Distance between the Second and First Contour Lines

This step is shown in the following example.

EXAMPLE 16-8

Determine the cuts and fills for the site plan shown in Figures 16-7 and 16-8. The bottom of the building excavation is at 99.00 feet.

Solution: For the intersection at B2, the existing grade is 100.0 and the proposed grade is 99.0. The cut or fill is calculated using Eq. (16-19) as follows:

$$\text{Cut/Fill}_{B2} = 99.0 - 100.0 = -1.0$$

Because the answer is negative, there is a cut of 1.0 foot at B2. For B3 the existing grade is 101.0 and the proposed grade

is 99.0. For B4 the existing grade is 102.0 and the proposed grade is 99.0. For C2 the existing grade is 98.0 and the proposed grade is 99.0. For C3 the existing grade is 101.0 and the proposed grade is 99.0. For C4 the existing grade must be interpolated using Eq. (16-20). The distance between the 101.0 contour line and the intersection is 50 feet, and the distance between the 101.0 and 102.0 contour lines is about 75 feet. The grade at the intersection is calculated as follows:

$$\text{Grade}_p = (102.0 - 101.0)\frac{(50 \text{ ft})}{(75 \text{ ft})} + 101.0 = 101.7$$

The proposed grade at C4 is 99.0. The cuts and fills for the intersections are calculated using Eq. (16-19) as follows:

$$\text{Cut/Fill}_{B3} = 99.0 - 101.0 = -2.0$$
$$\text{Cut/Fill}_{B4} = 99.0 - 102.0 = -3.0$$
$$\text{Cut/Fill}_{C2} = 99.0 - 98.0 = 1.0$$
$$\text{Cut/Fill}_{C3} = 99.0 - 101.0 = -2.0$$
$$\text{Cut/Fill}_{C4} = 99.0 - 101.7 = -2.7$$

The existing grades, proposed grades, and cut or fill for each intersection are shown in Figure 16-9. The existing grades are above and left of the intersection, the proposed grades are below and left of the intersection, and the cuts or fills are below and right of the intersections. The fills begin with the letter "F" and the cuts begin with the letter "C."

Step 3: Separate the Cuts and Fills with a Zero Line

The next step is to draw a line between the cuts and fills. This is referred to as a zero line because it represents the location where there is neither a cut nor a fill.

CHAPTER SIXTEEN

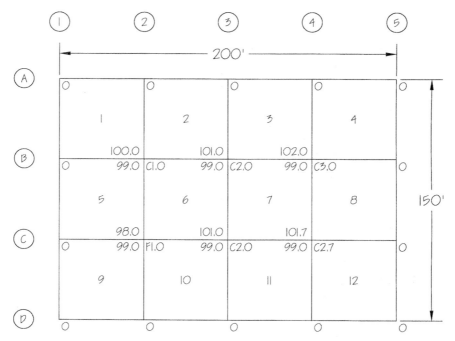

FIGURE 16-9 Cuts and Fills

Before drawing the zero line, one needs to determine the points where cuts change into fills. A change will occur any time there is a cut and a fill next to each other on the grid. For the site plan in Figure 16-9, this will occur between points B2 and C2 and between points C2 and C3. The location of the zero is determined by interpolating between two adjacent intersections using the following equation:

$$\text{Dist}_{C-0} = (\text{Dist}_{C-F})\frac{(\text{Cut})}{(\text{Cut} + \text{Fill})} \quad (16\text{-}21)$$

where

Dist_{C-0} = Distance from the Intersection with a Cut to the Zero Line
Dist_{C-F} = Distance from the Intersection with a Cut to the Intersection with a Fill
Cut = Cut at the Intersection with a Cut (must be positive)
Fill = Fill at the Intersection with a Fill

When the rectangle on a grid has three zeros, as is the case with rectangles 5 and 10 in Figure 16-9, the zero

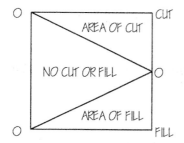

FIGURE 16-10 Zero Lines at Boundaries

line is drawn as shown in Figure 16-10. This condition often occurs at boundaries.

Locating the zero line is shown in the following example.

EXAMPLE 16-9

Determine the location of the zero line for the site plan shown in Figure 16-9.

Solution: The distance from B2 to the zero line in the direction of C2 is calculated using Eq. (16-21) as follows:

$$\text{Dist}_{C2-0} = (50 \text{ ft})\frac{(1.0)}{(1.0 + 1.0)} = 25 \text{ ft}$$

The distance from C3 to the zero line in the direction of C2 is calculated using Eq. (16-21) as follows:

$$\text{Dist}_{C2-0} = (50 \text{ ft})\frac{(2.0)}{(2.0 + 1.0)} = 33 \text{ ft}$$

The zero line for Figure 16-9 is drawn as shown in Figure 16-11.

Step 4: Calculate the Volume of the Cuts and Fills

The next step is to calculate the volume of the cuts and the fills. The volume of the cuts is calculated separately from that for the fills. When a rectangle contains both cuts and fills, the rectangle is divided into separate areas (a cut area and a fill area) by the zero line as shown in Figure 16-11. The volume of the cut for a rectangle equals the average of the cuts and zeros surrounding the cut area within the rectangle multiplied by the area of the

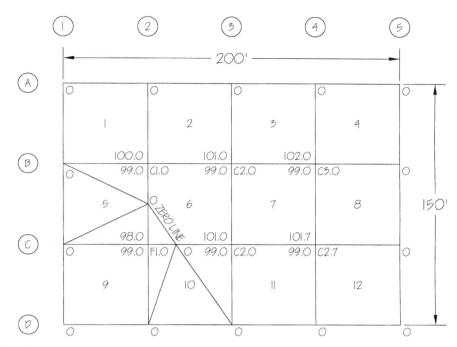

FIGURE 16-11 The Zero Line

cut area of the rectangle. Similarly, the volume of the fill for a rectangle equals the average of the fills and zeros surrounding the fill area within the rectangle multiplied by the area of the fill area of the rectangle. These calculations are shown in the following example.

EXAMPLE 16-10

Determine the cuts and fills for each of the 12 rectangles shown in Figure 16-11. What is the total cut for all of the areas? What is the total fill for all of the areas?

Solution: First we will calculate the cuts. All of rectangle 1 is a cut comprising zero cuts at three corners and a cut of 1.0 foot at the fourth. The cut volume is calculated as follows:

$$\text{Volume}_1 = \left(\frac{0\text{ ft} + 0\text{ ft} + 0\text{ ft} + 1.0\text{ ft}}{4}\right)(50\text{ ft})(50\text{ ft})\left(\frac{1\text{ yd}^3}{27\text{ ft}^3}\right)$$
$$= 23\text{ yd}^3$$

The cut volumes for rectangles 2 through 4 are calculated in a similar manner. They are as follows:

$$\text{Volume}_2 = \left(\frac{0\text{ ft} + 0\text{ ft} + 2.0\text{ ft} + 1.0\text{ ft}}{4}\right)(50\text{ ft})(50\text{ ft})\left(\frac{1\text{ yd}^3}{27\text{ ft}^3}\right)$$
$$= 69\text{ yd}^3$$

$$\text{Volume}_3 = \left(\frac{0\text{ ft} + 0\text{ ft} + 2.0\text{ ft} + 3.0\text{ ft}}{4}\right)(50\text{ ft})(50\text{ ft})\left(\frac{1\text{ yd}^3}{27\text{ ft}^3}\right)$$
$$= 116\text{ yd}^3$$

$$\text{Volume}_4 = \left(\frac{0\text{ ft} + 0\text{ ft} + 0\text{ ft} + 3.0\text{ ft}}{4}\right)(50\text{ ft})(50\text{ ft})\left(\frac{1\text{ yd}^3}{27\text{ ft}^3}\right)$$
$$= 69\text{ yd}^3$$

For rectangle 5, the area must be divided into three areas: a cut area, a fill area, and an area that is neither a cut nor a fill. From Example 16-9, the area of the cut is a triangle with a base of 25 feet and a height of 50 feet. The triangle is surrounded by two zeros and a cut of 1.0 foot. The cut volume is calculated as follows:

$$\text{Volume}_5 = \left(\frac{0\text{ ft} + 0\text{ ft} + 1.0\text{ ft}}{3}\right)\left[\frac{(50\text{ ft})(25\text{ ft})}{2}\right]\left(\frac{1\text{ yd}^3}{27\text{ ft}^3}\right)$$
$$= 8\text{ yd}^3$$

For rectangle 6, the area must be divided into two areas: a cut area and a fill area. From Example 16-9, the area of the fill is a triangle with a base of 25 feet (50 ft − 25 ft) and a height of 17 feet (50 ft − 33 ft). The area of the cut is the area of the rectangle (50 ft × 50 ft) less the area of the triangle. The cut area is surrounded by two zeros, a cut of 1.0 foot, and two cuts of 2.0 feet. The cut volume is calculated as follows:

$$\text{Volume}_6 = \left(\frac{0\text{ ft} + 0\text{ ft} + 1.0\text{ ft} + 2.0\text{ ft} + 2.0\text{ ft}}{5}\right)$$
$$\times \left[(50\text{ ft})(50\text{ ft}) - \frac{(17\text{ ft})(25\text{ ft})}{2}\right]$$
$$= \left(\frac{5\text{ ft}}{5}\right)\left[(50\text{ ft})(50\text{ ft}) - \frac{(17\text{ ft})(25\text{ ft})}{2}\right]\left(\frac{1\text{ yd}^3}{27\text{ ft}^3}\right)$$
$$= 85\text{ yd}^3$$

The cut volumes for rectangles 7 and 8 are as follows:

$$\text{Volume}_7 = \left(\frac{2.0\text{ ft} + 2.0\text{ ft} + 3.0\text{ ft} + 2.7\text{ ft}}{4}\right)$$
$$\times (50\text{ ft})(50\text{ ft})\left(\frac{1\text{ yd}^3}{27\text{ ft}^3}\right) = 225\text{ yd}^3$$

$$\text{Volume}_8 = \left(\frac{0\text{ ft} + 0\text{ ft} + 2.7\text{ ft} + 3.0\text{ ft}}{4}\right)(50\text{ ft})(50\text{ ft})\left(\frac{1\text{ yd}^3}{27\text{ ft}^3}\right)$$
$$= 132\text{ yd}^3$$

Rectangle 9 does not have any cuts. The cut volume for rectangle 10 is calculated in the same manner as for rectangle 5 and is calculated as follows:

$$\text{Volume}_{10} = \left(\frac{0 \text{ ft} + 0 \text{ ft} + 2.0 \text{ ft}}{3}\right)\left[\frac{(50 \text{ ft})(33 \text{ ft})}{2}\right]\left(\frac{1 \text{ yd}^3}{27 \text{ ft}^3}\right)$$
$$= 20 \text{ yd}^3$$

The cut volumes for rectangles 11 and 12 are as follows:

$$\text{Volume}_{11} = \left(\frac{0 \text{ ft} + 0 \text{ ft} + 2.0 \text{ ft} + 2.7 \text{ ft}}{4}\right)(50 \text{ ft})(50 \text{ ft})\left(\frac{1 \text{ yd}^3}{27 \text{ ft}^3}\right)$$
$$= 109 \text{ yd}^3$$

$$\text{Volume}_{12} = \left(\frac{0 \text{ ft} + 0 \text{ ft} + 0 \text{ ft} + 2.7 \text{ ft}}{4}\right)(50 \text{ ft})(50 \text{ ft})\left(\frac{1 \text{ yd}^3}{27 \text{ ft}^3}\right)$$
$$= 63 \text{ yd}^3$$

The total volume of the cut is calculated by summing the individual cut volumes as follows:

Volume = 23 yd³ + 69 yd³ + 116 yd³ + 69 yd³ + 8 yd³ + 85 yd³
+ 225 yd³ + 132 yd³ + 20 yd³ + 109 yd³ + 63 yd³
= 919 yd³

The fill volumns are calculated as follows:

$$\text{Volume}_5 = \left(\frac{0 \text{ ft} + 0 \text{ ft} + 1.0 \text{ ft}}{3}\right)\left[\frac{(50 \text{ ft})(25 \text{ ft})}{2}\right]\left(\frac{1 \text{ yd}^3}{27 \text{ ft}^3}\right)$$
$$= 8 \text{ yd}^3$$

$$\text{Volume}_6 = \left(\frac{0 \text{ ft} + 0 \text{ ft} + 1.0 \text{ ft}}{3}\right)\left[\frac{(17 \text{ ft})(25 \text{ ft})}{2}\right]\left(\frac{1 \text{ yd}^3}{27 \text{ ft}^3}\right)$$
$$= 3 \text{ yd}^3$$

$$\text{Volume}_9 = \left(\frac{0 \text{ ft} + 0 \text{ ft} + 0 \text{ ft} + 1.0 \text{ ft}}{4}\right)(50 \text{ ft})(50 \text{ ft})\left(\frac{1 \text{ yd}^3}{27 \text{ ft}^3}\right)$$
$$= 23 \text{ yd}^3$$

$$\text{Volume}_{10} = \left(\frac{0 \text{ ft} + 0 \text{ ft} + 1.0 \text{ ft}}{3}\right)\left[\frac{(50 \text{ ft})(17 \text{ ft})}{2}\right]\left(\frac{1 \text{ yd}^3}{27 \text{ ft}^3}\right)$$
$$= 5 \text{ yd}^3$$

The total fill volume is calculated by summing the individual fill volumes as follows:

Volume = 8 yd³ + 3 yd³ + 23 yd³ + 5 yd³ = 39 yd³ ■

The cross-sectional method may be used on building excavation even when contours do not exist, as long as cuts and fills can be calculated. The use of the cross-sectional method is shown in the following example.

EXAMPLE 16-11

Determine the excavation volume for Example 16-3 using the cross-sectional method.

Solution: The grids are aligned with the top and bottom of the slopes. The grids and the cuts are shown in Figure 16-12. The cuts are shown below and to the right of each intersection. The cuts for each of the rectangles are calculated as follows:

$$\text{Volume}_1 = \left(\frac{0 \text{ ft} + 0 \text{ ft} + 0 \text{ ft} + 10 \text{ ft}}{4}\right)(10 \text{ ft})(10 \text{ ft})\left(\frac{1 \text{ yd}^3}{27 \text{ ft}^3}\right)$$
$$= 9 \text{ yd}^3$$

$$\text{Volume}_2 = \left(\frac{0 \text{ ft} + 0 \text{ ft} + 10 \text{ ft} + 10 \text{ ft}}{4}\right)(10 \text{ ft})(30 \text{ ft})\left(\frac{1 \text{ yd}^3}{27 \text{ ft}^3}\right)$$
$$= 56 \text{ yd}^3$$

$$\text{Volume}_3 = \left(\frac{0 \text{ ft} + 0 \text{ ft} + 0 \text{ ft} + 10 \text{ ft}}{4}\right)(10 \text{ ft})(10 \text{ ft})\left(\frac{1 \text{ yd}^3}{27 \text{ ft}^3}\right)$$
$$= 9 \text{ yd}^3$$

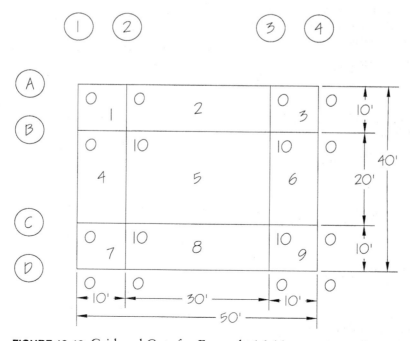

FIGURE 16-12 Grid and Cuts for Example 16-11

$$\text{Volume}_4 = \left(\frac{0\text{ ft} + 0\text{ ft} + 10\text{ ft} + 10\text{ ft}}{4}\right)(10\text{ ft})(20\text{ ft})\left(\frac{1\text{ yd}^3}{27\text{ ft}^3}\right)$$
$$= 37\text{ yd}^3$$
$$\text{Volume}_5 = \left(\frac{10\text{ ft} + 10\text{ ft} + 10\text{ ft} + 10\text{ ft}}{4}\right)(30\text{ ft})(20\text{ ft})\left(\frac{1\text{ yd}^3}{27\text{ ft}^3}\right)$$
$$= 222\text{ yd}^3$$
$$\text{Volume}_6 = \left(\frac{0\text{ ft} + 0\text{ ft} + 10\text{ ft} + 10\text{ ft}}{4}\right)(10\text{ ft})(20\text{ ft})\left(\frac{1\text{ yd}^3}{27\text{ ft}^3}\right)$$
$$= 37\text{ yd}^3$$
$$\text{Volume}_7 = \left(\frac{0\text{ ft} + 0\text{ ft} + 0\text{ ft} + 10\text{ ft}}{4}\right)(10\text{ ft})(10\text{ ft})\left(\frac{1\text{ yd}^3}{27\text{ ft}^3}\right)$$
$$= 9\text{ yd}^3$$
$$\text{Volume}_8 = \left(\frac{0\text{ ft} + 0\text{ ft} + 10\text{ ft} + 10\text{ ft}}{4}\right)(10\text{ ft})(30\text{ ft})\left(\frac{1\text{ yd}^3}{27\text{ ft}^3}\right)$$
$$= 56\text{ yd}^3$$
$$\text{Volume}_9 = \left(\frac{0\text{ ft} + 0\text{ ft} + 0\text{ ft} + 10\text{ ft}}{4}\right)(10\text{ ft})(10\text{ ft})\left(\frac{1\text{ yd}^3}{27\text{ ft}^3}\right)$$
$$= 9\text{ yd}^3$$

The total cut is calculated as follows:

$$\text{Volume} = 9\text{ yd}^3 + 56\text{ yd}^3 + 9\text{ yd}^3 + 37\text{ yd}^3 + 222\text{ yd}^3$$
$$+ 37\text{ yd}^3 + 9\text{ yd}^3 + 56\text{ yd}^3 + 9\text{ yd}^3$$
$$\text{Volume} = 444\text{ yd}^3$$

COMPARISON OF METHODS

The excavation in Example 16-3 has been calculated using the five methods presented in this chapter. The geometric method produces an exact quantity for the excavation. A comparison among the methods is shown in Table 16-3. Although the geometric method and the modified-average-end method both produce 456 cubic yards, when the calculations are done without rounding, there is a 0.1% difference in their answers. In some cases, these two methods can produce vastly different numbers. With five methods for calculating volume, it can be hard to know which method is best. The most accurate method is the geometric method, but in many cases it will be very hard, if not impossible, to use. The method most often used is the method that is easiest to use based on the grading and excavation data available.

TABLE 16-3   Comparison of Volumes

Method	Volume (yd³)	Error (%)
Geometric	456	0
Average-width-length-depth	444	−2.6
Average-end	481	+5.5
Modified-average-end	456	−0.1
Cross-sectional	444	−2.6

BACKFILL

The easiest way to calculate the quantity of backfill for footing, foundations, and so forth is to determine the volume of excavation and subtract the volume of the footing or foundation. This is shown in the following example.

EXAMPLE 16-12

The footing in Example 16-3 supports a 2-foot by 12-foot column. Using the quantity of excavation from Example 16-3, determine the volume of backfill that needs to be placed after the footing has been constructed.

Solution: The volume of excavation from Example 16-3 is 456 cubic yards. The entire footing and the bottom 2 feet (10 ft − 8 ft) of the column are below the finished grade. The volume of the footing and column that displaces soil is calculated as follows:

$$\text{Volume}_{\text{Footing}} = (16\text{ ft})(26\text{ ft})(8\text{ ft})\left(\frac{1\text{ yd}^3}{27\text{ ft}^3}\right) = 123\text{ yd}^3$$
$$\text{Volume}_{\text{Column}} = (2\text{ ft})(12\text{ ft})(2\text{ ft})\left(\frac{1\text{ yd}^3}{27\text{ ft}^3}\right) = 2\text{ yd}^3$$

The volume of backfill is calculated as follows:

$$\text{Volume} = 456\text{ yd}^3 - 123\text{ yd}^3 - 2\text{ yd}^3 = 331\text{ yd}^3$$

Once the quantities of excavation and backfill have been calculated, these quantities along with the quantity of soil imported or exported need to be related to each other. If the excavated material is suitable, it may be used for the fill. If it is unsuitable for fill, it must be hauled off (exported) and an acceptable fill material brought in (imported). The calculation of soil import and export is shown in the following example.

EXAMPLE 16-13

The soils in Example 16-12 have a swell percentage of 30% and a shrinkage percentage of 5%. The excavated soils may be used as backfill. Determine the volume of the soils that needs to be exported.

Solution: From Example 16-12 the volume of excavation is 456 cubic yards and the volume of backfill is 331 cubic yards. The compacted volume of the backfill is converted to a bank quantity using Eq. (16-6) as follows:

$$V_B = \frac{(331\text{ yd}^3)}{\left(1 - \frac{5}{100}\right)} = 348\text{ yd}^3$$

Three hundred and forty-eight yards of the excavated material are needed for the backfill. The quantity exported is 108 bank cubic yards (456 yd³ − 348 yd³). The loose volume of the exported soils is converted from a bank quantity to a loose quantity using Eq. (16-3) as follows:

$$V_L = (108\text{ yd}^3)\left(1 + \frac{30}{100}\right) = 140\text{ yd}^3$$

One hundred and forty loose cubic yards of soil will need to be exported.

SOILS REPORT

Estimators should read soils report carefully. A soils report provides valuable information, such as the presence of groundwater or rocky soil. Both of these conditions increase the cost of excavation. If a soils report is not provided with the bid documents, the estimator should contact the design professionals to see if a copy is available.

SAMPLE TAKEOFF FOR THE RESIDENTIAL GARAGE

A sample takeoff for excavation from a set of plans is shown in the following example.

EXAMPLE 16-14

Determine the excavation, under-slab gravel, and backfill needed to complete the residential garage given in Appendix G. The excavated soils may be used as backfill.

Solution: The excavation will consist of excavating the entire footprint of the building, the sidewalk, and the driveway and excavating and backfilling the footings. The pad excavation will extend 1 foot beyond the sidewalk and driveway and 2 feet beyond the building. The excavation quantities are calculated using the cross-sectional method. Begin by calculating the excavation for the building, including 2 feet beyond in all directions. The excavation grade is 8 inches (0.67 ft) below the grade of the building slab shown on the floor plan. The cuts for the building corners and total cut volume are as follows:

$$\text{Cut}_{\text{NW Corner}} = 99.51 \text{ ft} - 98.83 \text{ ft} = 0.68 \text{ ft}$$
$$\text{Cut}_{\text{SW Corner}} = 99.47 \text{ ft} - 98.83 \text{ ft} = 0.64 \text{ ft}$$
$$\text{Cut}_{\text{SE Corner}} = 99.95 \text{ ft} - 99.33 \text{ ft} = 0.62 \text{ ft}$$
$$\text{Cut}_{\text{NE Corner}} = 100.00 \text{ ft} - 99.33 \text{ ft} = 0.67 \text{ ft}$$
$$\text{Volume} = \left(\frac{0.68 \text{ ft} + 0.64 \text{ ft} + 0.62 \text{ ft} + 0.67 \text{ ft}}{4}\right)$$
$$\times (28 \text{ ft})(30 \text{ ft})\left(\frac{1 \text{ yd}^3}{27 \text{ ft}^3}\right)$$
$$= 20.3 \text{ yd}^3$$

Next, the volume of the driveway is calculated. The east side of the drive has already been excavated to a point 2 feet west of the garage. At the west end, the drive will need to match the grade of the existing driveway. The cuts for the slab corners and total cut volume are as follows:

$$\text{Cut}_{\text{NW Corner}} = 98.60 \text{ ft} - 97.93 \text{ ft} = 0.67 \text{ ft}$$
$$\text{Cut}_{\text{SW Corner}} = 98.70 \text{ ft} - 98.03 \text{ ft} = 0.67 \text{ ft}$$
$$\text{Cut}_{\text{NE Corner}} = 99.50 \text{ ft} - 98.78 \text{ ft} = 0.72 \text{ ft}$$
$$\text{Cut}_{\text{SE Corner}} = 99.52 \text{ ft} - 98.78 \text{ ft} = 0.74 \text{ ft}$$
$$\text{Volume} = \left(\frac{0.67 \text{ ft} + 0.67 \text{ ft} + 0.72 \text{ ft} + 0.74 \text{ ft}}{4}\right)$$
$$\times (20 \text{ ft})(24.58 \text{ ft})\left(\frac{1 \text{ yd}^3}{27 \text{ ft}^3}\right)$$
$$= 12.7 \text{ yd}^3$$

Next, the volume of the sidewalk is calculated. The width of the sidewalk to be excavated is 4 feet plus the 1 foot over-excavation less the 2 feet of over-excavation already accomplished when excavating the garage. The length of the sidewalk to the south of the building is 32 feet (1 ft + 4 ft + 26 ft + 1 ft), which includes 1 foot over-excavation at both ends. The length of sidewalk to the west is 5 feet (3 ft + 2 ft). For the grades of the sidewalk we use the cut of the southeast building corner and southeast corner of the driveway. The total cut volume is calculated as follows:

$$\text{Volume} = \left(\frac{0.74 \text{ ft} + 0.62 \text{ ft}}{2}\right)(32 \text{ ft} + 7 \text{ ft})(3 \text{ ft})\left(\frac{1 \text{ yd}^3}{27 \text{ ft}^3}\right)$$
$$= 2.9 \text{ yd}^3$$

Next, the excavation for the foundation wall is calculated. The cuts to the bottom of the foundation in the four corners of the building are as follows:

$$\text{Cut}_{\text{NW Corner}} = 98.83 \text{ ft} - 97.66 \text{ ft} = 1.17 \text{ ft}$$
$$\text{Cut}_{\text{SW Corner}} = 98.83 \text{ ft} - 97.66 \text{ ft} = 1.17 \text{ ft}$$
$$\text{Cut}_{\text{SE Corner}} = 99.33 \text{ ft} - 97.66 \text{ ft} = 1.67 \text{ ft}$$
$$\text{Cut}_{\text{NE Corner}} = 99.33 \text{ ft} - 97.66 \text{ ft} = 1.67 \text{ ft}$$

Allow for a 4-foot-wide trench for the footings. Extend the trench out 1 foot beyond the outside of the footings. The footings extend 0.5 feet outside the foundation wall. The north and south trenches are 29 feet (1 ft + 0.5 ft + 26 ft + 0.5 ft + 1 ft) long and include the corners. The east and west trenches are 19 feet long and will run from the inside of the north trench to the inside of the south trench. Because the excavations are shallow, the walls of the excavation can be vertical. The excavation is calculated using the average-width-length-depth method as follows:

$$\text{Volume}_N = (29 \text{ ft})\left(\frac{1.67 \text{ ft} + 1.17 \text{ ft}}{2}\right)(4)\left(\frac{1 \text{ yd}^3}{27 \text{ ft}^3}\right) = 6.1 \text{ yd}^3$$
$$\text{Volume}_E = (19 \text{ ft})\left(\frac{1.67 \text{ ft} + 1.67 \text{ ft}}{2}\right)(4)\left(\frac{1 \text{ yd}^3}{27 \text{ ft}^3}\right) = 4.7 \text{ yd}^3$$
$$\text{Volume}_S = (29 \text{ ft})\left(\frac{1.67 \text{ ft} + 1.17 \text{ ft}}{2}\right)(4)\left(\frac{1 \text{ yd}^3}{27 \text{ ft}^3}\right) = 6.1 \text{ yd}^3$$
$$\text{Volume}_W = (19 \text{ ft})\left(\frac{1.17 \text{ ft} + 1.17 \text{ ft}}{2}\right)(4)\left(\frac{1 \text{ yd}^3}{27 \text{ ft}^3}\right) = 3.3 \text{ yd}^3$$
$$\text{Volume} = 6.1 \text{ yd}^3 + 4.7 \text{ yd}^3 + 6.1 \text{ yd}^3 + 3.3 \text{ yd}^3 = 20.2 \text{ yd}^3$$

The total excavation is calculated as follows:

$$\text{Volume} = 20.3 \text{ yd}^3 + 12.7 \text{ yd}^3 + 2.9 \text{ yd}^3 + 20.2 \text{ yd}^3 = 56 \text{ yd}^3$$

Backfill is needed around the footings and in the areas over-excavated during the excavation of the pad. The backfill of the footings is calculated by determining the volume of the concrete in the excavation and subtracting it from the volume of excavation. The height of the wall below the excavation level on the west side is 0.34 feet (1.17 ft − 10 in) and on the east side is 0.83 feet (1.67 ft − 10 in). The west wall includes the wall on the footing and the frost wall. The footing volume is calculated as follows:

$$\text{Length}_{\text{Footing}} = 3\text{ ft }4\text{ in} + 27\text{ ft} + 21\text{ ft }8\text{ in} + 27\text{ ft} + 3\text{ ft }4\text{ in}$$
$$= 82\text{ ft }3\text{ in} = 82.33\text{ ft}$$

$$\text{Volume}_{\text{Footing}} = (82.33\text{ ft})\frac{(10\text{ in})(20\text{ in})}{(12\text{ in/ft})^2}\left(\frac{1\text{ yd}^3}{27\text{ ft}^3}\right) = 4.24\text{ yd}^3$$

$$\text{Volume}_{\text{West Wall}} = [2(3.33\text{ ft})(0.34\text{ ft})(0.67\text{ ft})$$
$$+ (16\text{ ft})(1.17\text{ ft})(0.67\text{ ft})]$$
$$= 14.0\text{ ft}^3\left(\frac{1\text{ yd}^3}{27\text{ ft}^3}\right) = 0.52\text{ yd}^3$$

$$\text{Volume}_{\text{North Wall}} = (26\text{ ft})\left(\frac{0.34\text{ ft} + 0.83}{2}\right)(0.67\text{ ft})\left(\frac{1\text{ yd}^3}{27\text{ ft}^3}\right)$$
$$= 0.38\text{ yd}^3$$

$$\text{Volume}_{\text{East Wall}} = (22.66\text{ ft})(0.83)(0.67\text{ ft})\left(\frac{1\text{ yd}^3}{27\text{ ft}^3}\right) = 0.47\text{ yd}^3$$

$$\text{Volume}_{\text{South Wall}} = (26\text{ ft})\left(\frac{0.34\text{ ft} + 0.83}{2}\right)(0.67\text{ ft})\left(\frac{1\text{ yd}^3}{27\text{ ft}^3}\right)$$
$$= 0.38\text{ yd}^3$$

$$\text{Volume}_{\text{Concrete}} = 4.24\text{ yd}^3 + 0.52\text{ yd}^3 + 0.38\text{ yd}^3$$
$$+ 0.47\text{ yd}^3 + 0.38\text{ yd}^3$$
$$= 6.0\text{ yd}^3$$

$$\text{Volume}_{\text{Backfill}} = 20.2\text{ yd}^3 - 6.0\text{ yd}^3 = 14.2\text{ yd}^3$$

The volume of backfill needed for the over-excavation is calculated as follows:

$$\text{Volume} = [(1\text{ ft})(24.58\text{ ft}) + (2\text{ ft})(3\text{ ft}) + (2\text{ ft})(30\text{ ft})$$
$$+ (2\text{ ft})(24\text{ ft}) + (1\text{ ft})(4\text{ ft}) + (1\text{ ft})(32\text{ ft})$$
$$+ (1\text{ ft})(7\text{ ft}) + (1\text{ ft})(21.58\text{ ft})](0.67\text{ ft})(1\text{ yd}^3/27\text{ ft}^3)$$
$$= 5.0\text{ yd}^3$$

The total backfill is 19.2 cubic yards (14.2 yd³ + 5.0 yd³). Assuming a shrinkage percentage of 5%, we calculate the quantity of excavated material needed for the fill using Eq. (16-6) as follows:

$$V_B = \frac{(19.2\text{ yd}^3)}{\left(1 - \frac{5}{100}\right)} = 20.2\text{ yd}^3$$

The amount of export is 36 bank cubic yards (56 yd³ − 20 yd³).

The last item to calculate is the volume of gravel needed under the concrete slab, driveway, and sidewalk. The volume is calculated as follows:

$$\text{Area}_{\text{Slab}} = (24\text{ ft }8\text{ in})(22\text{ ft }8\text{ in}) + (16\text{ ft})(8\text{ in}) = 570\text{ ft}^2$$
$$\text{Area}_{\text{Drive}} = (18\text{ ft})(26\text{ ft }7\text{ in}) = 479\text{ ft}^2$$
$$\text{Area}_{\text{Sidewalk}} = (33\text{ ft})(4\text{ ft}) = 132\text{ ft}^2$$
$$\text{Area} = 570\text{ ft}^2 + 479\text{ ft}^2 + 132\text{ ft}^2 = 1{,}181\text{ ft}^2$$
$$\text{Volume} = (1{,}181\text{ ft}^2)\frac{(4\text{ in})}{(12\text{ in/ft})}\left(\frac{1\text{ yd}^3}{27\text{ ft}^3}\right) = 14.6\text{ yd}^3$$

The quantities needed for the garage, grouped by the cost codes in Appendix B, are shown in Table 16-4.

TABLE 16-4 Quantities for Residential Garage

31-200 Grading and Excavation		
Excavation	56	bank cyd
Backfill	19	bank cyd
Export	36	bank cyd
3/4" gravel	15	compacted cyd

CONCLUSION

The volume of excavation can be calculated in a number of ways including the geometric method, the average-width-length-depth method, the average-end method, the modified-average-end method, and the cross-sectional method. Because of variation in the surface area to be excavated and the inability of excavating equipment to quickly excavate exact geometric shapes, any calculation of the volume of excavation is only an approximation.

When soils are excavated their volume increases. This is known as swell. When soils are compacted their volume often decreases to a volume less than their volume before they were excavated. This is known as shrink. Swell and shrink must be taken into account when determining the volume of soil that needs to be transported and placed as compacted fill.

PROBLEMS

1. A soil has a bank dry density of 108 pounds per cubic foot, a loose dry density of 85 pounds per cubic foot, and a compacted dry density of 120 pounds per cubic foot. Determine the swell percentage and shrinkage percentage for the soils.

2. A soil has a bank dry density of 100 pounds per cubic foot, a loose dry density of 90 pounds per cubic foot, and a compacted dry density of 105 pounds per cubic foot. Determine the swell percentage and shrinkage percentage for the soils.

3. Using the geometric method, determine the amount of excavation required for a 24-foot by 50-foot basement. The measurements are from the outside of the foundation walls. The depth of the excavation is 7 feet. The footings extend 1 foot outside of

the foundation walls, and a 3-foot space between the footing and the sides of the excavation must be provided to form the footings. The soil is excavated at a 1:1 (horizontal:vertical) slope. Express your answer in cubic yards.

4. Using the geometric method, determine the amount of excavation required for the basement in Figure 16-13. The basement measurements are from the outside of the foundation walls. The average depth of the excavation is 6 feet. The footings extend 1 foot outside of the foundation walls, and a 2-foot space between the footing and the sides of the excavation must be provided to form the footings. The soil is excavated at a 1.5:1 (horizontal:vertical) slope. Express your answer in cubic yards.

5. Solve Problem 3 using the average-width-length-depth method.

6. Solve Problem 4 using the average-width-length-depth method.

7. Solve Problem 3 using the average-end method.

8. Solve Problem 4 using the average-end method.

9. Solve Problem 3 using the modified-average-end method.

10. Solve Problem 4 using the modified-average-end method.

11. Solve Problem 3 using the cross-sectional method.

12. Solve Problem 4 using the cross-sectional method.

13. Using the average-end method, determine the required excavation for a new road given the stations and their associated cut areas shown in Table 16-5. Express your answer in cubic yards.

14. Using the average-end method, determine the required volume of the cuts and fills for a new road given the stations and their associated cut and fill areas shown in Table 16-6. Express your answer in cubic yards.

TABLE 16-5 Cuts for Problem 13

Station	Cut (ft^2)
0+00	0
0+50	235
1+00	475
1+50	232
2+00	52
2+25	0

TABLE 16-6 Cuts and Fills for Problem 14

Station	Cut (ft^2)	Fill (ft^2)
0+00	0	0
0+50	175	10
1+00	250	45
1+50	75	225
2+00	0	450
2+25	0	0

15. Determine the volume of backfill needed for the basement in Problem 3. The footing height is 12 inches. Use the excavation quantities from Problem 3.

16. Determine the volume of backfill needed for the basement in Problem 4. The footing height is 12 inches. Use the excavation quantities from Problem 4.

17. The soils in Problem 15 have a bank dry unit weight of 105 pounds per cubic foot, a loose dry unit weight of 80 pounds per cubic foot, and a compacted dry unit weight of 110 pounds per cubic foot. Determine the volume of the soils that needs to be exported.

18. The soils in Problem 16 have a swell percentage of 25% and a shrinkage percentage of 8%. Determine the volume of the soils that needs to be exported.

19. For the site plan in Figure 16-14, determine the total volume of the cuts, the total volume of the fills, and the volume of the import or export using the cross-sectional method. The grids are 50 feet apart in both directions. The existing grade appears above the proposed grade. Express your answer in cubic yards. The shrinkage percentage is 5% and the swell percentage is 28%.

20. Determine the excavation and backfill needed for the Johnson Residence given in Appendix G. Assume that a 3-foot space between the footing and the sides of the excavation must be provided to form the footings. The sides are excavated at a 1:1 (horizontal:vertical) slope.

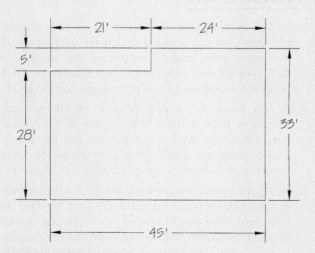

FIGURE 16-13 Basement Plan

21. Determine the excavation and backfill needed for the West Street Video project given in Appendix G. Assume that a 1-foot space between the footing and the sides of the excavation must be provided to form the footings. The sides are excavated at a 0.5:1 (horizontal:vertical) slope.

22. Determine the under-slab gravel needed for the West Street Video project given in Appendix G.
23. Set up Excel Quick Tip 16.1 in Excel.
24. Set up Excel Quick Tip 16.2 in Excel.
25. Set up Excel Quick Tip 16.3 in Excel.

	A	B	C	D	E	F	G	
1	104.5 / 104.5	104.7 / 104.7	105.1 / 105.1	105.3 / 105.3	105.9 / 105.9	106.1 / 106.1	106.5 / 106.5	1
		1	2	3	4	5	6	
2	104.2 / 104.2	104.6 / 104.7	104.8 / 104.7	105.3 / 104.7	105.5 / 104.7	105.8 / 104.7	106.0 / 106.0	2
		7	8	9	10	11	12	
3	103.9 / 103.9	104.3 / 104.7	104.7 / 104.7	104.9 / 104.7	105.2 / 104.7	105.4 / 104.7	105.5 / 105.5	3
		13	14	15	16	17	18	
4	103.4 / 103.4	104.0 / 104.7	104.5 / 104.7	104.8 / 104.7	105.1 / 104.7	105.1 / 104.7	105.3 / 105.3	4
		19	20	21	22	23	24	
5	103.1 / 103.1	103.3 / 104.7	104.1 / 104.7	104.3 / 104.7	104.6 / 104.7	105.0 / 104.7	105.0 / 105.0	5
		25	26	27	28	29	30	
6	102.8 / 102.8	103.2 / 104.7	103.6 / 104.7	103.9 / 104.7	104.5 / 104.7	104.7 / 104.7	104.9 / 104.9	6
		31	32	33	34	35	36	
7	102.7 / 102.7	102.9 / 102.9	103.1 / 103.1	103.5 / 103.5	103.9 / 103.9	104.4 / 104.4	104.7 / 104.7	7
	A	B	C	D	E	F	G	

FIGURE 16-14 Site Plan

CHAPTER SEVENTEEN

EXTERIOR IMPROVEMENTS

In this chapter you will learn how to apply the principles in Chapter 4 to exterior improvements including asphalt, site concrete, and landscaping. This chapter includes a sample spreadsheet that may be used in the quantity takeoff.

Exterior improvements include base, flexible and ridged paving, curbs and gutters, fences, athletic surfaces, retaining walls, bridges, irrigation systems, and landscaping. In this chapter we limit the discussion of exterior improvements to asphalt and base, site concrete, and landscaping. We begin by looking at asphalt and base.

ASPHALT AND BASE

Asphalt and base are bid by the ton. The volume of asphalt and base is determined by multiplying the area of the asphalt or base by the depth of the asphalt or base. The density of the asphalt or base is used to convert the volume to tons as shown in the following example.

EXAMPLE 17-1

Determine the tons of asphalt and base required to complete the parking lot in Figure 17-1. The base has an average thickness of 8 inches and a density of 125 pounds per cubic foot. The asphalt has an average thickness of 3.5 inches and a density of 150 pounds per cubic foot. The curb and gutter is 30 inches wide and has an inside radius of 5 feet.

Solution: The area of the parking lot will be divided into four areas: the large main area, the drive, and the two small pieces next to the radius of the curb and gutter. The main area is 151 feet long (156 ft − 30 in − 30 in) by 62 feet wide (67 ft − 30 in − 30 in). The drive is 25 feet wide by 13.5 feet long (11 ft + 30 in). The outside radius (R) of curb and gutter is 7.5 feet (5 ft + 30 in). The small areas next to the radius of the curbs are each equal to a square less a quarter circle. The areas are calculated as follows:

$$Area_{Main} = (151\ ft)(62\ ft) = 9{,}362\ ft^2$$

$$Area_{Dirve} = (25\ ft)(13.5\ ft) = 338\ ft^2$$

$$Area_{Curves} = 2\left[R^2\left(1-\frac{\pi}{4}\right)\right] = 2\left[(7.5\ ft)^2\left(1-\frac{\pi}{4}\right)\right] = 24\ ft^2$$

$$Area_{Total} = 9{,}362\ ft^2 + 338\ ft^2 + 24\ ft^2 = 9{,}724\ ft^2$$

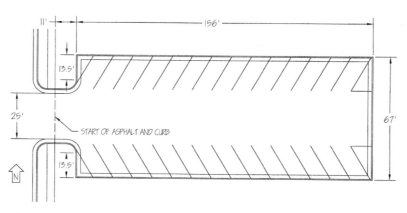

FIGURE 17-1 Parking Lot Layout

The volume of base and asphalt needed is calculated as follows:

$$\text{Volume}_{\text{Base}} = (9{,}724 \text{ ft})(8 \text{ in})\left(\frac{1 \text{ ft}}{12 \text{ in}}\right) = 6{,}483 \text{ ft}^3$$

$$\text{Volume}_{\text{Asphalt}} = (9{,}724 \text{ ft}^2)(3.5 \text{ in})\left(\frac{1 \text{ ft}}{12 \text{ in}}\right) = 2{,}836 \text{ ft}^3$$

The tons of base and asphalt needed are calculated as follows:

$$\text{Weight}_{\text{Base}} = (6{,}483 \text{ ft}^3)\left(\frac{125 \text{ lb}}{\text{ft}^3}\right)\left(\frac{1 \text{ ton}}{2{,}000 \text{ lb}}\right) = 405 \text{ tons}$$

$$\text{Weight}_{\text{Asphalt}} = (2{,}836 \text{ ft}^3)\left(\frac{150 \text{ lb}}{\text{ft}^3}\right)\left(\frac{1 \text{ ton}}{2{,}000 \text{ lb}}\right) = 213 \text{ tons}$$

EXCEL QUICK TIP 17-1
Asphalt or Base

The tons of asphalt or base needed for an area is set up in a spreadsheet by entering the data and formatting the cells as follows:

	A	B	C
1	Area	9,723	sft
2	Thickness	3.50	in
3	Density	150	lb/cft
4			
5	Weight	212.69	tons

The following formula needs to be entered in Cell B5:

=B1*(B2/12)*B3/2000

The data for asphalt or base is entered in Cells B1 through B3. The data shown in the foregoing figure is for the asphalt from Example 17-1.

SITE CONCRETE

Site concrete is bid using the same principles as for building concrete. The quantity takeoff of site concrete is shown in the following example.

EXAMPLE 17-2

Determine the quantity of concrete and lineal footage of forms needed for the parking lot in Figure 17-1. The cross section of the curb is shown in Figure 17-2.

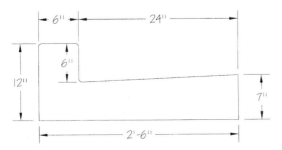

FIGURE 17-2 Curb Cross Section

Solution: The length of curb and gutter needed for each side of the entry is equal to 11 feet minus the inside radius (5 ft) of the curb and gutter, or 6 feet. The length of the curb and gutter on the north and south sides is 156 feet. The length of the curb and gutter on the east end is 67 feet less twice the width of the curb and gutter, or 62 feet (67 ft − 30 in − 30 in). This reduction in length is necessary to avoid double counting the corners. The length of the curb and gutter needed on the west ends is 13.5 feet less the width of the curb and gutter, or 11 feet each. The length of the curved portion of the gutter is based on the radius (R) of the centerline (average of the inside and outside radius), or 6.25 feet (5 ft/2 + 7.5 ft/2). The length of one of the curved portions of the curb and gutter is calculated as follows:

$$\text{Length} = \frac{2\pi R}{4} = \frac{2\pi(6.25)}{4} = 10 \text{ ft}$$

The total length of curve (clockwise from the driveway) is calculated as follows:

$$\text{Length} = 6 \text{ ft} + 10 \text{ ft} + 11 \text{ ft} + 156 \text{ ft} + 62 \text{ ft} + 156 \text{ ft}$$
$$+ 11 \text{ ft} + 10 \text{ ft} + 6 \text{ ft}$$
$$\text{Length} = 428 \text{ ft}$$

The cross-sectional area of the curb is calculated as follows:

$$\text{Area} = \left[(6 \text{ in})(12 \text{ in}) + (24 \text{ in})\frac{(6 \text{ in} + 7 \text{ in})}{2}\right]\left(\frac{1 \text{ ft}^2}{144 \text{ in}^2}\right)$$
$$= 1.58 \text{ ft}^2$$

The volume of concrete is calculated as follows:

$$\text{Volume} = (428 \text{ ft})(1.58 \text{ ft}^2)\left(\frac{1 \text{ yd}^3}{27 \text{ ft}^2}\right) = 25.0 \text{ yd}^3$$

The lengths of the forms needed to form the outside and the inside of the curb and gutter are calculated as follows:

$$\text{Length}_{\text{Outside}} = 6 \text{ ft} + \frac{2\pi(5 \text{ ft})}{4} + 13.5 \text{ ft} + 156 \text{ ft} + 67 \text{ ft}$$
$$+ 156 \text{ ft} + 13.5 \text{ ft} + \frac{2\pi(5 \text{ ft})}{4} + 6 \text{ ft}$$
$$\text{Length}_{\text{Outside}} = 434 \text{ ft}$$

$$\text{Length}_{\text{Inside}} = 6 \text{ ft} + \frac{2\pi(7.5 \text{ ft})}{4} + 11 \text{ ft} + 151 \text{ ft} + 62 \text{ ft}$$
$$+ 151 \text{ ft} + 11 \text{ ft} + \frac{2\pi(7.5 \text{ ft})}{4} + 6 \text{ ft}$$
$$\text{Length}_{\text{Inside}} = 422 \text{ ft}$$

LANDSCAPING

Landscaping includes providing and grading topsoil; providing an underground sprinkler system; and planting seed, sod, and plants. Providing topsoil is bid by the cubic yard, with the volume of topsoil being determined by the area times the thickness. The finish grading of the topsoil is bid based on the area. The underground sprinkler system is most often placed using a trenching machine to dig a narrow trench in which the pipe is placed. The trenching is bid based on the length of the trench, and the sprinkler system is bid based on the same principles used to determine the components of a plumbing system. Sod and seed are bid based on the area, and

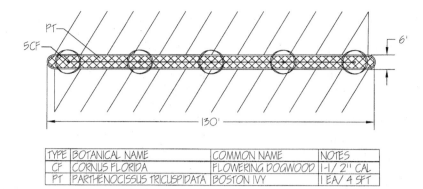

FIGURE 17-3 Landscaping Plan

plants are bid as a counted item. The quantity takeoff for landscaping is shown in the following example.

EXAMPLE 17-3

The planter shown in Figure 17-3 requires 2 inches of topsoil. Determine the topsoil and landscaping plants required to complete the planter. The dimensions are measured to the outside of a 6-inch-wide curb.

Solution: The area of the planter consists of one rectangle 124 feet (130 ft − 2 × 3 ft) long and 5 feet (6 ft − 2 × 6 in) wide and two half circles with a radius of 2.5 feet (3 ft − 6 in). The area is calculated as follows:

$$\text{Area} = (124 \text{ ft})(5 \text{ ft}) + \frac{2\pi(2.5 \text{ ft})^2}{2} = 640 \text{ ft}^2$$

The volume of topsoil is calculated as follows:

$$\text{Volume} = (640 \text{ ft}^2)\frac{(2 \text{ in})}{(12 \text{ in/ft})}\left(\frac{1 \text{ yd}^3}{27 \text{ ft}^3}\right) = 4 \text{ yd}^3$$

The number of flowering dogwood trees is five. The number of Boston ivy is calculated as follows:

$$\text{Number} = (640 \text{ ft}^2)\left(\frac{1 \text{ ea}}{4 \text{ ft}^2}\right) = 160 \text{ each}$$

SAMPLE TAKEOFF FOR THE RESIDENTIAL GARAGE

A sample takeoff for site improvements from a set of plans is shown in the following example.

EXAMPLE 17-4

Determine the site concrete and landscaping needed to complete the residential garage in Appendix G.

Solution: The site concrete consists of the driveway and the sidewalk. The area and volume of concrete needed for the driveway are calculated as follows:

$$\text{Area} = (18 \text{ ft})(26 \text{ ft } 7 \text{ in}) = 479 \text{ ft}^2$$

$$\text{Volume} = (479 \text{ ft}^2)\frac{(4 \text{ in})}{(12 \text{ in/ft})}\left(\frac{1 \text{ yd}^2}{27 \text{ ft}^2}\right) = 5.9 \text{ yd}^3$$

The driveway will require forms on the north and south side for a total of 54 feet of forms. In addition, a 27-foot-long screed will need to be set up in the middle of the driveway. There is 33 feet (26 ft + 4 ft + 3 ft) of 4-foot-wide sidewalk. The area and volume of concrete needed for the sidewalk are calculated as follows:

$$\text{Area} = (33 \text{ ft})(4 \text{ ft}) = 132 \text{ ft}^2$$

$$\text{Volume} = (132 \text{ ft}^2)\frac{(4 \text{ in})}{(12 \text{ in/ft})}\left(\frac{1 \text{ yd}^2}{27 \text{ ft}^2}\right) = 1.6 \text{ yd}^3$$

The sidewalk will require 37 feet (26 ft + 4 ft + 4 ft + 3 ft) of forms. Expansion joints are required between the driveway and the garage and the sidewalk and the garage, requiring 47 feet (18 ft + 3 ft + 26 ft) of expansion joint, or five 10-foot-long pieces.

The landscaping will consist of landscape repair around the garage. For this we will allow for an area 2 feet wide to be repaired. Beginning at the top left and going clockwise around the garage, we calculate the area of landscape repair as follows:

$$\begin{aligned}\text{Area} = (2 \text{ ft})(&26 \text{ ft } 7 \text{ in} + 3 \text{ ft} + 26 \text{ ft} + 24 \text{ ft} \\&+ 4 \text{ ft} + 26 \text{ ft} + 4 \text{ ft} + 7 \text{ ft} + 24 \text{ ft } 7 \text{ in}) \\= 291 \text{ ft}^2&\end{aligned}$$

The quantities needed for the garage, grouped by the cost codes in Appendix B, are shown in Table 17-1.

TABLE 17-1 Quantities for Residential Garage

32-110 Site Concrete—Labor		
Form driveway	54	ft
Screed	27	ft
Pour driveway	5.9	cyd
Finish driveway	479	sft
Form sidewalk	37	ft
Pour sidewalk	1.6	cyd
Finish sidewalk	132	sft
10′ expansion joint	5	ea
32-120 Site Concrete—Concrete		
Driveway	5.9	cyd
Sidewalk	1.6	cyd
32-900 Landscaping		
Sod Repair	291	sft

CONCLUSION

Asphalt and base are bid by the ton, which is determined by calculating the volume and multiplying it by the density of the asphalt or base. Site concrete is bid in the same way as building concrete. Topsoil is bid by the cubic yard; grading, seeding, and sod are bid by the square foot; plants are bid as a counted item; trenching is bid by the foot; and a sprinkler system is bid using the same principles used to bid plumbing systems.

PROBLEMS

1. Determine the number of tons of base required to complete 1,000 feet of road 22 feet wide. The base has a compacted weight of 125 pounds per cubic foot and is 12 inches thick.
2. Determine the number of tons of asphalt required to complete 1,000 feet of road 22 feet wide. The asphalt has a compacted weight of 145 pounds per cubic foot and is 4 inches thick.
3. How many cubic yards of concrete are required to construct an 80-foot by 5-foot by 4-inch-thick concrete sidewalk? Include 8% waste. Concrete must be ordered in quarter-yard increments.
4. Determine the number of tons of base required to complete the parking lot in Figure 17-4. The base has an average thickness of 8 inches and a density of 115 pounds per cubic foot. The curb is 6 inches wide and has an inside radius of 5 feet.
5. Determine the number of tons of asphalt required to complete the parking lot in Figure 17-4. The asphalt has an average thickness of 3 inches and a density of 145 pounds per cubic foot. The curb is 6 inches wide and has an inside radius of 5 feet.
6. Determine the volume of concrete needed to complete the curb in Figure 17-4. The curb is 6 inches wide by 18 inches high. Include 10% waste. Concrete must be ordered in quarter-yard increments.
7. Determine the number of tons of base required to complete the parking lot in Figure 17-5. The base has an average thickness of 6 inches and a density of 120 pounds per cubic foot. The curb is 6 inches wide and has an inside radius of 5 feet.
8. Determine the number of tons of asphalt required to complete the parking lot in Figure 17-5. The asphalt has an average thickness of 2.5 inches and a density of 150 pounds per cubic foot. The curb is 6 inches wide and has an inside radius of 5 feet.
9. Determine the volume of concrete needed to complete the curb in Figure 17-5. The curb is 6 inches wide by 18 inches high. Include 6% waste. Concrete must be ordered in quarter-yard increments.
10. The site shown in Figure 17-6 requires 4 inches of topsoil. The bark is 2 inches thick. Determine the topsoil, bark, and landscaping plants required to complete the planter.
11. The site shown in Figure 17-7 requires 2 inches of topsoil. Forty-five percent of the planter is covered with pansies. Determine the topsoil and landscaping plants required to complete the planter.

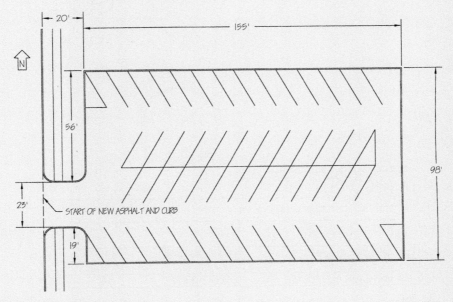

FIGURE 17-4 Parking Lot Layout

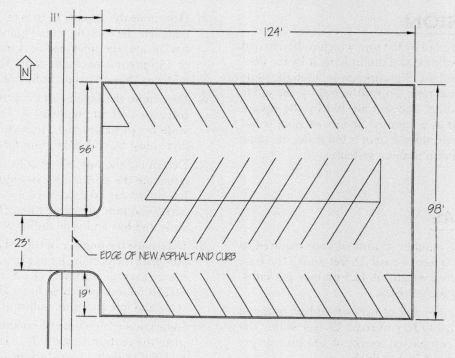

FIGURE 17-5 Parking Lot Layout

12. Determine the number of tons of asphalt and base required to complete the West Street Video project given in Appendix G. The asphalt has a density of 145 pounds per cubic foot and the base has a density of 130 pounds per cubic foot.

13. Determine the site concrete needed to complete the West Street Video project given in Appendix G.

14. Determine the landscaping needed to complete the West Street Video project given in Appendix G.

15. Set up Excel Quick Tip 17-1 in Excel.

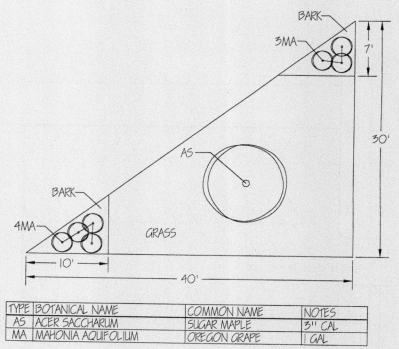

TYPE	BOTANICAL NAME	COMMON NAME	NOTES
AS	ACER SACCHARUM	SUGAR MAPLE	3" CAL
MA	MAHONIA AQUIFOLIUM	OREGON GRAPE	1 GAL

FIGURE 17-6 Landscaping Plan

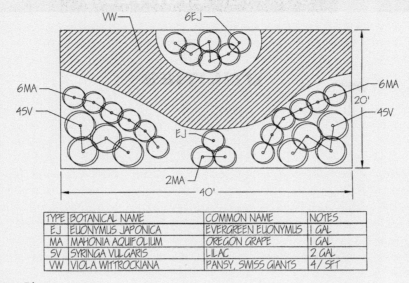

TYPE	BOTANICAL NAME	COMMON NAME	NOTES
EJ	EUONYMUS JAPONICA	EVERGREEN EUONYMUS	1 GAL
MA	MAHONIA AQUIFOLIUM	OREGON GRAPE	1 GAL
SV	SYRINGA VULGARIS	LILAC	2 GAL
VW	VIOLA WITTROCKIANA	PANSY, SWISS GIANTS	4/ SFT

FIGURE 17-7 Landscaping Plan

CHAPTER EIGHTEEN

UTILITIES

In this chapter you will learn how to apply the principles in Chapter 4 to utilities including sanitary sewers, water lines, storm drains, gas lines, underground power lines, and telephone lines. This chapter includes sample spreadsheets that may be used in the quantity takeoff.

Underground utilities are often placed in a similar manner, whether they are sanitary sewers, water lines, storm drains, gas lines, underground power lines, or telephone lines. They all consist of piping or cable placed in an underground trench. The lines may also include such features as manholes, boxes, and so forth. The work of placing underground utilities may be divided into four basic components: excavation, placement of bedding material (if required), placement of the utility line, and backfill. Let's begin by looking at excavation.

EXCAVATION

Excavation for manholes, boxes, and other such items is calculated in the same manner as excavation for a footing.

A typical utility trench is shown in Figure 18-1. The sides of utility trenches often consist of a vertical component (in this case with a constant height of d_v) and a sloped component with a slope of S horizontal feet per vertical foot. The sides of the trench are sloped to prevent cave-in and to meet safety regulations. The height of the sloped component of the sides of the trench varies as the depth of the trench varies. The height d_v and the slope of the sidewalls S are set by safety regulations and the soil conditions.

The volume of excavation may be estimated by the average-end method or the modified-average-end method presented in Chapter 16. As previously stated, the average-end method often understates the volume. The use of the average-end method to calculate the volume of excavation is shown in the following example.

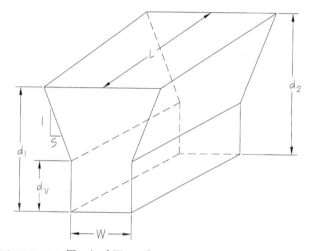

FIGURE 18-1 Typical Trench

EXAMPLE 18-1

A utility contractor is to install the sewer line whose plan and profile are shown in Figure 18-2. The trench is to have the cross section shown in Figure 18-3. Using the average-end method, determine the volume of excavation in cubic yards needed to install the sewer line. The excavation for the manhole is to be 6 feet by 6 feet and 2 feet deeper than the flow line at the manhole.

Solution: The depth of excavation at either manhole equals the elevation of the rim less the flow line (FL) of the sewer line plus the distance between the flow line and the bottom of the trench. The depth of excavation (d_1) at Station 14+30.20 and the depth of excavation (d_2) at Station 15+80.20 are calculated as follows:

$$d_1 = 1{,}148.20 \text{ ft} - 1{,}140.88 \text{ ft} + 0.50 \text{ ft} = 7.82 \text{ ft}$$

$$d_2 = 1{,}146.71 \text{ ft} - 1{,}143.88 \text{ ft} + 0.50 \text{ ft} = 3.33 \text{ ft}$$

At Station 14+30.20 the trench wall begins to slope at 4.82 feet (7.82 ft − 3 ft) below the top of the trench and widens each side of the trench by 3.62 feet (0.75 × 4.82 ft). At Station 15+80.20 the trench wall begins to slope at 0.33 feet (3.33 ft − 3 ft) below the top of the trench and widens each

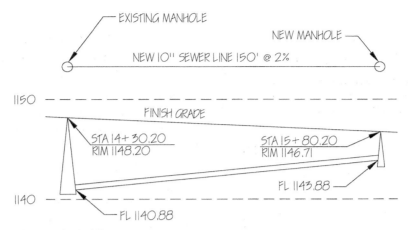

FIGURE 18-2 Sewer Line Plan and Profile

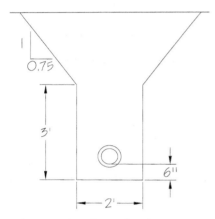

FIGURE 18-3 Sewer Line Cross Section

side of the trench by 0.25 feet (0.75 × 0.33 ft). The cross sections for the end of the trenches are shown in Figure 18-4.

The area of the end of the trench ($Area_1$) at Station 14+30.20 and the area of the end of the trench ($Area_2$) at Station 15+80.20 are calculated as follows:

$$Area_1 = (2 \text{ ft})(7.82 \text{ ft}) + \frac{2(4.82 \text{ ft})(3.62 \text{ ft})}{2} = 33.1 \text{ ft}^2$$

$$Area_2 = (2 \text{ ft})(3.33 \text{ ft}) + \frac{2(0.33 \text{ ft})(0.25 \text{ ft})}{2} = 6.7 \text{ ft}^2$$

The volume of excavation required for the trench is calculated using Eq. (16-16) as follows:

$$\text{Volume} = L\frac{(A_1 + A_2)}{2}$$

$$\text{Volume} = (150 \text{ ft})\left[\frac{(33.1 \text{ ft}^2 + 6.7 \text{ ft}^2)}{2}\right]\left(\frac{1 \text{ yd}^3}{27 \text{ ft}^3}\right) = 111 \text{ yd}^3$$

The depth of the excavation for the manhole is calculated as follows:

$$d = 1{,}146.71 \text{ ft} - 1{,}143.88 \text{ ft} + 2.00 \text{ ft} = 4.83 \text{ ft}$$

The volume of manhole excavation is calculated using the volume of column as follows:

$$\text{Volume} = (6 \text{ ft})(6 \text{ ft})(4.83 \text{ ft})\left(\frac{1 \text{ yd}^3}{27 \text{ ft}^3}\right) = 6 \text{ yd}^3$$

A small amount of the calculated quantity of excavation for the manhole is also included in the calculated quantity for the trench. The total excavation is 117 cubic yards (111 yd³ + 6 yd³).

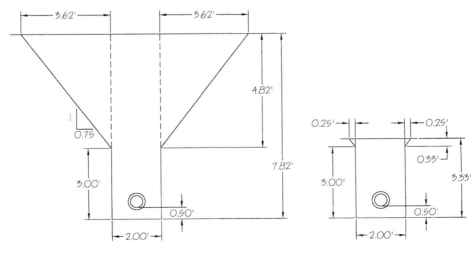

FIGURE 18-4 Trench Cross Sections

> **EXCEL QUICK TIP 18-1**
>
> **Trench Excavation Using Average-End Method**
>
> The volume of trench excavation using the average-end method is set up in a spreadsheet by entering the data and formatting the cells as follows:
>
> | | A | B | C |
> |---|---|---|---|
> | 1 | Length of Trench | 150 | ft |
> | 2 | Width of Trench Bottom | 2.00 | ft |
> | 3 | Depth at One End | 7.82 | ft |
> | 4 | Depth at Other End | 3.33 | ft |
> | 5 | Depth of Vertical Wall | 3.00 | ft |
> | 6 | Slope of Trench Walls | 0.75 | ft:ft |
> | 7 | | | |
> | 8 | Volume | 111 | cyd |
>
> The following formula needs to be entered in Cell B8:
>
> ```
> =B1*(B2*B3+IF(B5>B3,0,B6*(B3-B5)^2)
>  +B2*B4+IF(B5>B4,0,B6*(B4-B5)^2))/(2*27)
> ```
>
> The data for the trench is entered in Cells B1 through B6. The data shown in the foregoing figure is from Example 18-1.

The modified-average-end method often produces a more accurate answer but may also overstate the quantity. The use of the modified-average-end method to calculate the volume of excavation is shown in the following example.

EXAMPLE 18-2

Solve Example 18-1 using the modified-average-end method.

Solution: From Example 18-1, excavation volume for the manhole is 6 cy³ and the end areas are 33.1 ft² and 6.7 ft². The volume of excavation required for the trench is calculated using Eq. (16-18) as follows:

$$\text{Volume} = L\left(\frac{A_1 + A_2 + \sqrt{A_1 A_2}}{3}\right)$$

$$\text{Volume} = (150 \text{ ft})\left(\frac{33.1 \text{ ft}^2 + 6.7 \text{ ft}^2 + \sqrt{(33.1 \text{ ft}^2)(6.7 \text{ ft}^2)}}{3}\right)$$

$$\times \left(\frac{1 \text{ yd}^3}{27 \text{ ft}^3}\right)$$

$$\text{Volume} = 101 \text{ yd}^3$$

A small amount of the calculated quantity of excavation for the manhole is also included in the calculated quantity for the trench. The total excavation is 107 cubic yards (101 yd³ + 6 yd³).

A more accurate volume may be obtained by determining the volume of the geometric shape of the trench. This is done for the trench in Figure 18-1 by using the following equation:

> **EXCEL QUICK TIP 18-2**
>
> **Trench Excavation Using Modified-Average-End Method**
>
> The volume of trench excavation using the modified-average-end method is set up in a spreadsheet by entering the data and formatting the cells as follows:
>
> | | A | B | C |
> |---|---|---|---|
> | 1 | Length of Trench | 150 | ft |
> | 2 | Width of Trench Bottom | 2.00 | ft |
> | 3 | Depth at One End | 7.82 | ft |
> | 4 | Depth at Other End | 3.33 | ft |
> | 5 | Depth of Vertical Wall | 3.00 | ft |
> | 6 | Slope of Trench Walls | 0.75 | ft:ft |
> | 7 | | | |
> | 8 | Volume | 101 | cyd |
>
> The following formula needs to be entered in Cell B8:
>
> ```
> =B1*(B2*B3+IF(B5>B3,0,B6*(B3-B5)^2)
>  +B2*B4+IF(B5>B4,0,B6*(B4-B5)^2)
>  +((B2*B3+IF(B5>B3,0,B6*
>  (B3-B5)^2))*(B2*B4+IF(B5>B4,0,B6*
>  (B4-B5)^2)))^0.5)/(3*27)
> ```
>
> The data for the trench is entered in Cells B1 through B6. The data shown in the foregoing figure is from Example 18-2.

$$\text{Volume} = L\left[W\frac{(d_1 + d_2)}{2} + S\frac{(d_1^2 + d_1 d_2 + d_2^2 + 3d_v^2 - 3d_1 d_v - 3d_2 d_v)}{3}\right]$$

(18-1)

where

- L = Length of the Trench
- W = Width of the Trench at the Bottom
- d_1 = Depth at One End of the Trench
- d_2 = Depth at the Other End of the Trench
- d_v = Depth of the Vertical Trench Wall
- S = Slope of the Trench Walls

This equation is based on five assumptions. First, the width at the bottom of the trench is constant. Second, the height of the vertical sidewall of the trench is constant. Third, the height of the vertical sidewall is less than or equal to the depth at both ends ($d_v \le d_1$ and $d_v \le d_2$). Fourth, the slope of the sloped sidewalls is constant and, as a result, as the trench gets deeper, the width of the trench at the surface increases. Fifth, the change in depth of the trench is uniform over the length of the trench; therefore, the depth may be expressed as a linear function of the distance from one end of the trench to the other. The use of Eq. (18-1) is shown in the following example.

EXAMPLE 18-3

Solve Example 18-1 using Eq. (18-1).

Solution: From Example 18-1 the depths of excavation at the manholes are 7.82 feet and 3.33 feet. The volume of excavation required for the trench is calculated using Eq. (18-1) as follows:

$$\begin{aligned}\text{Volume} = (150 \text{ ft})\Big\{&(2 \text{ ft})\frac{(7.82 \text{ ft} + 3.33 \text{ ft})}{2} + 0.75\Big[(7.82 \text{ ft})^2 \\ &+ (7.82 \text{ ft})(3.33 \text{ ft}) + (3.33 \text{ ft})^2 + 3(3 \text{ ft})^2 \\ &- 3(7.82 \text{ ft})(3 \text{ ft}) - 3(3.33 \text{ ft})(3 \text{ ft})\Big]/3\Big\}\left(\frac{1 \text{ yd}^3}{27 \text{ ft}^3}\right) \\ = &\; 97 \text{ yd}^3\end{aligned}$$

From Example 18-1, the volume of excavation for the manhole is 6 cubic yards, bringing the total excavation volume to 103 cubic yards (97 yd³ + 6 yd³).

EXCEL QUICK TIP 18-3
Trench Excavation Using Eq. (18-1)

The volume of trench excavation using Eq. (18-1) is set up in a spreadsheet by entering the data and formatting the cells as follows:

	A	B	C
1	Length of Trench	150	ft
2	Width of Trench Bottom	2.00	ft
3	Depth at One End	7.82	ft
4	Depth at Other End	3.33	ft
5	Depth of Vertical Wall	3.00	ft
6	Slope of Trench Walls	0.75	ft:ft
7			
8	Volume	97	cyd

The following formula needs to be entered in Cell B8:

```
=IF(B5>B3,"ERROR",IF(B5>B4,"ERROR",
B1*(B2*(B3+B4)/2+B6*(B3^2+B3*B4+B4^2
+3*B5^2-3*B3*B5-3*B4*B5)/3)/27))
```

The use of the nested IF statements will display the word "ERROR" in Cell B8 should the depth of the vertical wall be greater than the depth at one of the ends. The data for the trench is entered in Cells B1 through B6. The data shown in the foregoing figure is from Example 18-3.

BEDDING

Often underground pipes are placed on a bedding material, such as gravel or sand, which provides a place for the pipe to rest in and may even surround the pipe. These two conditions are shown in Figure 18-5. The volume of bedding material is determined by calculating the cross-sectional area of the bedding material and multiplying it by the length of the pipe. Bedding material is also placed below manholes and boxes.

For pipe sitting on the bedding material, the angle a is determined by the following equation:

$$a = \cos^{-1}\left(1 - \frac{2Y}{D_P}\right) \quad (18\text{-}2)$$

where

Y = Depth of the Pipe in the Bedding
D_P = Outside Diameter of the Pipe

The quantity of bedding is estimated using Eq. (18-3) if the angle a is measured in radians and Eq. (18-4) if the angle a is measured in degrees:

$$\text{Volume} = \left\{W(D_B) - \frac{[a - (\sin a)(\cos a)]}{4}(D_P)^2\right\}L \quad (18\text{-}3)$$

$$\text{Volume} = \left\{W(D_B) - \frac{[2\pi a/360 - (\sin a)(\cos a)]}{4}(D_P)^2\right\}L \quad (18\text{-}4)$$

where

W = Width of the Trench at the Bottom
D_B = Depth of the Bedding
a = Angle from Eq. (18-2)
D_P = Outside Diameter of the Pipe
L = Length of the Trench

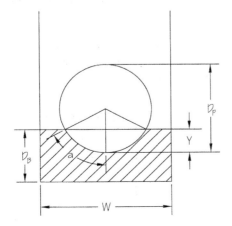
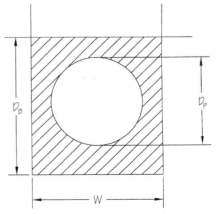

FIGURE 18-5 Bedding Conditions

For pipes barely in the bedding material, the area of the pipe is often ignored. The calculation of bedding material for a pipe partially buried in the bedding material is shown in the following example.

EXAMPLE 18-4

The pipe in Example 18-1 is bedded in gravel. The depth of the gravel is 1 foot, and the pipe is bedded into the gravel 6 inches. The outside diameter of the sewer pipe is 11 inches. Ignoring the manhole, determine the quantity of gravel needed for the sewer line.

Solution: Find the angle a in radians using Eq. (18-2) as follows:

$$a = \cos^{-1}\left[1 - \frac{2(6 \text{ in})}{11 \text{ in}}\right] = 1.66 \text{ radians}$$

The volume of bedding needed for the pipe is calculated using Eq. (18-4) as follows:

$$\text{Volume} = \left\{(2 \text{ ft})(1 \text{ ft}) - \frac{[1.66 - (\sin 1.66)(\cos 1.66)]}{4}\right.$$
$$\left. \times \left[\frac{11 \text{ in}}{(12 \text{ in/ft})}\right]^2\right\}(150 \text{ ft})\left(\frac{1 \text{ yd}^3}{27 \text{ ft}^3}\right)$$
$$= 9 \text{ yd}^3$$

EXCEL QUICK TIP 18-4
Bedding Volume for Pipe Resting on Bedding

The volume of bedding using Eqs. (18-2) and (18-3) is set up in a spreadsheet by entering the data and formatting the cells as follows:

	A	B	C
1	Width of Trench	2.00	ft
2	Depth of Bedding	1.00	ft
3	Depth of Pipe in Bedding	6.00	in
4	Diameter of Pipe	11.00	in
5	Length of Pipe	150	ft
6			
7	Angle a	1.66	rad
8	Volume	9.1	cyd

The following formulas need to be entered in the associated cells:

Cell	Formula
B7	=ACOS(1-2*B3/B4)
B8	=(B1*B2-(B7-SIN(B7)*COS(B7))* (B4/12)^2/4)*B5/27

The data for the pipe and bedding is entered in Cells B1 through B5. The data shown in the foregoing figure is from Example 18-4.

For pipe surrounded by bedding material, the quantity of bedding material is estimated using the following equation:

$$\text{Volume} = \left[W(D_B) - \frac{\pi}{4}(D_P)^2\right]L \quad (18.5)$$

where

W = Width of Trench at the Bottom
D_B = Depth of the Bedding
D_P = Outside Diameter of the Pipe
L = Length of the Trench

For bedding below manholes and boxes, the quantity of bedding is calculated using one of the excavation methods from Chapter 16. The calculation of bedding material is shown in the following example.

EXAMPLE 18-5

The pipe in Example 18-1 is surrounded by gravel. The depth of the gravel is 2 feet. The outside diameter of the sewer pipe is 11 inches. Twelve inches of bedding is also placed below the manhole. Determine the quantity of gravel needed for the sewer line.

Solution: The volume of bedding needed for the pipe is calculated using Eq. (18-5) as follows:

$$\text{Volume} = \left\{(2 \text{ ft})(2 \text{ ft}) - \frac{\pi}{4}\left[\frac{11 \text{ in}}{(12 \text{ in/ft})}\right]^2\right\}(150 \text{ ft})\left(\frac{1 \text{ yd}^3}{27 \text{ ft}^3}\right)$$
$$= 18.6 \text{ yd}^3$$

The volume of bedding needed for the manhole is calculated using the volume of a column as follows:

$$\text{Volume} = (6 \text{ ft})(6 \text{ ft})(1 \text{ ft})\left(\frac{1 \text{ yd}^3}{27 \text{ ft}^3}\right) = 1.3 \text{ yd}^3$$

The total volume needed is 20 cubic yards.

EXCEL QUICK TIP 18-5
Bedding Volume for Pipe Surrounded by Bedding

The volume of bedding using Eq. (18-5) is set up in a spreadsheet by entering the data and formatting the cells as follows:

	A	B	C
1	Width of Trench	2.00	ft
2	Depth of Bedding	2.00	ft
3	Diameter of Pipe	11.00	in
4	Length of Pipe	150	ft
5			
6	Volume	18.6	cyd

The following formula needs to be entered in Cell B6:

```
=IF(B3/12>B1, "ERROR",IF(B3/12>B2,
 "ERROR", (B1*B2-PI()*(B3/12)^2/4)*
 B4/27))
```

The use of the nested IF statements will display the word "ERROR" in Cell B6 should the diameter of the pipe exceed the width of the trench or the depth of bedding. The data for pipe and bedding is entered in Cells B1 through B4. The data shown in the foregoing figure is from Example 18-5.

UTILITY LINES

The quantities of utility components (such as pipe, manholes, valves, conduit, wire, and so forth) needed for the utility line are calculated in the same manner as plumbing and electrical components were calculated. The quantity takeoff for utility components is shown in the following example.

EXAMPLE 18-6

Determine the utility components needed for the sewer line in Example 18-1. The pipe is available in 20-foot lengths and includes a bell with a gasket at one end.

Solution: The new manhole will require a 3-foot-high manhole base with a hole and gasket for the sewer pipe, a flat manhole top, a metal ring, and a manhole cover. Eight (150 ft/20 ft) sections of 20-foot-long pipe are needed. The existing manhole will need to be core drilled for the new pipe, and the pipe will need to be sealed with a gasket or concrete.

BACKFILL

The quantity of backfill is determined by subtracting the volume of bedding and the volume of the utility components from the excavation volume. When determining the quantity of material that must be exported or imported, the estimator must take into account shrink and swell, as discussed in Chapter 16, and whether the excavated material can be used as backfill. The volume of backfill is shown in the following example.

EXAMPLE 18-7

Using the excavation volume from Example 18-3 and the bedding quantity from Example 18-5, determine the volume of backfill for the sewer line in Example 18-1. The manhole has an outside diameter of 4 feet and its bottom is 1 foot below the flow line.

Solution: From Example 18-3, the volume of excavation is 103 cubic yards. From Example 18-5, the volume of bedding is 20 cubic yards. The volume of the pipe is calculated as follows:

$$\text{Volume} = \frac{\pi}{4}\left[\frac{11 \text{ in}}{(12 \text{ in/ft})}\right]^2 (150 \text{ ft})\left(\frac{1 \text{ yd}^3}{27 \text{ ft}^3}\right) = 4 \text{ yd}^3$$

The height of the manhole is 3.83 feet, and it has a radius of 2 feet. The volume of the manhole is calculated as follows:

$$\text{Volume} = \pi (2 \text{ ft})^2 (3.83 \text{ ft})\left(\frac{1 \text{ yd}^3}{27 \text{ ft}^3}\right) = 2 \text{ yd}^3$$

The volume of the backfill is calculated as follows:

$$\text{Volume} = 103 \text{ yd}^3 - (20 \text{ yd}^3 + 4 \text{ yd}^3 + 2 \text{ yd}^3) = 77 \text{ yd}^3$$

CONCLUSION

Underground utilities consist of the excavation, placement of bedding material (if required), placement of the utility line, and backfill. The quantity of excavation may be calculated by the average-end method, the modified-average-end method, or Eq. (18-1). The quantity of bedding for pipes partially surrounded by bedding is calculated using Eq. (18-2) through (18-4), and the quantity of bedding for pipe fully surrounded by bedding is calculated using Eq. (18-5). The components of the utility lines are calculated in the same manner as plumbing and electrical components are calculated. The quantity of backfill needed equals the quantity of excavation less the quantity of backfill less the volume of the utility lines.

PROBLEMS

1. An 18-inch-wide by 3-foot-deep trench is dug in the ground and a water line with a 3.5 inch outside diameter is placed in the trench. The water line is placed on 6 inches of bedding, and bedding surrounds the pipe to a height of 2.5 inches above the water line. The length of the trench is 500 feet. Determine the volume of excavation needed for the water line.
2. How much bedding is needed for Problem 1?
3. The water line in Problem 1 is constructed of 20-foot-long pipes with a bell and gasket on one end. How many pieces of pipe are needed?
4. How much backfill is needed for the water line in Problem 1?
5. A utility contractor is to install the sewer line whose plan and profile are shown in Figure 18-6. The trench is to have the cross section shown in Figure 18-7. Using the average-end method, determine the volume of excavation in cubic yards needed to install the sewer line. The excavation for the manhole is to be 6 feet by 6 feet and 2 feet deeper than the flow line at the manhole.
6. Solve Problem 5 using the modified-average-end method.

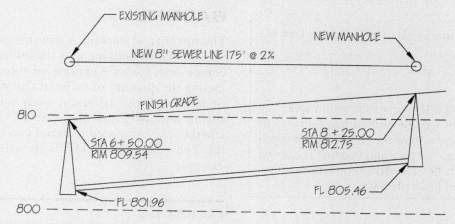

FIGURE 18-6 Sewer Line Plan and Profile

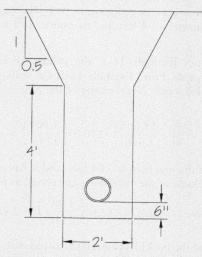

FIGURE 18-7 Sewer Line Cross Section

7. Solve Problem 5 using Eq. (18-1).
8. The pipe in Problem 5 is bedded in gravel. The depth of the gravel is 1 foot, and the pipe is bedded into the gravel 3 inches. The outside diameter of the sewer pipe is 9 inches. Ignoring the manhole, determine the quantity of gravel needed for the sewer line.
9. The pipe in Problem 5 is surrounded by gravel. The depth of the gravel is 18 inches. The outside diameter of the sewer pipe is 9 inches. Twelve inches of bedding is also placed below the manhole. Determine the quantity of gravel needed for the sewer line.
10. Determine the utility components needed for the sewer line in Problem 5. The pipe is available in 20-foot lengths and includes a bell with a gasket at one end.
11. Using the excavation volume from Problem 7 and the bedding quantity from Problem 9, determine the volume of backfill for the sewer line in Problem 5. The manhole has an outside diameter of 4 feet and its bottom is 1 foot below the flow line.
12. Determine the quantities of excavation, bedding material, utility components, and backfill needed for the Johnson Residence given in Appendix G.
13. Determine the quantities of excavation, bedding material, utility components, and backfill needed for the West Street Video project given in Appendix G.
14. Set up Excel Quick Tip 18-1 in Excel.
15. Set up Excel Quick Tip 18-2 in Excel.
16. Set up Excel Quick Tip 18-3 in Excel.
17. Set up Excel Quick Tip 18-4 in Excel.
18. Set up Excel Quick Tip 18-5 in Excel.

PART THREE

PUTTING COSTS TO THE ESTIMATE

Chapter 19   Material Pricing
Chapter 20   Labor Productivity and Hours
Chapter 21   Labor Rates
Chapter 22   Equipment Costs
Chapter 23   Crew Rates
Chapter 24   Subcontract Pricing
Chapter 25   Markups
Chapter 26   Pricing Extensions
Chapter 27   Avoiding Errors in Estimates

In this section, you will learn how to put costs to the estimate, including how to price materials, labor, and equipment. You will also learn to calculate markups and how to check for errors in the estimate. In Chapter 26 you will see how to apply these principles to the residential garage given in Appendix G.

CHAPTER NINTEEN

MATERIAL PRICING

In this chapter you will learn how to determine material pricing. Material costs include the cost of the materials, shipping costs, sales tax, storage costs, and escalation allowance. Material pricing can be obtained from suppliers, historical data, and national reference books.

The best way to determine pricing for materials is to obtain current pricing from suppliers. This is done by sending information on the quantity of materials needed to the supplier for pricing. Sometimes the supplier will do the quantity takeoff for you; however, when you have the supplier perform the quantity takeoff, you must make sure that the supplier is guaranteeing that there are sufficient quantities to perform the work. If the supplier does not guarantee that there are sufficient quantities, you should perform your own quantity takeoff to verify that there are sufficient quantities of materials to complete the work. Supplier pricing can be obtained by sending the supplier a Request for Materials Quote, such as the one shown in Figure 19-1, or by calling for pricing. It is usually better to get written pricing than a telephone quote because there is less of a chance that there will be a misunderstanding. When you receive a telephone quote, you should document it on a phone quote sheet such as the one shown in Figure 2-4.

In Figure 19-1, two columns are used for the quantity. The left column is used for the numeric portion of quantity (for example, 23, 56, and so forth) and the right column is used for units (for example, ft, cyd, miles, and so forth). The total cost for each item equals the numeric portion of the quantity multiplied by the cost per unit. This calculation is expressed by the following equation:

$$\text{Cost} = (\text{Quantity})(\text{Cost per Unit}) \quad (19\text{-}1)$$

Another way to obtain current pricing from suppliers is to set up contracts for materials with them. For example, you may set up an annual contract with a hardware supplier to supply residential door and bathroom hardware at a set unit price for a period of one year. You may then use these prices on all jobs that you bid during the year. One advantage of annual contracts is that suppliers often give better pricing because the contractor is committing to buy a large quantity of materials over the course of the year.

The next best way to determine pricing is to price materials based on historical costs. Historical costs may be obtained by determining past costs from the accounting system or by looking at previous materials invoices. Determining historical costs based on data from the accounting system is done by determining the material cost for a certain quantity of work and dividing it by the quantity of work performed. This calculation is performed by solving Eq. (19-1) for cost per unit and is expressed by the following equation:

$$\text{Cost per Unit} = \frac{\text{Cost}}{\text{Quantity}} \quad (19\text{-}2)$$

EXAMPLE 19-1

A recent project included 1,235 square feet of 4-inch-thick sidewalk. The cost of the concrete materials for the sidewalk was $1,372. Determine the material cost for 1 square foot of sidewalk.

Solution: The material cost per square foot is determined by using Eq. (19-2) as follows:

$$\text{Material Cost} = \frac{\$1,372}{1,235 \text{ ft}^2} = \$1.111/\text{ft}^2$$

For historical data to be accurate, the actual project costs and quantities must be accurately tracked in the accounting system. Poor tracking of costs or failure to update changes in the quantities will render the data useless.

The other method of determining historical cost is to look at previous invoices to see what prices the company has been paying. When looking at previous invoices one

REQUEST FOR MATERIALS QUOTE				
To:				
From:				
Project:				
QUANTITY	DESCRIPTION		$/UNIT	COST
			SUBTOTAL	
			SALES TAX	0.00%
			SUBTOTAL	
			DELIVERY	
			TOTAL	
The supplier agrees to supply the materials in a timely manner and for the prices stated above. This quote is good for 60 days.				
By:			Date:	

FIGURE 19-1 Request for Materials Quote

should use the most current invoices from a job with equivalent specifications and similar quantities.

The least accurate source of data is national reference books. Material prices vary greatly around the country. As a result, national reference books should only be used as a last resort for material pricing when supplier quotes or historical data are not available.

In addition to the cost of the materials, material pricing needs to include shipping costs, sales tax, and storage costs. When material prices are not guaranteed or materials are purchased over a long period of time, changes in the material prices (escalation) need to be included in the pricing. Let's look at these costs.

SHIPPING COSTS

Material prices are often quoted FOB the factory, FOB the supplier's warehouse, or FOB the jobsite. The term "FOB" is short for "free on board," which identifies the point where the supplier relinquishes responsibility for shipping and protection of the materials against loss to the purchaser of the materials. For example, if the price for a HVAC unit is quoted FOB the factory, the purchaser is responsible for paying for the shipping from the factory to the jobsite, and should the unit be damaged during shipping, the purchaser would be responsible for the cost of the damage. However, if the price for the HVAC unit is quoted FOB the jobsite, the supplier is responsible for shipping the unit to the jobsite, and should the unit be damaged during shipping, the supplier, not the purchaser, would be responsible for the cost of the damage.

Prices for shipping should be obtained from the supplier or a shipper. Shipping charges from local suppliers are often a flat rate per delivery or by the hour. For example, a lumber supplier may charge $25 per delivery and an aggregate supplier may charge $65 per hour for delivery. Long-distance shipping by truck is often charged by the hundredweight (cwt) or fraction thereof. One hundredweight equals 100 pounds. Shipping rates may also be based upon the size of the item being shipped. Long-distance shipping charges are calculated as shown in the following example.

EXAMPLE 19-2

The price for an 11,225-pound HVAC unit has been quoted FOB the factory. You have obtained a shipping price of $25.65/cwt to ship the unit from the factory to the jobsite. Determine the cost of shipping the HVAC unit.

Solution: The weight, in hundredweight, of the unit is calculated as follows:

$$\text{Weight} = \frac{11,225 \text{ lb}}{100 \text{ lb/cwt}} = 112.25 \text{ cwt}$$

The shipper will charge for 113 cwt. The shipping cost is calculated as follows:

Shipping Cost = (113 cwt)($25.65/cwt) = $2,898.45

SALES TAX

States often charge sales tax on material purchases. Some states may charge sales tax on permanent materials (materials incorporated into the finish project such as doors and windows) but not charge sales tax on construction materials (materials used in the construction process but not incorporated into the finished project, such as saw blades). The sales tax rate is expressed as a percentage of the sales price of the taxable materials. Some states allow counties and cities to levy additional sales tax. As a result, the sales tax rate may vary by city and county as well as by state. In addition, some public projects are exempt from sales tax. If this is the case, it will be identified in the construction documents. Because the sales tax rate can vary by project, city, county, and state, it is important for the estimator to determine the local sales tax rate for each project.

When purchasing materials from an in-state supplier, the supplier often collects the sales tax along with the cost of the materials. When purchasing materials from an out-of-state supplier, the supplier often does not pay sales tax on the materials, leaving purchasers to pay their local sales tax directly to the state by paying an use tax. When getting prices from suppliers, the estimator must find out whether a supplier has included sales tax in the pricing or if sales tax needs to be added.

The cost of the sales tax is determined by multiplying the cost of the taxable materials by the sales tax rate. This calculation is expressed by the following equation:

$$\text{Sales Tax} = (\text{Taxable Materials})(\text{Sales Tax Rate}) \quad (19\text{-}3)$$

EXAMPLE 19-3

You have been quoted $125,236.23 for an order of lumber, all of which is subject to sales tax. The price does not include sales tax. The local sales tax rate is 6.75%. How much sales tax must be paid on the lumber?

Solution: The sales tax is calculated using Eq. (19-3) as follows:

Sales Tax = ($125,236.23)(0.0675) = $8,453.45

STORAGE COSTS

When purchasing large quantities of materials, large equipment, or materials before they are needed, the contractor may have to provide a place to store the materials and equipment until they are installed. The storage can take place on the project, in which case the material must be stored in an out-of-the-way place and protected against theft and damage. Storage costs for on-site storage may include temporary fencing or containers in which to store the material or security to protect the material and equipment from theft. Alternatively, the materials can be stored off-site. When the materials are stored off-site, secure storage space for the materials may need to be rented, and the materials will need to be shipped to the site as they are needed, all of which adds to the cost of the materials. When storing materials or equipment, any additional handling (e.g., equipment to move the materials and equipment a second time) and additional shipping costs need to be included in the estimate.

ESCALATION

Sometimes material suppliers will not guarantee the material prices for the duration of the construction project, as is often the case with lumber. When the material prices are expected to increase over the duration of their use on the construction project, the estimator needs to include an escalation allowance in the material prices to cover the anticipated increases. When dealing with annual contracts or material suppliers that raise their prices annually, it is easy to anticipate price changes and incorporate these changes into the bids. With pricing that fluctuates with market prices, such as lumber, escalation is impossible to accurately predict, but escalation still needs to be included in the bid. Trade magazines, which report on the trends in the industry, can give a contractor insight into factors within the economy and industry that may lead to escalation of material costs. Material pricing can be adjusted for escalation using the following equation:

$$\text{Cost}_{\text{Escalated}} = \text{Cost}(1 + f) \quad (19\text{-}4)$$

where

$\text{Cost}_{\text{Escalated}}$ = Escalated Material Cost
Cost = Quoted Cost of Materials
f = Expected Escalation or Inflation Rate (in decimal format)

If material prices are expected to decline, a negative escalation rate may be used in this equation. The calculation of the escalated cost is shown in the following example.

EXAMPLE 19-4

You have been quoted $88,325 for an order of lumber. You expect lumber prices to increase by 3.5% between now and the time you place the order. How much should you include in your bid to cover the cost of the lumber?

Solution: The expected escalation rate is 3.5%, or 0.035. The escalated cost is calculated using Eq. (19-4) as follows:

$$\text{Cost}_{\text{Escalated}} = (\$88,325)(1 + 0.035) = \$91,416$$

To protect against material price increases, some contractors may purchase large quantities of materials at the beginning of the construction project and store the materials and pay the storage costs rather than risk price increases.

EXCEL QUICK TIP 19-1
Request for Materials Quote

The Request for Materials Quote shown in Figure 19-1 is set up in a spreadsheet by entering the data and formatting the cells as follows:

[Spreadsheet image of Request for Materials Quote form with columns A–E, rows 1–41, including fields for To, From, Project, Quantity, Description, $/Unit, Cost, Subtotal, Sales Tax, Delivery, Total, By, and Date]

The following formulas need to be entered into the associated cells:

Cell	Formula
E12	=A12*D12
E32	=SUM(E12:E31)
E33	=E32*D33
E34	=E32+E33
E36	=E34+E35

After entering the formula in Cell E12, copy the formula in Cell E12 to Cells E13 through E31. In Rows 12 through 31, the numeric portion of the quantity is entered in Column A, the units associated with the quantity in Column B, the material description in Column C, and the unit pricing in Column D. The sales tax rate is entered in Cell D33 and the shipping cost in Cell E35. One assumption in this spreadsheet is that all material costs (except for the delivery cost) are subject to sales tax. The cells containing the costs are formatted using the comma style, which replaces zeros with dashes.

CONCLUSION

Material costs include the cost of the materials, shipping costs, sales tax, storage costs, and escalation allowance. Material pricing may be obtained from suppliers, historical data, and national reference books.

PROBLEMS

1. A recent project included 11,000 square feet of 5-inch-thick concrete slab. The cost of the concrete material for the sidewalk was $15,950. Determine the material cost for 1 square foot of sidewalk.
2. A recent project included 6,450 square feet of painting. The cost of the paint for the painting was $609.10. Determine the material cost for 1 square foot of painting.
3. The price for a 5,595-pound HVAC unit has been quoted FOB the factory. You have obtained a shipping price of $15.75/cwt to ship the unit from the factory to the jobsite. Determine the cost of shipping the HVAC unit.
4. The price for a 3,023-pound transformer has been quoted FOB the factory. You have obtained a shipping price of $21.29/cwt to ship the unit from the factory to the jobsite. Determine the cost of shipping the transformer.
5. You have been quoted $13,568.22 for an order of lumber, all of which is subject to sales tax. The price does not include sales tax. The local sales tax rate is 4.25%. How much sales tax must be paid on the lumber?
6. You have been quoted $27,569.65 for electrical switchgear, which is subject to sales tax. The price does not include sales tax. The local sales tax rate is 7.60%. How much sales tax must be paid on the switchgear?
7. You have been quoted $165,665 for an order of lumber. You expect lumber prices to increase by 7.0% between now and the time you place the order. How much should you include in your bid to cover the cost of the lumber?
8. You have been quoted $2,359 for an order of drywall. You expect drywall prices to increase by 4.25% between now and the time you place the order. How much should you include in your bid to cover the cost of the drywall?
9. Prepare Request for Materials Quotes for the Johnson Residence given in "Appendix G" for the following Cost Codes:
   a. 03200 Rebar
   b. 03310 Footings and Foundation—Concrete, 03330 Slab/Floor—Concrete, and 32120 Site Concrete—Concrete
   c. 06110 Lumber
   d. 06120 Trusses
   e. 06210 Wood Trim
   f. 06400 Cabinetry and Countertops
   g. 08100 Metal Doors and 08110 Wood Doors
   h. 08300 Overhead Doors
   i. 08500 Windows
   j. 08700 Hardware
   k. 10210 Toilet and Bath Accessories
10. Prepare Request for Materials Quotes for the West Street Video project given in "Appendix G" for the following Cost Codes:
    a. 03200 Rebar and 32130 Rebar
    b. 03310 Footings and Foundation—Concrete, 03330 Slab/Floor—Concrete, and 32120 Site Concrete—Concrete
    c. 05100 Structural Steel
    d. 05200 Joist and Deck
    e. 06110 Lumber
    f. 06400 Cabinetry and Counter Tops
    g. 07700 Roof Specialties
    h. 08100 Metal Doors and Frames
    i. 08700 Hardware
    j. 10210 Toilet and Bath Accessories
    k. 10400 Fire Extinguishers and Cabinets
11. Set up Excel Quick Tip 19-1 in Excel.

CHAPTER TWENTY

LABOR PRODUCTIVITY AND HOURS

In this chapter you will learn what factors affect labor productivity and how to calculate labor productivity using historical data, field observations, and national reference books. You will also learn how to calculate labor hours from labor productivity.

Labor productivity is a measure of how fast construction tasks can be performed. Productivity can be reported as the number of units performed during a labor hour[1] or crew day, which is often referred to as output. For example, the productivity for a two-person framing crew installing roof sheathing may be 1,600 square feet of sheathing per 8-hour day. It also may be expressed as 100 square feet of sheathing per labor hour (lhr). When measuring output, the larger the number, the greater is the productivity.

Alternatively, it can be expressed as the number of labor hours required to complete one unit of work. For example, the productivity to install the roof sheathing may be expressed as 0.010 labor hour per square foot. In most estimating, labor hours per unit are used rather than output. When measuring productivity in labor hours per unit, the smaller the number, the greater is the productivity.

Daily output per crew day is converted to labor hours per unit by dividing the number of labor hours performed by the crew per day by the output, using the following equation:

$$\text{Labor Hours/Unit} = \frac{\text{Labor Hours}}{\text{Output}} \quad (20\text{-}1)$$

EXAMPLE 20-1

Convert the output of 1,600 square feet of sheathing per 8-hour day for a two-person framing crew to labor hours per square foot.

Solution: The number of labor hours per square foot is calculated using Eq. (20-1) as follows:

$$\text{Labor Hours/Unit} = \frac{(2 \text{ persons})(8 \text{ hr/person-day})}{(1,600 \text{ ft}^2/\text{day})}$$
$$= 0.010 \text{ lhr/ft}^2 \quad \blacksquare$$

Output per labor hour is converted to labor hours per unit by dividing 1 by the output, using the following equation:

$$\text{Labor Hours/Unit} = \frac{1}{\text{Output}} \quad (20\text{-}2)$$

EXAMPLE 20-2

Convert the output of 100 square feet of sheathing per labor hour to labor hours per square foot.

Solution: The number of labor hours per square foot is calculated using Eq. (20-2) as follows:

$$\text{Labor Hours/Unit} = \frac{1}{(100 \text{ ft}^2/\text{lhr})} = 0.010 \text{ lhr/ft}^2 \quad \blacksquare$$

FACTORS AFFECTING LABOR PRODUCTIVITY

Productivity is affected by many factors, including project size, overtime, size of crew, delays, interruptions, weather, project layout, safety, and so forth. Because these factors vary from project to project, it is impossible to assign one productivity rate to a given task. Each time an estimator bids a project, he or she must look at these factors and determine whether to expect a higher or lower productivity than on previous projects and adjust the productivity accordingly. Let's look at some of these factors.

Project Size

Larger projects tend to have higher productivities, which is particularly true for projects that are highly repetitive. This is because of a number of factors, including the spreading out of the fixed time over a greater number of units, workers learning to do the task faster, and specialization among the crew.

The time it takes to complete a task can be divided into a fixed component and a variable component. The fixed component includes things that must be done regardless of the number of units of work completed. Daily setup takes a fixed amount of time each day, regardless of the daily production. The variable component includes things that are a function of the number of units of work completed. To demonstrate fixed and variable time, let's look at an example of installing entry doors on a residence. To install one entry door, the carpenter needs to go to the residence, familiarize himself or herself with the plans, set up the equipment, install the door, clean up, and leave. Although the installation of the door by itself may only take 1 hour, the total process may take 4 hours. In this case, the fixed time is 3 hours and the variable time is 1 hour (the time actually used to install the door), with the fixed time being spread over one door. This results in a productivity of 4 labor hours per door. Should the carpenter install a second door at the same time, both doors could be installed in 5 hours, resulting in a productivity of 2 1/2 labor hours per door, spreading the fixed time over two doors. Three doors would result in a productivity of 2 labor hours per door.

As a person performs a task over and over, he or she should become more proficient in performing the task. This is often referred to as moving down the learning curve. An example of this is an electrician wiring hotel rooms. On the first room he or she has to consult the plans often to determine the location of the outlets, switches, and fixtures, which increases the time it takes to wire the room. On the second room, the electrician needs to consult the plans less than he or she did on the first room. By the time the electrician has done a dozen or so rooms, he or she can complete the rooms without consulting the plans, further increasing his or her productivity.

As the projects get larger, the crew can specialize in performing certain tasks. For example, one framing crew may frame the walls while another frames the floors and yet another frames the roof. This specialization moves the crews down the learning curve faster than if all three crews framed walls, floors, and roofs.

Overtime

Increasing the amount of time workers work each week decreases the productivity of the workers. The longer this goes on, the greater is the decrease in productivity. For example, workers may easily work an extra Saturday one week to complete a project with little or no loss in production. However, if the workers work every Saturday for 2 months, they will begin to tire and will need a rest. As a result, a greater loss in productivity will occur than it did when they only worked one Saturday. Because of the loss in productivity and the higher labor costs, it is uneconomical to use overtime to solve long-term labor problems. Overtime should only be used to solve short-term labor problems. If a long-term solution is needed, additional workers or crews need to be added.

Size of Crew

The size of the crew has a great effect on productivity. Having a crew that is too large or too small will lead to a decrease in productivity. With crews that are too large, members of the crew are often in each other's way, or there may not be sufficient work to keep all of the members of the crew busy. Crews that are too small often lack enough members to complete the tasks. For example, a two-person framing crew will find it difficult to stand walls because they lack a sufficient number of workers to lift all but the smallest walls in place. Because of this, there is an optimum size of crew, which is different for different projects.

Delays

One of the greatest enemies to productivity comes from delays on the construction project. For work to proceed on the project, the right materials, equipment, labor, and design information must be on the project when the project is ready for it. If the right materials and equipment are not available when they are needed, time must be spent obtaining them or the crews must leave the project until they have been obtained. This wastes time and decreases productivity. When the wrong class of labor is on the project, work may stop on the project until the right class of labor is available. If work proceeds with the wrong class of labor, it is often less productive than if it was performed by the correct class of labor. When the necessary design information is not available or is incorrect, work must stop until the correct design information is obtained, decreasing the productivity of the work.

Interruptions

Interruptions in work flow decrease productivity. When a worker is jumping from task to task, he or she is less productive than if he or she is allowed to work on a single task until it is complete because time is being wasted moving in from task to task.

Weather

Weather has a great effect on productivity. Hot weather tires workers out and drains their strength, requiring them to take frequent breaks for rest and to drink fluids, thus decreasing their productivity. Cold weather requires workers to wear weather protection that decreases their mobility and ability to handle small items (such as nails) and in turn decreases their productivity. In cold weather, workers tire easily and must take frequent breaks for rest and to warm up. Wet weather makes the site muddy and working surfaces slippery, decreasing productivity. Snow requires workers to remove the snow from the work area before beginning to work and covers up materials, making them harder to find. This results in a decrease in productivity. In addition, adverse weather requires that workers protect their work against the weather. For example, concrete must be protected against drying too quickly on hot days and freezing on cold days. Protecting construction work from weather also decreases productivity.

Project Layout

The location where materials can be delivered and stored in relationship to the construction project affects productivity. On sites where materials cannot be delivered or stored near the location at which they will be used, the time it takes to move the materials will increase and the productivity will decrease.

Safety

The setup and use of safety equipment often decreases productivity. In spite of the time it takes to set up and use safety equipment, safety should never be sacrificed for production. Estimators need to incorporate the time it takes to set up and use safety equipment into the productivity rates.

Production can be estimated using historical data, field observations, and national reference books.

HISTORICAL DATA

The best source for labor productivity is historical data. Historical labor productivity is obtained by determining the number of labor hours required to complete a certain quantity of work on past jobs. This information is obtained from the accounting system. The productivity is calculated by dividing the number of labor hours by the quantity of work performed as expressed by the following equation:

$$\text{Labor Hours/Unit} = \frac{\text{Labor Hours}}{\text{Quantity}} \quad (20\text{-}3)$$

EXAMPLE 20-3

A recent project included 1,235 square feet of 4-inch-thick sidewalk. It took 11 labor hours to complete the sidewalk. Determine the productivity in labor hours per square foot based on this historical data.

Solution: The number of labor hours per square foot is calculated using Eq. (20-3) as follows:

$$\text{Labor Hours/Unit} = \frac{11 \text{ lhr}}{1{,}235 \text{ ft}^2} = 0.0089 \text{ lhr/ft}^2$$

For historical data to be accurate, the actual labor hours must be accurately tracked in the accounting system and an accurate record of the quantities of work performed must be kept. Poor tracking of labor hours or failure to update changes in the quantities of work performed will render the data useless.

FIELD OBSERVATIONS

Labor productivity from field observations can be calculated in three ways: cycle time, rate of progress, and daily production. Let's look at each of these ways.

Cycle Time

Cycle time is used for repetitive tasks. Repetitive tasks are tasks that repeat over and over, such as installing sheathing on a wall. In the case of sheathing, the process of placing one piece of sheathing is known as a cycle. The cycle consists of picking up a sheet of sheathing from the stock pile, moving it to where it is going to be used, placing and fastening the sheathing, and returning to the stock pile. The productivity of repetitive tasks is determined by measuring the time it takes to complete a cycle, which is known as the cycle time. Because the time it takes to complete a cycle varies from cycle to cycle, at least 30 cycles should be measured when determining the cycle time. The average cycle time is determined by summing the cycle times and dividing by the number of observations, using the following equation:

$$\text{CT}_{\text{Ave}} = \frac{(\text{CT}_1 + \text{CT}_2 + \cdots + \text{CT}_n)}{n} \quad (20\text{-}4)$$

where

$\text{CT} = $ Cycle Time
$n = $ Number of Observations

EXAMPLE 20-4

The cycle times in Table 20-1 were observed for the process of installing plywood sheathing. Determine the average cycle time.

Solution: The average cycle time is calculated using Eq. (20-4) as follows:

Average CT = [(10.34 + 12.27 + 10.90 + 17.67 + 8.74 + 10.90
 + 10.48 + 8.32 + 11.84 + 15.42 + 10.11 + 11.09
 + 10.48 + 10.34 + 10.53 + 8.98 + 7.99
 + 12.22 + 13.82 + 14.10 + 10.25 + 15.60 + 10.25
 + 12.41 + 16.45 + 11.47 + 10.01 + 13.91
 + 8.08 + 9.35) min/cycle]/30

$$= \frac{(344.32 \text{ min/cycle})}{30} = 11.48 \text{ min/cycle}$$

TABLE 20-1 Cycle Times (CTs) for Example 20-4

Observation	CT (min)	Observation	CT (min)	Observation	CT (min)
1	10.34	11	10.11	21	10.25
2	12.27	12	11.09	22	15.60
3	10.90	13	10.48	23	10.25
4	17.67	14	10.34	24	12.41
5	8.74	15	10.53	25	16.45
6	10.90	16	8.98	26	11.47
7	10.48	17	7.99	27	10.01
8	8.32	18	12.22	28	13.91
9	11.84	19	13.82	29	8.08
10	15.42	20	14.10	30	9.35

To convert the average cycle time to labor hours per unit, one must take into account how the observed conditions compare to typical conditions, size of the crew, system efficiency, and the units constructed per cycle.

Often the observations are taken over a short period of time, and, as a result, they are not representative of the average conditions under which the work is performed. To account for this, an adjustment factor is used. If the cycle times were measured at a time when productivity was higher than average because of favorable conditions, an adjustment factor that is less than 1 is used. If the cycle times were measured at a time when productivity was lower than average because of unfavorable conditions, an adjustment factor that is greater than 1 is used. The adjustment factor must be estimated by the person making the observations, which is a judgment call.

No systems or processes are 100% efficient, and people are not always working on those items measured during a cycle time. During an hour the workers may only be spending 50 minutes working on tasks measured in the cycle time. During the remaining 10 minutes the workers may be taking breaks or engaged in other nonproductive tasks, such as going to the bathroom. The system efficiency is the number of productive minutes per hour. Typical system efficiencies are 45 to 50 minutes per hour.

When measuring the cycle time the number of workers that were required to complete one cycle was not taken into account. A one-hour cycle using two crew members will require twice the number of labor hours as a one-hour cycle using a single crew member. The size of the crew takes this into account.

Finally, one needs to account for the number of units produced during a single cycle because production is measured by the number of units produced.

All of these factors are accounted for in the following equation:

$$\text{Labor Hours/Unit} = \frac{(CT_{Ave})(AF)(\text{Size})}{(SE)(\text{Units})} \quad (20\text{-}5)$$

where

CT_{Ave} = Average Cycle Time in Minutes per Cycle
AF = Adjustment Factor
Size = Size of Crew
SE = System Efficiency (Minutes per 60-Minute Hour)
Units = Number of Units Produced per Cycle

EXAMPLE 20-5

Using the average cycle time from Example 20-4, a system efficiency of 50 minutes per hour, an adjustment factor of 0.90, and a crew size of two, determine the productivity in labor hours per square foot of sheathing.

Solution: One sheet is 32 square feet. The productivity is calculated using Eq. (20-5) as follows:

$$\text{Labor Hours/Unit} = \frac{(11.48 \text{ min/cycle})(0.9)(2)}{(50 \text{ min/hr})(32 \text{ ft}^2/\text{cycle})}$$
$$= 0.0129 \text{ lhr/ft}^2$$

Rate of Progress

Rate of progress is used for linear tasks. Linear tasks are tasks that progress along a line rather than in a cyclical motion. Linear tasks include paving roads, installing pipes, painting striping on roads, installing curbs, and so forth. The productivity for linear tasks is measured by the rate the task proceeds or rate of progress. Linear tasks consist of two components: production time and travel time. Production time includes the time spent producing

> **EXCEL QUICK TIP 20-1**
> **Cycle Time Worksheet**
>
> The spreadsheet shown in the following figure can be used to calculate the average cycle time and the labor hours per unit.
>
> | | A | B | C | D | E | F | G | H |
> |---|---|---|---|---|---|---|---|---|
> | 1 | | Cycle Time Calculator | | | | | | |
> | 2 | OBS. | CT (min) | | OBS. | CT (min) | | Adjustment Factor | 0.90 |
> | 3 | 1 | 10.34 | | 16 | 8.98 | | Crew Size | 2 |
> | 4 | 2 | 12.27 | | 17 | 7.99 | | System Efficiency | 50 |
> | 5 | 3 | 10.90 | | 18 | 12.22 | | Units | 32.00 |
> | 6 | 4 | 17.67 | | 19 | 13.82 | | | |
> | 7 | 5 | 8.74 | | 20 | 14.10 | | Average Cycle Time | 11.48 |
> | 8 | 6 | 10.90 | | 21 | 10.25 | | Labor-Hour/Unit | 0.01291 |
> | 9 | 7 | 10.48 | | 22 | 15.60 | | | |
> | 10 | 8 | 8.32 | | 23 | 10.25 | | | |
> | 11 | 9 | 11.84 | | 24 | 12.41 | | | |
> | 12 | 10 | 15.42 | | 25 | 16.45 | | | |
> | 13 | 11 | 10.11 | | 26 | 11.47 | | | |
> | 14 | 12 | 11.09 | | 27 | 10.01 | | | |
> | 15 | 13 | 10.48 | | 28 | 13.91 | | | |
> | 16 | 14 | 10.34 | | 29 | 8.08 | | | |
> | 17 | 15 | 10.53 | | 30 | 9.35 | | | |
>
> To complete this spreadsheet, enter the data and format the cells as shown in the figure. The following formulas need to be entered into the associated cells:
>
> | Cell | Formula |
> |------|---------|
> | H7 | =AVERAGE(B3:B17,E3:E17) |
> | H8 | =H7*H2*H3/(H4*H5) |
>
> The data is entered into Cells B3 through B17, E3 through E17, and H2 through H5. The data shown in the foregoing figure is from Examples 20-4 and 20-5.

the finished product. In the case of an employee operating a pavement-striping machine to paint parking stalls, this time includes the time the employee is painting the lines in the parking lot. The second component of a linear task is the travel time. In the case of the employee painting the parking stalls, the travel time is the time spent moving the striping machine from the end of one line to the beginning of the next line. The travel time represents breaks in the work and must be included in the production rate. As with repetitive tasks, when calculating the productivity of linear tasks, one must include the system efficiency. The productivity is calculated from the following equation:

$$\text{Labor Hours/Unit} = \frac{\left(\dfrac{\text{Quantity}}{\text{RoP}} + \text{TT}\right)(\text{Size})}{(\text{Quantity})(\text{SE})} \quad (20\text{-}6)$$

where

Quantity = Quantity of Work to Be Performed
RoP = Rate of Progress Measured in Units per Minute
TT = Travel Time (Minutes)
Size = Size of Crew
SE = System Efficiency (Minutes per 60-Minute Hour)

EXAMPLE 20-6

An operator of a paint-striping machine is painting parking stalls in a parking lot. The work consists of 4 lines 440 feet long, 200 lines 40 feet long, and 100 lines 20 feet long. When striping, the operator can paint 40 feet of line per minute. The travel time between lines is 0.5 minute. Using a system efficiency of 50 minutes per hour, determine the productivity for striping of the parking lot in labor hours per foot.

Solution: The quantity of work is calculated as follows:

$$\text{Quantity} = (4 \text{ ea})(440 \text{ ft/ea}) + (200 \text{ ea})(40 \text{ ft/ea})$$
$$+ (100 \text{ ea})(20 \text{ ft/ea})$$
$$= 11{,}760 \text{ ft}$$

The number of times the operator needs to travel is one less than the number of lines. The number of lines and travel time are calculated as follows:

$$\text{Number of Lines} = 4 \text{ ea} + 200 \text{ ea} + 100 \text{ ea} = 304 \text{ ea}$$

$$TT = (303 \text{ ea})(0.5 \text{ min/ea}) = 151.5 \text{ min}$$

The productivity is calculated using Eq. (20-6) as follows:

$$\text{Labor Hours/Unit} = \frac{\left(\dfrac{11{,}760 \text{ ft}}{40 \text{ ft/min}} + 151.5 \text{ min}\right)(1)}{(11{,}760 \text{ ft})(50 \text{ min/hr})}$$

$$\text{Labor Hours/Unit} = 0.000758 \text{ lhr/ft}$$

EXCEL QUICK TIP 20-2
Rate of Progress Worksheet

The spreadsheet shown in the following figure can be used to calculate the labor hours per unit using the rate of progress:

	A	B	C
1	Rate of Progress Calculator		
2	Quantity	11,760	units
3	Rate of Progress	40	units per min
4	No. of Moves	303	ea
5	Time per Moves	0.5	min
6	Crew Size	1	persons
7	System Efficiency	50	per 60 min hr
8	Labor-Hour/Unit	0.000758	lhr/unit

To complete this spreadsheet, enter the data and format the cells as shown in the figure. The following formula needs to be entered into Cell B8:

```
=(B2/B3+B4*B5)*B6/(B2*B7)
```

In this equation the travel time is broken down into two components: number of moves and the time per move. The travel time is calculated by multiplying the number of moves (Cell B4) by the time per move (Cell B5).

The data is entered into Cells B2 through B7. The data shown in the foregoing figure is from Example 20-6.

Daily Production

Alternatively, the production for all tasks can be determined by measuring the quantity of work performed during a day and dividing it into the number of labor hours required to complete the work. The productivity is determined by Eq. (20-3) in the same manner the productivity was determined from historical data. Because the whole day is used, the system efficiency is already incorporated in the calculation.

EXAMPLE 20-7

The measured output for a crew is 1,696 square feet of sheathing in 1 day. The crew consists of two crew members, who work 8 hours each during the day. Determine the productivity of the crew in labor hours per square foot of sheathing.

Solution: The number of labor hours per square foot is calculated using Eq. (20-3) as follows:

$$\text{Labor Hours/Unit} = \frac{(2 \text{ persons})(8 \text{ hr/person-day})}{(1{,}696 \text{ ft}^2/\text{day})}$$

$$= 0.0094 \text{ lhr/ft}^2$$

NATIONAL REFERENCE BOOKS

The least accurate source of data is that from national reference books. Labor productivity varies greatly around the country. As a result, national reference books should only be used when historical data or field observations are not available or as a check of the calculated productivity. When using national reference, the labor productivity should be adjusted for area, project size, time of year, and so forth. A sample page from *RSMeans Building Construction Cost Data 2015* is shown in Figure 20-1. From this figure, we can see that the crew includes two carpenters, the daily output is 1,600 square feet, and the labor hours per square foot is 0.010 for 5/16″-thick CDX plywood sheathing on roofs. Sample labor productivity rates for many of the tasks in Part II of this book are given in Appendix C.

LABOR HOURS

The number of labor hours required to complete a task is calculated by multiplying the labor productivity (1 hr/unit) by the quantity of work and is expressed by the following equation:

$$\text{Labor Hours} = (\text{Quantity})(\text{Labor Hours/Unit}) \quad (20\text{-}7)$$

EXAMPLE 20-8

A crew needs to install 2,650 square feet of 7/16-inch oriented strand board (OSB) on a roof. The OSB is to be pneumatically nailed. Using data from Figure 20-1, determine the number of labor hours needed to install the OSB.

Solution: From Figure 20-1, the number of labor hours needed to install one square foot of 7/16-inch OSB pneumatically nailed on a roof is 0.009 (see Line 4505). The number of labor hours needed is calculated using Eq. (20-7) as follows:

$$\text{Labor Hours} = (2{,}650 \text{ ft}^2)(0.009 \text{ lhr/ft}^2) = 24 \text{ lhr}$$

06 16 Sheathing

06 16 36 – Wood Panel Product Sheathing

06 16 36.10 Sheathing		Crew	Daily Output	Labor-Hours	Unit	Material	2015 Bare Costs Labor	Equipment	Total	Total Incl O&P
0010	**SHEATHING** R061636-20									
0012	Plywood on roofs, CDX									
0030	5/16" thick	2 Carp	1600	.010	S.F.	.56	.47		1.03	1.33
0035	Pneumatic nailed R061110-30		1952	.008		.56	.39		.95	1.20
0050	3/8" thick		1525	.010		.60	.49		1.09	1.42
0055	Pneumatic nailed		1860	.009		.60	.40		1	1.28
0100	1/2" thick		1400	.011		.66	.54		1.20	1.55
0105	Pneumatic nailed		1708	.009		.66	.44		1.10	1.40
0200	5/8" thick		1300	.012		.78	.58		1.36	1.75
0205	Pneumatic nailed		1586	.010		.78	.47		1.25	1.59
0300	3/4" thick		1200	.013		.93	.63		1.56	1.98
0305	Pneumatic nailed		1464	.011		.93	.51		1.44	1.81
0500	Plywood on walls, with exterior CDX, 3/8" thick		1200	.013		.60	.63		1.23	1.62
0505	Pneumatic nailed		1488	.011		.60	.50		1.10	1.44
0600	1/2" thick		1125	.014		.66	.67		1.33	1.75
0605	Pneumatic nailed		1395	.011		.66	.54		1.20	1.55
0700	5/8" thick		1050	.015		.78	.72		1.50	1.96
0705	Pneumatic nailed		1302	.012		.78	.58		1.36	1.75
0800	3/4" thick		975	.016		.93	.77		1.70	2.21
0805	Pneumatic nailed		1209	.013		.93	.62		1.55	1.98
1000	For shear wall construction, add						20%			
1200	For structural 1 exterior plywood, add				S.F.	10%				
3000	Wood fiber, regular, no vapor barrier, 1/2" thick	2 Carp	1200	.013		.61	.63		1.24	1.63
3100	5/8" thick		1200	.013		.77	.63		1.40	1.81
3300	No vapor barrier, in colors, 1/2" thick		1200	.013		.74	.63		1.37	1.77
3400	5/8" thick		1200	.013		.78	.63		1.41	1.82
3600	With vapor barrier one side, white, 1/2" thick		1200	.013		.60	.63		1.23	1.62
3700	Vapor barrier 2 sides, 1/2" thick		1200	.013		.83	.63		1.46	1.87
3800	Asphalt impregnated, 25/32" thick		1200	.013		.33	.63		.96	1.32
3850	Intermediate, 1/2" thick		1200	.013		.25	.63		.88	1.24
4500	Oriented strand board, on roof, 7/16" thick G		1460	.011		.50	.51		1.01	1.34
4505	Pneumatic nailed G		1780	.009		.50	.42		.92	1.20
4550	1/2" thick G		1400	.011		.50	.54		1.04	1.38
4555	Pneumatic nailed G		1736	.009		.50	.43		.93	1.22
4600	5/8" thick G		1300	.012		.69	.58		1.27	1.65
4605	Pneumatic nailed G		1586	.010		.69	.47		1.16	1.49
4610	On walls, 7/16" thick		1200	.013		.50	.63		1.13	1.51
4615	Pneumatic nailed		1488	.011		.50	.50		1	1.33
4620	1/2" thick		1195	.013		.50	.63		1.13	1.52
4625	Pneumatic nailed		1325	.012		.50	.57		1.07	1.42
4630	5/8" thick		1050	.015		.69	.72		1.41	1.86
4635	Pneumatic nailed		1302	.012		.69	.58		1.27	1.65
4700	Oriented strand board, factory laminated W.R. barrier, on roof, 1/2" thick G		1400	.011		.78	.54		1.32	1.69
4705	Pneumatic nailed G		1736	.009		.78	.43		1.21	1.53
4720	5/8" thick G		1300	.012		.93	.58		1.51	1.91
4725	Pneumatic nailed G		1586	.010		.93	.47		1.40	1.75
4730	5/8" thick, T&G G		1150	.014		.93	.65		1.58	2.03
4735	Pneumatic nailed, T&G G		1400	.011		.93	.54		1.47	1.85
4740	On walls, 7/16" thick G		1200	.013		.66	.63		1.29	1.69
4745	Pneumatic nailed G		1488	.011		.66	.50		1.16	1.51
4750	1/2" thick G		1195	.013		.78	.63		1.41	1.83
4755	Pneumatic nailed G		1325	.012		.78	.57		1.35	1.73

FIGURE 20-1 Construction Cost Data From RSMeans Building Construction Cost Data, 2015. Copyright RSMeans, Rockland, MA 781-422-5000; All rights reserved.

CONCLUSION

Productivity is affected by a number of factors, including project size, overtime, size of crew, delays, interruptions, weather, project layout, and safety. On each estimate, the estimator must take these factors into account when estimating labor productivity. Labor productivity can be calculated from historical data, field operations, and national reference books. Field observations can be based on measured cycle times, rate of progress, or daily production.

PROBLEMS

1. Convert the output of 3,000 square feet of 6-inch-thick blown fiberglass insulation per 8-hour day for a three-person crew to labor hours per square foot.
2. Convert the output of 165 square feet of brick veneer per 8-hour day for a five-person crew to labor hours per square foot.
3. Convert the output of 9.9 square feet of concrete masonry unit per labor hour to labor hours per square foot.
4. Convert the output of 35 lineal feet of trim per labor hour to labor hours per lineal foot.
5. A recent project included 40 lineal feet of counter top. It took 12.5 labor hours to complete the installation of the counter top. Determine the productivity in labor hours per linear foot based on this historical data.
6. A recent project included 2,015 square feet of suspended acoustical ceiling. It took 26 labor hours to complete the ceiling. Determine the productivity in labor hours per square foot based on this historical data.
7. The cycle times in Table 20-2 were observed for the installation of 8-foot by 15-inch batts of fiberglass insulation. Determine the average cycle time.
8. The cycle times in Table 20-3 were observed for the installation of residential windows. Determine the average cycle time.
9. Using the average cycle time from Problem 7, a system efficiency of 50 minutes per hour, an adjustment factor of 1.10, and a crew size of one, determine the productivity in labor hours per square foot of insulation.
10. Using the average cycle time from Problem 8, a system efficiency of 45 minutes per hour, an adjustment factor of 0.95, and a crew size of two, determine the productivity in labor hours per window.
11. An operator of a paint-striping machine is painting the centerline of a road. The road is 5 miles long and includes 30 intersections. Each intersection is 50 feet wide, and the centerline is not painted in the intersections. When striping, the operator can paint 50 feet of line per minute. The average time it takes to get through the intersection is 2 minutes. Using a system efficiency of 45 minutes per hour, determine the productivity for striping the road in labor hours per mile.
12. An asphalt-paving crew is paving a 36-foot-wide section of road 12 miles long. The road will be paved in widths of 12 feet, requiring the paving crew to make three passes along the road. At the end of each pass, it will take the paving crew 2 hours to move to the start of the road in preparation for the next pass. The paving crew consists of 11 persons. When paving, the paving crew can pave 12 feet of 12-foot-wide road per minute. Using a system efficiency of 50 minutes per 60-minute hour, determine the productivity for paving the road in labor hours per square yard.

TABLE 20-2 Cycle Times for Problem 7

Observation	CT (min)	Observation	CT (min)	Observation	CT (min)
1	2.47	11	4.26	21	3.85
2	3.00	12	2.70	22	2.50
3	2.67	13	2.39	23	3.58
4	3.11	14	2.64	24	2.11
5	3.18	15	3.11	25	3.60
6	3.08	16	2.59	26	2.69
7	3.89	17	3.00	27	2.54
8	3.10	18	4.90	28	2.70
9	4.38	19	3.72	29	2.88
10	4.62	20	2.93	30	2.77

TABLE 20-3 Cycle Times for Problem 8

Observation	CT (min)	Observation	CT (min)	Observation	CT (min)
1	37.26	11	56.55	21	39.57
2	28.55	12	37.90	22	29.82
3	52.57	13	55.91	23	46.70
4	34.85	14	38.51	24	30.95
5	42.27	15	41.47	25	35.41
6	42.41	16	27.97	26	61.83
7	30.87	17	32.15	27	31.12
8	46.36	18	60.10	28	33.37
9	36.77	19	36.27	29	50.72
10	32.89	20	39.33	30	37.64

13. The measured output for a crew finishing concrete is 2,050 square feet per day. The crew consists of three crew members, who work 8 hours each during the day. Determine the productivity of the crew in labor hours per square foot of concrete finished.

14. The measured output for a plumbing crew is 115 feet of 1 1/2-inch steel pipe per day. The crew consists of a plumber and an apprentice, who work 10 hours each during the day. Determine the productivity of the crew in labor hours per foot of pipe.

15. A crew needs to install 3,200 square feet of 1/2-inch CDX plywood on walls. The plywood is to be pneumatically nailed. Using the productivities in Figure 20-1, determine the number of labor hours required to install the sheathing.

16. A crew needs to paint 11,625 feet of parking lot striping. The estimated productivity is 0.8 labor hour per 1,000 feet. Determine the number of labor hours required to paint the striping.

17. Set up Excel Quick Tip 20-1 in Excel.

18. Set up Excel Quick Tip 20-2 in Excel.

REFERENCE

1. A labor hour is defined as one person working for 1 hour.

CHAPTER TWENTY-ONE

LABOR RATES

In this chapter you will learn how to determine hourly labor rates for a given class of employees. This chapter includes how to determine wage rates and labor burden, including cash equivalents and allowances, payroll taxes, unemployment insurance, workers' compensation insurance, general liability insurance, insurance benefits, retirement contributions, union payments, and other benefits.

The hourly wage rate for a given employee class includes the wages paid to the employee plus labor burden. Labor burden includes cash equivalents and allowances paid to the employee, payroll taxes, unemployment insurance, workers' compensation insurance, general liability insurance, insurance benefits, retirement contributions, union payments, vacation, sick leave, and other benefits paid by the employer. The average wage rate paid during the year is calculated by determining the annual costs of an employee and dividing it by the number of billable hours as shown in the following equation:

$$\text{Labor Rate} = \frac{\text{Annual Costs}}{\text{Billable Hours}} \quad (21\text{-}1)$$

Let's look at how to calculate the average hourly wage rate for an employee class.

BILLABLE HOURS

Billable hours are the number of hours an employee can be billed to projects during a year, which is usually different than the number of hours an employee is paid during the year. Often employees are paid for holidays, vacation, and sick leave, which are not billable to a project. The costs of holidays, vacation, and sick leave should be accrued throughout the year by billing it to the jobs as a burden cost and paid from the accruals at the time the holidays, vacation, and sick leave are taken.[1] In addition to holidays, vacation, and sick leave, travel time to, between, and from a job that is paid by the company may not be billable to the job.

WAGES

The annual wages should include the wages paid to employees including all bonuses and any anticipated raises that may occur during the project. In this section we will look at the two most common types of employees: nonexempt employees that are paid an hourly wage and exempt employees that are paid a salary.

The first group includes nonexempt employees who are paid an hourly wage based on the number of hours they work during the pay period. To estimate the wages for hourly employees, one must first estimate the number of hours each employee will work during an average pay period. The pay rates for hourly employees may be different for overtime, holidays, weekends, and swing shifts; and may be different based on the type of work begin preformed. Also the number of hours worked during a week may vary during different times of the year (summer versus winter). Nonexempt employees are often subject to the Fair Labor Standards Act, and may subject to the Davis-Bacon Act, contracts with labor unions, and state/city labor laws.

FAIR LABOR STANDARDS ACT

The Fair Labor Standards Act (FLSA) applies to all construction companies with revenues over $500,000 and many smaller companies. The FLSA is administered by the Wage and Hour Division (WHD) of the US Department of Labor. The FLSA requires that employers pay nonexempt employees at least the current minimum wage established by the federal government ($7.25 as of the time of this writing) and at least one and a half times the regular pay for hours worked in excess of 40 hours per work week (overtime). A work week is defined as 168 hours during 7 consecutive 24-hour periods, with the employer choosing when the work week starts and stops. The FLSA also defines the hours worked as all time the employee must be on duty; and includes travel time between jobs if the employee works on multiple jobs during the day. If an employee works on multiple jobs during the week, the hours from all jobs must be combined when determining compliance with the minimum wage and overtime requirements. Executive, administrative, professional, and outside sales employees are exempt from the minimum wage and overtime pay requirements. It is important to note that paying a nonexempt employee a salary does not relieve the employer from the requirement of the FLSA.

When a nonexempt employee works in excess of 40 hours per work week, half of the base wage rate must be added to the hours in excess of 40 hours per week. For example, if an employee's base wage rate is $20.66 per hour and the employee works 50 hour during a work week, the employee would be paid $1,033.00 (50 hours at $20.66 per hour) as a base wage plus an overtime premium of $10.33 per hour (half of $20.66 per hour) for the 10 hours in excess of 40 hours, for an additional $103.30. This would bring the employee's wages for the week to $1,136.30.

When the employer makes deductions from the employee's wages for tools, employer-required uniforms, or required safety equipment, the employer must use the wages after these deductions when determining if the minimum wage has been meet. If the employer pays a nonexempt employee a fixed salary or by the piece rate, the employer is responsible for verifying that the wages paid meet or exceed the minimum wage rate and that overtime is paid on the hours in excess of 40 hours per week. For more information on FLSA (including how to calculate overtime) see the *Handy Reference Guide to the Fair Labor Standards Act* by the U.S. Department of Labor.[2]

DAVIS-BACON ACT

The Davis–Bacon Act requires that prevailing wages be paid on jobs with federal funding. The prevailing wages vary by state, county, and the type of work being performed (building, heavy, highway, and residential). The prevailing wages are published Wage and Hour Division of the U.S. Department of Labor as general decisions. The general decision that is to be used for a project is usually included in the project manual. If it is not, the estimator should contact the contracting officer to obtain the correct general decision.

General decisions establish a minimum wage rate for different worker classifications and often includes a fringe benefit. Figure 21-1[3] is a sample general decision for general building construction in Salt Lake County in Utah. It requires that carpenters (other than carpenters working on acoustical ceiling installation, drywall hanging, formwork, and metal stud installation) be paid a minimum wage of $20.66 per hour and a fringe benefit of $7.47 per hour.

Employers comply with the minimum wage and overtime requirements using the same procedures they used to comply with FLSA. Overtime must be paid on the minimum wage rate, but not the fringe benefit. For example, the overtime rate for the carpenter would be $30.99 (1.5 times $20.66) per hour plus fringe benefits of $7.47 per hour. Employers may comply with the fringe benefit requirement by:

- Adding the fringe benefit to the base wage rate. In the case of the carpenter, the contractor may simply pay the carpenter $28.13 ($20.66 + $7.47) per hour.
- Providing "bona fide" fringe benefits with a value, when combined with the employee's base wage rate, exceeds the minimum wage rate plus the fringe benefit in the general decision. Examples of "bona fide" benefits include life insurance, health insurance, pension, vacations, holidays, and sick leave. Use of a vehicle, Thanksgiving turkey, Christmas bonus, workers' compensation insurance, unemployment compensations, and Social Security and Medicare contributions are not "bona fide" fringe benefits. In the case of the carpenter, the employer could pay the employee (1) wages of $20.66 per hour and provide "bona fide" fringe benefits with a value of $7.47 per hour, (2) wages of $25.00 per hour and provide "bona fide" fringe benefits with a value of $3.13 per hour, or (3) some other combination that exceeds $28.13 per hour.

The general contractor is responsible for documenting compliance with the minimum wage, fringe benefit, and overtime. They must collect this documentation from the subcontractors and submit it, along with their documentation, to the contracting officer. For more information on Davis–Bacon Act see the *Prevailing Wage Resource Book* by the U.S. Department of Labor.[4]

LABOR CONTRACTS

The wages for certain classes of workers may be governed by contracts with labor unions. These contracts may include: (1) wage rates, (2) wage premiums for shift work, holidays, weekends, and overtime, (3) fringe benefits including pension, health insurance, vacation, holidays, and sick leave, and (4) other provisions—such as number and length of breaks—that may affect employment costs and labor productivity. If a labor contract is due to expire before completion of the job, the estimator must take into account how any potential changes in the agreement will affect the costs of the project. If wages are expected to increase, the estimator will have to include

```
General Decision Number: UT150106 06/05/2015  UT106

Superseded General Decision Number: UT20140106

State: Utah

Construction Type: Building

County: Salt Lake County in Utah.

BUILDING CONSTRUCTION PROJECTS (does not include single family
homes or apartments up to and including 4 stories).

Note: Executive Order (EO) 13658 establishes an hourly minimum
wage of $10.10 for 2015 that applies to all contracts subject
to the Davis-Bacon Act for which the solicitation is issued on
or after January 1, 2015. If this contract is covered by the
EO, the contractor must pay all workers in any classification
listed on this wage determination at least $10.10 (or the
applicable wage rate listed on this wage determination, if it
is higher) for all hours spent performing on the contract. The
EO minimum wage rate will be adjusted annually. Additional
information on contractor requirements and worker protections
under the EO is available at www.dol.gov/whd/govcontracts.


Modification Number      Publication Date
         0                  01/02/2015
         1                  01/16/2015
         2                  06/05/2015

 CARP1507-001 01/01/2014

                                Rates          Fringes
CARPENTER (Drywall Hanging
and Metal Stud Installation
Only)..........................$ 21.00           8.66
----------------------------------------------------------------
* ELEC0354-001 06/01/2015

                                Rates          Fringes
ELECTRICIAN (Low Voltage
Wiring Only)...................$ 22.86           9.49
ELECTRICIAN....................$ 30.39          11.35
----------------------------------------------------------------
 ELEV0038-003 01/01/2014

                                Rates          Fringes
ELEVATOR MECHANIC..............$ 39.28          26.785
----------------------------------------------------------------
 PAIN0077-003 01/01/2013

                                Rates          Fringes
DRYWALL FINISHER/TAPER.........$ 19.50           6.28
----------------------------------------------------------------
 PAIN0077-004 08/01/2013
```

FIGURE 21–1a General Decision

```
                                Rates          Fringes
PAINTER (Brush, Roller, and
Spray, excluding
Drywall/Finisher and Taper)....$ 18.25           6.65
----------------------------------------------------------------
 PLUM0140-001 08/01/2013

                                Rates          Fringes
PLUMBER/PIPEFITTER.............$ 30.05          12.93
----------------------------------------------------------------
 SFUT0669-003 07/01/2013

                                Rates          Fringes
SPRINKLER FITTER (Fire
Sprinklers)....................$ 29.93          16.87
----------------------------------------------------------------
 SHEE0312-002 07/01/2014

                                Rates          Fringes
SHEET METAL WORKER (Including
HVAC Duct Installation)........$ 32.11          11.40
----------------------------------------------------------------
 SUUT2012-017 07/29/2014

                                Rates          Fringes
CARPENTER (Acoustical Ceiling
Installation Only).............$ 21.25           2.15

CARPENTER (Form Work Only).....$ 16.93           1.93

CARPENTER, Excludes
Acoustical Ceiling
Installation, Drywall
Hanging, Form Work, and Metal
Stud Installation..............$ 20.66           7.47

CEMENT MASON/CONCRETE FINISHER.$ 15.00           0.00

IRONWORKER, STRUCTURAL.........$ 20.21           3.22

LABORER:  Common or General....$ 13.84           0.00

LABORER:  Mason Tender - Brick.$ 16.38           1.00

LABORER:  Mason Tender -
Cement/Concrete................$ 14.94           0.00

LABORER:  Pipelayer............$ 13.57           0.00

LABORER: Landscape and
Irrigation.....................$  9.50           0.00

OPERATOR:
Backhoe/Excavator/Trackhoe.....$ 14.48           0.00

OPERATOR:  Loader..............$ 19.34           0.00
```

FIGURE 21–1b General Decision

enough in the estimate to cover the higher cost of all work done after the expiration of the agreement.

STATE AND LOCAL EMPLOYMENT LAWS

State and local cities may have employment laws that affect the cost of labor. States, and sometimes cities, will establish a minimum wage rate, which may be higher or lower than the minimum wage established by the federal government. When determining the minimum wage rate, the estimator must use the larger of the federal, state or local minimum wage rate.

The total wages for an employee or class of employees is calculated by multiplying the number of hours by the wage rate for these hours and adding any anticipated bonuses.

EXAMPLE 21-1

A typical carpenter works 50 hours per week for 20 weeks of the year and 40 hours per week for 29 weeks of the year, and the remaining 3 weeks per year comprises holiday leave, vacation, and sick leave, for which the carpenters are paid for 40 hours per week. Carpenters are paid $20.66 per hour and are paid time-and-a-half on any hours worked over 40 hours per week. In the past the carpenters received a $100 Christmas bonus. Determine the wages paid to the typical carpenter and the number of billable hours during a one-year period. Assume that all hours are billable except holidays, vacation, and sick leave.

Solution: The typical carpenter is paid $20.66 per hour for 40 hours per week for 52 weeks a year and is paid $30.99 ($20.66 × 1.5) for 10 hours per week for 20 weeks. The wages, including bonus, are calculated as follows:

$$\text{Wages} = (\$20.66/\text{hr})(40\text{ hr/wk})(52\text{ wk/yr})$$
$$+ (\$30.99/\text{hr})(10\text{ hr/wk})(20\text{ wk/yr}) + \$100/\text{yr}$$
$$= \$49,271/\text{yr}$$

The number of billable hours per year is calculated as follows:

$$\text{Billable Hours} = (50\,\text{hr/wk})(20\,\text{wk/yr}) + (40\,\text{hr/wk})(29\,\text{wk/yr})$$
$$= 2,160\text{ hr/yr}$$

```
PLASTERER.....................$ 18.36        0.00

ROOFER........................$ 13.22        0.00

TILE FINISHER.................$ 13.54        0.00

TILE SETTER...................$ 23.50        0.00

TRUCK DRIVER:  Dump Truck.....$ 15.50        0.00
----------------------------------------------------------------
WELDERS - Receive rate prescribed for craft performing
operation to which welding is incidental.
================================================================
Unlisted classifications needed for work not included within
the scope of the classifications listed may be added after
award only as provided in the labor standards contract clauses
(29CFR 5.5 (a) (1) (ii)).
----------------------------------------------------------------
The body of each wage determination lists the classification
and wage rates that have been found to be prevailing for the
cited type(s) of construction in the area covered by the wage
determination. The classifications are listed in alphabetical
order of "identifiers" that indicate whether the particular
rate is a union rate (current union negotiated rate for local),
a survey rate (weighted average rate) or a union average rate
(weighted union average rate).

Union Rate Identifiers

A four letter classification abbreviation identifier enclosed
in dotted lines beginning with characters other than "SU" or
"UAVG" denotes that the union classification and rate were
prevailing for that classification in the survey. Example:
PLUM0198-005 07/01/2014. PLUM is an abbreviation identifier of
the union which prevailed in the survey for this
classification, which in this example would be Plumbers. 0198
indicates the local union number or district council number
where applicable, i.e., Plumbers Local 0198. The next number,
005 in the example, is an internal number used in processing
the wage determination. 07/01/2014 is the effective date of the
most current negotiated rate, which in this example is July 1,
2014.

Union prevailing wage rates are updated to reflect all rate
changes in the collective bargaining agreement (CBA) governing
this classification and rate.

Survey Rate Identifiers

Classifications listed under the "SU" identifier indicate that
no one rate prevailed for this classification in the survey and
the published rate is derived by computing a weighted average
rate based on all the rates reported in the survey for that
classification.  As this weighted average rate includes all
rates reported in the survey, it may include both union and
non-union rates. Example: SULA2012-007 5/13/2014. SU indicates
the rates are survey rates based on a weighted average
```

FIGURE 21-1c General Decision

```
calculation of rates and are not majority rates. LA indicates
the State of Louisiana. 2012 is the year of survey on which
these classifications and rates are based. The next number, 007
in the example, is an internal number used in producing the
wage determination. 5/13/2014 indicates the survey completion
date for the classifications and rates under that identifier.

Survey wage rates are not updated and remain in effect until a
new survey is conducted.

Union Average Rate Identifiers

Classification(s) listed under the UAVG identifier indicate
that no single majority rate prevailed for those
classifications; however, 100% of the data reported for the
classifications was union data. EXAMPLE: UAVG-OH-0010
08/29/2014. UAVG indicates that the rate is a weighted union
average rate. OH indicates the state. The next number, 0010 in
the example, is an internal number used in producing the wage
determination. 08/29/2014 indicates the survey completion date
for the classifications and rates under that identifier.

A UAVG rate will be updated once a year, usually in January of
each year, to reflect a weighted average of the current
negotiated/CBA rate of the union locals from which the rate is
based.

----------------------------------------------------------------
              WAGE DETERMINATION APPEALS PROCESS

1.) Has there been an initial decision in the matter? This can
be:

*   an existing published wage determination
*   a survey underlying a wage determination
*   a Wage and Hour Division letter setting forth a position on
    a wage determination matter
*   a conformance (additional classification and rate) ruling

On survey related matters, initial contact, including requests
for summaries of surveys, should be with the Wage and Hour
Regional Office for the area in which the survey was conducted
because those Regional Offices have responsibility for the
Davis-Bacon survey program. If the response from this initial
contact is not satisfactory, then the process described in 2.)
and 3.) should be followed.

With regard to any other matter not yet ripe for the formal
process described here, initial contact should be with the
Branch of Construction Wage Determinations.  Write to:

             Branch of Construction Wage Determinations
             Wage and Hour Division
             U.S. Department of Labor
             200 Constitution Avenue, N.W.
             Washington, DC 20210
```

FIGURE 21-1d General Decision

```
2.) If the answer to the question in 1.) is yes, then an
interested party (those affected by the action) can request
review and reconsideration from the Wage and Hour Administrator
(See 29 CFR Part 1.8 and 29 CFR Part 7). Write to:

             Wage and Hour Administrator
             U.S. Department of Labor
             200 Constitution Avenue, N.W.
             Washington, DC 20210

The request should be accompanied by a full statement of the
interested party's position and by any information (wage
payment data, project description, area practice material,
etc.) that the requestor considers relevant to the issue.

3.) If the decision of the Administrator is not favorable, an
interested party may appeal directly to the Administrative
Review Board (formerly the Wage Appeals Board).  Write to:

             Administrative Review Board
             U.S. Department of Labor
             200 Constitution Avenue, N.W.
             Washington, DC 20210

4.) All decisions by the Administrative Review Board are final.
================================================================
                    END OF GENERAL DECISION
```

FIGURE 21-1e General Decision

The second group we will look at includes all exempt employees who are paid a flat rate (salary) per pay period. To estimate the wages for these employees, you do not need to know how many hours each employee works. The total wages for an employee or class of employees is calculated by summing the salaries of the employees and adding any anticipated bonuses. The costs for these employees are calculated by the week or month rather than by the hour.

Burden includes all payroll taxes, unemployment insurance, workers' compensation insurance, general liability insurance, and fringe benefits or their cash equivalent paid for by the employer. Holidays, vacation, and sick leave are also part of labor burden; however, they have already been accounted for by not including these hours in the billable hours. Let's look at how labor burden is calculated.

CASH EQUIVALENTS AND ALLOWANCES

Cash equivalents and allowances include cash paid in lieu of providing fringe benefits or cash paid as an allowance. In some cases where a company is required to provide fringe benefits, the employer may be allowed to pay the cash equivalent of those benefits and let employees use the money to purchase their own benefits, if they so desire. Allowances are used to help the employee cover out-of-pocket expenses associated with their job. For example, a company located in an out-of-the-way place or a company that requires their employees to use their personal vehicles during work hours may pay employees a vehicle allowance. An allowance is different than reimbursing the employee for expenses or mileage because it is not based on the actual cost or mileage. Reimbursements are not considered part of the labor wages and burden but are costs that can be billed to the appropriate project.

Cash equivalents and allowances are treated as regular wages for the purpose of payroll taxes, unemployment insurance, workers' compensation insurance, and general liability insurance.

EXAMPLE 21-2

The typical carpenter in Example 21-1 is paid a tool allowance of $20 per month. Determine the annual wages including allowances paid to the typical carpenter during a one-year period.

Solution: From Example 21-1, the typical carpenter is paid $49,271 per year plus allowances. The wages, including allowances, are calculated as follows:

$$\text{Allowances} = (\$20/\text{mo})(12 \text{ mo/yr}) = \$240/\text{yr}$$
$$\text{Wages and Allowances} = \$49,271/\text{yr} + \$240/\text{yr}$$
$$= \$49,511/\text{yr}$$

PAYROLL TAXES

The Federal Insurance Contributions Act (FICA) requires employers to pay Social Security and Medicare taxes for each employee. For the year 2016, the employer pays a Social Security tax of 6.2% of each employee's first $118,500 in wages.[5] Although the tax rate has not changed for many years, the amount of wages on which the Social Security tax is paid increases almost every year. Be sure to check with your tax advisor or the Internal Revenue Service for the current rates. For the year 2016, the employer pays a Medicare tax of 1.45% of the employee's entire wages.[6] The employee is required to match these payments. The Social Security and Medicare taxes paid by an employee are deducted from his or her wages and are not a cost to the employer. Cash equivalents and allowances must be included when calculating these taxes; however, in some cases the amount the employee pays for his or her benefits may be excluded.

EXAMPLE 21-3

Determine the Social Security and Medicare taxes paid by the employer during the year for the typical carpenter in Example 21-2. The typical carpenter pays $150 per month for health insurance, which is not subject to Social Security and Medicare taxes. Use the tax rates for the year 2016.

Solution: The wages used in calculating the Social Security and Medicare tax must include the allowances paid to the typical carpenter and exclude the $150 per month paid by the employee for health insurance. From Example 21-2 the annual wages, including allowances, for the typical carpenter are $49,511. This is reduced by the $1,800 ($150/mo × 12 mo/yr) per year the typical carpenter pays for health insurance. The typical carpenter's wages that are subject to Social Security and Medicare taxes are $47,711 ($49,511 − $1,800). Because the typical carpenter's wages are less than $118,500, the employer must pay 6.2% for Social Security tax on all of his or her wages. The Social Security tax is calculated as follows:

$$\text{Social Security Tax} = (0.062)(\$47,711/\text{yr}) = \$2,958/\text{yr}$$

The Medicare tax is 1.45% of the carpenter's wages and is calculated as follows:

$$\text{Medicare Tax} = (0.0145)(\$47,711/\text{yr}) = \$692/\text{yr}$$

These same amounts are deducted from the carpenter's paychecks.

UNEMPLOYMENT INSURANCE

The Federal Unemployment Tax Act (FUTA) requires employers to pay federal unemployment tax. If a state program exists, the State Unemployment Tax Act (SUTA) requires employers to pay state unemployment tax.

For the year 2016 the federal unemployment tax rate is 6.0% on the first $7,000 of each employee's wages paid during the year.[7] If a company pays into a state unemployment program, the company may reduce its FUTA liability by the amount paid into the program. This reduction is limited to 5.4% of first $7,000 of each employee's wages. To receive the entire 5.4%, the company must pay its state unemployment tax on time. When a company is eligible for the full credit, its federal unemployment tax rate is reduced to 0.6% (6.0% − 5.4%) of the first $7,000 of each employee's wages paid during the year.

A company's state unemployment tax rate is based in part on its claims history and must be obtained from the state agency that administers the state unemployment tax. The maximum rate a state charges and the amount of the employees' wages that are subject to tax vary from state to state. For example, for the year 2014, Texas' minimum rate was 0.54%, the maximum rate was 7.35%, and companies paid SUTA on the first $9,000 of each employee's wages paid during the year. Arizona's minimum rate was 0.02%, the maximum rate was 6.67%, and companies paid SUTA on the first $7,000 of each employee's wages paid during the year. Utah's minimum rate was 0.40%, the maximum rate was 7.4%, and companies paid SUTA on the first $30,800 of each employee's wages paid during the year.[8]

EXAMPLE 21-4

Determine the state unemployment and federal unemployment taxes paid by the employer during the 2016 year for the typical carpenter in Example 21-2. The company's state unemployment rate is 2.0% on the first $9,000 of the employee's wages. The company can take the full 5.4% credit against its federal unemployment tax.

Solution: The wages used in calculating the unemployment taxes must include the allowances paid to the carpenter. From Example 21-2 the annual wages, including allowances, for the carpenter are $49,511. Because the carpenter's wages are more than $9,000, the employer must pay 2.0% on $9,000 for state unemployment tax. The state unemployment tax is calculated as follows:

$$SUTA = 0.02(\$9,000/yr) = \$180/yr$$

The federal unemployment tax is calculated as follows:

$$FUTA = 0.006(\$7,000/yr) = \$42/yr$$

The total unemployment tax paid is $222 ($180 + $42).

The underlying assumption in the calculations for Example 21-4 is that the same person fills the carpenter position for the entire year. This may not be the case. If more than one person fills the carpenter's position during the year because of employee turnover, the company must pay FUTA on the first $7,000 of wages from each employee as shown by the following example.

EXAMPLE 21-5

During the year, a carpenter's position is filled by three different carpenters, two of whom left the company during the year. For the year, the first carpenter was paid $5,130, the second carpenter was paid $11,000, and the third carpenter was paid $9,200. Determine the federal unemployment taxes paid by the employer on this carpenter's position. The company can take the full 5.4% credit against its federal unemployment tax.

Solution: The employer must pay 0.6% on the first $7,000 paid to each employee. The federal unemployment tax is calculated as follows:

$$FUTA = (\$5,130/yr)(0.006) + (\$7,000/yr)(0.006)$$
$$+ (\$7,000/yr)(0.006)$$
$$= \$115/yr$$

Here employee turnover increased the costs of FUTA by 273%. The same situation occurs with SUTA. One way to take this into account is to determine the average FUTA and SUTA paid during the last year as a percentage of labor costs based on data from the company's accounting system. These percentages then can be applied to all labor costs when determining labor burden. The calculations are shown in the following example:

EXAMPLE 21-6

Last year your company paid $1,191,300 in wages, $2,486 in FUTA, and $7,992 in SUTA. Determine the company's FUTA and SUTA rate as a percentage of wages paid. Also determine the state unemployment and federal unemployment taxes paid by the employer for the typical carpenter in Example 21-2 based on these rates.

Solution: The FUTA rate as a percentage of wages is calculated as follows:

$$\text{FUTA Percentage} = \frac{\$2,486/yr}{\$1,191,300/yr} = 0.00209 \text{ or } 0.209\%$$

The FUTA paid on a typical carpenter is calculated as follows:

$$FUTA = (\$49,511/yr)(0.00209) = \$103/yr$$

The SUTA rate as a percentage of wages is calculated as follows:

$$\text{SUTA Percentage} = \frac{\$7,992/yr}{\$1,191,300/yr} = 0.00671 \text{ or } 0.671\%$$

The SUTA paid on a typical carpenter is calculated as follows:

$$SUTA = (\$49,511/yr)(0.00671) = \$332/yr$$

The total unemployment tax is $435 ($103 + $332).

Looking at Examples 21-4 and 21-6, one notices that the amount calculated by Example 21-6 is more than the legally required amount calculated by Example 21-4. This is because the method used in Example 21-6 uses an average percentage for the FUTA and SUTA calculations and bases these calculations on the total wages paid. As a result it understates the FUTA and SUTA on lower-cost employees (such as laborers) and overstates the cost on higher-paid employees. For all of the field employees, the company in the foregoing examples paid 20% more FUTA and SUTA in Example 21-6 than would have been calculated using the procedures used to calculate FUTA and SUTA in Example 21-4 because of employee turnover.

WORKERS' COMPENSATION INSURANCE

By law all employers are required to provide their employees with workers' compensation insurance. Workers' compensation insurance is governed by the individual states, and the requirements vary from state to state.

Workers' compensation insurance covers reasonable medical costs as well as some of the lost wages for employees who are injured on the job or who contract an occupational illness. For employees who are killed on the job, workers' compensation insurance may pay part of the burial expense and may provide surviving family members a weekly or monthly benefit.

The cost of the insurance is to be paid entirely by the employer. The premium is based on the gross payroll, the type of work performed by the employees, the company's accident history, and other factors. Employees are grouped into a standard set of classifications established by the National Council on Compensation Insurance (NCCI) based on the type of work they do. The NCCI sets a standard lost cost factor for each job classification, which is modified by the individual states to take into account local variances in losses and regulations. The premium rate is based on the lost cost factor and is expressed in dollars per $100 of payroll.

The premium rate may be modified by multiplying the premium rate by a so-called experience modifier. Experience modifiers are calculated by NCCI and reflect the relationship between the company's actual losses and the expected losses for similar companies. An experience modifier greater than 1 indicates that a company had higher-than-expected losses, whereas an experience modifier of less than 1 indicates that a company had lower-than-expected losses. The experience modifiers are based on the past three years of losses, not including the most recent policy year. For example, the experience modifier for the year 2017 would be based on the years 2013, 2014, and 2015. For companies to receive an experience modification they must meet a minimum level of premiums, and thus companies with small payrolls are often not given experience modifiers. Experience modifiers may be as low as 0.6 and as high as 2.0.

The workers' compensation insurance premiums may also be adjusted based on the safety practices of the company. Other discounts may be offered for such items as policy size.

Throughout the year, companies pay workers' compensation insurance premiums based on the estimated payroll. At the end of the year or more frequently the insurance carrier will audit the payroll of the company and make adjustments in the premiums to reflect the actual payroll. It is important to note that this is an adjustment in the total premium, not the premium rate.

EXAMPLE 21-7

Determine the cost of workers' compensation insurance for the typical carpenter in Example 21-2. The company's workers' compensation insurance rate for carpenters is $6.25 per $100 in wages.

Solution: The wages used in calculating the workers' compensation insurance must include the allowances paid to the carpenter. From Example 21-2 the annual wages, including allowances, for the typical carpenter are $49,511 per year. The workers' compensation insurance cost is calculated as follows:

$$\text{Workers' Compensation} = (\$6.25/\$100)(\$49,511/\text{yr})$$
$$= \$3,094/\text{yr} \quad \blacksquare$$

GENERAL LIABILITY INSURANCE

General liability insurance protects the company against claims because of negligent business activities and failure to use reasonable care. The types of claims include bodily injury, property damage or loss, and other personal injury such as slander or damage to reputation. Like workers' compensation insurance, the cost of the insurance is based on gross revenues and varies by worker class.

Throughout the year, companies pay general liability insurance premiums based on the estimated payroll. At the end of the year, or more frequently, the insurance carrier will audit the payroll of the company and make adjustments in the premiums to reflect the actual payroll. This is an adjustment to the total premium and not the premium rates.

EXAMPLE 21-8

Determine the cost of general liability insurance for the typical carpenter in Example 21-2. The company's workers' compensation insurance rate for carpenters is 0.97% of wages.

Solution: The wages used in calculating the general liability insurance must include the allowances paid to the typical carpenter. From Example 21-2 the annual wages, including allowances, for the typical carpenter are $49,511 per year. The general liability insurance cost is calculated as follows:

$$\text{Liability Insurance} = (0.0097)(\$49,511/\text{yr}) = \$480/\text{yr} \quad \blacksquare$$

INSURANCE BENEFITS

As part of the employee benefit package, the employer may provide health, dental, life, and disability insurance for which the employee and their families are beneficiaries. The employer may pay the entire cost of the benefits, split the cost with the employee, or require the employee to pay the entire amount. The amount the employees pay is deducted from their wages and does not represent a cost to the employer. The part of the cost that is paid by the employer is a real cost to the employer and needs to be included in the cost of the benefits.

EXAMPLE 21-9

Determine the cost of health insurance for the typical carpenter in Example 21-2. The company pays $1,050 per month toward health insurance for its employees. An additional $150 is deducted from the employee's paycheck.

Solution: The cost of the health insurance includes only those costs paid by the employer and is calculated as follows:

$$\text{Health Insurance} = (\$1,050/\text{mo})(12 \text{ mo/yr}) = \$12,600/\text{yr} \quad \blacksquare$$

RETIREMENT CONTRIBUTIONS

As part of the employee benefit package, the employer may provide traditional pension plans, pay funds to a union to provide pension benefits, or participate in a profit-sharing plan such as a 401(k). Like insurance benefits, the employer may pay all, part, or none of the cost of the retirement. The amount the employees pay is deducted from their wages and does not represent a cost to the employer. The part of the cost that is paid by the employer is a real cost to the employer and needs to be included in the cost of the benefits.

EXAMPLE 21-10

Determine the cost of retirement for the typical carpenter in Example 21-2. For retirement, the company has provided the employee with access to a 401(k) plan and matches the employee's contributions to the plan at a rate of $0.50 per $1.00 contributed by the employee on the first 6% of the employee's wages—including allowances—for a maximum matching contribution of 3% of the employee's wages. The typical carpenter is expected to contribute at least 6% of his or her wages to the 401(k) plan.

Solution: From Example 21-2 the annual wages, including allowances, for the typical carpenter are $49,511. The cost of retirement is calculated as follows:

$$\text{Retirement} = \left(\frac{\$0.50}{\$1.00}\right)(0.06)(\$49,511/\text{yr}) = \$1,485/\text{yr}$$

UNION PAYMENTS

Employee unions often require the employer to make payments directly to the union for the union to provide benefits to the employees. Payments to unions are governed by the contract between the company and the union. Union costs paid by the employer should be included in the cost of the benefits. Unions may also require the employer to deduct union dues from the employees' paycheck. The amounts the employees pay do not represent a cost to the employer and should not be included in the cost of the benefits.

OTHER BENEFITS

The employer may provide other benefits not covered in one of the foregoing categories. Where possible the costs of these benefits should be included.

ANNUAL COSTS AND BURDEN MARKUP

The annual cost of the employee is determined by summing the individual burden items and adding the wages paid to the employee. The burden markup is calculated as follows:

$$\text{Burden Markup} = \frac{\text{Annual Cost}}{\text{Wages}_{\text{Billable Hours}}} - 1 \quad (21\text{-}2)$$

where the wages include only those wages paid for billable hours and should include allowances. By doing this the wages paid on unbillable hours, vacation, and sick leave are included as part of the burden.

EXAMPLE 21-11

Determine the annual cost, average hourly cost, and burden markup for the typical carpenter in Example 21-1 using Examples 21-2 through 21-4 and 21-7 through 21-10. Assume that the same person fills the position during the entire year.

Solution: From Examples 21-1 through 21-4, and 21-7 through 21-10 the following costs were calculated:

Wages and Allowances	= $49,511/yr
Social Security Tax	= $2,958/yr
Medicare Tax	= $692/yr
SUTA	= $180/yr
FUTA	= $42/yr
Workers Compensation	= $3,094/yr
Liability Insurance	= $480/yr
Health Insurance	= $12,600/yr
Retirement	= $1,485/yr

The annual cost is calculated as follows:

$$\begin{aligned}\text{Annual Cost} = {} & \$49,511/\text{yr} + \$2,958/\text{yr} + \$692/\text{yr} \\ & + \$180/\text{yr} + \$42/\text{yr} + \$3,094/\text{yr} \\ & + \$480/\text{yr} + \$12,600/\text{yr} + \$1,485/\text{yr}\end{aligned}$$

$$\text{Annual Cost} = \$71,042/\text{yr}$$

Using the billable hours from Example 21-1, we calculate the average hourly wage rate using Eq. (21-1) as follows:

$$\text{Average Hourly Wage Rate} = \frac{\$71,042/\text{yr}}{2,160 \text{ hr/yr}} = \$32.89/\text{hr}$$

The carpenter is paid $2,479 (3 weeks × 40 hours/week × $20.66 per hour) for vacation, holidays, and sick leave. The burden markup is calculated using Eq. (21-2) as follows:

$$\text{Burden Markup} = \frac{\$71,042/\text{yr}}{(\$49,511/\text{yr} - \$2,479/\text{yr})} - 1$$
$$= 0.5105 \text{ or } 51.05\%$$

It is important to note that one cannot take the base hourly rate of $20.66 and add the burden markup to get the hourly cost because some of the employee's hours are paid at time-and-a-half.

When using "bona fide" fringe benefits to meet the Davis-Bacon fringe benefit requirements, one needs to verify that the actual fringe benefits provided meet the minimum required. This is done by dividing the fringe benefits by the billable hours. This is shown in the following example:

EXAMPLE 21-12

Determine if the fringe benefits provided to the carpenter in Example 21-11 meet the Davis-Bacon requirements.

Solution: From Figure 21-1, the required fringe benefits for a carpenter are $7.47 per hour. From Example 21-11, the fringe benefits provided to the employee include vacation, holidays, and sick leave ($2,479); the employer's portion of the health insurance premiums ($12,600); and the employer's contributions to employee's retirement ($1,485). The actual fringe benefit provided is calculated as follows:

$$\text{Fringe} = \frac{\$2,479 + \$12,600 + \$1,485}{2,160 \text{ hours}} = \$7.67/\text{hr}$$

The fringe benefits meet the require minimum.

EXCEL QUICK TIP 21-1
Labor Rate Worksheet

The spreadsheet shown in the following figure can be used to calculate the average hourly wage and labor burden for a class of employees.

	A	B	C	D	E	F
1		Wage Calculation Worksheet				
2		Hours per	Weeks	Base		
3		Week	per Year	Rate	Wages	
4	Week Type 1	50.00	20.00	20.66	22,726	
5	Week Type 2	40.00	29.00	20.66	23,966	
6	Week Type 3				-	
7	Week Type 4				-	
8	Vacation & Sick Leave	40.00	3.00	20.66	2,479	
9	Bonus				100	
10	Billable Hours					2,160
11	Wages					49,271
12	Allowances	20.00	per mo	12	mo	240
13	Employee Paid Health Ins.	150.00	per mo	12	mo	1,800
14	Social Security/Med. Wages					47,711
15						
16	Wages and Allowances					49,511
17	Social Security	6.20%	on first	118800	of wages	2,958
18	Medicare	1.45%	on first		of wages	692
19	SUTA	2.00%	on first	9,000	of wages	180
20	FUTA	0.80%	on first	7,000	of wages	42
21	Workers' Compensation	6.25	$/$100 of wages			3,094
22	General Liability	0.97%	of wages			480
23	Employer Paid Health Ins.	100.00	per mo	12	mo	12,600
24	Retirement	50.00%	match on	6.00%	of wages	1,485
25	Union Payments	0.00%	of wages			-
26				Annual Cost of Employee		71,043
27				Average Hourly Wage Rate		32.89
28				Fringe Benefit Rate		7.67
29				Labor Burden Markup		51.05%

To complete this spreadsheet, enter the data and format the cells as shown in the figure. The following formulas need to be entered into the associated cells:

Cell	Formula
E4	=IF(B4>40,40*D4+(B4-40)*D4*1.5,B4*D4)*C4
F10	=B4*C4+B5*C5+B6*C6+B7*C7
F11	=SUM(E4:E9)
F12	=B12*D12
F13	=B13*D13
F14	=F11+F12-F13
F16	=F11+F12
F17	=IF(D17="",F14*B17,IF(F14<D17,F14*B17,D17*B17))
F18	=IF(D18="",F14*B18,IF(F14<D18,F14*B18,D18*B18))
F19	=IF(D19="",F16*B19,IF(F16<D19,F16*B19,D19*B19))
F20	=IF(D20="",F16*B20,IF(F16<D20,F16*B20,D20*B20))

Cell	Formula
F21	=(B21/100)*F16
F22	=B22*F16
F23	=B23*D23
F24	=B24*D24*F16
F25	=F16*B25
F26	=SUM(F16:F25)
F27	=F26/F10
F28	=(F24+F23+E8)/F10
F29	=F26/(F16-E8)-1

An IF statement is used in cell E4 to calculate the weekly wages while including time-and-a-half for hours worked in excess of 40 hours per week. The IF statement checks to see if the hours worked is greater than 40 hours per week. If this is the case, the weekly wages equal the hours over 40 hours per week multiplied by 1.5 times the wage rate plus 40 hours at the wage rate. If the hours worked are less than or equal to 40 hours, the weekly wages are equal to the hours per week times the wage rate. The weekly wages are then multiplied by the number of weeks per year to get total wages.

Cells F17 through F20 use a nested IF statement, which works as follows for Row 17. If Cell D17 is blank, the wages in Cell F14 are multiplied by the FICA rate in Cell B17; if the cell is not blank, the next IF statement checks to see if the wages in Cell F14 are less than the wage limit in Cell D17. If this is true, the wages in Cell F14 are multiplied by the FICA rate in Cell B17; if the wages are greater than the limit, the FICA limit in D17 is multiplied by the FICA rate in Cell B17. If there is not a limit, then this equation applies the FICA rate to all of the wages.

After entering these formulas, one needs to copy Cell E4 into Cells E5 through E8. The data for the employee class is entered into Cells B4 through D8, E9, B12, B13, D12, D13, B17 through B25, D17 through D20, D23, and D24. The data shown in the foregoing figure is from Examples 21–1 through 21–4 and 21–7 through 21–12 and is formatted using the comma style, which replaces zeros with dashes.

CONCLUSION

When projecting costs that include labor, it is important to include all of the costs associated with the employee. The cost of an employee includes wages and the associated labor burden. The employee's wages may be determined by market rates, union contracts, or federal Davis-Bacon wage decisions. Labor burden includes cash equivalents and allowances paid to the employee, payroll taxes, unemployment insurance, workers' compensation insurance, general liability insurance, insurance benefits, retirement contributions, union dues, and other benefits.

PROBLEMS

1. An employee is paid $0.45 per mile for use of his or her personal vehicle on the job. During the month of June, the employee drove 342 miles to perform job-related errands. How much of this mileage pay is subject to Social Security and Medicare taxes?

2. A typical drywall finisher works 45 hours per week for 16 weeks of the year and 40 hour per week for 32 weeks of the year, and gets 4 weeks off in the form of vacation, holidays, and sick leave (for which the drywall finishers are paid for 40 hours per week). The typical drywall finisher is paid $25.78 per hour. The employer must comply with the overtime requirements in the Fair Labor Standards Act. Determine the wages paid to a typical drywall finisher and the number of billable hours during a one-year period. Assume that all hours are billable except holidays, vacation, and sick leave.

3. Plumbers on a federal project are required by the Davis-Bacon act to be paid $30.05 per hour in wages and $12.93 per hour in fringe benefits. The typical plumber works 50 hour per week for 47 weeks of the year and is given 5 weeks off in the form of vacation, holidays, and sick leave (for which the plumbers are paid for 40 hours per week). The employer must comply with the overtime requirements in the Fair Labor Standards Act. The employer plans on covering the $12.93 for fringe benefits by providing "Bona fide" fringe benefits. Determine the wages paid to the typical plumber during the year, the number of billable hours, and the amount of fringe benefits (in addition to holidays, vacation, and sick leave) that the employer must provide.

4. Plumbers on a federal project are required by the Davis-Bacon act to be paid $30.05 per hour in wages and $12.93 per hour in fringe benefits. The typical plumber works 50 hour per week for 52 weeks of the year. The employer must comply with the overtime requirements in the Fair Labor Standards Act. The employer plans on covering the $12.93 for fringe benefits by paying them as wages. Determine the wages paid to a typical plumber and the number of billable hours during a one-year period.

5. Using the 2016 rates, determine the Social Security and Medicare taxes paid by the employer on an employee that is paid $75,800 in wages for the year.

6. Using the 2016 rates, determine the Social Security and Medicare taxes paid by the employer on an employee that is paid $135,000 in wages for the year.

7. Determine the state and federal unemployment taxes paid by the employer on an employee that is paid $75,800 in wages for the year. The company's state unemployment rate is 2.83% on the first $14,000 of the employee's wages. The company can take the full 5.4% credit against its federal unemployment tax.

8. Determine the state and federal unemployment taxes paid by the employer on an employee that is paid $20,300 in wages for the year. The company's state unemployment rate is 2.75% on the first $35,200 of the employee's wages. The company can

take the full 5.4% credit against its federal unemployment tax.

9. Determine the cost of workers' compensation insurance for a carpenter that is paid $52,400 per year. The workers' compensation insurance rate for carpentry is $12.62 per $100 of wages.

10. Determine the cost of workers' compensation insurance for a superintendent that is paid $132,000 per year. The workers' compensation insurance rate for supervisors is $2.21 per $100 of wages.

11. Determine the cost of general liability insurance for a carpenter that is paid $52,400 per year. The general liability insurance rate for carpentry is 1.09% of wages.

12. Determine the cost of general liability insurance for a superintendent that is paid $132,000 per year. The general liability insurance rate for supervisors is 2.11% of wages.

13. Health insurance premium for an employee and his family is $1,025 per month. The employer pays 75% of the premium and the remaining portion is deducted from the employee's paycheck. Determine the cost of the health insurance to the employer.

14. Determine the cost of health insurance for the typical plumber making $97,800 per year. The company pays $675 per month toward health insurance for its employees. An additional $325 is deducted from the employee's paycheck.

15. Determine the cost of retirement for a plumber making $97,800 per year. For retirement, the company has provided the employee with access to a 401(k) plan and matches the employee's contributions to the plan at a rate of $0.50 per $1.00 contributed by the employee on the first 6% of the employee's wages for a maximum matching contribution of 3% of the employee's wages. The plumber is expected to contribute at least 6% of his or her wages to the 401(k) plan.

16. Determine the cost of retirement for a carpenter making $52,400 per year. For retirement, the company has provided the employee with access to a 401(k) plan and matches the employee's contributions to the plan at a rate of $0.75 per $1.00 contributed by the employee on the first 6% of the employee's wages for a maximum matching contribution of 4.5% of the employee's wages. The carpenter is expected to contribute at least 6% of his or her wages to the 401(k) plan.

17. A contract with the local labor union requires that plumbers be paid $30.05 per hour in wages, with time-and-a-half being paid on overtime (over 40 hours per week). It also requires that the employee pay 5% of his or her wages as union dues and requires that the employer pay $12.93 per hour worked by the plumbers to the union. The union uses these payments from the employees and employer to provide training, insurance, and retirement benefits to the plumbers. Determine the annual amount paid to the union by the employer and by the employee, assuming that the plumbers work 45 hours per week, 50 weeks a year.

18. During the year, the average carpenter works 50 hours per week for 30 weeks, 40 hours per week for 15 weeks, and 30 hours per week for 5 weeks. He or she is paid time-and-a-half for any hours over 40 hours per week and receives 10 paid days off per year in the form of vacation, sick leave, and holidays. The employee's health insurance is paid for entirely by the employer. Assuming that the employee takes full advantage of the 401(k) benefit, determine the average hourly cost and burden markup for the average carpenter given in Table 1.

TABLE 1

Item	Cost
Wages	$18.86/hr
Bonus	$500
Allowances	None
Social Security	6.2% of wages to $118,500
Medicare	1.45% of wages
FUTA	0.6% of wages to $7,000
SUTA	1.85% of wages to $20,000
Workers' Compensation	$7.85 per $100 of wages
General Liability	0.65% of wages
401(k)	50% match up to 6% of wages
Health Insurance	$500/month

19. During the year, the average equipment operator works 50 hours per week for 35 weeks, 40 hours per week for 5 weeks, and 20 hours per week for 9 weeks. He or she is paid time-and-a-half for any hours over 40 hours per week and receives 15 paid days off per year in the form of vacation, sick leave, and holidays. The employee pays $200 per month for health insurance, which is not subject to Social Security and Medicare taxes. Assuming that the employee takes full advantage of the 401(k) benefit, determine the average hourly cost and burden markup for the average equipment operator given in Table 2.

20. Prepare burdened labor rates for the trades in Table 3, which are to be used on the Johnson Residence given in Appendix G. Assume the employees work 40 hours per week 50 weeks a year and do not get any paid time off, bonuses, or allowances. Use the Social Security, Medicare, and FUTA rates for the current year. The company's SUTA rate is 3.12% on the first $12,000 of an employee's wages. The company pays 1.2% of the employee's wages for general liability insurance and $350 per month towards

TABLE 2

Item	Cost
Wages	$32.11/hr
Bonus	$1,000
Allowances	$25/month
Social Security	6.2% of wages to $118,500
Medicare	1.45% of wages
FUTA	0.6% of wages to $7,000
SUTA	2.85% of wages to $12,000
Workers' Compensation	$4.60 per $100 of wages
General Liability	1.05% of wages
401(k)	50% match up to 6% of wages
Health Insurance	$750/month

each employee's health insurance. For retirement, the company has provided the employee with access to a 401(k) plan and matches the employee's contributions to the plan at a rate of $1.00 per $1.00 contributed by the employee on the first 6% of the employee's wages for a maximum matching contribution of 6% of the employee's wages.

21. Prepare burdened labor rates for the trades in Table 4, which are to be used on the West Street Video project given in Appendix G. Assume the employees work 40 hours per week 50 weeks a year and do not get any paid time off, bonuses, or allowances. Use the Social Security, Medicare, and FUTA rates for the current year. The company's SUTA rate is 1.25% on the first $33,600 of an employee's wages. The company pays 1.05% of the employee's wages for general liability insurance and $625 per

TABLE 3 Wage and Workers' Compensation Rates

Trade	Wages	Workers' Compensation
Laborer	$14.14/hr	$7.42/$100
Cement Mason	$18.86/hr	$8.72/$100
Rough Carpenter	$28.13/hr	$10.08/$100
Finish Carpenter	$29.66/hr	$7.40/$100

TABLE 4 Wage and Workers' Compensation Rates

Trade	Wages	Workers' Compensation
Laborer	$16.55/hr	$8.42/$100
Cement Mason	$21.65/hr	$11.54/$100
Ironworker	$32.50/hr	$14.36/$100
Rough Carpenter	$24.17/hr	$10.66/$100
Finish Carpenter	$28.24/hr	$6.95/$100

month towards each employee's health insurance. For retirement, the company has provided the employee with access to a 401(k) plan and matches the employee's contributions to the plan at a rate of $0.50 per $1.00 contributed by the employee on the first 6% of the employee's wages for a maximum matching contribution of 3% of the employee's wages.

22. Set up the worksheet in Excel Quick Tip 21-1 in Excel.

REFERENCES

1. For more information on how to accrue vacation and sick leave costs, see Steven J. Peterson, *Construction Accounting and Financial Management,* Prentice Hall, Upper Saddle River, NJ.
2. Wage and Hour Division, U.S. Department of Labor, *Handy Reference Guide to the Fair Labor Standards Act,* accessed from http://www.dol.gov/whd/regs/compliance/hrg.htm.
3. Wage and Hour Division, U.S. Department of Labor, 2015, *General Decision Number UT150106,* accessed from http://www.wdol.gov/dba.aspx.
4. Wage and Hour Division, U.S. Department of Labor, *Prevailing Wage Resource Book,* accessed from http://www.dol.gov/whd/recovery/pwrb/toc.htm.
5. Internal Revenue Service, *Circular E, Employer's Tax Guide,* Publication 15, 2016, p. 1.
6. Internal Revenue Service, *Circular E, Employer's Tax Guide,* Publication 15, 2016, p. 2.
7. Internal Revenue Service, *Circular E, Employer's Tax Guide,* Publication 15, 2016, p.34.
8. Tax Policy Center, *State Unemployment Tax Rates, 2014,* accessed from http://www.taxpolicycenter.org/taxfacts/displayafact.cfm?Docid=541.

CHAPTER TWENTY-TWO

EQUIPMENT COSTS

In this chapter you will learn how to determine equipment costs for owned, leased, and rented equipment. The cost of equipment includes depreciation and interest, taxes and licenses, insurance, storage, tires and other wear items, fuel, lubricants and filters, and repair reserves.

The equipment used on construction projects can be divided into three types of equipment: owned, leased, and rented. Let's begin by looking at determining the cost of owned equipment.

Equipment costs for owned equipment can be divided into two categories: ownership costs and operating costs. Ownership costs are those costs that are most closely related to how long a piece of equipment is owned rather than how much a piece of equipment is used. Ownership costs include depreciation and interest, taxes and licensing, insurance, and storage. Operation costs are those costs that are more closely related to how long a piece of equipment is operated rather than how long it has been owned. Operation costs include tires and other wear items, fuel, lubricants and filters, and repair reserves. During the life of a piece of equipment, the ownership costs tend to decrease, whereas the operation costs increase. For the purpose of estimating, the costs are often equally spread throughout the expected life of the piece of equipment.

DEPRECIATION AND INTEREST

One of the costs the company must recoup on equipment is its loss in value over its useful life. This is known as depreciation. Depreciation over the useful life of a piece of equipment equals the purchase price less the salvage value. When working with equipment, the purchase price of a piece of equipment should include the cost of the piece of equipment, sales tax, transportation cost, set up of the equipment, and any other costs the equipment dealer charges in conjunction with its purchase. Tires and other wear items, such as cutting edges, are considered operation costs and are often excluded from the purchase price and treated as a separate item. The salvage value for a piece of equipment should include all expected revenues from its sale or disposal. The depreciation discussed here should not be confused with depreciation for tax purposes, which is handled in a different manner.

By investing in equipment a company in effect takes out a loan at a specified interest rate or forgoes the opportunity to invest its capital in other opportunities that would increase the value of its invested funds. In either case, there is an interest cost (either paid or lost) that the company must recoup as part of the cost of the equipment.

The annual cost of depreciation and interest is incorporated into the equipment costs by the following equation:

$$\text{Cost}_{D\&I} = \frac{[P(1 + i)^n - F]i}{[(1 + i)^n - 1]} \qquad (22\text{-}1)$$

where

$\text{Cost}_{D\&I}$ = Annual Depreciation and Interest Costs
P = Purchase Price
F = Salvage Value
i = Interest Rate Expressed in Decimal Format
n = Life of the Equipment in Years

EXAMPLE 22-1

The purchase price of a front-end loader, including the tires, sales tax, and all other costs associated with its purchase, is $135,000. The life of the loader is 7 years, at which time it will have an estimated salvage value of $20,000. The replacement cost of the tires is $5,000 for all tires on the loader and is not to be included in the depreciation and interest cost. Determine the annual depreciation and interest cost of the loader using an interest rate of 9%.

Solution: The purchase price for the loader must be reduced by the cost of the tires, for a cost of $130,000 ($135,000 − $5,000). The annual depreciation and interest cost of the loader is determined by Eq. (22-1) as follows:

$$\text{Cost}_{D\&I} = \frac{[\$130{,}000(1 + 0.09)^7 - \$20{,}000](0.09)}{[(1 + 0.09)^7 - 1]} = \$23{,}656/\text{yr}$$

TAXES AND LICENSING

Taxes and licensing includes all taxes and licensing fees assessed by government agencies. Companies that own equipment usually have to pay a property tax based on the value of the equipment. The property taxes are often calculated by multiplying an assessed value for the equipment by a specified tax rate. The property tax due each year is the same for a piece of equipment that sits in the equipment yard all year as for a piece of equipment that is used for most of the year; therefore, it is considered an ownership cost. Additional licensing fees are often required for vehicles that travel over public roads.

Taxes and licensing can be incorporated into the cost of the equipment in two ways. The first uses the actual cost of the taxes and licensing, which is obtained from the company's accountant or estimated from the previous year's tax and licensing bills. The second way is to add the tax rate to the interest rate used in Eq. (22-1) to cover the cost of taxes.

INSURANCE

Construction equipment is covered by three general types of insurance. Automotive insurance covers vehicles used on public roads including cars, trucks, portable office trailers used on the job site, and construction equipment that drives on public roads, such as dump trucks. Marine insurance covers equipment used on waterways, such as barges and boats. Inland marine insurance covers off-road construction equipment, such as backhoes, scrapers, and dump trucks not licensed for use on public roads. These insurances cover the company against liability arising out of the operation of the equipment, damage to the equipment, theft of the equipment, and damage caused by the equipment.

Insurance is based on potential loss because of damage to the insured equipment and potential loss because of damage to other equipment and to people caused by the equipment. The potential loss because of damage to the insured equipment is a function of the value of the equipment. The loss of a $10,000 dump truck in an accident is less than the loss of a $60,000 dump truck. The potential loss because of damage to other equipment and to people caused by the equipment is independent of the cost of the equipment. A $10,000 dump truck can cause the same amount of damage as a $60,000 dump truck.

Insurance costs can be incorporated into the cost of the equipment in two ways. The first uses the actual cost of the insurance, which is obtained from the company's insurance agent or estimated from the previous year's insurance bills. The second way is to add a percentage to the interest rate used in Eq. (22-1) to cover the cost of insurance.

STORAGE

Companies often include equipment storage in the cost of the equipment. Storage costs include the cost of property used to store equipment, security for the property, and so forth. Storage costs can be incorporated into the cost of the equipment in two ways. The first uses actual cost of the storage, which is done by allocating the storage costs to all of the pieces of equipment owned by the company. The second way is to add a percentage to the interest rate used in Eq. (22-1) to cover the cost of storage.

HOURLY OWNERSHIP COST

Thus far the ownership cost of the piece of equipment has been determined on an annual basis. Because most equipment is not used on a project for an entire year, the costs must be converted to an hourly, weekly, or monthly cost. This is done by estimating the number of billable hours, weeks, or months the equipment will be used during the year and dividing the annual cost by the number of billable hours, weeks, or months. The hourly cost is determined by the following equation:

$$\text{Cost}_{\text{Hourly}} = \frac{\text{Cost}_{\text{Annually}}}{\text{Hours}} \qquad (22\text{-}2)$$

where

$\text{Cost}_{\text{Hourly}}$ = Hourly Cost
$\text{Cost}_{\text{Annually}}$ = Annual Cost
Hours = Billable Hours

The weekly and monthly costs are determined by substituting billable weeks or billable months for hours into Eq. (22-2), respectively. Because the ownership costs are the same regardless of the number of hours the equipment is used each year, the hourly, weekly, and monthly costs are highly sensitive to the number of hours, weeks, and months the equipment is used. If the usage of the equipment doubles, the ownership costs are cut in half.

EXAMPLE 22-2

For the front-end loader in Example 22-1, the annual taxes and licensing are $2,600, the annual insurance is $2,100, and the storage cost is $1,200. It is estimated that the loader will operate 1,200 hours per year. Determine the hourly ownership cost of the loader.

Solution: From Example 22-1, the annual depreciation and interest cost is $23,656. The annual ownership cost is as follows:

$$\text{Cost}_{\text{Ownership}} = \text{Cost}_{\text{D\&I}} + \text{Cost}_{\text{Taxes \& Licensing}} + \text{Cost}_{\text{Insurance}}$$
$$+ \text{Cost}_{\text{Storage}}$$
$$= \$23,656/\text{yr} + \$2,600/\text{yr} + \$2,100/\text{yr} + \$1,200/\text{yr}$$
$$= \$29,556/\text{yr}$$

The hourly ownership cost of the loader is determined by Eq. (22-2) as follows:

$$\text{Cost}_{\text{Ownership}} = \frac{(\$29,556/\text{yr})}{(1,200\ \text{hr/yr})} = \$24.63/\text{hr}$$

TIRES AND OTHER WEAR ITEMS

Wear items include tires, cutting edges, bucket teeth, and other items that frequently wear out. The cost of tires and other wear items includes all costs associated with the purchase or replacement of the item. Because the costs of tires and other wear items are operating costs and their consumption is more closely based on the hours of operation than the duration of ownership, their life is often expressed in operating hours.

If the life of the tires and wear items is one year or less, the hourly cost is determined by dividing the cost of the tires or wear items by their life. The hourly cost is calculated by the following equation:

$$\text{Cost}_{\text{Tires/Wear}} = \frac{\text{Cost}}{\text{Life}} \quad (22\text{-}3)$$

where

$\text{Cost}_{\text{Tires/Wear}}$ = Hourly Cost of Tires or Wear Items
Cost = Cost of the Tires or Wear Items
Life = Useful Life of the Tires or Wear Items in Hours

When the useful life of the tires or wear items is greater than one year, the costs associated with these items are then spread out over the life of the items, just as the price of a piece of equipment is spread over the life of the piece of equipment. The annual costs of the tires or wear items are calculated using Eq. (22-1) and substituting the life of the tires or wear items for the life of the equipment. The hourly cost is then determined by using Eq. (22-2). Because the tires and other wear items have a different useful life than the equipment, they cannot simply be included in the depreciation and interest costs.

Although the life of the tires and other wear items is based on operating hours, their life varies based on the conditions under which the equipment operates. For example, a bucket tooth digging in hard, blasted rock will wear out faster than the same bucket tooth digging in clay. When determining the life of tires and other wear items, the operation conditions must be taken into account. The equipment manufacturer is a good source of information on costs and life of tires and other wear items.

Tire repair is included in the cost of the tires by using the following equation:

$$\text{Cost}_{\text{Tire Repair}} = \frac{(\text{Repair Multiplier})(\text{Cost}_{\text{Tires}})}{\text{Life}} \quad (22\text{-}4)$$

where

$\text{Cost}_{\text{Tire Repair}}$ = Hourly Cost of Tire Repair
$\text{Cost}_{\text{Tires}}$ = Cost of the Tires
Life = Useful Life of the Tires in Hours

The repair multiplier is the historical repair cost of the tire as a percentage of the cost of the tires and is calculated by the following equation:

$$\text{Repair Multiplier} = \frac{\text{Repair Cost}}{\text{Cost}_{\text{Tires}}} \quad (22\text{-}5)$$

A common repair multiplier is 15%.

EXAMPLE 22-3

For the front-end loader in Example 22-1, the estimated life of the tires is 3,000 hours and the repair multiplier is 15%. Determine the hourly cost of the tires, including tire repair.

Solution: From Example 22-1, the cost of the tires is $5,000. The life of the tires in years is calculated as follows:

$$n = \frac{(3,000\ \text{hr})}{(1,200\ \text{hr/yr})} = 2.5\ \text{yr}$$

The annual cost of the tires is calculated using Eq. (22-1) as follows:

$$\text{Cost}_{\text{Annual}} = \frac{[\$5,000(1 + 0.09)^{2.5} - \$0](0.09)}{[(1 + 0.09)^{2.5} - 1]} = \$2,322/\text{yr}$$

The hourly cost of the tires is calculated using Eq. (22-2) as follows:

$$\text{Cost}_{\text{Hourly}} = \frac{(\$2,322/\text{yr})}{(1,200\ \text{hr/yr})} = \$1.94/\text{hr}$$

The hourly repair cost is calculated using Eq. (22-4) as follows:

$$\text{Cost}_{\text{Tire Repair}} = \frac{(0.15)(\$5,000)}{(3,000\ \text{hr})} = \$0.25/\text{hr}$$

The total hourly cost of tires and tire repair is $2.19 per hour ($1.94/hr + $0.25/hr).

FUEL

Fuel consumption is a function of the number of hours of operation. Like tires and other wear items, the consumption of fuel varies based on the conditions under which the equipment operates. For example, an excavator digging in hard-to-dig shale will consume more fuel per hour than will the same piece of equipment digging in easy-to-dig sand. When determining fuel consumption, the operation conditions must be taken into account. The equipment manufacturer is a good source of information for the rate of fuel consumption. The cost of fuel is calculated by multiplying the fuel consumption by the estimated fuel cost per gallon.

EXAMPLE 22-4

For the front-end loader in Example 22-1, it is estimated that under normal conditions it will consume 3 gallons of diesel per hour. Using a cost of $3.70 per gallon of diesel, determine the fuel cost for the front-end loader.

Solution: The cost of the fuel is calculated as follows:

$$\text{Cost}_{\text{Fuel}} = (3 \text{ gal/hr})(\$3.70/\text{gal}) = \$11.10/\text{hr}$$

LUBRICANTS AND FILTERS

Lubricants and filters include oils for use in the engine, transmission, and final drives; hydraulic fluids; grease; air filters; oil filters; hydraulic filters; transmission filters; and fuel filters. Lubricant consumption includes the replacement of the lubricants at regular intervals and the loss of lubricants because of leaks. The cost of lubricants and filters is a function of the number of hours of operation. Like other operating costs, the cost of lubricants and filters varies based on the conditions under which the equipment operates. For example, a truck working in a dusty, off-road environment will require more frequent changing of the air filter than would a truck operating on a paved highway. The operating conditions must be taken into account when determining the cost of lubricants and filters. Equipment manufacturers often provide information on the rate of consumption of lubricants and filters, which, combined with local pricing, can be used to estimate the cost of lubricants and filters.

EXAMPLE 22-5

For the front-end loader in Example 22-1, it is estimated that under normal conditions its hourly lubricant consumptions are as follows: crankcase, 0.010 gal/hr; transmission, 0.006 gal/hr; final drives including differential, 0.006 gal/hr; and hydraulic controls, 0.007 gal/hr. A tube of grease is used during each 8-hour day to lubricate the grease fittings. The local costs of lubricants are as follows: oil for the crankcase, $28.00/gal; transmission fluid, $20.00/gal; oil for the final drives, $29.50/gal; hydraulic fluid, $9.20/gal; and grease, $4.70/tube. The filters for the loader cost $0.18 per hour of operation including the labor to install them. Determine the cost of lubricants and filters for the loader.

Solution: The cost of lubricants and filters is calculated as follows:

$$\begin{aligned}\text{Cost}_{\text{L\&F}} = &(0.010 \text{ gal/hr})(\$28.00/\text{gal}) + (0.006 \text{ gal/hr})(\$20.00/\text{gal}) \\&+ (0.006 \text{ gal/hr})(\$28.50/\text{gal}) + 0.007 \text{ gal/hr})(\$9.20/\text{gal}) \\&+ (1 \text{ tube}/8 \text{ hr})(\$4.70/\text{tube}) + \$0.18/\text{hr} \\= &\$1.41/\text{hr}\end{aligned}$$

REPAIR RESERVE

Repairs include all operation costs not included in one of the foregoing items and including preventative maintenance (not included in lubricants and filters), repairs from damage and wear (not included in tires and other wear items), and overhauls. As a piece of equipment gets older, the need for repairs and the associated costs increase, resulting in the need to increase the operational cost over the life of the piece of equipment. Rather than increasing the cost over time, the average hourly cost over the life of the equipment is used. Using the average hourly cost produces excess funds during the early life of the piece of the equipment, which is placed in reserve to cover the higher cost of maintenance during the latter part of a piece of equipment's life. These excess funds are known as a repair reserve. Equipment manufacturers often provide information on the repair reserve required for a piece of equipment.

EXAMPLE 22-6

For the front-end loader in Example 22-1, the repair reserve is $6.50 per hour. Determine the hourly operation cost and hourly ownership and operation cost of the loader.

Solution: Using data from Examples 22-3 through 22-5, we calculate the hourly operation cost as follows:

$$\begin{aligned}\text{Cost}_{\text{Operation}} &= \text{Cost}_{\text{Tire}} + \text{Cost}_{\text{Fuel}} + \text{Cost}_{\text{L\&F}} + \text{Cost}_{\text{Repair Reserve}} \\&= \$2.19/\text{hr} + \$11.10/\text{hr} + \$1.41/\text{hr} + \$6.50/\text{hr} \\&= \$21.20/\text{hr}\end{aligned}$$

Using the hourly cost of ownership from Example 22-2, we calculate the hourly ownership and operation cost as follows:

$$\begin{aligned}\text{Cost} &= \text{Cost}_{\text{Ownership}} + \text{Cost}_{\text{Operation}} = \$24.63/\text{hr} + \$21.20/\text{hr} \\&= \$45.83/\text{hr}\end{aligned}$$

LEASED EQUIPMENT

With leased equipment the depreciation and interest costs are replaced with the lease payment. With a typical lease, the company leasing the equipment is still responsible for paying the cost of the taxes and licenses, insurance, storage, tires and wear items, fuel, lubricants and filters, and repairs. These costs are calculated in the same manner as they are calculated for owned equipment.

RENTED EQUIPMENT

With most rented equipment the depreciation and interest, taxes and licenses, insurance, storage, tires and wear items, lubricants and filters, and repairs are covered by the company that rents the equipment. This leaves the contractor to pay the rental cost and the cost of fuel. However, rental agreements may vary, and the estimator

EXCEL QUICK TIP 22-1
Equipment Cost Worksheet

A spreadsheet can be set up to calculate the ownership and operation cost for equipment. The spreadsheet shown in the following figure allows the user to account for tires; alternatively, the user may replace the tires with another wear item. This spreadsheet assumes that the tires have a life of greater than one year.

	A	B	C	D	E
1	Ownership Cost			Operation Cost	
2	Purchase Price ($)	135,000		Tires ($)	5,000
3	Salvage Value ($)	20,000		Tire Life (hr)	3,000
4	Equipment Life (yr)	7		Repair Multiplier	15.00%
5	Billable Hours (hr/yr)	1,200		Fuel Consumption (gal/hr)	3.00
6	Interest Rate	9.00%		Fuel Cost ($/gal)	3.70
7	Taxes & Licensing ($/yr)	2,600		Crankcase (gal/hr)	0.010
8	Insurance ($/yr)	2,100		Transmission (gal/hr)	0.006
9	Storage ($/yr)	1,200		Final Drives (gal/hr)	0.006
10	Ownership Cost ($/hr)	24.63		Hydraulic Controls (gal/hr)	0.007
11				Grease (tubes/hr)	0.125
12	Total ($/hr)	45.82		Filters ($/hr)	0.18
13				Crankcase Oil ($/gal)	28.00
14				Transmission Fluid ($/gal)	20.00
15				Final Drive Oil ($/gal)	29.50
16				Hydraulic Fluid ($/gal)	9.20
17				Grease ($/tube)	4.70
18				Repair Reserve ($/hr)	6.50
19					
20				Tires ($/hr)	2.18
21				Fuel ($/hr)	11.10
22				Lubricants and Filters ($/hr)	1.41
23				Operation Cost ($/hr)	21.19

To complete this spreadsheet, enter the data and format the cells as shown in the figure. The following formulas need to be entered into the associated cells:

Cell	Formula
B10	=(((B2-E2)*(1+B6)^B4-B3)*B6/((1+B6)^B4-1)+B7+B8+B9)/B5
B12	=B10+E23
E20	=E2*(1+B6)^(E3/B5)*B6/((1+B6)^(E3/B5)-1)/B5+E4*E2/E3
E21	=E5*E6
E22	=E7*E13+E8*E14+E9*E15+E10*E16+E11*E17+E12
E23	=E18+E20+E21+E22

The data for the equipment is entered into Cells B2 through B9 and Cells E2 through E18. The data shown in the foregoing figure is from Examples 22-1 through 22-6 and is formatted using the comma style, which replaces zeros with dashes.

should verify which costs are covered by the rental cost. The cost of the equipment equals the rental cost plus any costs not covered by the rental agreement. The costs not covered by the rental agreement are calculated in the same manner as they are calculated for owned equipment.

CONCLUSION

Equipment costs include the ownership costs of depreciation and interest, taxes and licenses, insurance, and storage; and the operation costs of tires and other wear items, fuel, lubricants and filters, and repair reserves. Data to calculate these costs are available from historical data and the equipment manufacturer.

PROBLEMS

1. The purchase price of an excavator is $185,950, excluding sales tax and deliver. Delivery is $3,500 and sales tax is 6.5% of the purchase price (excluding delivery). The expected useful life of the excavator is 8 years. At the end of its useful life, the excavator is expect to have a salvage value of $19,000. It is expected that the excavator will be used 1,350 hours per year. Using an interest rate of 8.25% determine the hourly depreciation and interest cost for the excavator.

2. The purchase price of a front-end loader, including the tires, sales tax, and all other costs associated with its purchase, is $117,450. The life of the loader is 6 years, at which time it will have an estimated salvage value of $11,500. The replacement cost of the tires is $7,500 for all tires on the loader and is not to be included in the depreciation and interest cost. The loaders is expected to be used 1,200 hours per year. Determine the annual depreciation and interest cost of the loader using an interest rate of 7%.

3. For the excavator in Problem 1, the owner decides to incorporate the taxes and license, insurance, and storage costs into the interest rate by adding 1.25% for taxes and license, 1.75% for insurance, and 1.1% for storage. Determine the ownership cost for the excavator.

4. The annual taxes for a front-end loader are $1,700, the annual insurance is $2,475, and the monthly storage costs is $200. Determine the hourly taxes, insurance, and storage costs for the loader if it is used 1,500 hours per year.

5. A set of cutters for a cold planer cost $2,500 and last an average of 80 hours. Determine the hourly cost of the cutters for the cold planer.

6. The tires for a front-end loader cost $6,700 for all four tires. The loader is used 1,200 hour per year and the estimated life of the tires is 4,000 hours. Using a repair multiplier of 20% and an interest rate of 8%, determine the hourly cost of the tires, including tire repair.

7. An excavator consumes 4.3 gallons of fuel per hour. Determine the hourly fuel cost for the excavator if fuel cost $3.95 per gallon.

8. For an off-road dump truck, it is estimated that under normal conditions its hourly lubricant consumptions are as follows: crankcase, 0.009 gal/hr; transmission, 0.005 gal/hr; final drives including differential, 0.004 gal/hr; and hydraulic controls, 0.006 gal/hr. A tube of grease is used during each 8-hour day to lubricate the grease fittings. The local costs of lubricants are as follows: oil for the crankcase, $27.50/gal; transmission fluid, $19.30/gal; oil for the final drives, $30.45/gal; hydraulic fluid, $9.90/gal; and grease, $4.95/tube. The filters for the truck cost $0.29 per hour of operation including the labor to install.

9. For a grader, it is estimated that under normal conditions its hourly lubricant consumptions are as follows: crankcase, 0.020 gal/hr; transmission, 0.012 gal/hr; final drives including differential, 0.023 gal/hr; and hydraulic controls, 0.014 gal/hr. Two tubes of grease is used during each 8-hour day to lubricate the grease fittings. The local costs of lubricants are as follows: oil for the crankcase, $28.75/gal; transmission fluid, $19.85/gal; oil for the final drives, $31.10/gal; hydraulic fluid, $10.20/gal; and grease, $4.75/tube. The filters for the truck cost $0.32 per hour of operation including the labor to install.

10. The purchase price of a loader including the tires, sales tax, and all other costs associated with its purchase is $210,000. The life of the loader is 9 years, at which time it will have an estimated salvage value of $23,000. The replacement cost of the tires is $11,000 for all tires on the loader. The life of the tires is estimated to be 3,600 hours. Tire repairs are expected to be 15% of the cost of the tires. The interest rate is 7.5%. The company adds to the interest rate 1.5% for property taxes, 2.0% for insurance, and 1.0% for storage. The fuel consumption is estimated to be 10.75 gal/hr, at an estimated cost of $3.45/gal. It is estimated that under normal conditions its hourly lubricant consumptions are as follows: crankcase, 0.064 gal/hr; transmission, 0.018 gal/hr; final drives including differential, 0.039 gal/hr; and hydraulic controls, 0.035 gal/hr. Two tubes of grease are used during each 8-hour day to lubricate the grease fittings. The local costs of lubricants are as follows: oil for the crankcase, $28.75/gal; transmission fluid, $20.40/gal; oil for the final drives, $30.70/gal; hydraulic fluid, $9.90/gal; and grease, $4.99/tube. The filters for the loader cost $0.44/hr of operation, including the labor to install them. The estimated repair reserve is $14.40. Using 1,500 billable hours per year; determine the hourly ownership and operation cost of the loader.

11. The purchase price of the hydraulic excavator, including sales tax and all other costs associated with its purchase, is $150,000. The life of the

excavator is 6 years, at which time it will have an estimated salvage value of $20,000. The interest rate is 8.25%. The taxes and licensing costs are $3,220/yr, the insurance costs are $2,930/yr, and the storage costs are $2,110/yr. The fuel consumption is estimated to be 3.2 gal/hr, at an estimated cost of $3.55/gal. It is estimated that under normal conditions, its hourly lubricant consumptions are as follows: crankcase, $0.008 gal/hr; final drives, $0.001 gal/hr; and hydraulic controls, $0.021 gal/hr. One tube of grease is used during each 8-hour day to lubricate the grease fittings. The local costs of lubricants are as follows: oil for the crankcase, $27.85/gal; oil for the final drives, $29.95/gal; hydraulic fluid, $9.75/gal; and grease, $4.45/tube. The filters for the excavator cost $0.25/hr of operation, including the labor to install them. The estimated repair reserve is $4.20/hr. Using 1,400 billable hours per year, determine the hourly ownership and operation cost of the excavator.

12. Contact a local equipment rental company and get a rental price for a telescoping forklift. Find out which costs are included in the rental costs; and which costs are excluded. Prepare an estimate for the costs excluded from the rental agreement. What is the hourly cost of the forklift?

13. Set up Excel Quick Tip 22-1 in Excel.

CHAPTER TWENTY-THREE

CREW RATES

In this chapter you will learn how to calculate crew labor and equipment rates, the average cost of labor per labor hour, and the equipment cost per labor hour.

Most work on a construction project requires the use of more than one class of employee and may involve multiple pieces of equipment. For example, the framing of a building often requires a lead carpenter or foreperson, journeyman carpenters, apprentice carpenters, and laborers, as well as compressors, nail guns, and a forklift. An estimator needs to be able to calculate the average cost of labor per labor hour for a specific crew and the equipment cost per labor hour. This is because the productivity is based on the number of labor hours needed to perform a unit of work.

The average labor cost is calculated by summing the cost of labor for the entire crew for one day and then dividing it by the number of labor hours performed by the crew during one day. The labor cost is determined by multiplying the number of workers in the class by the labor rate for the class by the number of hours worked per day, as shown in the following equation:

$$\text{Cost}_{\text{Labor}} = (\text{Workers})(\text{Labor Rate})(\text{Hours Worked per Day}) \quad (23\text{-}1)$$

Determining the labor rate is discussed in Chapter 21.

The equipment cost per labor hour is determined by summing the cost of all of the equipment for one day and then dividing it by the number of labor hours performed by the crew during one day. The equipment cost is calculated from hourly rates by multiplying the number of pieces of the same type of equipment by the equipment cost by the number of hours worked per day, as shown in the following equation:

$$\text{Cost}_{\text{Equipment}} = (\text{No. of Pieces})(\text{Equipment Cost}) \times (\text{Hours Worked per Day}) \quad (23\text{-}2)$$

Alternatively, the equipment cost may be determined from daily rates by multiplying the number of pieces of the same type of equipment by the equipment cost, as shown in the following equation:

$$\text{Cost}_{\text{Equipment}} = (\text{No. of Pieces})(\text{Equipment Cost}) \quad (23\text{-}3)$$

Determining the equipment cost is discussed in Chapter 22. Determining the average cost of labor per labor hour and equipment cost per labor hour is shown in the following example.

EXAMPLE 23-1

A pipe-laying crew consists of two hydraulic excavators, a front-end loader, a trench box, a gravel box, a foreperson, a pipe layer, two equipment operators, and a laborer. The cost of a hydraulic excavator is $90.00 per hour, a front-end loader is $75.00 per hour, a trench box is $25.00 per day, and a gravel box is $18.00 per day. The foreman costs $44.85 per hour, an operator costs $33.91 per hour, the pipe layer costs $42.85 per hour, and the laborer costs $23.08 an hour. Labor costs include labor burdens. All employees work an 8-hour day except the foreperson, who works a 9-hour day. Determine the average cost of labor per labor hour and the equipment cost per labor hour for the pipe-laying crew.

Solution: The daily labor costs for each class of worker are calculated using Eq. (23-1) as follows:

$$\text{Cost}_{\text{Foreperson}} = (1)(\$44.85/\text{hr})(9 \text{ hr/day}) = \$403.65/\text{day}$$
$$\text{Cost}_{\text{Pipe Layer}} = (1)(\$42.85/\text{hr})(8 \text{ hr/day}) = \$342.80/\text{day}$$
$$\text{Cost}_{\text{Operators}} = (2)(\$33.91/\text{hr})(8 \text{ hr/day}) = \$542.56/\text{day}$$
$$\text{Cost}_{\text{Laborer}} = (1)(\$23.08/\text{hr})(8 \text{ hr/day}) = \$184.64/\text{day}$$

The daily labor cost for the entire crew is calculated as follows:

$$\text{Cost}_{\text{Labor}} = \$403.65/\text{day} + \$342.80/\text{day} + \$542.56/\text{day} + \$184.64/\text{day}$$
$$= \$1{,}473.65/\text{day}$$

The number of labor hours (lhr) worked in one day is calculated as follows:

$$\text{Hours} = (1)(9 \text{ hr/day}) + (1)(8 \text{ hr/day}) + (2)(8 \text{ hr/day})$$
$$+ (1)(8 \text{ hr/day})$$
$$= 41 \text{ lhr/day}$$

The average cost per labor hour is calculated as follows:

$$\text{Cost}_{\text{Labor Hour}} = \frac{(\$1{,}473.65/\text{day})}{(41 \text{ lhr/day})} = \$35.94/\text{lhr}$$

The daily equipment costs for each type of equipment are calculated using Eq. (23-2) and (23-3) as follows:

$$\text{Cost}_{\text{Excavators}} = (2)(\$90.00/\text{hr})(8 \text{ hr/day}) = \$1{,}440.00/\text{day}$$
$$\text{Cost}_{\text{Loader}} = (1)(\$75.00/\text{hr})(8 \text{ hr/day}) = +\$600.00/\text{day}$$
$$\text{Cost}_{\text{Trench Box}} = (1)(\$25.00/\text{day}) = \$25.00/\text{day}$$
$$\text{Cost}_{\text{Gravel Box}} = (1)(\$18.00/\text{day}) = \$18.00/\text{day}$$

The daily equipment cost for the entire crew is calculated as follows:

$$\text{Cost}_{\text{Equipment}} = \$1{,}440.00/\text{day} + \$600.00/\text{day}$$
$$+ \$25.00/\text{day} + \$18.00/\text{day}$$
$$= \$2{,}083.00/\text{day}$$

The equipment cost per labor hour (lhr) is calculated as follows:

$$\text{Cost}_{\text{Equipment}} = \frac{(\$2{,}083.00/\text{day})}{(41 \text{ lhr/day})} = \$50.80/\text{lhr}$$

The crew makeup can be determined by a number of factors. Sometimes, limits on the crew makeup are set by contract or government regulations. For example, state law may require there be at least one journeyman plumber for every two apprentice plumbers on a job. Crew makeup is also determined by the size of the job. A small framing job may only require a journeyman carpenter and a laborer, whereas a large job may require a lead carpenter, three journeyman carpenters, four apprentice carpenters, and three laborers.

EXCEL QUICK TIP 23-1
Crew Rates Worksheet

The following spreadsheet can be used to calculate the average cost of labor per labor hour and the equipment cost per labor hour for a crew.

	A	B	C	D	E	F
1			Labor			
2					Daily	
3	Class	Number	Rate	Hr/Day	Wages	
4	Foreperson	1	44.85	9.00	403.65	
5	Pipe Layer	1	42.85	8.00	342.80	
6	Operators	2	33.91	8.00	542.56	
7	Labor	1	23.08	8.00	184.64	
8					-	
9					-	
10				41.00	1,473.65	
11						
12			Cost Per Labor Hour		35.94	
13						
14			Equipment			
15			Hourly			Daily
16	Class	Number	Rate	Hr/Day	Daily Rate	Wages
17	Excavator	2	90.00	8.00		1,440
18	Loader	1	75.00	8.00		600
19	Trench Box	1			25.00	25
20	Gravel Box	1			18.00	18
21						-
22						-
23						2,083
24						
25			Cost Per Labor Hour			50.80

To complete this spreadsheet, enter the data and format the cells as shown in the figure. The following formulas need to be entered into the associated cells:

Cell	Formula
E4	=B4*C4*D4
D10	=B4*D4+B5*D5+B6*D6+B7*D7+B8*D8+B9*D9
E10	=SUM(E4:E9)
E12	=E10/D10
F17	=B17*C17*D17+B17*E17
F23	=SUM(F17:F22)
F25	=F23/D10

After entering these formulas, copy Cell E4 to Cells E5 through E9 and copy Cell F17 to Cells F18 through F22.

The data for the labor and equipment is entered into Cells A4 through E9 and Cells A17 through F22. Costs for equipment should only be entered in Column D (the hourly rate) or Column E (the daily rate) but not both. The data shown in the foregoing figure is from Example 23-1 and is formatted using the comma style, which replaces zeros with dashes.

CONCLUSION

The average labor rate per labor hour for a crew is determined by calculating the daily cost of the crew and dividing it by the labor hours performed by the crew. The equipment cost per labor hour is determined by calculating the daily cost of the equipment and dividing it by the number of labor hours worked by the crew.

PROBLEMS

1. An excavation crew consists of a hydraulic excavator, three dump trucks, a foreperson, and three truck drivers. The foreperson operates the hydraulic excavator. The cost of a hydraulic excavator is $90.00/hr and a truck costs $55.00/hr. The foreperson costs $35.00/hr and a truck driver costs $21.00/hr. Labor costs include labor burdens. All employees work an 8-hour day. Determine the average cost per labor hour and the equipment cost per labor hour for the excavation crew.

2. A carpentry crew consists of a lead carpenter, three journeyman carpenters, four apprentice carpenters, and three laborers. The lead carpenter costs $36.00/hr, a journeyman carpenter costs $29.39/hr, an apprentice carpenter costs $23.63/hr, and a laborer costs $17.28/hr. Labor costs include labor burdens. All employees work an 8-hour day, except the foreperson, who works a 9-hour day. The equipment to support the carpenters includes six nail guns, three saws, two compressors, and a forklift. A nail gun costs $5/day, a saw costs $4/day, a compressor costs $10/day, and a forklift costs $35.00/hr. Determine the average cost per labor hour and the equipment cost per labor hour for the carpentry crew.

3. Prepare crew rates for the crews in Table 1, which are to be used on the Johnson Residence given in Appendix G. Assume all employees work an 8 hour day. Use the labor rates calculated from Problem 20 of Chapter 21.

TABLE 1 Crews for Johnson Residence

Crew	Labor	Equipment ($/day)
Concrete	2 each cement masons 1 each laborer	$20.00
Rough Carpentry	3 each rough carpenters 2 each laborers	$80.00
Finish Carpentry	1 each finish carpenter 1 each laborer	$25.00

4. Prepare crew rates for the crews in Table 2, which are to be used on the West Street Video project given in Appendix G. Assume all employees work an 8 hour day. Use the labor rates calculated from Problem 21 of Chapter 21.

TABLE 2 Crews for West Street Video

Crew	Labor	Equipment ($/day)
Concrete	3 each cement masons 2 each laborer	$40.00
Steel Installation	5 each ironworkers 2 each laborers	$1,200.00
Rough Carpentry	5 each rough carpenters 3 each laborers	$350.00
Finish Carpentry	2 each finish carpenter 1 each laborer	$45.00

5. Set up Excel Quick Tip 23-1 in Excel.

CHAPTER TWENTY-FOUR

SUBCONTRACT PRICING

In this chapter you will learn how to obtain subcontract pricing, including how to prepare a request for a quote and a scope of work, as well as how to select the best bid.

Subcontractors provide labor and often materials and equipment to the project for a fixed price or for a unit price. The best source of subcontractor pricing is the subcontractor. When pricing from the subcontractor is not available, the subcontractor pricing can be estimated from historical data.

When obtaining pricing from the subcontractor, it is important for the subcontractor and the contractor to clearly communicate the work that is to be performed, the price for the work, and any other conditions associated with the work. This is best done by communicating these parameters in writing in the form of a Request for Quote (RFQ) and having the subcontractor respond with a written proposal. The Request for Quote, sometimes referred to as a Request for Proposal (RFP), is sent to the subcontractor. The preparation and submission of quotes is discussed in Chapter 28.

REQUEST FOR QUOTE

Prior to preparing the RFQ, the estimator needs to break the project down into work packages. Each work package should consist of work that would be done by a specific class of subcontractor or in-house crew. For example, a project may contain an electrical work package that consists of the electrical work needed to complete the project. Once the work packages have been identified, the estimator needs to identify which subcontractors he or she wants to bid on each of the work packages to be subcontracted out. Next, the estimator needs to contact each of the subcontractors and obtain a commitment that the subcontractor will provide a bid for the project. Finally, the estimator should send each subcontractor the RFQ for the work package that the subcontractor has agreed to bid. This is done to prevent miscommunications as to the project, its bid time, and the scope of the work that needs to be performed to complete the work package. The RFQ should contain the name of the subcontractor from whom the quote is being requested, the name of the contractor requesting the quote, the project for which the quote is being requested, the bid date and time for the project, and the scope of the work included in the work package. A sample RFQ is shown in Figure 24-1.

REQUEST FOR QUOTE

TO: A&M Structural Steel
            856 West Jefferson Street
            Ogden, Utah 84403

FROM: Eagle Gate Contractors
            1556 East Adams Ave.
            Ogden, Utah 84403

PROJECT: West Street Video
            4755 South West Street
            Ogden, Utah 84403

DATE & TIME OF BID: June 15, 2006 @ 1:00 p.m.

We invite your company to bid on the following work for the above project:

Furnish all material, labor, and equipment to complete Sections 05120—Structural Steel and 05500—Metal Fabrications to complete the West Street Video project as per the plans and specifications prepared by Steven J. Peterson, MBA, PE, and dated August 14, 2004. Placement of anchor bolts is to be done by others.

Plans are available at our office for your review.

Sincerely,

Steven Peterson
Chief Estimator

FIGURE 24-1 Request for Quote

WRITING A SCOPE OF WORK

For the RFQ to be useful it must clearly define the scope or limits of work to be performed; it will become the basis of the contract. Writing contracts is discussed in Chapter 29. The scope of work should clearly delineate what work is included in the work package and what is excluded. The scope of work should include any clarifications, exclusions, and other assumptions necessary to clearly define the scope of the work package.

The different scopes of work should cover all materials, labor, and equipment needed to complete the project. There should be no gaps among different scopes of work, otherwise the contractor might leave some costs out of the pricing and be open to a change order. Care must also be taken to ensure that there are no overlaps between multiple scopes of work. This would not only create confusion by duplicating work in two different scopes of work, but might lead the contractor to pay for the work twice if the duplication goes unnoticed. If an overlap between scopes of work is identified, the contractor must issue a change order or have changes made to the quotes to rectify the mistake.

Well-prepared scopes of work are particularly important when the lines delineating the work performed by two subcontractors are not clearly defined. This often occurs when there are two subcontractors working on a piece of equipment. For example, when installing a gas-fired hot-water heater provided by the plumbing subcontractor and a gas-fired furnace provided by the HVAC subcontractor, which subcontractor is responsible for providing the gas lines? If the HVAC subcontractor provides the gas lines, is the HVAC subcontractor responsible for connecting the gas lines to the hot-water heater, or does the HVAC subcontractor just provide a valve next to the hot-water heater, with the plumbing subcontractor responsible for connecting the hot-water heater to the gas lines? The reverse is true for the furnace when the plumbing subcontractor is providing the gas lines.

The scope of work can be prepared by referencing all or part of a set of contract documents, for example, the plans and project manual. When referencing a set of contract documents, the reference should include the title of the project, the architect or engineer who prepared the documents, and the document dates. It is very important to include the date for the drawings because the drawings may be revised during the bidding process and it is important to know on which drawings the scope of work is based. This may require that a table of drawings, with their dates, be included in the scope of work. A scope of work based on the contract document may be written as shown in Figure 24-1

Alternatively, the scope of work can be prepared by specifically spelling out what is included and what is excluded, as shown in Figure 24-2. In this scope of work the framing subcontractor provides the labor, equipment,

---

Provide all labor, equipment, tools, and supervision required to furnish and install all framing and all appurtenances required for a complete installation in accordance with the Contract Documents (the plans entitled "Westland Residence" prepared by John Murray, AIA, and dated August 15, 2003), applicable codes, and governing agencies. The work includes, but is not limited to the following:

1. Subcontractor shall install all rough carpentry members, including but not limited to sills, studs, plates, sleepers, columns, posts, beams, girders, floor joists, header joists, stringers, risers, treads, trimmers, backing, headers, blocking, lintels, bridging, firestopping, rafters, purlins, trusses, floor sheathing, roof sheathing, and drops.
2. Subcontractor shall install all framing hardware, including but not limited to nails, shots, pins, expansion bolts, machine bolts, lag screws, post-to-beam connections, joist and beam hinges, and straps.
3. Subcontractor shall provide all nails required for the completion of this contract. All other framing hardware to be provided by the contractor.
4. Subcontractor shall install all exterior doors, sliding glass doors, and windows. All doors and windows shall be hung true and plumb. Windows that line up shall be hung such that the vertical and horizontal lines are straight and true. Doors shall be hung such that they do not bind nor swing open or closed by themselves.
5. Subcontractor shall furnish a forklift and/or crane for use in transporting and placing materials installed as a part of this contract.
6. Subcontractor shall furnish all tools, saws, cords, and miscellaneous equipment required for the completion of this contract.
7. Contractor shall supply temporary power for the subcontractor.
8. Subcontractor shall coordinate with the project superintendent to:
   a. Ensure correct material sizes and quantities are ordered.
   b. Ensure prompt and timely material deliveries.
   c. Coordinate material distribution and storage.
   d. Coordinate installation procedures with other trades.
   e. Setup and maintain inventory control for receiving and distribution of materials.
   f. Spot-check rough openings and backing requirements.
9. The following items are specifically excluded from the subcontractor's contract and will be furnished by the Contractor: lumber, engineered wood products (for example, glue-laminated beams), trusses, steel columns, framing hardware, fasteners (except nails), drywall, doors, sliding glass doors, and windows.

FIGURE 24-2 Sample Scope of Work

and some of the materials to construct a home and the general contractor provides the rest of the materials.

Sample scopes of work are given in Appendix E and may be used as the basis for writing your own scopes of work.

HISTORICAL

In the absence of subcontractor pricing for the project, the estimated subcontractor cost can be calculated from historical data. Historical costs are determined by obtaining past costs from the accounting system or contracts for a certain quantity of work and dividing it by the quantity of work performed. This calculation is expressed by the following equation:

$$\text{Cost per Unit} = \frac{\text{Cost}}{\text{Quantity}} \qquad (24\text{-}1)$$

EXAMPLE 24-1

A recent project included 1,235 square feet of wood-framed building. The framing was subcontracted out for $5,700. Determine the framing cost per square foot of building.

Solution: The framing cost per square foot is calculated using Eq. (24-1) as follows:

$$\text{Cost per Unit} = \frac{(\$5,700)}{(1,235 \text{ ft}^2)} = \$4.62/\text{ft}^2 \qquad \blacksquare$$

The historical cost is then multiplied by the quantity of work that needs to be performed on the project to obtain an estimate of the cost of the subcontract work. This calculation is expressed by the following equation:

$$\text{Cost} = (\text{Quantity})(\text{Cost per Unit}) \qquad (24\text{-}2)$$

EXAMPLE 24-2

Using the data from Example 24-1, estimate the cost to subcontract out the framing for a 2,200-square foot building. The framing of the building in Example 24-1 is similar to this building.

Solution: The estimated framing cost is calculated using Eq. (24-2) as follows:

$$\text{Costs} = (\$4.62/\text{ft}^2)(2,200 \text{ ft}^2) = \$10,164 \qquad \blacksquare$$

With materials, one is able to determine accurate quantities for each component. With subcontracting, it is impossible to do this because the pricing data is often a single price for an entire work package. Because of this, the estimator must make sure that the historical data used in estimating subcontractor work is for a project that is very similar to the project he or she is bidding; otherwise, there could be a big discrepancy between the actual cost of the work and the cost used in the estimate. The comparison should be based on the size of the project, complexity of the work, and materials used.

BID SELECTION

When receiving bids from multiple subcontractors, the estimator must select the subcontractor with the best quote. Selection should be based on completeness of the scope of work, price, schedule, quality, and performance of the subcontractor.

The most important thing to do when one receives a quote from a subcontractor is to verify that the subcontractor has bid a complete work package without any extras. If the subcontractor has not bid a complete work package, the estimator needs to fill in the missing items or have the subcontractor provide costs for the missing items. If the subcontractor has extra items in the bid, the estimator needs to have the subcontractor deduct them.

Historically, contractor and subcontractor selection has been based on price alone. Many owners and contractors are beginning to select their contractors and subcontractors based on which subcontractor provides the best bid based on price, the ability of the subcontractor to meet the schedule and quality standards, and the past performance of the subcontractor. It may well be worth the extra cost to pay more for a subcontractor who can shorten the schedule and meet the quality standards the first time. Failure to meet the quality standards increases the duration of a project because it takes time to rework the materials. It also increases costs because the employees who are to perform these tasks have to wait for the quality defect to be repaired before proceeding. Scheduling delays, either because of poor schedule performance or poor quality, increase the duration of the project, which increases the project's overhead cost. The best way to judge the quality and schedule performance of subcontractors is based on their past performance.

CONCLUSION

Subcontractor pricing is best obtained from the subcontractors who are interested in performing the work on the project. When getting bids from subcontractors it is important that the estimator clearly communicate the work package that he or she wants the subcontractor to bid, the project to be bid, and the bid date and time. This is best done by sending a Request for Quote (RFQ) to the subcontractor. The RFQ should contain a complete scope of work for the work package. The scope of work should clearly delineate what work is included in the work package and what is excluded. The scope of work should be clearly defined, including any clarifications, exclusions, and other assumptions. When pricing from the subcontractor is not available, the estimated subcontractor cost can be calculated from historical data. Historical costs are determined by obtaining past costs from the accounting system or contracts for a certain quantity of work and dividing it by the quantity of work performed. When selecting among subcontractors, selection should be based on completeness of the scope of work, price, schedule, quality, and performance.

PROBLEMS

1. A recent project included a 21,000-square-foot warehouse with tilt-up concrete panel exterior. The construction of the tilt-up panels was subcontracted for $74,000. Determine the cost of the tilt-up panels per square foot of warehouse.

2. In a recent project, a 2,200-square-foot, single-story office building was constructed with a 30-year architectural shingle. The roofing was subcontracted for $3,900. Determine the cost of the roofing per square foot of office building.

3. Using the data from Problem 1, estimate the cost to subcontract out the tilt-up concrete panels for an 18,000-square-foot warehouse. The warehouse design is similar to that of the warehouse in Problem 1.

4. Using the data from Problem 2, estimate the cost to subcontract out the roofing for a 2,800-square-foot office building. The roofing and roof design are similar to those of the office building in Problem 2.

5. Prepare scope of works for the drawings of the residential garage given in Appendix G.

6. Prepare Request for Quotes, which includes a scope of work, for the Johnson Residence given in Appendix G for the following Cost Codes:
    a. 04200 Masonry
    b. 07200 Insulation
    c. 07220 Stucco
    d. 07400 Siding
    e. 07500 Roofing
    f. 07710 Rain Gutters
    g. 09200 Drywall
    h. 09300 Ceramic Tile
    i. 09600 Flooring
    j. 09900 Paint
    k. 22000 Plumbing
    l. 23000 HVAC
    m. 26000 Electrical
    n. 31200 Grading and Excavation
    o. 33100 Water Line and 33300 Sewer Line

7. Prepare Request for Quotes, which includes a scope of work, for the West Street Video project given in Appendix G for the following Cost Codes:
    a. 02400 Demolition
    b. 04200 Masonry
    c. 07200 Insulation
    d. 07500 Roofing and 07600 Sheet Metal
    e. 08400 Store Fronts
    f. 09200 Drywall
    g. 09300 Ceramic Tile
    h. 09500 Acoustical Ceilings
    i. 09600 Flooring
    j. 09900 Paint
    k. 21000 Fire Sprinklers
    l. 22000 Plumbing
    m. 23000 HVAC
    n. 26000 Electrical
    o. 31200 Grading and Excavation
    p. 32100 Asphalt
    q. 33100 Water Line, 33300 Sanitary Sewer, and 33400 Storm Drain

CHAPTER TWENTY-FIVE

MARKUPS

In this chapter you will learn how to determine markups for the building permit, payment and performance bonds, and profit and overhead.

Markups are costs that are added to the cost of the bid. The most common markups include a building permit, payment and performance bonds, and profit and overhead. Let's begin by looking at the building permit costs.

BUILDING PERMITS

Building permits are required on many construction projects. Highway and federal projects do not require building permits nor do many state and municipal projects. Building permits are used to cover the costs of inspections to verify that the project is being built in accordance with the applicable building codes. The cost of the building permit is based on the valuation of the project, which may be based on the actual bid for the project or a standardized cost per square foot established by the code enforcement agency. Sample building permit costs are shown in Table 25-1. The estimator should check with the contracting officer or code enforcement agency to determine: (1) if a building permit is required, (2) how the valuation of the building is established, and (3) the building permit cost schedule used by the code enforcement agency.

The building permit costs are calculated by selecting the appropriate range for the valuation of the project in the left column and determining the cost in accordance with the right column. These calculations are shown in the following example.

EXAMPLE 25-1

Using the building permit costs in Table 25-1, determine the building permit cost for a project valued at $257,061.

TABLE 25-1   Sample Building Permit Costs

Total Valuation ($)[1]	Rate
1 to 1,000	$50
1,001 to 5,000	$50 plus $45 per $1,000
5,001 to 25,000	$230 plus $25 per $1,000
25,001 to 50,000	$730 plus $15 per $1,000
50,001 to 100,000	$1,105 plus $10 per $1,000
100,001 to 250,000	$1,605 plus $8 per $1,000
250,001 to 500,000	$2,805 plus $7 per $1,000
500,001 to 1,000,000	$4,555 plus $6 per $1,000
Over 1,000,000	$7,555 plus $5 per $1,000

[1]All valuations are to be rounded up to the next $1,000 increment.

Solution: The valuation falls between $250,001 and $500,000; therefore, the cost of the building permit is $2,805 for the first $250,000 plus $7 for each additional $1,000 or fraction thereof. The valuation exceeds $250,000 by $7,061, or 8 additional $1,000s. The building permit cost is calculated as follows:

Building Permit Cost = $2,805 + (8)($7) = $2,861

In this example, it was assumed that the building permit valuation was known and did not change when the cost of the building permit was added to the bid. This is the case when the valuation is based on a standardized cost per square foot established by the code enforcement agency rather than the bid. When the cost of the building permit is based on the bid, the cost of the building permit will increase the cost of the bid, which will in turn change the cost of the building permit, which will increase the cost of the bid, and so on until the changes

EXCEL QUICK TIP 25-1
Building Permit Cost Worksheet

The following spreadsheet can be used to calculate the cost of the building permit:

	A	B	C	D	E	F
1	From	To	Base Rate			Per Additional $1,000
2	1	1,000	50			
3	1,001	5,000	50	for first	1,000	45.00
4	5,001	25,000	230	for first	5,000	25.00
5	25,001	50,000	730	for first	25,000	15.00
6	50,001	100,000	1,105	for first	50,000	10.00
7	100,001	250,000	1,605	for first	100,000	8.00
8	250,001	500,000	2,805	for first	250,000	7.00
9	500,001	1,000,000	4,555	for first	500,000	6.00
10	1,000,001		7,555	for first	1,000,000	5.00
11						
12			Input Valuation		257,061	
13			Valuation Rounded		258,000	
14			Buiding Permit Cost		2,861	

To complete this spreadsheet, enter the data and format the cells as shown in the above figure. The following formulas need to be entered into the associated cells:

Cell	Formula
B2	=A3-1
E3	=A3-1
E13	=ROUNDUP(E12,-3)
E14	=IF(E13<A3,C2,
	IF(E13<A4,C3+(E13-E3)*F3/1000,
	IF(E13<A5,C4+(E13-E4)*F4/1000,
	IF(E13<A6,C5+(E13-E5)*F5/1000,
	IF(E13<A7,C6+(E13-E6)*F6/1000,
	IF(E13<A8,C7+(E13-E7)*F7/1000,
	IF(E13<A9,C8+(E13-E8)*F8/1000,
	IF(E13<A10,C9+(E13-E9)*F9/1000,
	C10+(E13-E10)*F10/1000))))))))

After entering these formulas, copy Cell B2 to Cells B3 through B9 and Cell E3 to Cells E4 through E10. Cell E13 uses the ROUNDUP function to round up the valuation to the nearest $1,000 increment. Cell E14 uses nested IF functions to calculate the permit cost.

The building permit cost data is entered into Cells A2 through A10, C2 through C10, and F3 through F10, and the valuation is entered in Cell E12. The data shown in the foregoing figure is from Table 25-1 and Example 25-1.

are insignificant. When this is the case, either a rough estimate for the cost of the building permit must be included in the bid prior to the calculation of the cost of the building permit, or the cost of the building permit must be calculated by iteration. Chapter 32 will look at using iteration in a spreadsheet.

PAYMENT AND PERFORMANCE BONDS

Many jobs require payment and performance bonds. The cost of the bond is based on the total bid. The bonding schedule is divided into different brackets with a

TABLE 25-2 Sample Bond Schedule

Total Estimate ($)	Bond Rate (%)
0 to 50,000	1.50
50,001 to 100,000	1.25
100,001 to 250,000	1.00
250,001 to 500,000	0.90
500,001 to 1,000,000	0.80
1,000,001 PLUS	0.75

different bonding rate being applied to each bracket. A sample bonding schedule is shown in Table 25-2. A contractor using the bonding schedule in Table 25-2 would pay 1.5% on the first $50,000 of a bid, 1.25% on the next $50,000 ($100,000 − $50,000), 1% on the next $150,000 ($250,000 − $100,000), and so forth.

EXAMPLE 25-2

Using the bond costs in TABLE 25-2, determine the bonding cost for a project bid at $458,265.

Solution: The company will pay 1.5% on the first $50,000, 1.25% on the next $50,000, 1% on the next $150,000, and 0.9% on the remaining $208,265. The bonding cost is calculated as follows:

Bond Cost = ($50,000)(0.015) + ($50,000)(0.0125)
  + ($150,000)(0.01) + ($208,265)(0.009)

Bond Cost = $4,749.39

Because the cost of the bonds is based on the bid, the cost of the bonds will increase the bid, which will in turn increase the cost of the bonds, which will increase the cost of the bid, and so on until the changes are insignificant. To deal with this, either a rough estimate for the cost of the bonds must be included in the bid prior to the calculation of the cost of the bonds, or the cost of the bonds must be calculated by iteration. Chapter 32 will look at using iteration in a spreadsheet to determine the cost of the bonds.

PROFIT AND OVERHEAD

The profit and overhead (P&O) markup is used to cover the general overhead required by the main office and to provide a profit to the company owners. The profit and overhead markup is expressed as a percentage of the construction cost, which often includes the cost of the building permit and bond. The profit and overhead markup is determined by multiplying the construction cost by the profit and overhead markup rate as shown in the following equation:

EXCEL QUICK TIP 25-2
Bond Cost Worksheet

The following spreadsheet can be used to calculate the cost of a bond:

	A	B	C
1	From	To	Rate(%)
2	-	50,000	1.50%
3	50,001	100,000	1.25%
4	100,001	250,000	1.00%
5	250,001	500,000	0.90%
6	500,001	1,000,000	0.80%
7	1,000,001		0.75%
8			
9		Bid	458,265
10		Bond	4,749

To complete this spreadsheet, enter the data and format the cells as shown. The following formulas need to be entered into the associated cells:

Cell	Formula
B2	=A3-1
C10	=IF(C9<=B2,C9*C2,
	IF(C9<=B3, (C9-B2)*C3+B2*C2,
	IF(C9<=B4, (C9-B3)*C4+(B3-B2)*
	C3+B2*C2,
	IF(C9<=B5, (C9-B4)*C5+(B4-B3)*C4+
	(B3-B2)*C3+B2*C2,
	IF(C9<=B6, (C9-B5)*C6+(B5-B4)*C5+
	(B4-B3)*C4+(B3-B2)*C3+B2*C2,
	(C9-B6)*C7+(B6-B5)*C6+
	(B5-B4)*C5+(B4-B3)*
	C4+(B3-B2)*C3+B2*C2)))))

After entering these formulas, copy Cell B2 to Cells B3 through B6. Cell C10 uses nested IF functions to calculate the cost of the bond.

The bond ranges and rates are entered into Cells A2 through A7 and C2 through C7. The bid is entered into Cell C9. The data shown in the foregoing figure is from Table 25-2 and Example 25-2.

$$\text{P\&O Markup} = \frac{(\text{Construction Cost})(\text{P\&O Markup \%})}{100\%}$$

(25-1)

EXAMPLE 25-3

The estimated construction cost for a project is $452,632. The company's profit and overhead markup is 5% of construction cost. Determine the profit and overhead markup.

Solution: The profit and overhead markup is determined as follows using Eq. (25-1):

$$\text{P\&O Markup} = \frac{(\$452{,}632)(5\%)}{100\%} = \$22{,}631.60$$

Not all projects can or should be bid at the same profit and overhead markup. A company should set a minimum profit and overhead markup and then increase the markup when conditions warrant.[1] The difficult part is to determine when the conditions warrant an increase. The following are some common reasons for increasing the profit and overhead markup.

First, the company is submitting a bid as a courtesy to the customer but really doesn't want the project unless it is very profitable. This may happen when a company is asked to bid on a project that is outside its geographical area or area of specialization or when it already has plenty of work.

Second, the project is a difficult project, has a high degree of risk, or the project owners are difficult to work with. The level of risk is increased by poor document (plans and project manual) quality, short construction schedules, high liquid damages, and uncertainty that may lead to cost or schedule overruns. In this case the risk and headaches of the project are acceptable to the construction company only if there is a higher level of profit.

Third, the company is bidding on the project to check their prices but really doesn't want the project unless it is very profitable. When starting to bid after not bidding for a few months or bidding in a new market, it is wise to bid on two or three projects to get a feel for the market and to get back into a bidding rhythm. This gives the company's management a chance to get a feel for the level of profit and overhead markup that they can add to their bids, as well as give their estimators a chance to warm up. Should they win the job by accident, the project usually has a good profit margin.

Fourth, other companies bidding on the project are expected to be charging a higher profit and overhead markup or they have higher construction costs. When competition in the market is stiff, companies can seldom charge more than their minimum profit and overhead markup; however, when competition is meager, companies can often increase their profit and overhead markup.

To assist in determining when to charge a higher profit and overhead markup, a company must track how its competitors' prices compare to its own. The easiest way to do this is to keep a record of all of the competitors who have bid against your company along with each of their bids and the profit and overhead markup that your company would have had to add to your construction costs for your bid to equal the competitors' bids. This is done by using the following equation:

$$\text{P\&O Markup} = \frac{\text{Bid}}{\text{Construction Costs}} - 1 \quad (25\text{-}2)$$

EXAMPLE 25-4

Your construction company recently bid against ABC Construction Company. Your construction costs were \$157,260 and you added a 15% profit and overhead markup for a total bid of \$180,849. ABC's bid was \$179,249. What profit and overhead markup would you need to add to your construction costs to match ABC's bid?

Solution: The profit and overhead markup is determined as follows using Eq. (25-2):

$$\text{P\&O Markup} = \frac{\$179{,}249}{\$157{,}260} - 1 = 0.1398 \text{ or } 13.98\%$$

When tracking the competitors' bids it is important to keep track of the name of the project and the bid date as well as the competitors' bid and your construction costs. Competitors may be tracked using the simple spreadsheet format shown in Figure 25-1.

When deciding whether to increase the profit and overhead markup, you need to take into account the size of the project, type of project, and bid date. It is important to take size into account because as the project's size increases, the profit and overhead markup tends to decrease. It is important to take the type of project into account because some companies may bid different types of projects at different profit and overhead markups. Finally, it is important to take the bid date into account because projects bid early in the construction season—when companies are looking for the year's work—are often bid at a lower rate than projects later in the construction season when contractors' schedules begin to fill up.

ABBCO				
Project	Date	Bid	Costs	P&O
West City Park	6/24/2002	$ 875,256	$ 798,952	9.55%
Platt Park Restrooms	8/14/2002	$ 52,326	$ 42,165	24.10%
ABC Construction				
Project	Date	Bid	Costs	P&O
South Street Improvements	3/15/2002	$ 179,249	$ 157,260	13.98%
West City Park	6/24/2002	$ 859,462	$ 798,952	7.57%
East Side Community Center	7/22/2002	$ 1,152,634	$ 1,092,215	5.53%

FIGURE 25-1 Spreadsheet for Tracking Competitors' Bids

EXAMPLE 25-5

Your construction company is bidding against the two construction companies shown in Figure 25-1 on a municipal project with an engineer's estimate of $750,000 to $850,000. Your company's minimum profit and overhead markup is 8%. What are the chances of increasing your profit and overhead markup above the minimum 8%?

Solution: Your company bid against ABC Construction on the West City Park—a municipal project of similar size—where their bid was 7.57% above your costs. It is unlikely that you will be able to raise your profit and overhead markup and still be competitive with ABC Construction. You also bid against ABBCO on the same project, where their bid was 9.55% above your costs. If ABC Construction were not to bid, you might be able to raise your profit and overhead a little. ∎

The construction bidding market is constantly changing. Contractors who are winning work and feeling less pressure to get work may raise their profit and overhead markups, whereas other contractors who are completing projects and need more work may lower their profit and overhead markups to get this work. At the same time, other contractors may be adjusting the profit and overhead on courtesy bids or when they perceive that a project has a higher degree of risk. All of this makes it difficult to predict where contractors are going to bid. By tracking your competitors you increase your odds of reading the market right.

EXCEL QUICK TIP 25-3
Competitor-Tracking Worksheet

The competitor-tracking worksheet in Figure 25-1 is set up as a spreadsheet by entering the data and formatting the cell as follows:

	A	B	C	D	E
1	**ABBCO**				
2	Project	Date	Bid	Costs	P&O
3	West City Park	6/24/02	$ 875,256	$ 798,952	9.55%
4	Platt Park Restrooms	8/14/02	$ 52,326	$ 42,165	24.10%
5					
6					
7	**ABC Construction**				
8	Project	Date	Bid	Costs	P&O
9	South Street Improvements	3/15/02	$ 179,249	$ 157,260	13.98%
10	West City Park	6/24/02	$ 859,462	$ 798,952	7.57%
11	East Side Community Center	7/22/02	$ 1,152,634	$ 1,092,215	5.53%

The following formula needs to be entered into Cell E3:

    =(C3/D3-1)

This formula will need to be copied to all other applicable rows in Column E. In the foregoing example Cell E3 needs to be copied to Cells E4, E9, E10, and E11. Data for the bids is entered in Columns A through D. The data shown in the foregoing figure is from Figure 25-1.

CONCLUSION

Common markups include building permits, bonds, and the profit and overhead markup. The building permit may be based on the bid or a standardized cost per square foot established by the code enforcement agency. The bond is based on the bid and a bond schedule. The profit and overhead markup is added to construction costs and may be added to the cost of the building permit and bond. Estimators should track the bidding history of their competitors to identify where it may be possible to increase their profit and overhead markup.

PROBLEMS

1. Using the building permit costs in Table 25-1, determine the building permit cost for a project valued at $1,102,365.

2. Using the building permit costs in Table 25-1, determine the building permit cost for a project valued at $55,268.

3. Using the bond costs in Table 25-2, determine the bonding cost for a project bid at $1,102,365.

4. Using the bond costs in Table 25-2, determine the bonding cost for a project bid at $55,268.

5. The estimated construction cost for a project is $875,264. The company's profit and overhead markup is 8% of construction cost. Determine the profit and overhead markup.

6. The estimated construction cost for a project is $25,654. The company's profit and overhead markup is 20% of construction cost. Determine the profit and overhead markup.

7. Your construction company recently bid against ABC Construction Company. Your construction costs were $265,815, and you added an 11% profit and overhead markup for a total bid of $295,055. ABC's bid was $301,251. What profit and overhead markup would you need to add to your construction costs to match ABC's bid?

8. Your construction company recently bid against ABC Construction Company. Your construction costs were $1,125,572, and you added a 6% profit and overhead markup for a total bid of $1,193,106. ABC's bid was $1,179,999. What profit and overhead markup would you need to add to your construction costs to match ABC's bid?

9. For the purposes of determining the cost of the building permit, the local municipality values residential buildings at $120 per square foot for living space and $35 per square foot for garages. Using the building permit costs in Table 25-1, determine the building permit cost for the Johnson Residence given in Appendix G.

10. For the purposes of determining the cost of the building permit, the local municipality values retail space at $160 per square foot. Using the building permit costs in Table 25-1, determine the building permit cost for the West Street Video project given in Appendix G.

11. Set up Excel Quick Tip 25-1 in Excel.

12. Set up Excel Quick Tip 25-2 in Excel.

13. Set up Excel Quick Tip 25-3 in Excel.

REFERENCE

1. For information on how to set the profit and overhead markup, see Steven J. Peterson, *Construction Accounting and Financial Management,* Prentice Hall, Upper Saddle River, NJ.

CHAPTER TWENTY-SIX

PRICING EXTENSIONS

In this chapter you will learn how to calculate the price for each takeoff item and the total cost for the bid.

The previous chapters covered how to determine material pricing, labor productivity, labor hours, labor rates, equipment costs, crew rates, subcontractor pricing, and markups. It is now time to combine all of these concepts into preparing a total cost for the estimate. This is done by using the Detail worksheet created in the exercises in Chapter 3 and the Summary worksheet (Figure 2-2) in Chapter 2.

DETAIL WORKSHEET

The Detail worksheet is used for performing detailed estimates of the materials, labor, and equipment needed to complete the work in a cost code. Let's look at each of these costs.

MATERIAL COSTS

Material costs are calculated by multiplying material unit costs by the quantity of materials needed to get the cost of the materials, as expressed by the following equation:

Material Cost = (Material Unit Cost)(Quantity) (26-1)

The material unit cost should include shipping costs, sales tax, storage costs, and escalation. These calculations are shown in the following example.

EXAMPLE 26-1

A project requires the installation of 150 feet of water line. The work includes 70 cubic yards of excavation, 23 cubic yards of bedding to be placed around the pipe, and 42 cubic yards of compacted backfill. The cost of the pipe is $14.05 per foot, and the cost of the bedding is $23.50 per cubic yard. The sales tax rate is 5% and is paid only on materials. No materials are needed for the excavation, and the excavated materials are used for the compacted backfill. Excess materials from the excavation may be left on site. Determine the material costs for the installation of the water line.

Solution: The material costs are calculated using Eq. (19-3) as follows:

$$\text{Material Cost}_{\text{Bedding}} = (23 \text{ yd}^3)(\$23.50/\text{yd}^3) = \$541$$

$$\text{Material Cost}_{\text{Pipe}} = (150 \text{ ft})(\$14.05/\text{ft}) = \$2{,}108$$

$$\text{Material Cost}_{\text{Subtotal}} = \$541 + \$2{,}108 = \$2{,}649$$

The sales tax is calculated using Eq. (19-3) as follows:

$$\text{Sales Tax} = (\$2{,}649)(0.05) = \$132$$

The total cost is calculated as follows:

$$\text{Material Cost}_{\text{Total}} = \$2{,}649 + \$132 = \$2{,}781$$

	A	B	C	D	E	F	G	H	I	J	K	L
1				MATERIALS			LABOR			EQUIPMENT		
2	ITEM	QUANTITY		$/UNIT	COST	LHR/UNIT	LHR	$/LHR	COST	$/LHR	COST	TOTAL
3	33-100 Water Line											
4	Trench excavation	70	cyd	-	-							
5	Bedding	23	cyd	23.50	541							
6	10" water line	150	ft	14.05	2,108							
7	Backfill	42	cyd	-	-							
8	Subtotal				2,648							
9	Sales Tax				132							
10	Total				2,780							

FIGURE 26-1 Detail Worksheet for Example 26-1

This can also be done by entering the unit costs in Column D of the Detail worksheet created in Chapter 3. The material cost in Column E equals the quantity in Column B multiplied by the unit costs in Column D. These calculations are shown in Figure 26-1. The difference in the total labor cost in Example 26-1 and Figure 26-1 is because of rounding. The cells are formatted using the comma style, which replaces zeros with dashes.

Alternatively, the Request for Materials Quote form discussed in Chapter 19 could be used to calculate the materials costs.

LABOR COSTS

Labor costs are obtained by multiplying the productivity by the quantity to get the number of labor hours using Eq. (20-7). The number of labor hours is then multiplied by the average labor rate for the crew performing the task as expressed by the following equation:

$$\text{Labor Cost} = (\text{Labor Hours})(\text{Labor Rate}) \quad (26\text{-}2)$$

Labor productivity is discussed in Chapter 20, and the average labor rate for a crew is discussed in Chapter 23. These calculations are shown in the following example.

EXAMPLE 26-2

Using the average labor rate per labor hour from Example 23-1, determine the labor costs for the installation of the water line in Example 26-1. For excavation of the trench the productivity is 0.50 labor hour per cubic yard, for the placement of bedding the productivity is 0.30 labor hour per cubic yard, for the placement of the pipe the productivity is 0.26 labor hour per foot, and for the backfill of the trench the productivity is 3.20 labor hour per cubic yard.

Solution: The number of labor hours for excavation is calculated using Eq. (20-7) as follows:

$$\text{Labor Hours}_{\text{Excavation}} = (0.50 \text{ lhr/yd}^3)(70 \text{ yd}^3) = 35 \text{ lhr}$$

The labor cost for the excavation is calculated using Eq. (26-2) as follows:

$$\text{Labor Cost}_{\text{Excavation}} = (35 \text{ lhr})(\$35.94/\text{lhr}) = \$1,258$$

The labor hours and labor costs for the remaining items are calculated as follows:

$$\text{Labor Hours}_{\text{Bedding}} = (0.30 \text{ lhr/yd}^3)(23 \text{ yd}^3) = 6.9 \text{ lhr}$$
$$\text{Labor Cost}_{\text{Bedding}} = (6.9 \text{ lhr})(\$35.94/\text{lhr}) = \$248$$
$$\text{Labor Hours}_{\text{Pipe}} = (0.26 \text{ lhr/ft})(150 \text{ ft}) = 39 \text{ lhr}$$
$$\text{Labor Cost}_{\text{Pipe}} = (39 \text{ lhr})(\$35.94/\text{lhr}) = \$1,402$$
$$\text{Labor Hours}_{\text{Backfill}} = (3.20 \text{ lhr/yd}^3)(42 \text{ yd}^3) = 134.4 \text{ lhr}$$
$$\text{Labor Cost}_{\text{Backfill}} = (134.4 \text{ lhr})(\$35.94/\text{lhr}) = \$4,830$$

The total labor cost is calculated as follows:

$$\text{Labor Cost}_{\text{Total}} = \$1,258 + \$248 + \$1,402 + \$4,830$$
$$= \$7,738$$

This can also be done by entering the productivity in labor hours per unit in Column F and the average labor rate for the crew in Column H of the Detail worksheet created in Chapter 3. The labor hours in Column G equals the quantity in Column B multiplied by the productivity in Column F. The labor cost in Column I equals the labor hours in Column G multiplied by the average labor rate for the crew in Column H. These calculations are shown in Figure 26-2.

EQUIPMENT COSTS

Equipment costs are obtained by multiplying equipment cost per labor hour by the number of labor hours as expressed by the following equation:

$$\text{Equipment Cost} = (\text{Equipment Cost per Labor Hour}) \times (\text{Labor Hours}) \quad (26\text{-}3)$$

The labor hours used in this equation are the labor hours calculated from Eq. (20-7). Determining the equipment cost per labor hour is discussed in Chapter 23. These calculations are shown in the following example.

EXAMPLE 26-3

Using the equipment cost per labor hour from Example 23-1 and the labor hours from Example 26-2, determine the equipment costs for the installation of the water line in Examples 26-1 and 26-2.

Solution: The equipment costs are calculated using Eq. (26-3) as follows:

$$\text{Equipment Cost}_{\text{Excavation}} = (35 \text{ lhr})(\$50.80/\text{lhr}) = \$1,778$$
$$\text{Equipment Cost}_{\text{Bedding}} = (6.9 \text{ lhr})(\$50.80/\text{lhr}) = \$351$$
$$\text{Equipment Cost}_{\text{Pipe}} = (39 \text{ lhr})(\$50.80/\text{lhr}) = \$1,981$$
$$\text{Equipment Cost}_{\text{Backfill}} = (134.4 \text{ lhr})(\$50.80/\text{lhr}) = \$6,828$$

	A	B	C	D	E	F	G	H	I	J	K	L
1				MATERIALS			LABOR			EQUIPMENT		
2	ITEM	QUANTITY		$/UNIT	COST	LHR/UNIT	LHR	$/LHR	COST	$/LHR	COST	TOTAL
3	33-100 Water Line											
4	Trench excavation	70	cyd	-	-	0.50	35.00	35.94	1,258			
5	Bedding	23	cyd	23.50	541	0.30	6.90	35.94	248			
6	10" water line	150	ft	14.05	2,108	0.26	39.00	35.94	1,402			
7	Backfill	42	cyd	-	-	3.20	134.40	35.94	4,830			
8	Subtotal				2,648				7,738			
9	Sales Tax				132				-			
10	Total				2,780				7,738			

FIGURE 26-2 Detail Worksheet for Example 26-2

	A	B	C	D	E	F	G	H	I	J	K	L
1				MATERIALS			LABOR			EQUIPMENT		
2	ITEM	QUANTITY		$/UNIT	COST	LHR/UNIT	LHR	$/LHR	COST	$/LHR	COST	TOTAL
3	33-100 Water Line											
4	Trench excavation	70	cyd	-	-	0.50	35.00	35.94	1,258	50.80	1,778	
5	Bedding	23	cyd	23.50	541	0.30	6.90	35.94	248	50.80	351	
6	10" water line	150	ft	14.05	2,108	0.26	39.00	35.94	1,402	50.80	1,981	
7	Backfill	42	cyd	-	-	3.20	134.40	35.94	4,830	50.80	6,828	
8	Subtotal				2,648				7,738		10,937	
9	Sales Tax				132				-		-	
10	Total				2,780				7,738		10,937	

FIGURE 26-3 Detail Worksheet for Example 26-3

The total equipment cost is calculated as follows:

$$\text{Equipment Cost}_{\text{Total}} = \$1{,}778 + \$351 + \$1{,}981 + \$6{,}828$$
$$= \$10{,}938$$

This can also be done by entering the equipment cost per labor hour in Column J of the Detail worksheet created in Chapter 3. The equipment cost in Column K equals the quantity in Column G multiplied by the equipment cost per labor hour in Column J. These calculations are shown in Figure 26-3. The difference in the total equipment cost in Example 26-3 and Figure 26-3 is because of rounding.

TOTAL COST

The total cost for each bid item equals the material cost plus the labor cost plus the equipment cost, as shown in the following example.

EXAMPLE 26-4

Using the costs for Examples 26-1 through 26-3, determine the total cost for each bid item and for the installation of the water line in Examples 26-1 through 26-3.

Solution: The total cost for each bid item is calculated as follows:

$$\text{Total Cost}_{\text{Excavation}} = \$0 + \$1{,}258 + \$1{,}778 = \$3{,}036$$
$$\text{Total Cost}_{\text{Bedding}} = \$541 + \$248 + \$351 = \$1{,}140$$
$$\text{Total Cost}_{\text{Pipe}} = \$2{,}108 + \$1{,}402 + \$1{,}981 = \$5{,}491$$
$$\text{Total Cost}_{\text{Backfill}} = \$0 + \$4{,}830 + \$6{,}828 = \$11{,}658$$
$$\text{Total Cost}_{\text{Sales Tax}} = \$132 + \$0 + \$0 = \$132$$

The total cost of the water line is calculated as follows:

$$\text{Total Cost}_{\text{Total}} = \$3{,}036 + \$1{,}140 + \$5{,}491 + \$11{,}658 + \$132$$
$$= \$21{,}457$$

This can also be done by summing Columns E, I, and K of the Detail worksheet created in Chapter 3. These calculations are shown in Figure 26-4. The difference in the total cost in Example 26-4 and Figure 26-4 is because of rounding.

EXCEL QUICK TIP 26-1
Detail Worksheet

The calculations for the Detail worksheet from Chapter 3, the worksheet shown in Figure 26-4, is set up by entering the data and formatting the cells as shown in that figure. The following formulas need to be entered into the associated cells:

Cell	Formula
E4	=B4*D4
G4	=F4*B4
I4	=G4*H4
K4	=G4*J4
L4	=E4+I4+K4
L8	=SUM(L4:L7)

The formula in Cell E4 will need to be copied to all other applicable rows in Column E, which in the case of the foregoing example are Cells E5 through E7. The same is true of Columns G, I, K, and L. The formula in Cell L8 needs to be copied to Cells E8, I8, and K8. Chapter 32 looks at automating the writing of the formulas for Row 4 using a macro.

Data for the costs are entered in Columns A through D, F, H, and J.

SUMMARY WORKSHEET

The Summary worksheet is used to record costs from the Request for Material Quotes, the Detail worksheet, and subcontractor bids. The costs from the material quote sheet are entered into the materials column next to their respective cost codes. The costs from the Detail worksheet are entered into the materials, labor, and equipment columns next to their respective cost codes. Subcontract

	A	B	C	D	E	F	G	H	I	J	K	L
1				MATERIALS			LABOR			EQUIPMENT		
2	ITEM	QUANTITY		$/UNIT	COST	LHR/UNIT	LHR	$/LHR	COST	$/LHR	COST	TOTAL
3	33-100 Water Line											
4	Trench excavation	70	cyd	-	-	0.50	35.00	35.94	1,258	50.80	1,778	3,036
5	Bedding	23	cyd	23.50	541	0.30	6.90	35.94	248	50.80	351	1,139
6	10" water line	150	ft	14.05	2,108	0.26	39.00	35.94	1,402	50.80	1,981	5,490
7	Backfill	42	cyd	-	-	3.20	134.40	35.94	4,830	50.80	6,828	11,658
8	Subtotal				2,648				7,738		10,937	21,323
9	Sales Tax				132				-		-	132
10	Total				2,780				7,738		10,937	21,456

FIGURE 26-4 Detail Worksheet for Examples 26-1 through 26-3

	Code	Description	Materials	Labor	Equipment	Subcontract	Total
	32-000	EXTERIOR IMPROVEMENTS					
✓	32-100	Asphalt	-	-	-	25,256	25,256
✓	32-110	Site Concrete—Labor	-	-	-	11,999	11,999
✓	32-120	Site Concrete—Concrete	10,234	-	-	-	10,234
	32-130	Rebar					
	32-300	Fencing					
	32-310	Retaining Walls					
	32-320	Dumpster Enclosures					
	32-330	Signage					
	32-340	Outside Lighting					
	32-900	Landscaping					
	33-000	UTILITIES					
✓	33-100	Water Line	2,780	7,738	10,937	-	21,456
	33-300	Sanitary Sewer					
	33-400	Storm Drain					
	33-500	Gas Lines					
	33-700	Power Lines					
	33-800	Telephone Lines					
		SUBTOTAL	13,014	7,738	10,937	37,255	68,945

FIGURE 26-5 Summary Worksheet for Example 26-5

pricing is entered into the subcontract column next to the respective cost codes. The total of each cost code is then summed. The materials, labor, equipment, subcontract, and total cost for each cost code are then totaled to get the total construction cost for the project. The markups (building permit, bond, and profit and overhead) will need to be added to these costs.

EXAMPLE 26-5

Using the Summary worksheet from Example 3-9 and the costs from Figure 26-4, determine the construction costs for a site improvement project consisting of the water line, asphalt paving, and concrete sidewalk and curb. In addition to the cost of the water line from Examples 26-1 through 26-4, you have received a bid to complete the asphalt for $25,256 and a bid to form, pour, finish, and protect the concrete site work for $11,999. Your company will need to provide the concrete for the site work, for which you have a material quote from a local concrete supplier to supply the specified quantities of concrete for $10,234.

Solution: The Summary worksheet is filled out and the columns are totaled as shown in Figure 26-5.

Once the construction costs have been totaled, the cost of the building permit, the bonds, and the profit and overhead markup can be added to them. This is shown by the following example.

EXAMPLE 26-6

Determine the bid for the project in Example 26-5. A building permit is not required, and the cost of the bond is $1,025, which is treated as a material cost. The profit and overhead markup on materials, labor, and equipment costs is 15%, and the profit and overhead markup on subcontracted items is 5%.

Solution: The Summary worksheet is filled out and the columns are totaled as shown in Figure 26-6.

	Code	Description	Materials	Labor	Equipment	Subcontract	Total
	32-000	EXTERIOR IMPROVEMENTS					
✓	32-100	Asphalt	-	-	-	25,256	25,256
✓	32-110	Site Concrete—Labor	-	-	-	11,999	11,999
✓	32-120	Site Concrete—Concrete	10,234	-	-	-	10,234
	32-130	Rebar					
	32-300	Fencing					
	32-310	Retaining Walls					
	32-320	Dumpster Enclosures					
	32-330	Signage					
	32-340	Outside Lighting					
	32-900	Landscaping					
	33-000	UTILITIES					
✓	33-100	Water Line	2,780	7,738	10,937	-	21,456
	33-300	Sanitary Sewer					
	33-400	Storm Drain					
	33-500	Gas Lines					
	33-700	Power Lines					
	33-800	Telephone Lines					
		SUBTOTAL	13,014	7,738	10,937	37,255	68,945
		Building Permit					
✓		Bond	1,025				1,025
		SUBTOTAL	14,039	7,738	10,937	37,255	69,970
		Profit and Overhead Markup	15.0%	15.0%	15.0%	5.0%	
		Profit and Overhead	2,106	1,161	1,641	1,863	6,770
		TOTAL	16,145	8,899	12,578	39,118	76,740

FIGURE 26-6 Summary Worksheet for Example 26-6

The profit and overhead markups are calculated using Eq. (25-1) as follows:

$$P\&OMarkup_{Materials} = \frac{(\$14,039)(15\%)}{100\%} = \$2,106$$

$$P\&OMarkup_{Labor} = \frac{(\$7,738)(15\%)}{100\%} = \$1,161$$

$$P\&OMarkup_{Equipment} = \frac{(\$10,937)(15\%)}{100\%} = \$1,641$$

$$P\&OMarkup_{Subcontract} = \frac{(\$37,255)(5\%)}{100\%} = \$1,863$$

$$P\&OMarkup_{Total} = \$2,106 + \$1,161 + \$1,641 + \$1,863$$
$$= \$6,771$$

The difference in the total cost in Example 26-6 and Figure 26-6 is because of rounding.

EXCEL QUICK TIP 26-2
Summary Worksheet

The Summary worksheet in Figure 26-6 is set up as a spreadsheet by completing Example 3-9. The Summary worksheet is as follows:

	A	B	C	D	E	F	G	H
1		Code	Description	Materials	Labor	Equipment	Subcontract	Total
104		32000	**EXTERIOR IMPROVEMENTS**					
105	✓	32100	Asphalt	-	-	-	25,256	25,256
106	✓	32110	Site Concrete—Labor	-	-	-	11,999	11,999
107	✓	32120	Site Concrete—Concrete	10,234	-	-	-	10,234
108		32130	Rebar					
109		32300	Fencing					
116		33000	**UTILITIES**					
117	✓	33100	Water Line	2,780	7,738	10,937	-	21,456
118		33300	Sanitary Sewer					
119		33400	Storm Drain					
120		33500	Gas Lines					
121		33700	Power Lines					
122		33800	Telephone Lines					
123			**SUBTOTAL**	13,014	7,738	10,937	37,255	68,945
124			Building Permit					
125	✓		Bond	1,025				1,025
126			**SUBTOTAL**	14,039	7,738	10,937	37,255	69,970
127			Profit and Overhead Markup	15.0%	15.0%	15.0%	5.0%	
128			Profit and Overhead	2,106	1,161	1,641	1,863	6,770
129			**TOTAL**	16,145	8,899	12,578	39,118	76,740
130								76,740

In this figure, Rows 2 through 103 have been hidden. The following formulas need to be entered into the associated cells:

Cell	Formula
D123	=SUM(D2:D122)
H124	=SUM(D124:G124)
H125	=SUM(D125:G125)
D126	=SUM(D123:D125)
D128	=D126*D127
H128	=SUM(D128:G128)

Cell	Formula
D129	=D126+D128
H130	=SUM(D129:G129)

Cell D123 needs to be copied to Cells E123 through H123. Cell D126 needs to be copied to Cells E126 through H126. Cell D128 needs to be copied to Cells E128 through G128. Cell D129 needs to be copied to Cells E129 through H129. Cells H129 and H130 should be the same value and are used as a check for errors. The data shown in the foregoing figure is from Example 26-6 and Figure 26-6. The data is formatted using the comma style, which replaces zeros with dashes.

SAMPLE ESTIMATE: THE RESIDENTIAL GARAGE

A sample estimate for the residential garage drawings given in Appendix G is shown in Figures 26-7 and 26-8. The small differences in the numbers in Figures 26-7 and 26-8 are due to rounding. This estimate uses the quantities from Chapters 5, 8 through 11, and 15 through 17, and the burdened labor rates in Table 26-1.

TABLE 26-1   Labor Rates for Residential Garage

Cost Code	Rate($/lhr)
03-000	24.75
06-100 to 06-120	34.64
06-200 to 06-210	36.92
07-000	33.88
08-000	36.92
09-200	31.78
09-900	30.71
26-100	51.20
31-200	27.19
32-000	24.75

ITEM	QUANTITY		MATERIALS $/UNIT	MATERIALS COST	LABOR LHR/UNIT	LABOR LHR	LABOR $/LHR	LABOR COST	EQUIPMENT $/LHR	EQUIPMENT COST	TOTAL
				03-000 CONCRETE							
03-200 Rebar											
#4 × 20′ rebar (footings)	14	ea	9.09	127	-	-	24.75	-	-	-	127
7″ × 29″ L-dowels (footings)	66	ea	1.81	119	-	-	24.75	-	-	-	119
#4 × 20′ rebar (found.)	21	ea	9.09	191	-	-	24.75	-	-	-	191
18″ × 18″ L-dowels (found.)	13	ea	1.81	24	-	-	24.75	-	-	-	24
1/2″ × 10″ anchor bolt (found.)	45	ea	0.97	44	-	-	24.75	-	-	-	44
1/2″ nut (found.)	45	ea	0.39	18	-	-	24.75	-	-	-	18
1/2″ washer (found.)	45	ea	0.12	5	-	-	24.75	-	-	-	5
HPAHD22 holddowns	4	ea	12.79	51	-	-	24.75	-	-	-	51
				579		-		-		-	579
Sales Tax			6.50	38							38
				617		-		-		-	617
03-300 Footings and Foundation—Labor											
10″ high footing forms	168	ft	-	-	0.08	13.44	24.75	333	10.00	134	467
Install continuous rebar 20′	14	ea	-	-	0.25	3.50	24.75	87	10.00	35	122
Install dowel	66	ea	-	-	0.06	3.96	24.75	98	10.00	40	138
Pour footings	4.50	cyd	-	-	0.50	2.25	24.75	56	1.50	3	59
2′ high foundation forms	390	sft	-	-	0.15	58.50	24.75	1,448	15.00	878	2,325
Install continuous rebar 20′	21	ea	-	-	0.25	5.25	24.75	130	15.00	79	209
Install dowel	13	ea	-	-	0.06	0.78	24.75	19	15.00	12	31
Install anchor bolt	45	ea	-	-	0.08	3.60	24.75	89	15.00	54	143
Pour foundation	4.75	cyd	-	-	0.50	2.38	24.75	59	1.50	4	62
				-		94		2,318		1,238	3,556
03-310 Footings and Foundation—Concrete											
3500 psi concrete (footings)	4.50	cyd	95.00	428	-	-	24.75	-	-	-	428
3500 psi concrete (found.)	4.75	cyd	95.00	451	-	-	24.75	-	-	-	451
				879		-		-		-	879
Sales Tax			6.50	57							57
				936		-		-		-	936
03-320 Slab/Floor—Labor											
Pour slab	8.00	cyd	-	-	0.50	4.00	24.75	99	1.50	6	105
Finish slab	570	sft	-	-	0.01	5.70	24.75	141	1.50	9	150
				-		10		240		15	255

FIGURE 26-7a Costs for Residential Garage

CHAPTER TWENTY-SIX

ITEM	QUANTITY		MATERIALS $/UNIT	MATERIALS COST	LABOR LHR/UNIT	LABOR LHR	LABOR $/LHR	LABOR COST	EQUIPMENT $/LHR	EQUIPMENT COST	TOTAL
03-330 Slab/Floor–Concrete											
4000 psi concrete (slab)	8.00	cyd	103.00	824	–	–	24.75	–	–	–	824
				824		–		–		–	824
Sales Tax			6.50	54							54
				878		–		–		–	878
06-000 WOODS AND PLASTICS											
06-100 Rough Carpentry											
2 × 4 wall	83	ft	–	–	0.15	12.45	34.64	431	1.28	16	447
Single-hung door opening	1	ea	–	–	0.40	0.40	34.64	14	1.28	1	14
Overhead door opening	1	ea	–	–	1.50	1.50	34.64	52	1.28	2	54
Install trusses	14	ea	–	–	1.00	14.00	34.64	485	1.28	18	503
Install ledger	110	ft	–	–	0.05	5.50	34.64	191	1.28	7	198
Install fascia board	116	ft	–	–	0.06	6.96	34.64	241	1.28	9	250
Install 15″ blocking/outlooks	62	ea	–	–	0.05	3.10	34.64	107	1.28	4	111
Install ceiling blocking	48	ft	–	–	0.04	1.92	34.64	67	1.28	2	69
Install roof sheathing	829	sft	–	–	0.014	11.61	34.64	402	1.28	15	417
				–		57		1,990		74	2,063
06-110 Lumber											
Walls											
2 × 4 – 8 redwood	1	ea	6.77	7	–	–	34.64	–	–	–	7
2 × 4 – 12 redwood	4	ea	10.37	41	–	–	34.64	–	–	–	41
2 × 4 – 14 redwood	2	ea	12.20	24	–	–	34.64	–	–	–	24
Seal sill 50′	2	ea	8.77	18	–	–	34.64	–	–	–	18
Z-flashing 10′ long	9	ea	4.36	39	–	–	34.64	–	–	–	39
2 × 4 – 92 5/8″ stud	80	ea	2.48	198	–	–	34.64	–	–	–	198
2 × 6 – 8′	1	ea	5.92	6	–	–	34.64	–	–	–	6
7/16″ × 4′ × 8′ OSB	1	ea	8.85	9	–	–	34.64	–	–	–	9
3 – 1/2″ × 12″ GLB	16.5	ft	9.15	151	–	–	34.64	–	–	–	151
2 × 4 – 12	12	ea	5.25	63	–	–	34.64	–	–	–	63
2 × 4 – 14	4	ea	6.13	25	–	–	34.64	–	–	–	25
Roof											
Hurricane ties	24	ea	0.58	14	–	–	34.64	–	–	–	14
2 × 4 – 8	46	ea	2.83	130	–	–	34.64	–	–	–	130
7/16″ × 4′ × 8′ OSB	32	ea	8.85	283	–	–	34.64	–	–	–	283
Plywood clips	78	ea	0.13	10	–	–	34.64	–	–	–	10
				1,019		–		–		–	1,019
Sales Tax			6.50	66							66
				1,085		–		–		–	1,085
06-120 Trusses											
24′ 4:12 standard trusses w/18″ tails	12	ea	63.97	768	–	–	34.64	–	–	–	768
24′ 4:12 gable end trusses w/18″ tails	2	ea	69.00	138	–	–	34.64	–	–	–	138
				906		–		–		–	906
Sales Tax			6.50	59							59
				965		–		–		–	965
06-200 Finish Carpentry											
Install shaped fascia	116	ft	–	–	0.10	11.60	36.92	428	2.50	29	457
Install door casing-set	1	ea	–	–	0.50	0.50	36.92	18	1.28	1	19
				–		12		447		30	476
06-210 Wood Trim											
1 × 6 – 10′ cedar	6	ea	6.70	40	0.30	1.80	36.92	66	1.28	2	109
1 × 6 – 16′ cedar	4	ea	11.95	48	0.48	1.92	36.92	71	1.28	2	121
7′ foot casing	3	ea	3.71	11	0.21	0.63	36.92	23	1.28	1	35
				99		4		161		6	265
Sales Tax			6.50	6							6
				106		4		161		6	272
07-000 THERMAL & MOISTURE PROTECTION											
07-200 Insulation											
R-13 15″ × 32′ unfaced	16	ea	14.00	224	–	–	33.88	–	–	–	224
3 mil plastic 8′ by 12′	8	ea	9.94	80	–	–	33.88	–	–	–	80

FIGURE 26-7b Costs for Residential Garage

Pricing Extensions

ITEM	QUANTITY		MATERIALS $/UNIT	MATERIALS COST	LABOR LHR/UNIT	LABOR LHR	LABOR $/LHR	LABOR COST	EQUIPMENT $/LHR	EQUIPMENT COST	TOTAL
Install R-13 insulation	628	sft	-	-	0.007	4.40	33.88	149	-	-	149
R-25 23″ × 32′ unfaced	22	ea	30.67	675	-	-	33.88	-	-	-	675
Install R-25 insulation	598	sft	-	-	0.006	3.59	33.88	122	-	-	122
				978		8		270		-	1,249
Sales Tax			6.50	64							64
				1,042		8		270		-	1,312
07-400 Siding											
T1-11	26	ea	31.47	818	-	-	33.88	-	1.28	-	818
Install siding	792	sft	-	-	0.020	15.84	33.88	537	1.28	20	557
T1-11 sofit	6	ea	31.47	189	-	-	33.88	-	1.28	-	189
Install soffit	164	sft	-	-	0.06	9.84	33.88	333	1.28	13	346
3″ wide screen	58	ft	1.00	58	0.06	3.48	33.88	118	1.28	4	180
12 × 14 vent	2	ea	28.00	56	0.50	1.00	33.88	34	1.28	1	91
1 × 4 cedar 8′	16	ea	5.30	85	-	-	33.88	-	1.28	-	85
1 × 4 cedar 10′	13	ea	6.70	87	-	-	33.88	-	1.28	-	87
1 × 4 cedar 12′	4	ea	8.18	33	-	-	33.88	-	1.28	-	33
Cut and install trim	75	ft	-	-	0.10	7.50	33.88	254	1.28	10	264
Install trim	212	ft	-	-	0.03	6.36	33.88	215	1.28	8	224
				1,326		44		1,491		56	2,873
Sales Tax			6.50	86							86
				1,412		44		1,491		56	2,960
07-500 Roofing											
Drip edge 10′	12	ea	6.71	81	0.03	0.36	33.88	12	4.95	2	94
15 pound felt, 4 square roll	3	ea	22.96	69	1.20	3.60	33.88	122	4.95	18	209
20 yr 3-tab shingles	9	sq	30.15	271	2.00	18.00	33.88	610	4.95	89	970
				421		22		744		109	1,273
Sales Tax			6.50	27							27
				448		22		744		109	1,301
08-000 DOORS & WINDOWS											
08-100 Metal Doors											
3 − 0 × 6 − 8 6-panel metal prehung door w/4-5/8 wood jambs, 1-1/2 pair hinges, weather stripping, and threshold, drilled for deadbolt and lockset	1	ea	99.00	99	2.00	2.00	36.92	74	1.28	3	175
				99		2		74		3	175
Sales Tax			6.50	6							6
				105		2		74		3	182
08-300 Overhead Doors											
16′ × 7′ pre finished, insulated, sectional overhead door	1	ea	616.00	616	4.00	4.00	36.92	148	1.28	5	769
1/2 hp screw-drive opener	1	ea	183.00	183	1.00	1.00	36.92	37	1.28	1	221
Keyless entry	1	ea	39.97	40	1.00	1.00	36.92	37	1.28	1	78
Remotes	2	ea	-	-	-	-	36.92	-	1.28	-	-
				839		6		222		8	1,068
Sales Tax			6.50	55							55
				894		6		222		8	1,123
08-700 Hardware											
Single-keyed deadbolt	1	ea	29.92	30	1.00	1.00	36.92	37	1.28	1	68
Keyed lockset	1	ea	20.00	20	1.00	1.00	36.92	37	1.28	1	58
Floor mounted doorstop	1	ea	1.97	2	0.25	0.25	36.92	9	1.28	0	12
				52		2		83		3	138
Sales Tax			6.50	3							3
				55		2		83		3	141
09-000 FINISHES											
09-200 Drywall											
Hang and finish drywall	1238	sft	-	-	0.020	24.76	31.78	787	1.28	32	819
4′ × 8′ × 1/2″ gypsum board	1	ea	10.46	10	-	-	31.78	-	1.28	-	10
4′ × 12′ × 1/2″ gypsum board	17	ea	15.69	267	-	-	31.78	-	1.28	-	267

FIGURE 26-7c Costs for Residential Garage

ITEM	QUANTITY		MATERIALS		LABOR			EQUIPMENT		TOTAL
			$/UNIT	COST	LHR/UNIT	LHR	$/LHR	COST	$/LHR	COST	
4' × 14' × 1/2" gypsum board	9	ea	18.31	165	-	-	31.78	-	1.28	-	165
1 – 5/8" screws	14	lbs	6.47	91	-	-	31.78	-	1.28	-	91
Joint compound	5	box	12.98	65	-	-	31.78	-	1.28	-	65
Joint tape rolls	2	ea	1.75	4	-	-	31.78	-	1.28	-	4
				601		25		787		32	1,420
Sales Tax			6.50	39							39
				640		25		787		32	1,459
09-900 Paint											
Interior primer	5	gal	14.98	75	-	-	30.71	-	1.50	-	75
Interior latex	7	gal	28.97	203	-	-	30.71	-	1.50	-	203
Paint drywall w/3 coats	1238	sft	-	-	0.020	24.76	30.71	760	1.50	37	798
Exterior primer	4	gal	26.98	108	-	-	30.71	-	1.50	-	108
Exterior latex	6	gal	39.98	240	-	-	30.71	-	1.50	-	240
Paint T1-11 w/3 coats	956	sft	-	-	0.020	19.12	30.71	587	1.50	29	616
Exterior latex	0.25	gal	18.47	5	-	-	30.71	-	1.50	-	5
Paint single hung door	1	ea	-	-	2.00	2.00	30.71	61	1.50	3	64
				630		46		1,409		69	2,108
Sales Tax			6.50	41							41
				671		46		1,409		69	2,149
26-000 ELECTRICAL											
26-000 Electrical											
Meter tap	1	ea	4.95	5	-	-	51.20	-	1.28	-	5
1 – 1/4" set screw/threaded connector	1	ea	1.04	1	-	-	51.20	-	1.28	-	1
1 – 1/4" conduit	10	ft	1.47	15	-	-	51.20	-	1.28	-	15
1 – 1/4" comp./male threaded coupling	1	ea	1.40	1	-	-	51.20	-	1.28	-	1
1 – 1/4" slip/female threaded PVC cup.	1	ea	1.12	1	-	-	51.20	-	1.28	-	1
1 – 1/4" PVC 90° sweep	2	ea	1.92	4	-	-	51.20	-	1.28	-	4
1 – 1/4" PVC conduit	40	ft	0.49	20	-	-	51.20	-	1.28	-	20
Install Conduit	40	ft	-	-	0.14	5.60	51.20	287	1.28	7	294
3 ea #4 copper wire with ground	50	ft	2.07	104	0.03	1.50	51.20	77	1.28	2	182
70 amp single-phase panel w/main	1	ea	87.50	88	-	-	51.20	-	1.28	-	88
70 amp main breaker	1	ea	25.46	25	-	-	51.20	-	1.28	-	25
20 amp breaker	1	ea	3.72	4	-	-	51.20	-	1.28	-	4
20 amp GFI breaker	2	ea	35.97	72	-	-	51.20	-	1.28	-	72
6' copper ground rod	1	ea	11.28	11	-	-	51.20	-	1.28	-	11
Install panel	1	ea	-	-	8.00	8.00	51.20	410	1.28	10	420
Octagon box	3	ea	1.98	6	-	-	51.20	-	1.28	-	6
Single gang box	7	ea	0.52	4	-	-	51.20	-	1.28	-	4
Duplex outlets	5	ea	0.49	2	-	-	51.20	-	1.28	-	2
Outlet covers	4	ea	0.69	3	-	-	51.20	-	1.28	-	3
Weather proof outlet cover	1	ea	3.49	3	-	-	51.20	-	1.28	-	3
Quad-gang box	1	ea	3.95	4	-	-	51.20	-	1.28	-	4
Quad-gang switch cover	1	ea	2.98	3	-	-	51.20	-	1.28	-	3
Single-pole switches	4	ea	0.69	3	-	-	51.20	-	1.28	-	3
4-tube fluorescent fixtures	2	ea	32.97	66	-	-	51.20	-	1.28	-	66
4' fluorescent light bulbs	8	ea	2.00	16	-	-	51.20	-	1.28	-	16
Brass coach lights	3	ea	16.97	51	-	-	51.20	-	1.28	-	51
100 watt light bulbs	3	ea	1.50	5	-	-	51.20	-	1.28	-	5
250' roll 2-#12 w/ground Romex	1	ea	51.74	52	-	-	51.20	-	1.28	-	52
NM/SE cable connectors	3	ea	0.50	2	-	-	51.20	-	1.28	-	2
Electrical staples	1	box	4.39	4	-	-	51.20	-	1.28	-	4
Red wire nuts	1	bg	2.18	2	-	-	51.20	-	1.28	-	2
Install light fixture	5	ea	-	-	1.00	5.00	51.20	256	1.28	6	262
Install outlet	5	ea	-	-	0.60	3.00	51.20	154	1.28	4	157
Install Switch	4	ea	-	-	0.60	2.40	51.20	123	1.28	3	126

FIGURE 26-7d Costs for Residential Garage

ITEM	QUANTITY	MATERIALS $/UNIT	MATERIALS COST	LHR/UNIT	LABOR LHR	LABOR $/LHR	LABOR COST	EQUIPMENT $/LHR	EQUIPMENT COST	TOTAL
			575		26		1,306		33	1,913
Sales Tax		6.50	37							37
			613		26		1,306		33	1,951

31-000 Earthwork

31-200 Grading and Excavation

ITEM	QUANTITY		$/UNIT	COST	LHR/UNIT	LHR	$/LHR	COST	$/LHR	COST	TOTAL
Excavation	56	cyd	-	-	0.10	5.60	27.19	152	85.00	476	628
Backfill	19	cyd	-	-	0.20	3.80	27.19	103	85.00	323	426
Export	36	cyd	-	-	0.10	3.60	27.19	98	65.00	234	332
3/4" gravel	15	cyd	32.5	488	0.20	3.00	27.19	82	5.00	15	584
				488		16		435		1,048	1,970
Sales Tax			6.50	32							32
				519		16		435		1,048	2,002

32-000 Exterior Improvements

32-110 Site Concrete-Labor

ITEM	QUANTITY		$/UNIT	COST	LHR/UNIT	LHR	$/LHR	COST	$/LHR	COST	TOTAL
Form driveway	54	ft	-	-	0.09	4.86	24.75	120	5.00	24	145
Screed	27	ft	-	-	0.09	2.43	24.75	60	5.00	12	72
Pour driveway	5.9	cyd	-	-	0.60	3.54	24.75	88	1.50	5	93
Finish driveway	479	sft	-	-	0.03	14.37	24.75	356	1.50	22	377
Form sidewalk	37	ft	-	-	0.09	3.33	24.75	82	5.00	17	99
Pour sidewalk	1.6	cyd	-	-	0.60	0.96	24.75	24	1.50	1	25
Finish sidewalk	132	sft	-	-	0.03	3.96	24.75	98	1.50	6	104
10' expansion joint	5	ea	-	-	0.10	0.50	24.75	12	1.50	1	13
				-		34		840		88	928

32-120 Site Concrete-Concrete

ITEM	QUANTITY		$/UNIT	COST	LHR/UNIT	LHR	$/LHR	COST	$/LHR	COST	TOTAL
Driveway	5.9	cyd	105.00	620	-	-	24.75	-	-	-	620
Sidewalk	1.6	cyd	105.00	168	-	-	24.75	-	-	-	168
				788		-		-		-	788
Sales Tax			6.50	51							51
				839		-		-		-	839

32-900 Landscaping

ITEM	QUANTITY		$/UNIT	COST	LHR/UNIT	LHR	$/LHR	COST	$/LHR	COST	TOTAL
Sod Repair	291	sft	0.45	131	0.02	5.82	24.75	144	1.50	9	284
				131		6		144		9	284
Sales Tax			6.50	9							9
				139		6		144		9	292

FIGURE 26-7e Costs for Residential Garage

CHAPTER TWENTY-SIX

Code	Description	Materials	Labor	Equipment	Subcontract	Total
01-000	**GENERAL REQUIREMENTS**					
01-300	Supervision	-	1,000	-	-	1,000
01-500	Temporary Utilities	-	-	-	-	-
01-510	Temporary Phone	-	-	-	-	-
01-520	Temporary Facilities	-	-	-	-	-
01-700	Cleanup	-	-	-	-	-
02-000	**EXISTING CONDITIONS**					
02-400	Demolition	-	-	-	-	-
03-000	**CONCRETE**					
03-200	Rebar	617	-	-	-	617
03-300	Footing and Foundation—Labor	-	2,318	1,238	-	3,556
03-310	Footing and Foundation—Concrete	936	-	-	-	936
03-320	Slab/Floor—Labor	-	240	15	-	255
03-330	Slab/Floor—Concrete	878	-	-	-	878
03-340	Concrete Pump	-	-	-	-	-
03-400	Pre-cast Concrete	-	-	-	-	-
03-500	Light-weight Concrete	-	-	-	-	-
04-000	**MASONRY**					
04-200	Masonry	-	-	-	-	-
05-000	**METALS**					
05-100	Structural Steel	-	-	-	-	-
05-200	Joist and Deck	-	-	-	-	-
05-500	Metal Fabrications	-	-	-	-	-
05-900	Erection	-	-	-	-	-
06-000	**WOOD, PLASTICS, AND COMPOSITES**					
06-100	Rough Carpentry	-	1,990	74	-	2,064
06-110	Lumber	1,085	-	-	-	1,085
06-120	Trusses	965	-	-	-	965
06-200	Finish Carpentry	-	447	30	-	477
06-210	Wood Trim	106	161	6	-	273
06-400	Cabinetry and Counter Tops	-	-	-	-	-
06-410	Counter Tops	-	-	-	-	-
07-000	**THERMAL AND MOISTURE PROTECTION**					
07-100	Waterproofing	-	-	-	-	-
07-200	Insulation	1,042	270	-	-	1,312
07-210	Rigid Insulation	-	-	-	-	-
07-220	Stucco	-	-	-	-	-
07-400	Siding	1,412	1,491	56	-	2,959
07-500	Roofing	448	744	109	-	1,301
07-600	Sheet Metal	-	-	-	-	-
07-700	Roof Specialties	-	-	-	-	-
07-710	Rain Gutters	-	-	-	-	-
07-800	Fireproofing	-	-	-	-	-
07-900	Caulking and Sealants	-	-	-	-	-
08-000	**OPENINGS**					
08-100	Metal Doors and Frames	105	74	3	-	182
08-110	Wood Doors	-	-	-	-	-
08-300	Overhead Doors	894	222	8	-	1,124
08-400	Store Fronts	-	-	-	-	-
08-500	Windows	-	-	-	-	-
08-700	Hardware	55	83	3	-	141
08-800	Glass and Glazing	-	-	-	-	-

FIGURE 26-8a Summary of Costs for Residential Garage

Code	Description	Materials	Labor	Equipment	Subcontract	Total
09-000	**FINISHES**					
09-200	Drywall	640	787	32	-	1,459
09-210	Metal Studs	-	-	-	-	-
09-300	Ceramic Tile	-	-	-	-	-
09-500	Acoustical Ceilings	-	-	-	-	-
09-600	Flooring	-	-	-	-	-
09-700	Wall Coverings	-	-	-	-	-
09-900	Paint	671	1,409	69	-	2,149
10-000	**SPECIALTIES**					
10-100	Signage	-	-	-	-	-
10-200	Toilet Partitions	-	-	-	-	-
10-210	Toilet and Bath Accessories	-	-	-	-	-
10-400	Fire Extinguishers and Cabinets	-	-	-	-	-
11-000	**EQUIPMENT**					
11-300	Appliances	-	-	-	-	-
12-000	**FURNISHINGS**					
12-200	Window Treatments	-	-	-	-	-
14-000	**CONVEYING EQUIPMENT**					
14-200	Elevators	-	-	-	-	-
21-000	**FIRE SUPPRESSION**					
21-100	Fire Sprinklers	-	-	-	-	-
22-000	**PLUMBING**					
22-100	Plumbing	-	-	-	-	-
23-000	**HVAC**					
23-100	HVAC	-	-	-	-	-
26-000	**ELECTRICAL**					
26-100	Electrical	613	1,306	33	-	1,952
27-000	**COMMUNICATIONS**					
27-100	Communications	-	-	-	-	-
31-000	**EARTHWORK**					
31-100	Clearing and Grubbing	-	-	-	-	-
31-200	Grading and Excavation	519	435	1,048	-	2,002
32-000	**EXTERIOR IMPROVEMENTS**					
32-100	Asphalt	-	-	-	-	-
32-110	Site Concrete—Labor	-	840	88	-	928
32-120	Site Concrete—Concrete	839	-	-	-	839
32-130	Rebar	-	-	-	-	-
32-300	Fencing	-	-	-	-	-
32-310	Retaining Walls	-	-	-	-	-
32-320	Dumpster Enclosures	-	-	-	-	-
32-330	Signage	-	-	-	-	-
32-340	Outside Lighting	-	-	-	-	-
32-900	Landscaping	139	144	9	-	292
33-000	**UTILITIES**					
33-100	Water Line	-	-	-	-	-
33-300	Sanitary Sewer	-	-	-	-	-
33-400	Storm Drain	-	-	-	-	-
33-500	Gas Lines	-	-	-	-	-
33-700	Power Lines	-	-	-	-	-
33-800	Telephone Lines	-	-	-	-	-
	SUBTOTAL	11,964	13,961	2,821	-	28,746
	Building Permit	523				523
	Bond	559				559
	SUBTOTAL	13,046	13,961	2,821	-	29,828
	Profit and Overhead Markup	25.0%	25.0%	25.0%	25.0%	
	Profit and Overhead	3,262	3,490	705	-	7,457
	TOTAL	16,308	17,451	3,526	-	37,285

FIGURE 26-8b Summary of Costs for Residential Garage

CONCLUSION

The Detail worksheet is used for performing detailed estimates of the materials, labor, and equipment needed to complete the work in a cost code. The Summary worksheet is used to total the costs for materials quotes, the Detail worksheet, and subcontractor bids and to include the markups into the bid.

PROBLEMS

1. Prepare a detailed estimate and a total estimate for the residential plans given in Appendix G.
2. Prepare a detailed estimate and a total estimate for the West Street Video plans given in Appendix G.
3. Set up Excel Quick Tip 26-2 in Excel.

CHAPTER TWENTY-SEVEN

AVOIDING ERRORS IN ESTIMATES

In this chapter you will learn how to check the bid for errors and learn the difference between accuracy and completeness. Estimating is the lifeblood of any construction company. Produce good competitive estimates and the company thrives. Produce bad estimates and the company loses work or, worse, wins work that it cannot afford to complete. Errors are the greatest threat to producing good estimates.

Nothing can ruin a construction company faster than bad estimating. Bad estimating will either lose the company work or win work for the company that it cannot complete while making a reasonable profit. The opposite of a bad estimate is a good estimate. Good estimates have two characteristics: They are accurate and complete.

The accuracy of an estimate is a measure of how accurate or correct the numbers in the estimate are. Problems in accuracy can occur in the quantity takeoff (the quantity is wrong), productivity (too high or too low), labor rates (wrong labor rates have been used), or pricing (wrong prices have been used).

The completeness of an estimate is a measure of whether the bid has all the items needed for the project without duplicating items. Problems in completeness include failing to include all items needed to complete the project (for example, forgetting to add the cost of a bond to the bid) or duplicating items (for example, including the cost of the gas lines for the project in both the HVAC and plumbing costs).

Bad estimates are estimates that contain excessive errors in accuracy or completeness. It is impossible to eliminate all errors from bids; however, there are a number of ways errors can be minimized. Let's look at some of these ways.

LIST COST CODES

Before preparing the estimate make a list of all cost codes to be included in the estimate. This is best done by marking off the needed cost codes on a standardized form, such as the form given in Figure 2-2. If a checked item on Figure 2-2 does not have pricing, it is easy for the estimator to see that the estimate is incomplete, which helps to ensure the completeness of the estimate.

SPEND MORE TIME ON LARGE COSTS

Large costs carry more risk of error than small costs. For example, guessing the price for a large HVAC unit is riskier than guessing the price of a bathroom faucet. Being off by 10% on the cost of the HVAC unit will create a larger cost error than would being off by 300% on the cost of the faucet. Similarly, you should spend more time on small-dollar items that are used frequently in the building than on small-dollar items that are seldom used because the total cost of the frequently used items can be large. Underbidding a faucet by $10 becomes a big error when there are a few hundred faucets, as there would be in a hotel. This helps to ensure the accuracy of the estimate.

PREPARE DETAILED ESTIMATES

Broad estimates, such as bidding HVAC based on the square footage of a building, carry more risk of error than preparing a detailed estimate for the HVAC system. This is because the errors in the detailed estimate have the tendency to cancel each other out. For example, if one were to take a single item with an estimated cost of $300,000 and an estimated error of plus or minus 10% (±$30,000) and break it down into three items with an estimated cost of $100,000 each with the same estimated error of plus or minus 10% (±$10,000), on average the combined error for all three items would be plus or minus $17,320 or 5.77% provided the errors are independent of each other. This is because the errors will tend to cancel each other out. When one item is priced too high, it is likely that another item is priced too low, reducing the overall total error. Estimators should not rely on errors canceling each other out as justification to ignore known errors. Preparing detailed estimates only works when you have sufficient data to prepare the estimate. Preparing detailed estimates helps to ensure their accuracy.

MARK ITEMS COUNTED DURING THE QUANTITY TAKEOFF

One of the hardest things for an estimator to do is to remember which items have been taken off and which items still need to be taken off. On a complex set of plans it is easy to miss items unless they are marked off as they are taken off. The best way to do this is with a light-colored highlighter through which the blueprints can still be read. As items are taken off they are marked off. When all items in a detail are taken off, the detail should be marked off by marking the title of the detail with the highlighter. When all items on a page are accounted for, the page should be marked off by marking the page number with the highlighter. At the end of the estimate, all pages in the plans should be marked off, which is easy for the estimator to check. Takeoff software packages mark the items as they are taken off. Marking off items and pages ensures both completeness (none of the construction details have been left out) and accuracy (the quantities are correct).

DOUBLE-CHECK ALL TAKEOFFS

It is easy for an estimator to lose count or miscount items. For this reason it is important to double-check all takeoff quantities by performing all takeoffs at least twice until the same quantity is obtained on two separate takeoffs. Any difference in the numbers should be because of rounding. In a pinch, a quick-and-dirty takeoff can be used to double-check complex takeoffs. Double-checking takeoffs helps to ensure the accuracy of the quantity takeoffs.

INCLUDE UNITS IN CALCULATIONS

A number of common mistakes can be avoided by including the units in the calculations. They include: (1) failing to convert the quantities to the proper units, (2) using the improper conversion factor, and (3) dividing when one should be multiplying, or multiplying when one should be dividing. By including the units in the calculations, you can cancel them out and verify that you end up with the desired units. For example, if you end up with cubic feet and you needed cubic yards, you know you forgot to convert from cubic feet to cubic yards. Consider another example of how including the units can avoid an error: If an estimator mistakenly divided by 9 ft^2/yd^2 to convert from cubic feet to cubic yards, he or she would find that the units after the conversion were ft-yd^2, indicating that he or she used the wrong conversion factor. This is shown in the following equation:

$$27 \text{ ft}^3 = \frac{27 \text{ ft}^3}{(9 \text{ ft}^2/\text{yd}^2)} = 3 \text{ ft-yd}^2$$

Similarly, if the estimator was converting cubic feet to cubic yards and multiplied by 27 ft^3/yd^3 rather than dividing, he or she would end up with ft^6/yd^3, indicating that there were problems with the conversion. This is shown in the following equation:

$$27 \text{ ft}^3 = (27 \text{ ft}^3)(27 \text{ ft}^3/\text{yd}^3) = 729 \frac{\text{ft}^6}{\text{yd}^3}$$

Including the units in the calculations helps to ensure the accuracy of the estimate, as shown in the following equation:

$$27 \text{ ft}^3 = \frac{(27 \text{ ft}^3)}{(27 \text{ ft}^3/\text{yd}^3)} = 1 \text{ yd}^3$$

Including units in the calculations helps to insure the accuracy of the estimate.

AUTOMATE WITH SPREADSHEETS

Another source of error is use of the wrong formula. The best way to avoid this problem is to set up standard formulas in spreadsheets and macros (computer functions that automate processes) to be used in the bidding process rather than using hand calculations or writing new spreadsheet formulas each time. For this to be successful, the formulas, spreadsheets, and macros must be carefully tested for errors. This helps to ensure the accuracy of the estimate.

USE WELL-TESTED AND CHECKED FORMULAS

Formulas used in spreadsheets and macros must be well tested to ensure that they do not contain errors. An error in a spreadsheet can promulgate the error to a number of bids before it is found, increasing rather than decreasing the number of errors. The formulas should be tested under a variety of conditions. It is not uncommon for a formula to produce errors only in some conditions. Testing spreadsheets and formulas is discussed further in Chapter 32.

When a new formula is written that will only be used once, it is important to verify that it is free of error and that it references the correct cells. A common mistake is to insert a row for an item at the top or bottom of a group of items that are summed without having the row included in the sum. In this case, the item is left out of the bid. Testing and checking formulas is needed to ensure the accuracy and completeness of the bid.

DOUBLE-CHECK ALL CALCULATIONS

It is easy for an estimator to make a mistake when entering a number into a calculator. For this reason, it is important to double-check all calculations by entering the numbers into the calculator at least twice until you get the same answer. Some estimators like to use a different calculator when double-checking calculations. This helps to ensure the accuracy of the estimate.

PERFORM CALCULATIONS IN TWO WAYS

Where possible, perform spreadsheet calculations in two ways. For example, the total construction cost for a project can be obtained by summing the total cost for each cost code, or the total can be obtained by summing the totals for material, labor, equipment, subcontract, and other costs and then summing these totals. Performing the calculations in both of these ways should result in the same answer. When the totals are different it is an indication that there is an error in the spreadsheet and it is likely a number has been left out of one of the totals. Performing calculations in two ways helps to ensure the completeness of the estimate.

DROP THE PENNIES

Another common error is to inadvertently leave out a decimal point when adding costs consisting of dollars and cents. For this reason, pennies should be dropped from all costs where dropping the pennies will not make a significant difference on the cost of the project. Pennies should be dropped from the material cost, labor cost, equipment cost, and total columns of Figure 26-4. Pennies and fractions thereof (if desired) should be included on unit costs and labor rates because dropping the pennies on these items will have a significant effect on the cost of the work. Dropping pennies, where appropriate, helps to ensure the accuracy of the estimate.

HAVE SOMEONE REVIEW THE ESTIMATE

Before submitting an estimate it should be reviewed by a peer or boss. As a person works intensively on a project over a long period of time he or she begins to see what he or she expects to see. Often, when someone else reviews the estimate he or she will spot problems that the estimator has missed. This helps to ensure both the accuracy and the completeness of the estimate.

REVIEW EACH COST CODE AS A PERCENTAGE OF THE TOTAL COST

Another way to prevent errors is to calculate the cost for each cost code as a percentage of the total cost of the estimate and review these percentages to see whether any of the percentages are higher or lower than expected. As one does this he or she will become familiar with the expected range for each cost code (for example, electrical costs may run between 6% and 9% of the bid) and the relationships among the percentages (for example, electrical costs may be about the same percentage as the HVAC and plumbing costs combined) for the type of work the company typically bids. Reviewing the percentages does not identify errors but may point to areas in the bid where one might find errors. For example, if framing is 5% of the bid and lumber is 20% of the bid, one may suspect there is an error in either the framing or lumber costs or both. Reviewing the percentages of the bid for each cost code helps to ensure the accuracy and completeness of the bid.

CHECK UNIT COSTS FOR EACH COST CODE

Another way to prevent errors is to calculate and review the unit costs for each cost code. The unit cost can be based on the cost per square foot, cost per parking space, cost per apartment unit, or any other relevant unit of measure. As in calculating the percentages, over time one becomes familiar with the expected range for each cost code and the relationships among the costs. Reviewing the unit costs of the bid helps to ensure the accuracy and completeness of the bid.

COMPARE COSTS TO THOSE FOR ANOTHER PROJECT

Another way to prevent errors is to compare the cost of the estimate to the cost of other similar projects that have been bid to see if the costs are similar. For example, if a company recently bid a 10,000-square-foot warehouse for $195,000 and they are bidding on a 9,500-square-foot warehouse of similar design, one would expect the costs for the warehouse they are bidding to be slightly less than those for the 10,000-square-foot warehouse. If they are vastly different, either higher or lower, the estimator should check the bid carefully for errors. The costs can be compared based on the total cost or by the cost code. Comparing the costs to another project helps to ensure the accuracy and completeness of the bid.

ALLOW PLENTY OF TIME

One of the most important things an estimator can do is allow plenty of time to complete the estimate, especially on bid day. Bid days can be quite hectic. It is better to be overprepared than to scramble to fill in missing prices at the last minute. This helps to ensure both the accuracy and the completeness of the bid.

CONCLUSION

It is impossible to eliminate all errors from a bid; however, with careful procedures you can greatly reduce the number and size of errors. Errors can be reduced by listing all of the needed cost codes on a standardized form at the start of the estimating process, spending more time on the larger-cost items, preparing more detailed estimates, marking off items on the drawings as they are taken off, double-checking all takeoffs, including units in calculations, automating the estimating process with spreadsheets, using well-tested formulas, double-checking all calculations, performing calculations in two ways, dropping pennies from numbers where it will make little difference in the total bid, having someone review the estimate, reviewing each cost code as a percentage of the total cost, checking the unit cost for each cost code, comparing the bid to bids from other projects, and allowing plenty of time to complete the bid.

PROBLEMS

1. Define accuracy and completeness. What is the difference between accuracy and completeness?
2. Why is it important to use a bid summary form similar to the one shown in Figure 2-2?
3. Why should you spend more time on items with large costs than on items with small costs?
4. Why are detailed estimates more accurate than square-footage estimates?
5. Why should items on the drawings be marked off as they are taken off?
6. Why should you perform takeoffs twice?
7. Why should you include units in all calculations?
8. What are the advantages of using formulas and macros in spreadsheets?
9. Why must spreadsheet formulas be tested?
10. Why should you always double-check your calculations?
11. Why should calculations, where possible, be performed in two different ways?
12. Why should pennies be dropped from the material cost, labor cost, equipment cost, and total columns from Figure 26-4?
13. Why should the estimate be reviewed by a peer or boss?
14. Why is it important to check the percentage of the total bid for each cost code?
15. Why is it important to check the unit cost for each cost code?
16. Why should one compare the costs of the bid to the costs of other, similar projects?
17. Why is it important to allow plenty of time to prepare the bid?

PART FOUR

FINALIZING THE BID

Chapter 28   Submitting the Bid
Chapter 29   Project Buyout
Chapter 30   The Estimate as the Basis of the Schedule
Chapter 31   Ethics

In this section, you will learn how to complete and submit the bid, purchase the materials and hire subcontractors to complete the work, and incorporate data from the bid into the schedule. This section also includes a discussion of ethics as it relates to bidding.

CHAPTER TWENTY-EIGHT

SUBMITTING THE BID

In this chapter you will learn how to submit a bid using standardized bid documents and also how to write a proposal letter.

One of the most important things to do when submitting a bid is to have the bid in on time. For some owners, particularly public agencies, bids that are late, even by 1 minute, are not considered. The types of bids submitted may be divided into two broad categories: submitting the bid using standardized bid documents provided by the design professional and owner, or submitting a proposal. Let's first look at submitting the bid with standardized bid documents.

BID SUBMISSION WITH STANDARDIZED DOCUMENTS

Many owners or design professionals have a set of standardized bid documents that they use on all projects. The documents may include a bid form (such as the one shown in Figure 1-1), a schedule of values (such as the one shown in Figure 1-2), a bid bond, certifications, and so forth. When bidding using standardized bid forms, it is important to make sure that the form is completely filled out and that the pricing is accurate. For many municipalities, incomplete or missing forms result in the bid being classified as a nonresponsive bid, and the bid ends up being discarded. It would be unfortunate for the estimator to spend days preparing a bid and having the low price only to have the bid discarded because a form was missing or a line had been left blank on a form. Before submitting a bid, the estimator should carefully check the bid documents to make sure they are complete and accurate. Any blank items that must be filled in just before the bid is submitted, such as the price, should be marked with Post-It notes to minimize the possibility of

Bid Check List

Bid Form
- ☐ Are all blanks, except for the price(s), filled out?
- ☐ Are the blanks that need a price marked with a PostIt?
- ☐ Has the bid document been signed?
- ☐ Has the corporate seal, if required, been affixed?

Schedule of Values
- ☐ Are all blanks, except for the price(s), filled out?
- ☐ Are the blanks that need a price marked with a PostIt®?

Bid Bond
- ☐ Has the bond been properly filled out and signed by the surety?
- ☐ Has the bond been signed by our company's representative?
- ☐ Has the corporate seal, if required, been affixed?

Certifications
- ☐ Has the certification been properly filled out?
- ☐ Has the certifications been signed?

FIGURE 28-1 Sample Bid Submission Checklist

forgetting to fill in a blank on the day of the bid. It is also a good idea to use a checklist to make sure that nothing is forgotten. A sample checklist is shown in Figure 28-1.

WRITING A PROPOSAL

The second category is submitting a proposal. With a proposal, rather than using documents provided by the owner or the design professional, the estimator must prepare his or her own proposal. The owner may specify what needs to be included in the documents, but the

> 1802 University Circle
> Ogden, Utah 84408-1802
> May 27, 2006
>
> Mr. John Robertson
> West 10th Street Development
> 485 South Deer Born Lane
> South Ogden, Utah
>
> RE: Bid for West 10th Street Office Complex
>
> Dear Mr. Robertson:
>
> Thank you for giving us the opportunity to bid on the West 10th Street Office Complex.
>
> The cost to complete the building core and shell is $956,250. Progress payments will be billed at the end of each month and are due 25 days after receipt of the bill. The work will be completed as per the plans and specifications prepared by John Lake, AIA, and dated April 26, 2006. Included in this price is the cost of the building permit. We excluded all other fees assessed by the local municipality. This price is good for 60 days and is based upon the project being started before August 15, 2006, and will be completed in 160 calendar days.
>
> Enclosed is a copy of our company's brochure. If you have any questions please contact me at 801-626-0000.
>
> Sincerely,
>
>
> Steven Peterson, MBA, PE
>
> Enclosure
>
> CC: file

FIGURE 28-2 Sample Proposal Letter

actual format and organization of the documents is left up to the company submitting the bid. The documents may include a proposal letter, breakdown of the bid price, bid bonds, schedule, catalogue cut sheets, samples, and so forth.

One of the most important documents in the proposal is the proposal letter. The proposal letter should identify the price for the work, payment terms, the scope of the work being bid, work that is excluded from the bid, any clarifications that need to be made, the latest date the proposal may be accepted, the schedule for the start and completion of the work, suggested alternatives, severe-weather clauses, and other items that help to clarify the proposal. The proposal letter should be kept to under five or six pages. When all of this information cannot be contained in a short letter, these items should be included as attachments to the proposal. All attachments to the proposal should be referenced in the proposal letter. Let's look at the specific components of the proposal letter.

Price: The proposal letter must clearly communicate to the client the price to complete all or a specified portion of a construction project. The letter should contain a price or price range. Price ranges are often used when dealing with preliminary bids when the design of the project is still being established to indicate that the price is not a hard-and-fast number. If the price is to be broken down into a number of items, then the breakdown of the pricing should be submitted on a schedule of value form (similar to Figure 1-2) that is attached to the letter. When a schedule of values is included, the estimator must make sure that the costs on the schedule of values add up to the total bid price. Any backup to the pricing required by the owner, such as bids from subcontractors, should also be attached. In the letter in Figure 28-2, the price for the work is $956,250.

Payment Terms: Along with the price, the terms of payment (for example, retention rate and when payments are due) should be included in the letter. In the letter in Figure 28-2, the payments terms are that the owner will be billed progress payments at the end of each month, and payment is due from the owner 25 days after receipt of the bill.

Scope of Work: The proposal letter should clearly state the scope of work being bid. Often, with a proposal there is some flexibility to specify or modify the scope of work. Some owners may allow the contractor to bid on all or part of the work, may leave the specifying of materials up to the contractor, or leave design of all or part of the project to the contractor, as is the case in a design-build project. The scope of work should be prepared as discussed in Chapter 24 and should identify work that is included in the bid, work that is excluded from the bid, and any clarifications that need to be made. This may require the attachment of a sample specification (such as the one shown in Figure 28-3),

Item:	Specification:
Asphalt:	3" asphalt with 6" engineered road base
Concrete:	Footing and Foundation: 3000 psi engineered mix Floor Slabs: 3500 psi engineered mix
Rebar:	As per plans and specifications
Masonry:	4"x 2 2/3"x8" running bond brick veneer
Framing:	All wood framing to be Douglas Fir #2 or better All sheathing to be OSB
Waterproofing:	Membrane waterproofing
Insulation:	Attic: R-30 fiberglass batt Exterior Walls: R-19 fiberglass batt Interior Walls: R-11 fiberglass sound batts
Stucco:	Two coats (scratch and finish) Dryvit or Synergy synthetic stucco
Roofing:	Built-up gravel roof with 20-year warranty
Doors:	Wood Doors: Solid-core with oak veneer and hollow metal frame Glass Doors: Glass door with aluminum frame
Overhead Doors:	Ceco RSW 14" barrel with gray metal finish and insulation
Trim:	Stain-grade oak trim
Door Hardware:	Hardware Finish: Brushed aluminum Hardware Grade: Mid-grade commercial Keying: All locks to be master keyed Exterior: Keyed lockset with closer, 1 1/2 pair hinges, and spring stop Interior: Keyed lockset, 1 1/2 pair hinges and spring stop
Glass and Glazing:	Double pane, low-e glass in aluminum frame
Drywall:	5/8" type X with smooth finish with round corner bead
Carpeting:	Carpet: 32 oz. commercial grade carpet, glued down
Acoustical Ceiling:	2'x2' dropped acoustical ceiling with revealed edge and white metal grid
Ceramic Tile:	12"x12" ceramic tile to be selected by the owner
Paint:	1 coat PVA primer 2 coats latex paint with final coat rolled
Window Sills:	Ceramic tile
Bathroom Accessories:	Finish: Brushed chrome Toilet Paper Holder: Double roll, one per stall Towel Dispenser: Two per restroom Soap Dispenser: One per lavatory Grab Bars: One pair per bathroom Mirrors: Square edge with full chrome channel, 36" high and width to match vanity
Toilet Partitions:	Baked enamel
Cabinetry:	Light oak
Blinds:	2" PVC blinds with wood grained appearance
HVAC:	Furnace: Gas Air Conditioner: Central air Thermostat: Programmable thermostat Exhaust Fans: Bathroom HVAC is to be designed by the HVAC subcontractor
Plumbing:	Water Piping: Copper Waste Piping: ABS Water Heater: 50-gallon gas Bathroom Lavatory: Drop in china Bathroom Faucets: Single handle Moen Toilets: White china Hose Bibs: Two with keys
Electrical:	Outlets: White, decor Switches: White, decor Telephone Wire: Cat 5 cable Pre-wired Phone Jacks: Two per office or conference room. In open work areas allow for one for every 50 square feet of floor space Pre-wired CTV Jacks: Conference rooms Light Fixtures: 2x4 florescent

FIGURE 28-3 Sample Specifications

catalogue cut sheets, and so forth. In the letter in Figure 28-2, the work will be completed as per the plans and specification prepared by John Lake, AIA, and dated April 26, 2006. Included in the price is the cost of the building permit, and all other fees assessed by the local municipality are excluded.

Time Frame: The proposal letter should identify how long the client has to accept the proposal and the anticipated start and completion of the work. Because bids that have not been rejected can tie up the bonding capacity for a company and leave the company wondering whether they are going to do the work, a company needs to include a time frame for accepting the bid in the proposal letter. Because the time of the year the work is being done can greatly affect the price, the time frame for starting and completing the work should be included in the proposal letter. In the letter in Figure 28-2, the owner has 60 days from the date of the proposal to accept the proposal, and the proposal is based on the project being started before August 15, 2006, and being completed in 160 calendar days.

Alternatives: The proposal letter may also include a list of suggested alternatives for consideration by the owner. For example, the contractor may be able to obtain a piece of equipment that has not been approved but is similar to the equipment specified at a lower price. In this case the contractor should include the specified piece of equipment in the company's bid and offer the owner the opportunity to reduce the bid by changing to the proposed piece of equipment.

Other: The proposal should contain any severe-weather clauses or other items that help to clarify the proposal.

Contact Information: Finally, the proposal should contain the name of the person to contact and how to get a hold of him or her should the client have any questions or desire to make any changes to the proposal.

WRITING A BUSINESS LETTER

The proposal should be written in a standard business letter format. All business letters should leave the client with a good impression of the company. This is done by preparing a clearly written letter free of spelling and typographical errors and written in a standard business format. When working for a company, one should determine whether the company has a standard format or template for business letters, and if so, it should be used. The following is an overview of the parts and layout of a standard business letter.

Sender's Address: The sender's address should include the company name, street address, city, state, and zip code. The senders address may appear at the top of the letter as shown in Figure 28-2 or be included as a footer to the first page of the letter. When the letter is printed on letterhead that includes the sender's address, the sender's address is not included in the letter.

Date: The date of the letter should be the date the letter was written. When writing a letter that takes a number of days to complete, the completion date of the letter is used instead of the date the letter was started to avoid the appearance that the letter has been sitting around for a long time before someone got around to mailing it. The date should contain the month (written out completely), the day of the month, and the year. A comma should separate the day of the month and the year. The date should be directly below the sender's address, if there is one.

Inside Address: The inside address is the address of the person to whom the letter is being sent. Every letter should be addressed to a person and not just a company. The inside address should include the name, personal title (Ms., Mrs., Mr., or Dr.), and job title (for example, president). If one is not sure of the spelling of a person's name, preferred personal title, or job title, one should call his or her secretary or the company's receptionist and find out this information. When the sender's address is above the date in the letter, there should be one blank line between the date and the inside address. When the sender's address is not above the date in the letter, the inside address should be 1 inch below the date.

Subject Line: Every letter should have a subject line. The subject line allows the reader of the letter to quickly identify the main topic of the letter. A subject line is very useful when trying to locate an old letter about a certain subject when one is not sure of the date the letter was sent. During the course of a project, hundreds of letters may be written between the contractor and the owner. Without a subject line, it can take hours to locate old correspondence related to a specific topic. Subject lines often begin with RE:, short for regarding. Most often the subject line is not a complete sentence; therefore, it should not end with a period. The subject of the letter in Figure 28-2 is the submission of the bid for the West 10th Street Office Complex. There should be one blank line between the inside address and the subject line.

Salutation: The salutation is the greeting to the person to whom the letter is being written. The salutation should begin with the word Dear and end with a colon. The person who is greeted in the salutation is the person whose name appears in the inside address. Unless the writer is a close personal business acquaintance or friend, the person should be addressed by personal title and last name. If the writer is a close business acquaintance or friend, he or she may choose to address the recipient of the letter in a less formal manner, such as Dear Bob. There should be one blank line between the subject line and the salutation.

Body: The body of the letter should be written in a clear, concise, and orderly manner. For the proposal letter, the body of the letter should identify the price of the work, payment terms, and so forth. The first paragraph should contain a friendly opening, such as "Thank you for giving us the opportunity to bid on the West 10th Street Office Complex." Next, the body of the letter should state the main point of the letter. This is done in the first or second paragraph. After stating the main point, the body of the letter should continue to logically justify or clarify the main point of the letter. These justifications and clarifications should be organized in the order of importance, with the most important points first and the least important points last. Near the end of the body of the letter, the main points may be summarized, and, if desired, the writer of the letter should urge the recipient to take some specific action. Finally, the writer of the letter should identify whom to contact and how to contact him or her for answers to questions or for more information. This is very important because often letters are signed by someone other than the person who prepared the letter. There should be one blank line between the salutation and the body of the letter. The paragraphs of the letter should be single spaced with a space between the paragraphs.

Closing: The closing of a letter contains a closing statement (such as "Sincerely" or "Best regards"), a space for the signature, and the name and title of the person who signed the letter. There should be a blank line between the body of the letter and the closing statement. Only the first word of the closing statement should be capitalized, and the closing statement should be followed by a comma. Three blank lines should be provided for the signature before the name and title of the person who signs the letter. The name and title of the person who signs the letter follows the same rules as the name and title of the recipient.

Enclosures: If documents are included with the letter, the word Enclosure (for a single document) or Enclosures (for multiple documents) is included after the closing. There should be one blank line between the closing and the enclosure. If desired, the writer of the letter may list the documents enclosed with the letter.

Copies: Next, the letter should identify who received copies of the letter. This is done by typing CC: (short for carbon copies) followed by the name of the recipients. In Figure 28-2 a copy of the letter was sent to the file. There should be one blank line between the copies and the enclosure or closing if there are not any enclosures.

Writer's and Typist's Initials: If the letter was written or typed by someone other than the signer of the letter, the writer's initials and the typist's initials are often included after the copies. The writer's initials are capitalized and the typist's initials are written in lowercase. When both sets of initials are included, a colon separates the initials. There should be a blank line between the initials and the text that appears directly above the initials.

File Name and Location: With the advent of the computer, people often want to locate an electronic copy of a document. To make letters easier to find, the writer often includes the file name and file location (folder) at the end of the letter. There should be a blank line between the file name and location and the text that appears directly above it.

Letters Longer Than One Page: Two things must be done for letters longer than one page. First, the closing should not begin at the top of the page. The page breaks of the letter should be changed so that there are at least two lines of body on the page before the closing. Second, in case the pages of the letter get separated from each other, the pages need to be individually identified so they can be reunited. This is done by including a header on the second through last pages of the letter. The header should include the page number, whom the letter was written to, and the date of the letter.

LETTER FORMATS

There are three common formats for letters: block, modified block, and semiblock. In the block format, all contents in the letter are left justified and the first line of each paragraph is not indented. The letter in Figure 28-2 is written in the block format. For a letter written in the modified block format, the sender's address, date, and closing (signature block) begin at the center of the page (not centered), whereas all other items of the letter are left justified. Figure 28-4 shows a letter written in modified block format. The semiblock format is the same as the modified format except that the first line of each paragraph is indented. Figure 28-5 shows a letter written in the semiblock format.

WRITING E-MAILS

E-mail is widely used by contractors to communicate with design professionals, subcontractors, and suppliers, as well as with the company's own personnel. E-mail is

FIGURE 28-4 Modified Block Letter

> 1802 University Circle
> Ogden, Utah 84408-1802
> May 27, 2006
>
> Mr. John Robertson
> West 10th Street Development
> 485 South Deer Born Lane
> South Ogden, Utah
>
> RE: Bid for West 10th Street Office Complex
>
> Dear Mr. Robertson:
>
> Thank you for giving us the opportunity to bid on the West 10th Street Office Complex.
>
> The cost to complete the building core and shell is $956,250. Progress payments will be billed at the end of each month and are due 25 days after receipt of the bill. The work will be completed as per the plans and specifications prepared by John Lake, AIA, and dated April 26, 2006. Included in this price is the cost of the building permit. We excluded all other fees assessed by the local municipality. This price is good for 60 days and is based upon the project being started before August 15, 2006, and will be completed in 160 calendar days.
>
> Enclosed is a copy of our company's brochure. If you have any questions please contact me at 801-626-0000.
>
> Sincerely,
>
> Steven Peterson, MBA, PE
>
> Enclosure
>
> CC: file

FIGURE 28-5 Semiblock Letter

> 1802 University Circle
> Ogden, Utah 84408-1802
> May 27, 2006
>
> Mr. John Robertson
> West 10th Street Development
> 485 South Deer Born Lane
> South Ogden, Utah
>
> RE: Bid for West 10th Street Office Complex
>
> Dear Mr. Robertson:
>
>   Thank you for giving us the opportunity to bid on the West 10th Street Office Complex.
>
>   The cost to complete the building core and shell is $956,250. Progress payments will be billed at the end of each month and are due 25 days after receipt of the bill. The work will be completed as per the plans and specifications prepared by John Lake, AIA, and dated April 26, 2006. Included in this price is the cost of the building permit. We excluded all other fees assessed by the local municipality. This price is good for 60 days and is based upon the project being started before August 15, 2006, and will be completed in 160 calendar days.
>
>   Enclosed is a copy of our company's brochure. If you have any questions please contact me at 801-626-0000.
>
> Sincerely,
>
> Steven Peterson, MBA, PE
>
> Enclosure
>
> CC: file

```
From:     ABC Construction
To:       XYZ Lumber
Date:     8/22/2009  10:55 AM
Subject:  Lumber Pricing for West 10th Street Office Complex

Attached is the list of the lumber needed to construction
the West 10th Street Office Complex. Please provide
us with a quote for this lumber by 10:00 am on
September 1, 2009.

Steven Peterson, MBA, PE
Phone 801-626-0000
Fax 801-626-0001
```

FIGURE 28-6 Sample E-mail

often used to invite subcontractors to bid on a project, obtain pricing from suppliers, and submit bids to the general contractor. During both the bidding process and the construction of the project, e-mail is often used to request additional information and obtain clarifications regarding the drawings and the project manual.

It is important that e-mails leave the client, design professional, subcontractor, or supplier with a good impression of the company. As with letters, this is done by preparing a clearly written e-mail, free from spelling and typographical errors. A sample e-mail is shown in Figure 28-6. E-mails should abide by the following rules:

- E-mails should only be sent to the parties that need the information. Sending e-mails to a large number of people to cover your bases wastes their time and leaves them with a bad impression. Junk e-mails, cute stories, and other similar e-mails should not be forwarded unless you have permission from those you are sending them to.

- E-mails should have a subject line so that it can be quickly found. A common mistake is to reply to an e-mail with an e-mail that discusses a different topic without changing the subject line. If the topic changes, the subject line should change to reflect the change. If an e-mail is important enough to send, it is important enough to have a well written subject line. E-mails with poor subject lines are often placed in the trash without even being opened. In Figure 28-6, the subject of the e-mail is "Lumber Pricing for West 10th Street Office Complex."

- E-mails should be brief and to the point. They should clearly state the desired action the sender wants the recipient to take. The e-mail should not leave them wondering why it was sent. In Figure 28-6, the desired action is for the lumber supplier to provide ABC Construction a lumber quote for the West 10th Street Office Complex by 10:00 am on September 1, 2009.

- E-mails should be checked for proper spelling, punctuation, and capitalization. If your e-mail software does not check for spelling and grammar, the e-mail should be written in a word processer that does and copied to the e-mail software. E-mails should not be written using all capital letters.

- Emotion icons should not be used in business communication. Too often they do not come across as expected.

- The signature block should contain the contact information for the sender, such as his or her phone number(s) and fax number.

CONCLUSION

It is important for bids to be submitted on time because owners may reject late bids. Sometimes bids are submitted on standard bid documents that may include the bid form, a schedule of values, a bid bond, certifications, and so forth. Other times, the contractor may be required to submit a proposal that may include a proposal letter, breakdown of the bid price, bid bonds, schedule, catalogue cut sheets, samples, and so forth. The proposal letter should include the price for the work, payment terms, the scope of the work being bid, work that is excluded from the bid, any clarifications that need to be made, the latest date the proposal may be accepted, the schedule for the start and completion of the work, suggested alternatives, severe-weather clauses, and other items that help to clarify the proposal.

PROBLEMS

1. Why is it important to submit the bid on time?
2. Why is it important to make sure a complete bid package has been submitted?
3. Why should the proposal letter include the anticipated start and completion of the work?
4. Why should a letter or e-mail have a subject line?
5. Why should a letter contain the person to contact and how to get a hold of him or her should the recipient have questions?
6. Why should letters longer than one page have a header containing the page number, recipient's name, and date?
7. What is the difference between the block, modified block, and semiblock formats?
8. What are the rules for writing a good e-mail?
9. Prepare a proposal letter for the Johnson Residence given in Appendix G.
10. Prepare a proposal letter for the West Street Video project given in Appendix G.

CHAPTER TWENTY-NINE

PROJECT BUYOUT

In this chapter you will learn how to write subcontracts and purchase orders.

The buyout is the process of hiring subcontractors and procuring materials and equipment for a construction project. Subcontractors are hired by writing a subcontract between the contractor and the subcontractor. Materials and equipment are procured by issuing purchase orders or contracts to the materials and equipment suppliers. The estimator should write the subcontracts and the majority of the purchase orders before construction begins or during the early stages of construction. Let's begin by looking at subcontracts.

SUBCONTRACTS

Subcontracts are issued to subcontractors who provide labor at the construction site. These subcontractors may also provide materials and equipment. If a subcontractor does not provide labor at the construction site, a purchase order or material contract should be issued rather than a subcontract. During the buyout process, the estimator should prepare all of the subcontracts. The owner or designee should review and sign the contracts.

Many companies have standard subcontracts they use. Standard subcontracts are also available from the American Institute of Architects (AIA) and the Associated General Contractors of America (AGC). Construction companies should make sure that the subcontracts that they use meet their needs rather than assume that standard subcontracts are right for them. If contractors write their own subcontracts or make changes to a standard subcontract, the contract and changes should be reviewed by an attorney. When selecting or writing a subcontract to use, the contractor should consider the following issues:

Parties to the Contract: The subcontract is a legal agreement between the contractor and the subcontractor, or a subcontractor and a second-tier subcontractor (a subcontractor to a subcontractor). The parties to the contract should be clearly identified at the beginning of the contract. After identifying the parties, the parties to the contract are referred to as the "Contractor" and "Subcontractor." The contract should address the procedures for the assignment (the subcontracting) of part of or the entire subcontract to another party. If one or more of the parties are persons rather than corporations, what happens in the event of the death of one of the parties should be addressed.

Project Identification: The project name and location of the project should be clearly identified in the contract.

Scope of Work: The contract should include the scope of work covered by the subcontract. The scope of work is different for each subcontract. The scope of work should be written in accordance with Chapter 24 and should be tied to a set of construction documents along with their dates. This may be done by attaching a list of construction documents, such as the one given in Appendix G, to the contract and referencing the attachment in the subcontract.

Schedule of Values: The subcontract should clearly establish the price to be paid for the completion of the scope of the work. When work is done in phases, such as rough and finish electrical, the price for each phase needs to be established to avoid billing disputes during the construction process. When unit prices are included in the subcontract, the unit prices, as well as how the quantities will be determined, need to be included in the subcontract. If the contractor has the right to accept an alternative, such as constructing an additional building, during the construction of the project, the terms for acceptance and pricing need to be addressed in the contract. For example, the contractor may have the right to build a second identical building for the same price as the first building provided the construction of the second building is started within 6 months of the start of the first building.

Schedule: The subcontract should require the subcontractor to meet the schedule set by the general contractor. Often a preliminary schedule is attached to the subcontract. The number of days the subcontractor has to begin the work after receiving notice to proceed, the number of days the subcontractor has to complete the work, the minimum crew size the subcontractor agrees to maintain at the site during the period the work is being performed, and a requirement for the subcontractor to work overtime may be included in the subcontract. The subcontract should identify any coordination meetings the subcontractor is required to attend and any special coordination needed between the subcontractor and other trades. For example, if the mechanical subcontractor needs to coordinate the delivery of the roof-top units with the contractor so the contractor's crane is available to place the units, these requirements should be spelled out in the contract. The penalties for missing coordination meetings and missing deadlines should also be included in the subcontract.

Submittals: The subcontract should identify the submittals required, the process for their approval, and their due dates. If these requirements are included in the project's specifications, the specifications can simply be referenced in the subcontract.

Quality: The subcontract should identify the level of quality required along with any required testing and inspections. If these requirements are included in the project's specifications, the specifications can simply be referenced in the subcontract. The penalties for unacceptable quality and the procedures and time allowed for correcting the work need to be included in the subcontract.

Payment: The subcontract should clearly identify when and how payments will be made. The subcontract should include: (1) when invoices are due and when they will be paid, (2) whether partial or progress payment will be made versus payment on completion, (3) whether retention will be withheld from the payment, (4) whether the work is required to be billed within a specified number of days of its completion, (5) whether the payment will be held until payment is received from the owner (a paid-when-paid clause), (6) whether payment will be made for materials ordered or delivered to the site that have not been incorporated into the building, and (7) the requirements for final payment and release of retention. The subcontract should give the contractor the right to withhold payments for defective work; failure to meet contractual requirements, including providing lien waivers, insurance certificates, certified payrolls, as-built drawings, operation and maintenance manuals, and so forth; claims against the subcontractor; failure to pay for materials; and liens against the project caused by the subcontractor. The contractor should have the right to backcharge the subcontractor for damage to the work of other trades done by the subcontractor and charge the subcontractor the penalties set forth in the subcontract. The contractor should have the right to issue a joint check to the subcontractor and the subcontractor's suppliers if the contractor has concerns about the suppliers getting paid.

Change Orders: The subcontract should clearly identify the process for issuing change orders and indicate that payment will not be made for unauthorized change orders.

Insurance, License, and Bond Requirements: The subcontract should require the subcontractor to maintain specified insurance limits and name the general contractor as an additional insured on their insurance policy. It should also require the subcontractor to submit proof that they have workers' compensation insurance and a copy of their contractor's license (if licensing is required) and notify the contractor in the event either of these changes. The subcontract should identify any bonds required, and require the subcontractor to indemnify the contractor against any suits that are a result of the subcontractor's actions.

Warranty: The subcontract should identify the warranty requirements and spell out response times for different classes of warranty issues. For example, problems with heating systems should be handled within hours, whereas problems with paint should be handled within days.

Protection of Work: The subcontract should require the subcontractor to protect the work of other trades from damage caused by the subcontractor's work, protect their own work against weather, and protect their materials and equipment against theft and vandalism. It should also identify the subcontractor's rights and responsibility for using the existing or newly constructed facilities. For example, can the subcontractor park on the new parking lot? If so, what must the subcontractor do to protect the facilities?

Safety: The subcontract should require the subcontractor to comply with Occupational Safety and Health Act (OSHA) and other state and federal safety regulations, comply with the contractor's safety plan, attend or hold safety meetings, and provide Material Safety Data Sheets (MSDS). A copy of the contractor's safety plan should be attached to the subcontract.

Cleanup: The subcontract should require subcontractors to clean up after themselves on a daily basis and should identify the penalties for not cleaning up.

Termination: The subcontract should set the procedures for terminating the contract for nonperformance of the subcontractor or in the event the contractor's contract with the owner is canceled.

Dispute Resolution: The subcontract should set the procedures for dispute resolution. They may include mediation, arbitration, and going to court. The payment of attorney's fees in the event of a dispute should be addressed. If any of the parties to the contract are located in a different state than the state where the project is located, the jurisdiction under which the contract will be interpreted and where disputes will be resolved should be identified.

General Clauses: General provisions include how the contract is administered. They include: (1) stating the contractor's right to waive a provision without having to waive the provision in the future or without voiding the contract, (2) allowing unenforceable provisions thrown out by the courts to be severed from the contract, (3) describing under which state's law the contract will be enforced, (4) making sure that the entire agreement is contained within the contract, and (5) stipulating that the agreement supersedes all previous agreemes. All attachments to the contract should be identified in the contract.

PURCHASE ORDERS

Purchase orders are written to material and equipment suppliers. If these suppliers are providing labor at the construction site, a subcontract should be written rather than a purchase order. During the buyout process, the estimator should write the majority of purchase orders. The project manager and superintendent will write additional purchase orders as materials not purchased by the estimator are ordered.

The purchase orders are based on the Request for Materials Quote and should contain the following information: the number of the purchase order; the cost code to which the purchase order will be charged; the contractor's name, address, and contact information; the supplier's name, address, and contact information; the project's name, address, and contact information; the quantities, description, and pricing for the materials or equipment being ordered; delivery costs; sales tax; a total price for the purchase order; payment terms; the signature of the contractor's employee ordering the materials; and the date the purchase order was signed. A sample purchase order is shown in Figure 29-1.

PO # 1001	PURCHASE ORDER		Cost Code
To:			
From:			
Project:			
QUANTITY	DESCRIPTION	$/UNIT	COST
	SUBTOTAL		
	SALES TAX	0.00%	
	SUBTOTAL		
	DELIVERY		
	TOTAL		
Invoices received by the 25th day of the month will be paid on the 25th of the following month. Any application for payment received sixty (60) days or more after materials have been delivered shall not be honored. All invoices must be submitted to the contractor's office.			
By:		Date:	

FIGURE 29-1 Sample Purchase Order

Purchase Order Log				
Number	Project	Issued to	Description
1001			
1002			
1003			
1004			
1005			
...			

FIGURE 29-2 Sample Purchase Order Log

Purchase orders should be sequentially numbered and recorded in a purchase order log when they are issued. The purchase order log should contain the following information: the purchase order number, the project the purchase order was issued for, to whom the purchase order was issued, and a brief description of the materials or equipment ordered. A sample purchase order log is shown in Figure 29-2.

CONTRACTS FOR MATERIALS

When materials are purchased over a long period of time or annual agreements are made for materials, a contract for materials may be written rather than a purchase order. Material contracts differ from subcontracts in two ways. First, the contract may not be tied to a specific project as in the case of an annual contract for materials. Second, material suppliers do not perform work on the jobsite. Contracts for materials should be similar to subcontracts with the nonapplicable clauses eliminated. For example, material suppliers do not need to attend the weekly safety meetings.

CONCLUSION

Buyout is the process of hiring subcontractors and procuring materials and equipment for a construction project. Subcontractors are hired by writing a subcontract between the contractor and the subcontractor. Materials and equipment are procured by issuing purchase orders or contracts to the materials and equipment suppliers. The estimator should prepare the subcontracts and the majority of the purchase orders before construction begins or during the early stages of construction.

PROBLEMS

For Problems 1 through 12, obtain a copy of a subcontract used by a local contractor or a standard AIA or AGC subcontract.

1. How does the subcontract address scheduling issues?
2. How does the subcontract address quality issues?
3. What payment terms are established by the subcontract?
4. Under what circumstances can the contractor back-charge the subcontractor?
5. What are the procedures for processing a change order?
6. What are the insurance requirements?
7. What are the bonding requirements?
8. How does the subcontract address warranty issues?
9. How does the subcontract address protection of work?
10. How does the subcontract address safety?
11. Under what conditions can the subcontract be terminated?
12. What are the procedures for dispute resolution?
13. Prepare subcontracts and purchase orders for the Johnson Residence given in Appendix G.
14. Prepare subcontracts and purchase orders for the West Street Video project given in Appendix G.

CHAPTER THIRTY

THE ESTIMATE AS THE BASIS OF THE SCHEDULE

In this chapter you will learn how the data from the estimate relates to the project's schedule.

The decisions made during the estimate will affect the project's duration, and therefore, the decisions will directly affect those costs which are based on the project's duration—most notably the project's overhead. The selection of the means and methods and crew size for each task will affect the duration of the task, which will affect the duration of the project. The longer the project's duration, the higher the overhead costs. If the selected means and methods and crew sizes produces a project duration that is longer than allowed by the bid documents, the means and methods and crew size for specific tasks may need to be revised to meet the duration requirement, which will change the estimated cost of those tasks. Changing the crew size will change the average burdened labor rate, the equipment cost per labor hour, and the productivity of the crew, which will change the cost to complete the task. If overtime is needed to meet the duration requirement set forth in the bid documents, the cost of the overtime will need to be incorporated into the project's costs.

Because of this relationship between cost and project's duration, a preliminary schedule must be prepared during the bidding process to determine: (1) the estimated duration of the project and the project's overhead and (2) if the selected means and methods and crew sizes used in the estimate will meet the duration requirements set forth in the bid documents.

During the process of estimating the project, the estimator should have collected much of the data needed to prepare the schedule. For a task on the project to be completed, the proper labor, materials, and equipment must come together on the project at the same time the project is ready for the work to be done. For example, for a roof-top HVAC unit to be placed, the materials (the roof-top unit), the labor (the HVAC subcontractor), and the equipment (a crane) must come together when the building is ready for the roof-top unit. If the materials, labor, or equipment is not available or if the building is not ready for the work to be performed, the work cannot be performed. The following information from the estimate is needed to prepare the schedule.

Task Relationships: In preparing the estimate, the estimator should identify what tasks will need to be done before a task or group of tasks can be completed and what tasks must be done concurrently. Often, these relationships are very obvious; at other times they are not. For example, realizing that the footings must be placed before foundation walls and columns is obvious, whereas realizing that the HVAC unit must be placed as the trusses are placed, because there is limited access, is not obvious. It is such unobvious relationships that the estimator needs to pay particular attention to and should have identified as he or she builds the project in his or her mind.

Method of Construction: The method of construction affects the work flow. For example, if the estimator bid the project planning to reuse the foundation forms on the project four times, the schedule will need to reflect that the foundation will be poured in four phases.

Materials Required: The materials required to complete a task come from the estimate. Often, the purchase orders are written to coincide with the tasks on the schedule. For example, the purchase orders for lumber on a building may be divided into a wall package coinciding with the task of framing the walls and a roof package coinciding with the task of framing the roof.

TASK PLANNING FORM

Project:	Task:

Task Relationships:

Method of Construction:

Materials Required:

Labor Required:

Labor Hours: Crew Size: Duration:

Equipment Required:

Subcontractor Required:

Special Coordination, Inspections, and Long-Lead Items:

By:	Date:

FIGURE 30-1 Task Planning Form

Labor Required: The crew makeup and the number of labor hours to complete a task come from the estimate.

Equipment Required: The equipment required to complete a task comes from the estimate and is based on the method of construction used to determine the costs.

Subcontractor Required: The plans to use subcontractors to supply the labor for a task and possibly some of the equipment and materials are identified in the estimate.

Special Coordination, Inspections, and Long-Lead Items: In preparing the estimate, the estimator should identify special coordination, special inspections, and long-lead items that are needed to complete the project. It is vital that the estimator pass the information along to the persons responsible for preparing the schedule and completing the construction.

This information can be recorded on the form in Figure 30-1.

ESTIMATING DURATIONS

The estimated duration of a task should be related to the number of labor hours required to complete the task and the size of the crew assigned to it. The duration is estimated by dividing the number of labor hours from the estimate by the crew size, using the following equation:

$$\text{Duration} = \frac{\text{(Labor Hours)}}{\text{(Crew Size)}} \quad (30\text{-}1)$$

The schedule duration may need to be longer to reflect curing time and other time it takes to complete a task where labor is not working on the project. The calculation of duration is shown in the following example.

EXAMPLE 30-1

It is estimated it will take 475 labor hours to frame the interior partitions of an office building. How long will it take a five-person crew to complete the framing if they work 8 hours per day?

Solution: The duration is calculated using Eq. (30-1) as follows:

$$\text{Duration} = \frac{(475 \text{ lhr})}{(5 \text{ persons})} = (95 \text{ hr})\left(\frac{1 \text{ day}}{8 \text{ hr}}\right) = 12 \text{ days}$$

SAMPLE DURATIONS: THE RESIDENTIAL GARAGE

The estimated crew sizes and the durations for the residential garage drawings given in Appendix G are shown in Table 30-1. The labor hours are from Figure 26-7. These durations are based upon an 8-hour day and are rounded up to the next full day.

TABLE 30-1 Durations for the Residential Garage

	Labor Hours	Crew Size	Duration
Footings and foundation	94	3	4
Slab/floor	10	2	1
Rough carpentry	57	3	3
Finish carpentry	12	1	2
Wood trim	4	1	1
Insulation	8	1	1
Siding	44	2	3
Roofing	22	2	2
Metal doors	2	1	1
Overhead doors	6	2	1
Hardware	2	1	1
Drywall	25	2	2
Paint	46	2	3
Electrical	26	2	2
Grading and excavation	16	1	2
Site concrete	34	2	3
Landscaping	6	1	1

CONCLUSION

The estimate forms the basis of the schedule. The following information is obtained from the estimate: the order in which tasks need to take place; the method of construction; the required materials, labor, equipment, and subcontractors; special coordination and inspections; and long-lead items. The duration is estimated by dividing the number of labor hours from the estimate by the crew size.

PROBLEMS

1. It is estimated that it will take 265 labor hours to frame the interior partitions of an office building. How long will it take a three-person crew to complete the framing if they work 8 hours per day?

2. It is estimated that it will take 88 labor hours to set the tile in a residence. How long will it take a two-person crew to complete the tile if they work 9 hours per day?

3. Prepare task-planning forms for the residence given in Appendix G.

4. Prepare task-planning forms for the West Street Video project given in Appendix G.

CHAPTER THIRTY-ONE

ETHICS

In this chapter you will be introduced to ethics as it relates to estimating.

Ethics as it relates to estimating may be defined by two broad definitions: (1) a set of principles or rules that members of a group agree to abide by and (2) a set of values or a guiding philosophy of the individual.

Professional groups, such as the American Society of Professional Estimators,[1] have established a set of ethics by which their members agree to abide. These ethical principles, also known as canons, are shown in Figure 31-1. Principles such as these set minimum standards of behavior for members of the profession. Members of the industry help to improve the image and trustworthiness of the profession by following these standards.

Besides the ethical principles established by professional groups, each person has his or her own set of ethical values or guiding philosophy that affects how he or she behaves in a given situation. Let's look at a few of these.

WORK ETHIC

Work ethic relates to issues of how the estimator views the work he or she does and includes: (1) the completeness and accuracy of bids he or she prepares, (2) the quality of his or her work, (3) the commitment to expanding his or her skills, and (4) how hard he or she works.

Each estimator should commit to preparing an estimate that is as complete and as accurate as permitted by the time allowed and the quality of the construction documents. When there is insufficient time to prepare a complete and accurate bid for a project, the estimator has the duty to discuss with his or her supervisor the situation and how the time and document limitations will affect the bid.

Each estimator should commit to preparing work that positively reflects on the estimating profession. Work should be neat, with spelling and grammatical errors kept to a minimum. Estimates should be carefully documented and kept in an organized manner.

Each estimator should commit to expanding his or her area of expertise and keeping current with changes in the industry. This is done by attending classes and conferences, reading trade journals and publications, discussing estimating issues with other members of the profession, and discussing advances in materials and construction methods with field personnel, subcontractors, and suppliers.

Each estimator should commit to spending the time he or she is at work to promoting the goals of the company. In short, the employer should expect that the person will work hard to complete his or her duties as an estimator.

BIDDING PRACTICES

Bidding practices relate to how the estimator handles bids. When bids are handled in a manner that is fair to the client and all subcontractors and suppliers, relationships of trust are established among the parties. This trust results in better working relationships among the parties and often results in obtaining more work. When bids are handled in an unfair manner, or when it appears that bids were handled in an unfair manner even if they were handled fairly, mistrust is created among parties, and working relations are strained.

> **The American Society of Professional Estimators Code of Ethics**
>
> **Canon #1** – Professional estimators and those in training shall perform services in areas of their discipline and competence.
>
> **Canon #2** – Professional estimators and those in training shall continue to expand their professional capabilities through continuing education programs to better enable them to serve clients, employers and the industry.
>
> **Canon #3** – Professional estimators and those in training shall conduct themselves in a manner that will promote cooperation and good relations among members of our profession and those directly related to our profession.
>
> **Canon #4** – Professional estimators and those in training shall safeguard and keep in confidence all knowledge of the business affairs and technical procedures of an employer or client.
>
> **Canon #5** – Professional estimators and those in training shall conduct themselves with integrity at all times and not knowingly or willingly enter into agreements that violate the laws of the United States of America or of the states in which they practice. They shall establish guidelines for setting forth prices and receiving quotations that are fair and equitable to all parties.
>
> **Canon #6** – Professional estimators and those in training shall utilize their education, years of experience and acquired skills in the preparation of each estimate or assignment with full commitment to make each estimate or assignment as detailed and accurate as their talents and abilities allow.
>
> **Canon #7** – Professional estimators and those in training shall not engage in the practice of bid peddling as defined by this code. This is a breach of moral and ethical standards, and a member of this society shall not enter into this practice.
>
> **Canon #8** – Professional estimators and those in training to be estimators shall not enter into any agreement that may be considered acts of collusion or conspiracy (bid rigging) with the implied or express purpose of defrauding clients. Acts of this type are in direct violation of the code of ethics of the American Society of Professional Estimators.
>
> **Canon #9** – Professional estimators and those in training to be estimators shall not participate in acts such as the giving or receiving of gifts, which are intended to be or may be construed as being acts of bribery.
>
> Revision May, 2011

FIGURE 31-1 ASPE Code of Ethics (Courtesy of American Society of Professional Estimators)

The best way to ensure that bids are handled in a fair manner is to establish a policy regarding how bids are handled and follow it. In addition, the criteria for bid selection should be made available to all subcontractors and suppliers so they have an understanding of what they must do to be the selected bidder. The practices of bid rigging and bid shopping must be avoided.

LOYALTY TO EMPLOYER

Each estimator has the duty to be loyal to his or her employer, as long as the employer is behaving in an ethical manner. Estimators are not obligated to support or be involved in illegal or unethical behavior in the name of loyalty. Being loyal to an employer includes keeping the employer's proprietary information (such as bidding practices and historical cost data) confidential and making decisions that are in the best interest of the employer rather than of oneself.

When there is a potential conflict of interest that may lead to the estimator making a decision that is not in the best interest of the employer (for example, the estimator is purchasing materials from a family member), the estimator has the duty to disclose this conflict of interest and allow his or her work and decisions to be reviewed by someone else in the company. Conflicts of interest occur whenever an estimator does business with a family relation or a business he or she has an interest in or when he or she accepts gifts or favors from a company with which the estimator's company does business. It is not unethical to be in a situation in which a conflict of interest occurs provided everything is handled in an open, honest manner and the company's interests are held above the interest of the estimator. For the conflict of interest to be handled in an open manner, the potential conflict must be disclosed.

Employers should be able to expect that their employees are honest. Employers who expect an employee to steal, cheat, and lie for them should also expect the employee to steal from, lie to, and cheat them. Honesty is truly the best policy.

ETHICAL DILEMMAS

Ethical dilemmas occur when the company's and one's personal ethics conflict or when one's personal ethics as they relate to a specific situation are in conflict. Each person must decide how he or she will act when confronted with an ethical dilemma.

CONCLUSION

Ethics is both a set of principles or rules that members of a group agree to abide by and a set of values or a guiding philosophy of the individual. The personal ethics of the estimator include his or her work ethic (how he or she views the work), bidding practices (how bids are handled), and loyalty to the employer.

PROBLEM

1. A long-time supplier or subcontractor asks you to share a competitor's pricing or quantities before he or she submits a bid or before the bid closes. What do you do?

2. A long-time supplier or subcontractor asks you what price they have to beat to get the work. What do you do?
3. A salesperson offers you a gift in order to influence your purchasing decision. What do you do?
4. A long-time subcontractor is the second-lowest bidder, whereas an unknown subcontractor is the lowest bidder. Whom do you choose? Why?
5. A long-time subcontractor is the second-lowest bidder, whereas an unknown subcontractor is the lowest bidder. Your boss asks you to call the long-time subcontractor and ask him or her to meet the price of the lowest bidder. What do you do? How would having a company policy that established the procedure for selecting subcontractors affect your decision?
6. You are responsible for the budget on a cost-plus construction project. A supplier or subcontractor offers to overbill the client and split the difference with your company. What do you do?
7. You work for a subcontractor and are responsible for preparing the subcontractor's bids. A general contractor has a small project and asks you how much you would charge to do the work on the side. What do you do?

REFERENCE

1. For more information on the American Society of Professional Estimators, visit www.ASPEnational.org

PART FIVE

ADVANCED ESTIMATING

Chapter 32 Converting Existing Forms
Chapter 33 Creating New Forms
Chapter 34 Other Estimating Methods

In this section, you will learn how to take advantage of the full features of Excel, including how to convert existing forms to Excel, create new forms in Excel, prepare proposals using Excel, and incorporate error checking in your spreadsheets. In the final chapter, you will learn to prepare estimates using the project comparison, square-foot, and assembly estimating methods, which may be used to check a detailed estimate or to prepare an estimate during the design process.

CHAPTER THIRTY-TWO

CONVERTING EXISTING FORMS

In this chapter you will learn the procedures for converting existing paper forms to spreadsheets, how to automate a series of commands or steps using macros, and three methods of adding error protections to a worksheet. The chapter will take you step-by-step through the process of converting two paper forms (the Detail worksheet from Figure 3-25 and the Summary worksheet from Figure 2-2) *into computer spreadsheets and the creation of a macro.*

The easiest way to create a computerized estimating form is to create it from an existing paper form. This is because the user already has established a layout for the form and understands how to perform the calculations. There are five steps needed to convert existing forms. First, create the form's layout in Excel. Second, add the formulas into Excel. Third, automate the spreadsheet with macros. Fourth, test the spreadsheet to ensure that it is free of errors. Fifth, add protection against errors. The process is shown in Figure 32-1.

As a reminder, the following conventions are used in this book to describe actions that must be performed to set up a worksheet or change the settings for a worksheet:

Clicking, unless otherwise noted, means left clicking.
Bold words in the text represent buttons, tabs, icons, radio buttons, items in dropdown boxes, and check boxes that must be selected and keys that must be typed to complete the exercises.
Italic words in the text represent text or numbers that must be entered in a spreadsheet cell or text box to complete the exercises.
The <> brackets signify that the user must enter data represented by the text between the brackets, such as a file name. The appropriate data rather than the text should be entered between the brackets. For example, if you are instructed to type <your name> and your name is John Smith, you would type *John Smith*.
The shorthand notation "select **File** > **Alignment** > **Center** button" means that the user should select the **Center** Button from the **Alignment** Group from the **File** tab.
The shorthand "**Ctrl + N**" means that the user should type the **N** key while holding down the **Ctrl** key.

CREATING THE LAYOUT

The first step is to create the form's layout in Excel. For both the Detail and the Summary worksheets, this was done in Chapter 3 and they were saved as Chapter 3.xlsx in the Excel Exercise folder. If you have not completed Exercises 3-1 through 3-5, 3-8, and 3-9 from Chapter 3, you should do so now.

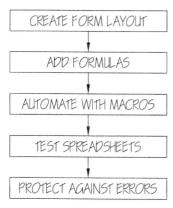

FIGURE 32-1 Process for Converting Forms to Excel

ADDING FORMULAS

The second step is to add formulas into the Excel worksheet. If you have completed Exercise 3-7 and the sidebars found in some of the preceding chapters, you have already had an opportunity to create some formulas. In the next exercise, you will add formulas to the Summary worksheet created in Chapter 3. The formulas for the Detail worksheet are added using macros. Macros are discussed in the next section.

Exercise 32-1

In this exercise you will add formulas to the Summary worksheet created in Chapter 3. Many of the formulas discussed in the individual steps of this exercise are shown in Figure 32-2.

The formulas are set up using the following steps:

1. Open Chapter 3, the workbook you created in Chapter 3.
2. Save the workbook as Chapter 32.xlsx by selecting the **File** tab, clicking on **Save As** in the left pane, double clicking on **This PC** to bring up the Save As dialogue box, selecting the **Excel Exercises** folder in the left pane, typing *Chapter 32* in the File Name text box, and clicking on the **Save** button.
3. Make sure the spreadsheet is open to the Summary worksheet. If not, click on the **Summary** tab to open the Summary worksheet.

Next, total the rows using the following steps:

4. Type =*Sum(D3:G3)* in Cell H3 and press the **Enter** key to sum Row 3. A dash will appear indicating that the sum is zero.
5. Copy Cell H3 to Cells H4 through H7, H10, H13 through H20, H23, H26 through H29, H32 through H38, H41 through H51, H54 through H60, H62 through H68, H71 through H74, H77, H80, H83, H86, H89, H92, H95, H98, H101, H102, H105 through H114, H117 through H122, H124, H125, and H128 to sum the total for each of these rows.
6. When you copied Cell H3 to Cells H122, H125, and H128, the borders under these cells were erased. Underline Cells H122, H125, and H128 by highlighting the cells and clicking the **Home** > **Font** > **Bottom Border** button (shown in Figure 3-47). If another border is shown in the Borders button, select the correct border by clicking on the small arrow to the right of the Borders button and selecting the **Bottom Border** from the Borders popup menu.

Next, enter the formulas used to calculate the subtotals found in Row 123 using the following steps:

7. Type =*Sum(D2:D122)* in Cell D123 and press the **Enter** key to sum the materials column.
8. Copy Cell D123 to Cells E123 through H123 to sum the labor, equipment, subcontract, and total columns.

	B	C	D	E	F	G	H
1	Code	Description	Materials	Labor	Equipment	Subcontract	Total
2	01-000	**GENERAL REQUIREMENTS**					
3	01-300	Supervision					=SUM(D3:G3)
4	01-500	Temporary Utilities					=SUM(D4:G4)
5	01-510	Temporary Phone					=SUM(D5:G5)
6	01-520	Temporary Facilities					=SUM(D6:G6)
7	01-700	Clean-Up					=SUM(D7:G7)
9	02-000	**EXISTING CONDITIONS**					
10	02-400	Demolition					=SUM(D10:G10)
12	33-000	**UTILITIES**					
13	33-100	Water Line					=SUM(D117:G117)
14	33-300	Sanitary Sewer					=SUM(D118:G118)
15	33-400	Storm Drain					=SUM(D119:G119)
16	33-500	Gas Lines					=SUM(D120:G120)
17	33-700	Power Lines					=SUM(D121:G121)
18	33-800	Telephone Lines					=SUM(D122:G122)
19		**SUBTOTAL**	=SUM(D2:D122)	=SUM(E2:E122)	=SUM(F2:F122)	=SUM(G2:G122)	=SUM(H2:H122)
20		Building Permit					=SUM(D124:G124)
21		Bond					=SUM(D125:G125)
22		**SUBTOTAL**	=SUM(D123:D125)	=SUM(E123:E125)	=SUM(F123:F125)	=SUM(G123:G125)	=SUM(H123:H125)
23		Profit and Overhead Markup					
24		Profit and Overhead	=D126*D127	=E126*E127	=F126*F127	=G126*G127	=SUM(D128:G128)
25		**TOTAL**	=D126+D128	=E126+E128	=F126+F128	=G126+G128	H126+H128
26							=SUM(D129:G129)

FIGURE 32-2 Formulas for Summary Worksheet

Next, enter the formulas used to calculate the subtotals found in Row 126 using the following steps:

9. Type =*Sum(D123:D125)* in Cell D126 and press the **Enter** key to calculate the subtotal.
10. Copy Cell D126 to Cells E126 through H126 to calculate the subtotal for the labor, equipment, subcontract, and total columns.

Next, enter the formulas used to calculate the profit and overhead markup found in Row 128 using the following steps:

11. Type =*D126*D127* in Cell D128 and press the **Enter** key to calculate the profit and overhead for the materials column.
12. Copy Cell D128 to Cells E128 through G128 to calculate the profit and overhead for the labor, equipment, and subcontract columns. Do not copy Cell D128 to Cell H128.

Next, you will enter the formulas used to calculate the total found in Row 129 using the following steps:

13. Type =*D126+D128* in Cell D129 and press the **Enter** key to calculate the total for the materials column.
14. Copy Cell D129 to Cells E129 through H129 to calculate the total for the labor, equipment, subcontract, and total columns.
15. As a check of the totals, type =*Sum(D129:G129)* in Cell H130 and press the **Enter** key to calculate the total. Cells H129 and H130 should contain the same value.
16. Save the workbook.

Chapter 25 looked at basing the building permit and bond costs on the total estimate of the bid using iteration. Because the building permit and bond costs are based on the total estimate and the total estimate is based on the building permit and bond costs, a circular reference is created. To solve a circular reference, Excel must be allowed to perform the calculations a number of times, each time getting closer to the correct answer. This is known as iteration. To allow Excel to use iteration to solve circular references, the user clicks on the **File** tab, clicks on **Options** to bring up the Excel Options dialogue box, selects **Formulas** in the left pane to display the Change options related to formulas calculation, performance, and error handling in the right pane (shown in Figure 32-3), and under the Calculation options heading checks the **Enable iterative calculation** check box. The maximum number of iterations is set in the Maximum iterations: text box and the maximum change in the Maximum change: text box. For example, if the maximum iterations is set to five and the maximum change is set to one, Excel will stop performing calculations after five iterations or after the change in values resulting from the last iteration was less than one, whichever comes first. For use on the Summary worksheet, five iterations and a maximum change of one is generally adequate for bids under $1,000,000. Setting the number of iterations higher and the maximum change lower will result in longer calculation times. In the next exercise you will incorporate the bond cost into the Summary worksheet.

Exercise 32-2

In this exercise you will add the bond rates from Table 25-2 to the Summary worksheet modified in exercise 32-1. To simplify the calculations, Table 25-2 is rewritten as shown in Table 32-1.

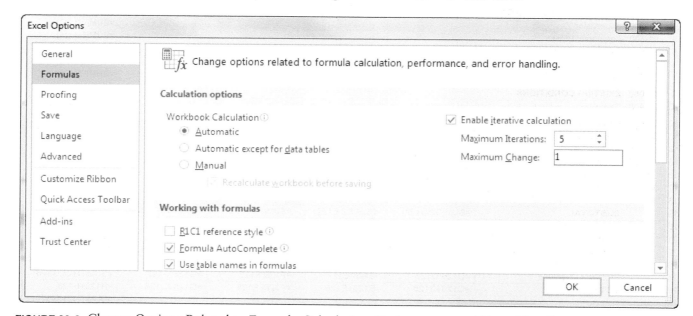

FIGURE 32-3 Change Options Related to Formula Calculation, Performance, and Error Handling

TABLE 32-1 Sample Bond Schedule

Total Estimate ($)	Bond Rate
0 to 50,000	1.50
50,001 to 100,000	1.25% on amount over $50,000 plus $750
100,001 to 250,000	1.00% on amount over $100,000 plus $1,375
250,001 to 500,000	0.90% on amount over $250,000 plus $2,875
500,001 to 1,000,000	0.80% on amount over $500,000 plus $5,125
1,000,001 plus	0.75% on amount over $1,000,000 plus $9,125

The table was generated using Table 25-2 and calculating the bonding costs for the upper end of the previous range. For example, a $500,000 bond costs $5,125 ($50,000 × 0.015 + $50,000 × 0.0125 + $150,000 × 0.01 + $250,000 × 0.009); therefore, the cost of a bond in the range of $500,001 to $1,000,000 is 0.80% on any amount over $500,000 plus $5,125 (the rate on the first $500,000).

This exercise is completed by the following steps:

1. Make sure that Chapter 32, the workbook created in Exercise 32-1, is open.
2. Make sure that the spreadsheet is open to the Summary worksheet.

Next, Excel needs to be set to allow it to perform iterations using the following step:

3. Click on the **File** tab, click on **Options** to bring up the Excel Options dialogue box, select **Formulas** in the left pane to display the Change options related to formula calculation, performance, and error handling in the right pane, under the Calculation options heading check the **Enable iterative calculation** check box, type 5 in the Maximum iterations: text box, type 1 in the Maximum change: text box, and click the **OK** button to close the Options dialogue box. This setting will need to be changed on any computer that runs this worksheet or when you get a circular reference warning.

The logic used to calculate the bond is shown in Figure 32-4.

Next, you will enter the bond formula as follows:

4. Type the following formula into Cell D125:
   =IF(H129<50000,H129*0.015,
   IF(H129<100000,(H129-50000)*0.0125+750,
   IF(H129<250000,(H129-100000)*0.01+1375,
   IF(H129<500000,(H129-250000)*0.009+2875,
   IF(H129<1000000,(H129-500000)*0.008+5125,
   (H129-1000000)*0.0075+9125)))))

5. Save the workbook.

AUTOMATING WITH MACROS

The third step is to automate the spreadsheets with macros. As we saw in Chapter 26 "Pricing Extensions," the following mathematical relationships exist on the Detail worksheet:

Column E = Column B × Column D
Column G = Column B × Column F
Column I = Column G × Column H
Column K = Column G × Column J
Column L = Column E + Column I + Column K

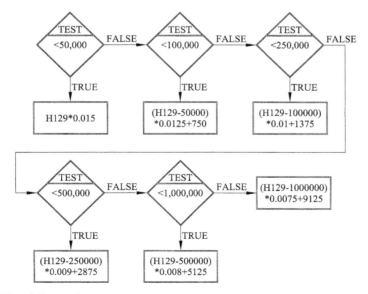

FIGURE 32-4 Logic for the Bond Formula

Each time the user adds a row to the Detail worksheet, he or she will need to add these formulas to Columns E, G, I, K, and L. Not only would this be time consuming, but would also require the user to test each formula. Alternatively, the user could copy the formulas from one row to another row. Although this is less time consuming, it is still slower than using a macro and is prone to errors. Macros greatly reduce the possibility of errors and increase efficiency.

When the user records a macro, he or she creates an executable code that can be infected by computer viruses. Running a macro with a virus can spread the virus and its damage to other files on the user's computer. To minimize the risk of infection, the user should only run macros he or she creates or macros from a trusted source.

Excel allows the user to set the security level for macros so that Excel: (1) disables all macros without notifying the user, (2) disables all macros but notifies the user and gives him or her the option to run the macro, (3) disables all macros except those digitally signed, or (4) enables all macros. When Excel is set to disable all macros with notification given to the user, a Security Warning indicating that the macros have been disabled appears below the menu ribbon, as shown in Figure 32-5. To enable the macro, the user clicks on **Enable Content**. Once you have enabled a document, Excel will remember it and you will not have to enable it again.

To set the level of security, the user clicks on the **File** tab, clicks on **Options** to bring up the Excel Options dialogue box, selects **Trust Center** in the left pane to display the Help keep your documents safe and your computer secure and healthy in the right pane, and under the Microsoft Office Excel Trust Center clicks on the **Trust Center Settings . . .** button to bring up the Trust Center dialogue box. From the Trust Center dialogue box, the user clicks on **Macro Settings** in the left pane to view the macro settings in the right pane, where he or she can select to: (1) Disable all macros without notification, (2) Disable all macros with notification, (3) Disable all macros except digitally signed macros, or (4) Enable all macros (not recommended; potentially dangerous code can run). The Trust Center dialogue box is shown in Figure 32-6.

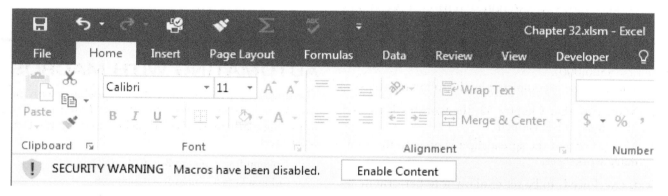

FIGURE 32-5 Excel Security Warning

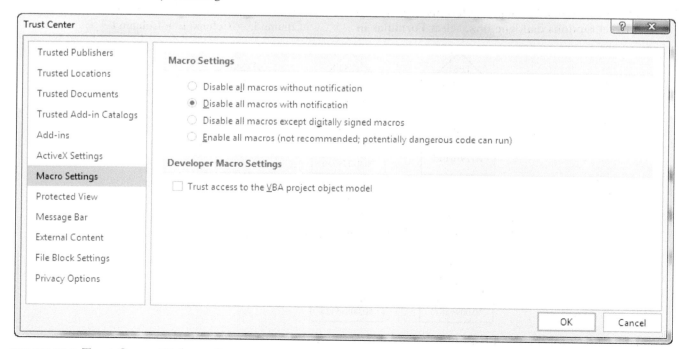

FIGURE 32-6 Trust Center Dialogue Box

The user closes the Trust Center dialogue box by clicking on the **OK** button.

Macros are written in a complex programming language; however, macros can be recorded as you would record your voice on a tape recorder, eliminating the need to learn a programming language. When recording a macro, you record the actions (key strokes, cursor movements, and so forth) taken while recording the macro. These actions can be repeated by playing back the macro.

The actions recorded by macros can be absolute or relative. Absolute macros are always executed in the same location on the spreadsheet, regardless of the location of the cursor when the macro is executed. For example, if an absolute macro was recorded starting in Cell A1, it will always move to Cell A1 before executing the macro. Relative macros are executed relative to the current position of the cursor. For example, if a relative macro was recorded with a start point in Cell A1 and then immediately moved the cursor to Cell A2, whenever the macro was executed it would start by moving the cursor one cell below the active cell. If the active cell were G17, the macro would move the active cell to G18. To select between absolute and relative references, the user clicks on the **View** menu tab, clicks on the arrow below the Macros button, and selects **Use Relative References** from the popup menu (shown in Figure 32-7).

When the Use Relative References is shaded, macros will be recorded with relative references. Relative references are turned off the same way they are turned on.

To record a macro, the user clicks on the **View** menu tab, clicks on the arrow below the Macros button, and selects **Record Macro...** from the popup menu (shown in Figure 32-8) to bring up the Record Macro dialogue box shown in Figure 32-9. The name of the macro is typed into the Macro name: text box. Each macro must have a unique name. Spaces may not be used in the name. A shortcut key can be assigned to a macro by typing a key into the Shortcut key: Ctrl+ text box. The macro can be stored in one of three places—this workbook (the workbook you are working on), a new workbook, or a personal macro workbook—by selecting the place to store the macro in the Store macro in: dropdown box. Macros that are used for a single workbook should be stored in the workbook, whereas macros that are used in multiple workbooks, such as inserting your name and title, should be stored in a personal workbook. A description of the macro is entered in the Description text box.

After completing the information on the record dialogue box, the user begins recording the macro by clicking the **OK** button. The user then enters the keystrokes, cursor movements, and so forth that are to be recorded. The user stops the recording of the macro by clicking

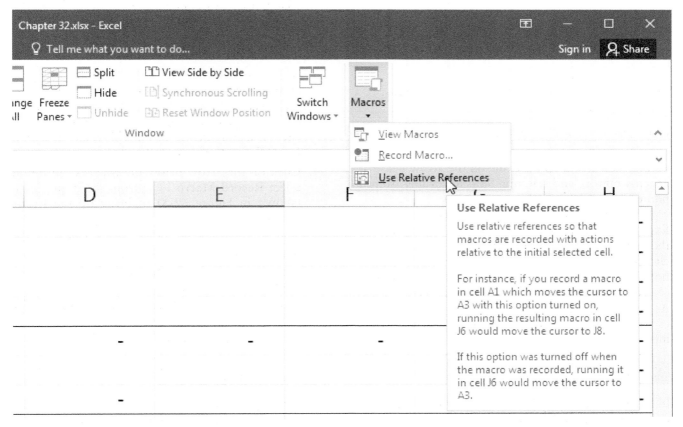

FIGURE 32-7 Use Relative References Button

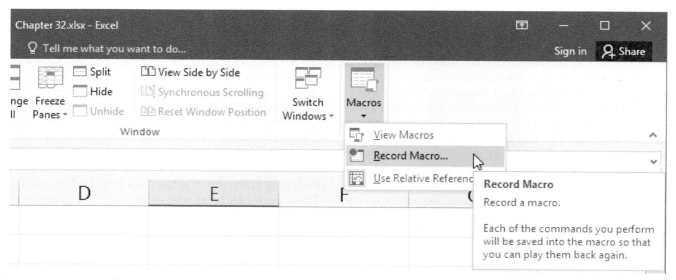

FIGURE 32-8 Record Macro Button

FIGURE 32-9 Record Macro Dialogue Box

on the **View** menu tab, clicking on the arrow below the Macros button, and selecting **Stop Recording** from the popup menu (shown in Figure 32-10).

Workbooks containing macros must be saved as an Excel macro-enable workbook. This is done by selecting the **File** tab, clicking on **Save As** in the left pane, double clicking on **This PC** to bring up the Save As dialogue box, and selecting Excel Macro-Enabled Workbook (*.xlsm) from the Save as Type dropdown box. The Save As dialogue box is shown in Figure 32-11.

The following exercise shows how to record a macro.

Exercise 32-3

In this exercise you will record a macro to add the formulas to a row on the Detail worksheet and assign it to the Ctrl + E keys so that it can be executed from the keyboard. Because this macro needs to work on all rows in the spreadsheet, the macro will need to be a relative macro. The macro is to be executed from Column D and will move the cursor down a row every time it is executed, so it can be executed over and over without the user having to move the cursor down a row. In addition, the macro will set the default labor rate to $28.00 per hour. The macro is created using the following steps:

1. Make sure that Chapter 32, the workbook modified in Exercise 32-2, is open.
2. Click on the **Detail** tab to move to the Detail worksheet and place the cursor on Cell D4.
3. Make sure that the macro is set to use relative references by clicking on the **View** menu tab, clicking on the arrow below the Macros button, and making sure that the box to the left of Use Relative References is highlighted. If it is not highlighted, click on **Use Relative Reference** to highlight it.
4. Click on the arrow below the Macros button and select **Record Macro . . .** from the popup menu (shown in Figure 32-8) to bring up the Record Macro dialogue box.
5. Type *Extension* in the Macro name: text box and *e* in the Shortcut key: Ctrl+ text box.
6. Type *This macro adds the formulas for an individual line on the Detail worksheet.* in the Description: text box. The Record Macro dialogue box should be similar to Figure 32-9.
7. Click the **OK** button to start recording the macro.

You will now record the key strokes and cursor movements that make up the macro.

8. Enter the formulas in Table 32-2 into the specified cells. After entering the formula for a cell, press the **Enter** key to complete the formula.

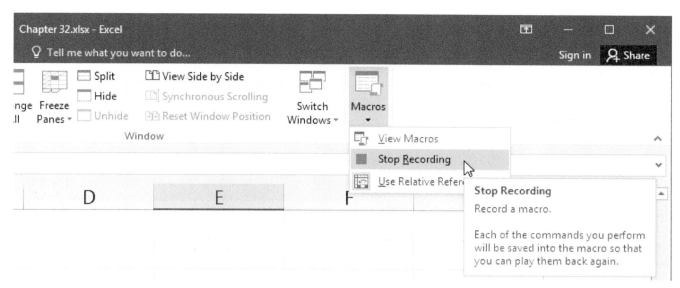

FIGURE 32-10 Stop Recording Button

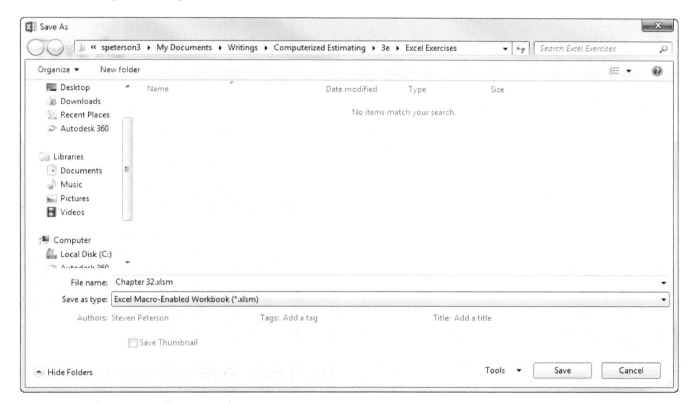

FIGURE 32-11 Save As Dialogue Box

9. After completing the entries in Table 32-2, move the cursor to Cell D5.
10. Click on the **View** menu tab, click on the arrow below the Macros button, and select **Stop Recording** from the popup menu (shown in Figure 32-10) to end the macro.
11. With the cursor on Cell D5, type **Ctrl + E**, and the formulas typed on Row 4 should be repeated on Row 5, except that they should reference Row 5 rather than Row 4. For example, the formula in Cell

TABLE 32-2 **Data for Cells**

E4	=B4*D4
G4	=B4*F4
H4	28
I4	=G4*H4
K4	=G4*J4
L4	=E4+I4+K4

	A	B	C	D	E	F	G	H	I	J	K	L
1				MATERIALS			LABOR			EQUIPMENT		
2	ITEM	QUANTITY		$/UNIT	COST	LHR/UNIT	LHR	$/LHR	COST	$/LHR	COST	TOTAL
3												
4					=B4*D4		=B4*F4	28.00	=G4*H4		=G4*J4	=E4+I4+K4
5					=B5*D5		=B5*F5	28.00	=G5*H5		=G5*J5	=E5+I5+K5

FIGURE 32-12 Formulas for Exercise 32-3

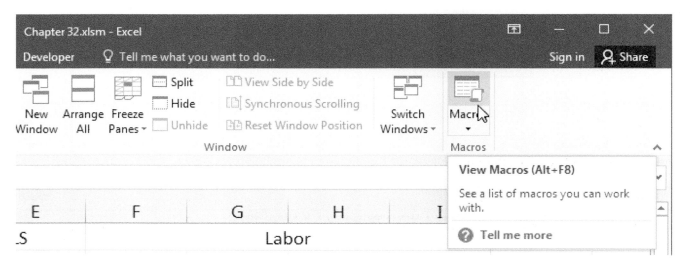

FIGURE 32-13 View: View Macros Button

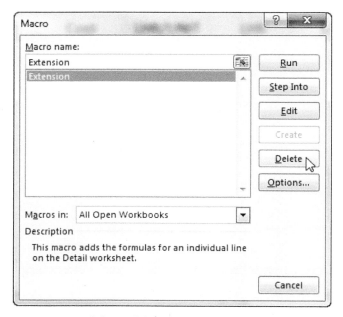

FIGURE 32-14 Macro Dialogue Box

E5 will read $=B5*D5$ rather than $=B4*D4$. The cursor should have moved to Cell D6. The formulas for this example are shown in Figure 32-12. The formulas on Row 4 were entered while creating the macro; whereas the formulas on Row 5 were entered by executing the macro.

12. Save the workbook as a macro-enabled workbook by selecting the **File** tab, clicking on **Save As** in the left pane, double click on **This PC** to bring up the Save As dialogue box, and selecting Excel Macro-Enabled Workbook (*.xlsm) from the Save as Type dropdown box. Do not change the name or location of the workbook. Click on **Save** to save the file.

Macros are deleted by clicking on the **View** menu tab and clicking on the **Macros** button (shown in Figure 32-13) to bring up the Macro dialogue box shown in Figure 32-14. The user selects the macro to be deleted and clicks on the **Delete** button. The user is asked if he or she wants to delete the macro. The macro is deleted by clicking on the **Yes** button.

TESTING THE WORKSHEETS

After creating formulas and macros, the user needs to test the formulas to verify that they are working correctly. To do this, the user enters data with a known result into the formula or macro to verify that it is producing the correct results. One at a time, each cell should be changed to verify that the formula works under all circumstances. This is shown in the following examples.

Exercise 32-4

In this exercise you will verify the formulas entered into the Summary worksheet from Exercises 32-1 and 32-2. For this worksheet, the formula created to sum the rows;

the formula created to sum the material, labor, equipment, subcontract, and total columns; and the formulas created to total the bid are verified. The verification process is done with the following steps.

1. Make sure that macro-enabled version of Chapter 32 (Chapter 32.xlsm), the workbook modified in Exercise 32-3, is open. If the file extensions are not displayed, hold your cursor over the file names and a popup box (similar to Figure 32-15) will appear that identifies the file type.
2. If the Security Warning appears (shown in Figure 32-5), click on the **Enable Content** button to enable the macros written in Exercise 32-3.
3. Click on the **Summary** tab.
4. Hide rows 9 through 115 by selecting (highlighting) the rows, right clicking on the selected rows, and selecting **Hide** from the popup menu shown in Figure 32-16.

First, verify the totals in Column H and Row 123 using the following steps:

5. Type *10* in Cell D3. Cells H3, D123, and H123 should all equal 10.
6. Type *20* in Cell E3. Cells H3 and H123 should equal 30, Cell D123 should equal 10, and Cell E123 should equal 20.
7. Type *30* in Cell F3. Cells H3 and H123 should equal 60, Cell D123 should equal 10, Cell E123 should equal 20, and Cell F123 should equal 30.

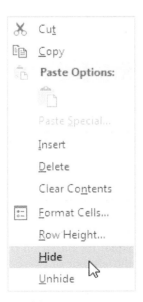

FIGURE 32-15 Popup Box

FIGURE 32-16 Popup Menu

8. Type *40* in Cell G3. Cells H3 and H123 should equal 100, Cell D123 should equal 10, Cell E123 should equal 20, Cell F123 should equal 30, and Cell G123 should equal 40.
9. Type *100* in Cells D122 through G122. Cell H3 should equal 100, Cell H122 should equal 400, Cell D123 should equal 110, Cell E123 should equal 120, Cell F123 should equal 130, Cell G123 should equal 140, and Cell H123 should equal 500.

By completing these actions we have verified the original formula that summed the rows and one of the copied formulas. We have also verified the formulas that summed the column by entering data in the top and bottom cells of the formula. Next, verify the formulas that create the total bid.

10. Type *150* in Cell D124. This will enter the cost of the building permit.
11. Type *40* in Cell D127, *30* in Cell E127, *20* in Cell F127, and *10* in Cell G 127. This will enter the profit markup for materials, labor, equipment, and subcontract.
12. Verify that the cells in the Summary worksheet have the same values shown in Figure 32-17.
13. Exit the workbook without saving the changes.

Exercise 32-5

In this exercise you will verify the macro created for the Detail worksheet in using the following steps:

1. Open Chapter 32.xlsm, the workbook modified in Exercise 32-3.
2. Click on the **Detail** tab to move to the Detail worksheet.
3. Enter the data shown in Table 32-3 into the specified cells.

After you complete the entries in Table 32-3, the spreadsheet should look like Figure 32-18.

4. Exit the workbook without saving the changes.

ADDING ERROR PROTECTION

This section looks at three ways of adding error protection. They are conditional formatting, data validation, and protecting the worksheet. Let's look at conditional formatting.

Conditional Formatting

Conditional formatting allows the user to set the font, border, color, and pattern for a cell by comparing the value or formula in a cell to a value, range of values, or other cells. Conditional formatting is very useful for

	A	B	C	D	E	F	G	H
1		Code	Description	Materials	Labor	Equipment	Subcontract	Total
2		01-000	**GENERAL REQUIREMENTS**					
3		01-300	Supervision	10	20	30	40	100
4		01-500	Temporary Utilities					-
5		01-510	Temporary Phone					-
6		01-520	Temporary Facilities					-
7		01-700	Clean-Up					-
9		33-000	**UTILITIES**					
10		33-100	Water Line					-
11		33-300	Sanitary Sewer					-
12		33-400	Storm Drain					-
13		33-500	Gas Lines					-
14		33-700	Power Lines					-
15		33-800	Telephone Lines	100	100	100	100	400
16			**SUBTOTAL**	110	120	130	140	500
17			Building Permit	150				150
18			Bond	13				13
19			**SUBTOTAL**	273	120	130	140	663
20			Profit and Overhead Markup	40.0%	30.0%	20.0%	10.0%	
21			Profit and Overhead	109	36	26	14	185
22			**TOTAL**	382	156	156	154	848
23								848

FIGURE 32-17 Results for Example 32-4

TABLE 32-3 Data for Cells

A3	*33-100 Water Line*
A4	*Bedding*
B4	23
C4	*cyd*
D4	23.50
F4	0.30
H4	35.94
J4	50.80
A5	*10" water line*
B5	150
C5	*ft*
D5	14.05
F5	0.26
H5	35.94
J5	50.80

marking cells that have unexpected values. For example, in the Summary worksheet, Cell H123 should equal the sum of Cells D123 through G123. If this is not the case, it would be very helpful to change the background color of Cell H123 to make it very apparent that the cells are not equal to each other.

The user sets the conditional formatting for a cell by placing the cursor in the cell or cells to be conditionally formatted, clicks on the **Home > Styles > Conditional Formatting** button (shown in Figure 32-19), places the cursor over the **Highlight Cell Rules** from the popup menu (shown in Figure 32-20), and selects the type of rules to be used in the conditional formatting from the popup menu (shown in Figure 32-21). The user may select from one of the preset rules or create their own rule by clicking on **More Rules...** to bring up the New Rule Formatting dialogue box (shown in Figure 32-22). We will look at creating our own rules.

From the New Rule Formatting dialogue box, the user selects the type of rule in the Select a Rule Type: box. We will limit our discussion to the Format only cells that contain rule, which is the rule type that is automatically selected when opening the New Rule Formatting dialogue box in the manner described above. In the left dropdown box, the user selects whether he or she wants to compare the cell's value or other property. For our example we are going to compare the cells value. In the next dropdown box (the second box from the left) the user selects the type of comparison to make by selecting from the following: **between, not between, equal to, not equal to, greater than, less than, greater than or equal to,** and **less than or equal to**. Based on the selection in the second box from the left, one or two boxes will appear to the right of this box. In these boxes the user enters cells, formulas, or values to be used in the comparison. Next, the user clicks on the **Format...** button to bring up the

Converting Existing Forms 343

	A	B	C	D	E	F	G	H	I	J	K	L
1				MATERIALS			Labor			Equipment		
2	Item	Quantity		$/Unit	Cost	LHR/UNIT	LHR	$/LHR	Cost	$/LHR	Cost	Total
3	33-100 Water Line											
4	Bedding	23	cyd	23.50	541	0.30	6.90	35.94	248	50.80	351	1,139
5	10" water line	150	ft	14.05	2,108	0.26	39.00	35.94	1,402	50.80	1,981	5,490

FIGURE 32-18 Results for Exercise 32-5

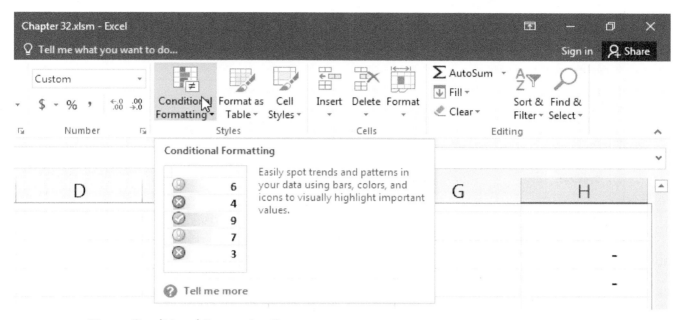

FIGURE 32-19 Home: Conditional Formatting Button

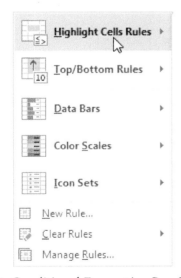

FIGURE 32-20 Conditional Formatting Pop Menu

FIGURE 32-21 Highlight Cell Rules Popup Menu

Format Cells dialogue box shown in Figure 32-23. The user then selects the desired formatting to use in conjunction with the conditional formatting and clicks on the **OK** button to close the Format Cells dialogue box.

Multiple rules for one cell can be written using the above procedures. To select the order the rules are applied, edit a rule, or to delete a rule, the user clicks on the **Home > Styles > Conditional Formatting** button (shown in Figure 32-19), and clicks on **Manage Rules . . .** from the popup menu (shown in Figure 32-24) to bring up the Conditional Formatting Rules Manager dialogue

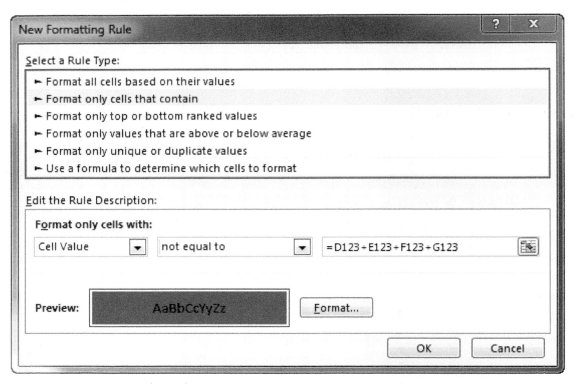

FIGURE 32-22 New Formatting Rule Dialogue Box

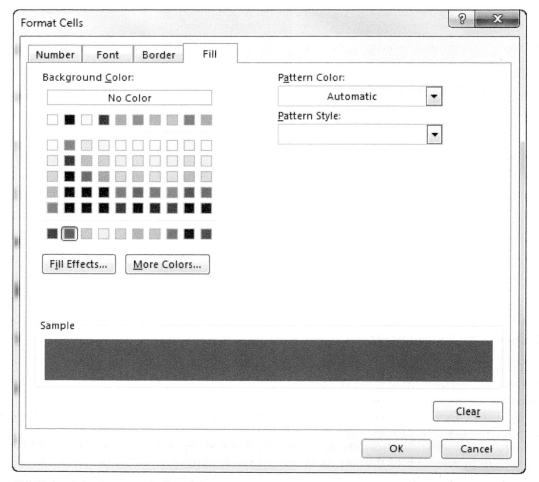

FIGURE 32-23 Fill Tab of the Format Cells Dialogue Box

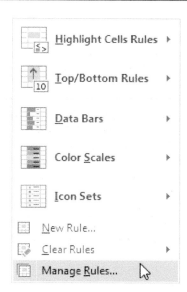

FIGURE 32-24 Conditional Formatting Popup Menu

box (shown in Figure 32-25). Only the rules that apply to the selected cell will be shown. From the Conditional Formatting Rules Manager dialogue box the user can add a new rule by clicking on the **New Rule . . .** button, edit the selected rule by clicking on the **Edit Rule . . .** button, or delete the selected rule by clicking on the **Delete Rule** button. When there are two or more rules, the user can select the order in which the rules are applied by highlighting the rule to move and clicking on the **Move Up** or **Move Down** buttons.

Exercise 32-6

In this exercise you will set the conditional formatting for Cell H123 so it turns red if the value in Cell H123 (the sum of Cells H2 through H122) does not equal the sum of Cells D123 through G123 using the following steps:

1. Make sure that Chapter 32.xlsm, the workbook modified in Exercise 32-3, is open. If the Security Warning appears, click on the **Enable Content** button to enable the macro.
2. Select the Summary worksheet.
3. Place the cursor in Cell H123, click on the the **Home > Styles > Conditional Formatting** button (shown in Figure 32-19), place the cursor over the **Highlight Cell Rules** from the popup menu (shown in Figure 32-20), and click on **More Rules . . .** (shown in Figure 32-21) to bring up the New Rule Formatting dialogue box (shown inFigure 32-22).
4. Select **Cell Value** from the first dropdown box and **not equal to** from the second dropdown box and type $=D123+E123+F123+G123$ in the text box.
5. Click on the **Format . . .** button to bring up the Format Cells dialogue box (shown in Figure 32-23), click the **Fill** tab, select the color red, and click the **OK** button.
6. The Conditional Formatting dialogue box should appear as shown in Figure 32-22, with a red background for the Preview box.
7. Click the **OK** button to complete the formatting.

In the Summary worksheet there are rows that represent the division headings. One example is Row 116, which is the division heading of the Utilities division. As such, costs should not be entered on this row. If costs are entered on this row, they will not be totaled in the Total column (Column H) because the row summing function was not copied to this row. We will enter a number onto this row to verify the conditional formatting.

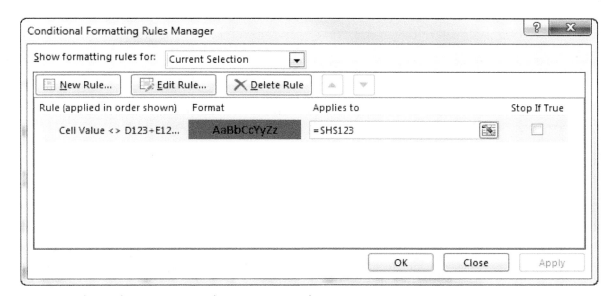

FIGURE 32-25 Conditional Formatting Rules Manager Dialogue Box

8. Enter *22* in Cell D116. Cell H123 should now be red and with a dash in the cell indicating that it is equal to zero. Cell D123 should be equal to 22. The spreadsheet should look like Figure 32-26.
9. Delete the contents of Cell D116 by selecting Cell D116 and pressing the **Delete** key.
10. Save the workbook.

Data Validation

Data validation allows the user to set limits on the data entered into a cell. For example, using data validation, the user may limit the values in a cell to a positive number. The user may limit the cell contents to a whole number, a decimal number, a list, dates, a time, or a text of a given length. Numeric data may be limited to values between two values, not between two values, equal to a value, not equal to a value, greater than a value, less than a value, greater than or equal to a value, or less than or equal to a value. Values may be fixed, be a value located in a cell, or calculated from a formula.

The user sets these limits by clicking on the **Data > Data Tools > Data Validation** button (shown in Figure 32-27) to bring up the Data Validation dialogue box, and selecting the **Settings** tab. The Settings tab of the Data Validation dialogue box is shown in Figure 32-28.

From the Data Validation dialogue box the user sets the type of validation to be used in the Allow: dropdown box, selects the type of comparisons to be made in the Data: dropdown box, and the values, functions, or cells to compare it to in the remaining boxes, which will change based on what is selected in the Allow: dropdown box.

⊿	A	B	C	D	E	F	G	H
1		Code	Description	Materials	Labor	Equipment	Subcontract	Total
2		33-000	UTILITIES	22				
3		33-100	Water Line					-
4		33-300	Sanitary Sewer					-
5		33-400	Storm Drain					-
6		33-500	Gas Lines					-
7		33-700	Power Lines					-
8		33-800	Telephone Lines					-
9			SUBTOTAL	22	-	-	-	-
10			Building Permit					-
11			Bond	-				-
12			SUBTOTAL	-	-	-	-	-
13			Profit and Overhead Markup					
14			Profit and Overhead	-	-	-	-	-
15			TOTAL	-	-	-	-	-
16								-

FIGURE 32-26 Spreadsheet for Exercise 32-6

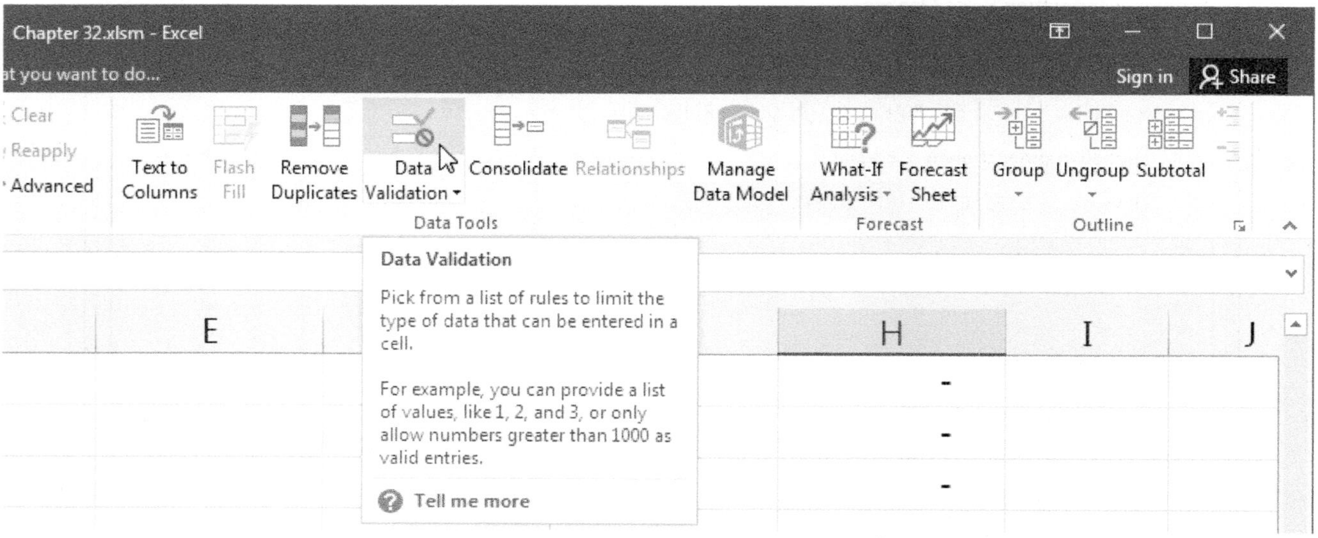

FIGURE 32-27 Data: Data Validation Button

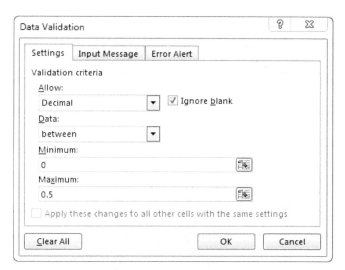

FIGURE 32-28 Settings Tab of the Data Validation Dialogue Box

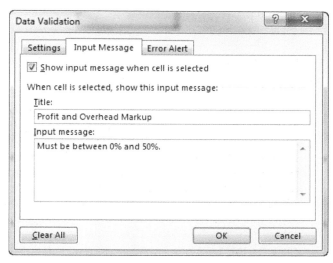

FIGURE 32-29 Input Message Tab of the Data Validation Dialogue Box

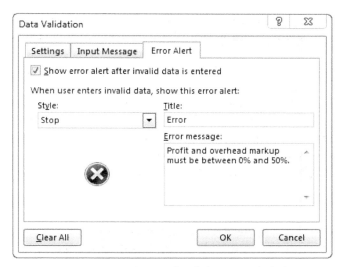

FIGURE 32-30 Error Alert Tab of the Data Validation Dialogue Box

In addition to making the comparison, the user may also create a custom prompt message on the Input Message tab and create a custom error message on the Error Alert tab. A custom prompt message, which will appear when the cell is selected, is created by selecting the **Input Message** tab, checking the **Show input message when cell is selected** check box, typing the title of the message in the Title: text box, and typing the message in the Input message: text box. The Input Message tab of the Data Validation dialogue box is shown in Figure 32-29.

A custom error message is created by selecting the **Error Alert** tab, checking the **Show error alert after invalid data is entered** check box, typing the title of the message in the Title: text box, and typing the message in the Error message: text box. The user may select between a stop, warning, and information error message in the Style: dropdown box. The stop error message does not let the user enter a value that does not meet the validation criteria. The warning error message will ask the user if he or she wants to continue even though the data does not meet the validation criteria. Although the information error message will inform the user that he or she has entered data that does not meet the validation criteria, it still makes the change. The Error Alert tab of the Data Validation dialogue box is shown in Figure 32-30.

Exercise 32-7

In this exercise you will limit the profit and overhead markups on the Summary worksheet to between 0% and 50% using data validation. This is done by the following steps:

1. Make sure that Chapter 32.xlsm, the workbook modified in Exercise 32-6, is open. If necessary, enable the macros.
2. Place the cursor in Cell D127 and click on the **Data>Data Tools>Data Validation** button (shown in Figure 32-27) to bring up the Data Validation dialogue box.
3. Select the **Settings** tab, select **Decimal** from the Allow: dropdown box, select **between** from the Data: dropdown box, and type *0* in the Minimum: box and *0.5* in the Maximum: box. The Data Validation box should now appear as in Figure 32-28.
4. Select the **Input Message** tab, check the **Show input message when cell is selected** check box, type *Profit and Overhead Markup* in the Title: text box, and type *Must be between 0% and 50%.* in the Input message: text box. The Data Validation box should now appear as Figure 32-29. When the cursor is placed in Cell D127, the message in Figure 32-31 will appear next to the cell.
5. Select the **Error Alert** tab, check the **Show error alert after invalid data is entered** check box, select **Stop** from the Style: dropdown box, type *Error* in

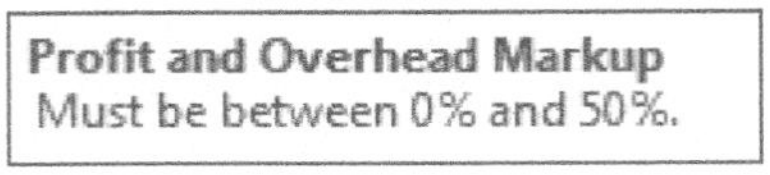

FIGURE 32-31 Input Message Created in Exercise 32-7

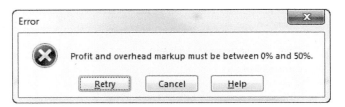

FIGURE 32-32 Error Alert Created in Exercise 32-7

the Title: text box, and type *Profit and overhead markup must be between 0% and 50%.* in the Error message: text box. The Data Validation box should now appear as in Figure 32-30. When numbers greater than 50% (0.50) or less than zero are entered into Cell D127, the error message in Figure 32-32 will appear and require the user to enter an approved value or cancel the attempt to change the value.

6. Click the **OK** button to complete the data validation.

Next, you will verify that the data validation is working correctly using the following steps:

7. Place the cursor in Cell D127 and verify that the popup box shown in Figure 32-31 appears next to the cell.
8. Type *55* in Cell D127 and verify that Figure 32-32 appears.
9. Click the **Retry** button, type *–5* (minus five) in Cell D127, and verify that Figure 32-32 again appears.
10. Click the **Retry** button, type 0 in Cell D127, and verify that the data is accepted.
11. Type 50 into Cell D127 and verify that the data is accepted.
12. Delete the contents of Cell D127.

Finally, copy the data validation from Cell D127 to Cells E127, F127, and G127 using the following steps:

13. Right click on Cell D127 and select **Copy** from the popup menu.
14. Select Cells E127, F127, and G127, right click on the selected cells and, select **Paste Special . . .** from the popup menu to bring up the Paste Special dialogue box.
15. Select the **Validation** radio button to paste only the data validation and click on the **OK** button to complete the paste.
16. Place the cursor in Cells E127, F127, and G127 one at a time and verify that the popup box shown Figure 32-31 appears next to each of the cells.
17. Save the workbook.

Protecting the Worksheet

The creator of a worksheet may protect specified cells of the worksheet against changes. Protecting the cells prevents the user of the worksheet from accidentally or intentionally changing a formula or entering data into a cell where data should not be entered. All estimating worksheets should be protected.

Before the creator of a worksheet can protect a worksheet, he or she must identify which cells the user will be allowed to change. This is done by selecting the cells that the user will be allowed to change, right clicking on the selected cells, selecting **Format Cells . . .** from the popup menu to bring up the Format Cells dialogue box, selecting the **Protection** tab, unchecking the **Locked** check box, and clicking on the **OK** button to close the dialogue box. The Format Cells dialogue box is shown in Figure 32-33. By default, all cells should have the Locked check box checked.

Once the creator of the worksheet has unchecked the Locked check box for all of the cells the user will be allowed to change, the creator may protect the worksheet. This is done by clicking on the **Review > Changes > Protect Sheet** button (shown in Figure 32-34) to bring up the Protect Sheet dialogue box shown in Figure 32-35. The user then makes sure that the **Protect worksheet and contents of locked cells** check box is checked, enters a password in the Password to unprotect sheet: text box, unchecks the **Select locked cells** check box (so the user can only select cell that can be changed), makes sure the **Select unlocked cells** check box is checked, makes sure all other check boxes are unchecked, and clicks on the **OK** button to bring up the Confirm Password dialogue box shown in Figure 32-36. The user retypes the password in the Reenter password to proceed. text box and clicks on the **OK** button to close the dialogue box and protect the worksheet. The user may now only select and change the cells that are not locked. A worksheet may be protected without a password by leaving the Password to unprotect sheet: text box blank, but it is not recommended because anyone can unprotect the worksheet and make changes.

To unprotect a worksheet, the creator clicks on the **Review > Changes > Unprotect Sheet** button shown in Figure 32-37, which has replaced the Protect Worksheet button, to bring up the Unprotect Sheet dialogue box shown in Figure 32-38. The user types the password used to protect the worksheet in the Password: text box and clicks on the **OK** button to close the dialogue box and unprotect the worksheet. The worksheet cannot be unprotected without the password, so it is a good idea to keep a record of the passwords used to protect worksheets in a safe place.

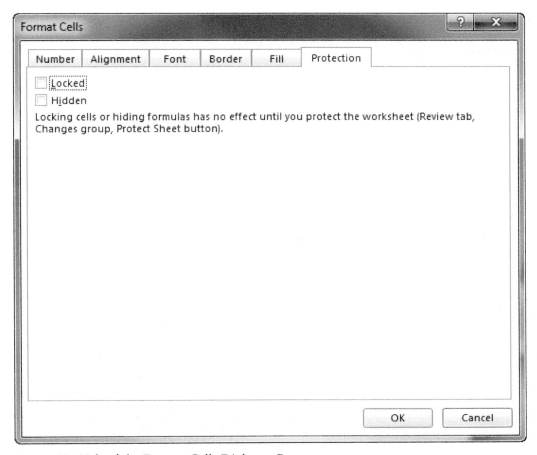

FIGURE 32-33 Protection Tab of the Format Cells Dialogue Box

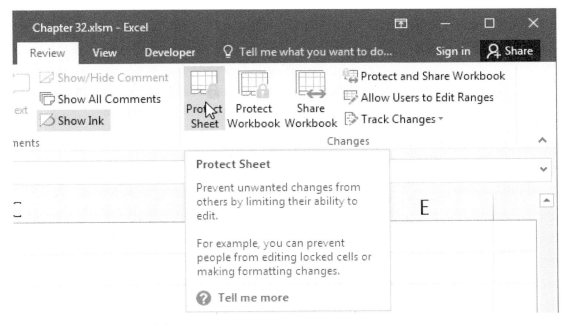

FIGURE 32-34 Review: Protect Worksheet Button

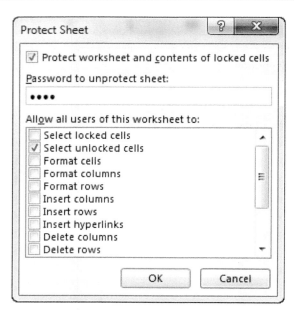

FIGURE 32-35 Protect Worksheet Dialogue Box

FIGURE 32-36 Confirm Password Dialogue Box

Exercise 32-8

In this exercise you will protect the Summary worksheet using the following steps:

1. Make sure Chapter 32.xlsm, the workbook modified in Exercise 32-7, is open. If necessary, enable the macros.
2. Select Cells A3 through A7, right click on the selected cells, select **Format Cells . . .** from the popup menu to bring up the Format Cells dialogue box (shown in Figure 32-33), select the **Protection** tab, uncheck the **Locked** check box, and click on the **OK** button to close the dialogue box.
3. Repeat Step 2 for the following: A10, A13 through A20, A23, A26 through A29, A32 through A38, A41 through A51, A54 through A60, A62 through A68, A71 through A74, A77, A80, A83, A86, A89, A92, A95, A98, A101, A102, A105 through A114, A117 through A122, A124, and A125. Hint: Multiple cells may be selected at one time using the Shift and the Ctrl keys.
4. Repeat Step 2 for the following: Cells D3 through G7, D10 through G10, D13 through G20, D23 through G23, D26 through G29, D32 through G38, D41 through G51, D54 through G60, D62 through G68, D71 through G74, D77 through G77, D80 through G80, D83 through G83, D86 through G86,

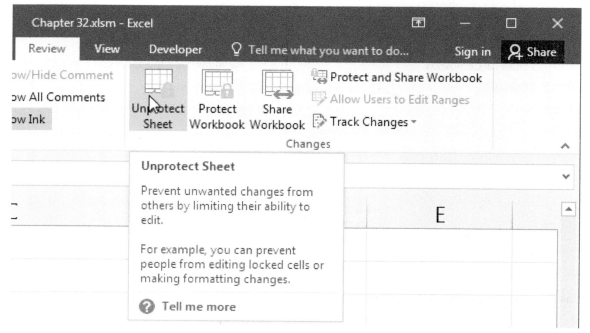

FIGURE 32-37 Review: Unprotect Worksheet Button

FIGURE 32-38 Unprotect Sheet Dialogue Box

D89 through G89, D92 through G92, D95 through G95, D98 through G98, D101 through G102, D105 through G114, D117 through G122, D124, and D127 through G127.

5. Protect the worksheet by clicking on the **Review > Changes > Protect Sheet** button (shown in Figure 32-34) to bring up the Protect Sheet dialogue box (shown in Figure 32-35).
6. Make sure the **Protect worksheet and contents of locked cells** check box is checked.
7. Type *1234* in the Password to unprotect sheet: text box.
8. Uncheck the **Select locked cells** check box.
9. Make sure the **Select unlocked cells** check box is checked.
10. Make sure that all other checks boxes are unchecked.
11. Click on the **OK** button to bring up the Confirm Password dialogue box (shown in Figure 32-36).
12. Type *1234* in the Reenter password to proceed. text box, and click on the **OK** button to close the dialogue box and protect the worksheet.
13. Verify that you can only select and change the unlocked cells by placing the cursor in Cell A3 and pressing the **Right Arrow** key. The cursor should move to the right and when it reaches Column G it should return to Column A. When it reaches Cell G127 it should return to Cell A3.
14. Save the workbook.

CONCLUSION

The steps for setting up an existing form are to create the form layout, enter the formulas, automate with macros, test the spreadsheets, and protect against errors. Macros are used to automate key strokes and cursor movements by recording the keystrokes and cursor movements as you would record your voice on a tape recorder. These actions will then be repeated each time the macro is used. Error protection may be added to the worksheet by (1) using conditional formatting to change the format for a cell when the specified conditions are met, (2) using data validation to only allow specific types of data to be entered into a cell, and (3) protecting the worksheet so only specified cells can be selected and modified.

PROBLEMS

1. What are the steps for converting an existing form?
2. What is a circular reference?
3. How can Excel solve circular references?
4. What is a macro?
5. What risk do you take if you use macros created by someone else?
6. How do you change the security level for macros?
7. What must you do to run a macro from the keyboard?
8. What is the purpose of conditional formatting?
9. How can conditional formatting be used to protect against errors?
10. What is the purpose of data validation?
11. How can data validation be used to protect against errors?
12. What is the purpose of protecting a worksheet?
13. How can protecting a worksheet be used to protect against errors?
14. For a worksheet that has been protected by a password, is there any way you can unprotect the worksheet without a password?

CHAPTER THIRTY-THREE

CREATING NEW FORMS

In this chapter you will learn how to create new Excel forms. This chapter will take you step by step through the creation of a spreadsheet to calculate the materials, labor, and costs for an asphalt shingle roof and create a ready-to-sign proposal from that spreadsheet.

Creating spreadsheet estimating forms from scratch is a more difficult task than creating spreadsheets from existing paper forms because the user has to decide what the form is going to accomplish and how it will be laid out. New forms should be carefully planned before starting to create them in Excel.

PLANNING NEW FORMS

When creating a spreadsheet form from scratch, it is best to carefully think through its design before beginning to set it up. There are five steps in planning a new spreadsheet: (1) Determine the desired output or results from the spreadsheet, (2) determine what input data will be entered into the spreadsheet, (3) solve different sample problems, (4) identify the steps and equations that were used to arrive at the solutions, and (5) prepare a sketch of the layout for the spreadsheet. Once the spreadsheet has been carefully planned, it can be set up using the same process used to convert existing forms. This process is shown in Figure 33-1. Let's look at each of the planning steps.

Identify Desired Results

The first step is to determine the desired output or result from the spreadsheet. This is what the estimator wants the spreadsheet to accomplish. This could be a materials list, a cost to complete an item, a proposal, or a preliminary cost estimate. The unit of measure associated with each of the results should be identified. For example, concrete materials should be in cubic yards, asphalt-impregnated felt in rolls, and so forth.

In this chapter we will prepare a spreadsheet for an asphalt shingle roof. The desired results of the spreadsheet are the quantities of roofing materials needed for the roof in standard ordering quantities (for example, whole bundles of shingles), the number of labor hours needed to complete the roof, and the cost to install the roof. The materials used on a typical asphalt shingle roof and that

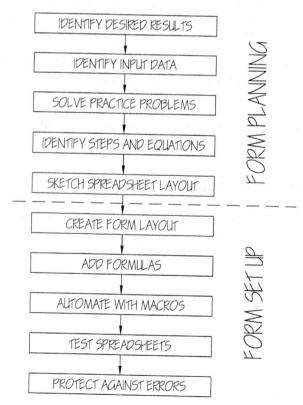

FIGURE 33-1 Process for Creating New Spreadsheets

need to be included in the spreadsheet are bundles of shingles, bundles of cap shingles, pounds of roofing nails, rolls of underlayment (felt), the number of underlayment nails in hundreds, pieces of 10-foot-long drip edge, pieces of 10-foot-long counter flashings, rolls of ridge vent, the number of turtle vents, the number of HVAC flashings, and the number of plumbing pipe flashings.

Identify Input Data

The second step is to determine the data the user of the spreadsheet will input (enter) into the spreadsheet for it to calculate the desired results. The units associated with each entry should be identified and should be the same as the units used in the takeoff. For example, most concrete footings are specified in a combination of feet and inches or just inches. Footing dimensions are never specified in yards, which is the unit of measure used in specifying the quantity of concrete. The spreadsheet should be set up to make the necessary conversions.

To bid an asphalt shingle roof, the following data needs to be collected from the plans and specifications: the slope of the roof (for example, 4:12 or 6:12), the type of shingle (for example, 25-year architectural), the type of underlayment (for example, 15-pound felt), the total length of the ridge(s) in feet, the number of ridges, the total length of the hip(s) and valley(s) in feet, the total length of the perimeter of the roof that is horizontal in feet, the total length of the perimeter of the roof that is sloped in feet, the total length of counter flashing placed horizontally in feet, the total length of counter flashing that is along a slope in feet, the number of HVAC flashings, the number of plumbing pipe flashings, and the type of attic ventilation (for example, gable end, ridge, or turtle vent). To determine the number of labor hours required to shingle the roof, the labor productivities for the different types of shingles are needed. To calculate total cost, the pricing for the materials and labor rates are needed.

For the spreadsheet developed in this chapter, there are four types of shingles: 20-year three tab, 25-year architectural, 30-year architectural, and 40-year architectural. There will also be two types of underlayment: 15-pound and 30-pound asphalt-impregnated felt. The pricing for the materials to be used in the spreadsheet

TABLE 33-1 **Material Pricing**

Item	Price	Conversion
Shingles		
20-year three tab	$12.39/bundle	bundle = 1/3 square
25-year architectural	$15.98/bundle	bundle = 1/3 square
30-year architectural	$16.65/bundle	bundle = 1/3 square
40-year architectural	$17.35/bundle	bundle = 1/3 square
Cap Shingles		
20-year three tab	$15.75/bundle	bundle = 21 ft
25-year architectural	$31.29/bundle	bundle = 21 ft
30-year architectural	$33.29/bundle	bundle = 21 ft
40-year architectural	$35.29/bundle	bundle = 21 ft
Underlayment		
15-pound felt	$14.95/roll	roll = 4 square
30-pound felt	$13.95/roll	roll = 2 square
Flashings and Vents		
Drip edge	$2.99/ea	each = 10 ft
Counter	$3.99/ea	each = 10 ft
Ridge vent	$42.00/ea	each = 20 ft
Turtle vents (61 sq in.)	$7.00/ea	
HVAC pipe flashing	$9.00/ea	
Plumbing flashing	$4.00/ea	
Nails		
Roofing nails	$1.39/lb	
Underlayment nails	$1.00/C (hundreds)	

TABLE 33-2 Labor Productivity

20-year three tab	2.50 lhr/square
25-year architectural	2.80 lhr/square
30-year architectural	3.00 lhr/square
40-Year architectural	3.20 lhr/square

is given in Table 33-1, and the labor productivity rates are given in Table 33-2. The crew rate per labor hour is $35. The pricing and productivity data are set up so they can be easily modified as material costs and labor rates change. Also given in Table 33-1 are conversion factors from standard order quantities (for example, rolls) to common units of measure (for example, feet).

Solve Practice Problems

The third step is to manually solve sample problems that cover all of the different types of problems the spreadsheet is to solve. During this process the developer of the spreadsheet will often discover additional input data that is needed to complete the calculations; this data should be added to the list prepared in the second step. For the roofing spreadsheet being set up in this chapter, sample problems similar to Examples 9-4 and 9-9 need to be solved. The reader should review these examples before continuing.

Identify Steps and Equations

The fourth step is to identify the steps and equations taken to solve the practice problems. Not all problems are solved in the same manner. For example, if the project specifies that turtle vents are needed, then the number of turtle vents needs to be calculated, whereas if the project specifies gable-end vents (which are not a part of the roofing package), then nothing would be done. As a result, the steps should identify any decisions made (such as selecting between two formulas) in the process of arriving at the solution.

For the roofing worksheet, the first step is to do a takeoff for the quantities from the roof. The slope of the roof, type of shingle, type of underlayment, and type of attic ventilation are read off the plans and specifications. The total length of the ridge(s), hip(s), valley(s), perimeter, and counter flashing are measured off the plans using a plan measurer, thus eliminating the need to add their lengths together. The drip edge and counter flashing on sloped edges are measured separately from the drip edge and counter flashing on horizontal edges so that the slope can be taken into account. The number of ridges, HVAC flashings, and plumbing pipe flashings are counted off the plans. The roof is divided into rectangles and the length and width of each rectangle are measured; the spreadsheet will need to calculate the total area of the roof. All of these measurements are measured in the plan view, with the spreadsheet accounting for slope.

The second step is to calculate the area of the roof in the plan view. This is done by multiplying the length by the width of each rectangle taken off the plans and then adding their areas together. If the user were to use a takeoff software package, the total area of the roof could be used in lieu of dividing the roof into rectangles and calculating the area.

The third step is to determine the quantities of materials. Unless noted, the spreadsheet will need to round all quantities up to the next whole number.

The quantity of shingles is found by summing the results of Eqs. (9-5), (9-7), and (9-8) using a standard exposure of 5 inches. The quantity is converted to bundles, where 3 bundles equal 1 square (100 square feet).

The quantity of cap is found by summing the results of Eqs. (9-9) and (9-10). The quantity is converted to bundles, where 1 bundle equals 21 lineal feet of 1-foot-wide cap or 0.21 squares of cap.

Based on historical data, 1 pound of roofing nails is needed for every 3 bundles of shingles, including cap shingles, and is calculated from the ordered quantities of shingles.

The quantity of underlayment is found using Eq. (9-3) and is converted to rolls. For 15-pound felt there are 4 squares of felt per roll, and for 30-pound felt there are 2 squares of felt per roll. The spreadsheet should decide which conversion factor to use.

From historical data, 125 nails are needed for 1 square of felt; therefore, 250 nails are needed for 1 roll of 30-pound-felt and 500 nails are needed for 1 roll of 15-pound felt. The number of nails is based on the order quantity of felt, and the spreadsheet will need to determine if 250 or 500 nails per roll will need to be ordered. Nails are ordered in units of 10 nails or 0.1 C, where C is the notation for 100.

The total drip edge equals the sum of the length of the drip edge placed horizontally plus the sum of the length of the drip edge placed on a sloped edge. The length of the drip edge placed on a sloped edge is determined by Eq. (9-1). The length of the drip edge is converted to 10-foot-long pieces by dividing the length by 9.8 feet, thus allowing 0.2 foot per lap. The counter flashing is calculated in the same manner.

If ridge vent is specified, the length of the ridge will equal the quantity of ridge vent needed. If ridge vent is not specified, then the length is zero. The quantity of ridge vent needs to be converted to 20-foot rolls.

If turtle vents are specified, 1 square foot of venting is required for every 150 square feet of attic, with half of this required ventilation in the upper portion of the roof and the other half located in the eaves; therefore, the turtle vents are required to provide 1 square foot of venting for every 300 square feet of attic space. Each vent provides 61 square inches (0.42 ft^2) of venting; thus 2.38 (1/0.42) vents are required to for every 300 square feet. As a result one vent is required for every 126 square feet (300 ft^2/2.38).

The quantities of HVAC and plumbing flashings are taken off from the plans.

All quantities will be rounded up to the next whole ordering unit (for example, bundle of shingles).

Materials	Quantity	Unit Price	Total
Shingle Type			
Shingle Type Cap			
Roofing Nails			
Underlayment Type			
Underlayment Nails			
10' Drip Edge			
10' Counter Flashing			
20' Ridge Vent			
Turtle Vents			
HVAC Flashing			
Plumbing Flashing			
		Subtotal	
Tax (6.5%)			
Roofing Crew	1hrs	35.00	
		Total	

FIGURE 33-2 First Tab of New Spreadsheet

The fourth step is to determine the material costs. The cost for each type of material (for example, shingles) is determined by using Eq. (26-1) and the previously determined quantities and the material unit cost for the item. The subtotal of the material cost is determined by summing the costs for each type of material. One adds to this subtotal the sales tax, which is calculated by using Eq. (19-3). The sales tax rate is 6.5%.

The fifth step is to determine the labor cost. The number of labor hours is based on the sloped area of the roof, which is calculated by using Eq. (9-7). The number of labor hours is calculated using Eq. (20-7) and the productivity associated with the type of shingle being installed.

Two hours of mobilization time are added to the number of labor hours, which are rounded to the tenth of an hour. The labor cost is calculated using Eq. (26-2) and a labor rate of $35 per hour.

Finally, the total cost is calculated by summing the material and labor costs, and the cost per square foot is calculated by dividing the total cost by the sloped area.

Sketch Spreadsheet Layout

The final step of planning is to sketch a preliminary layout for the spreadsheet. This helps the developer of the spreadsheet to plan how the final spreadsheet is to look.

The spreadsheet being developed in this chapter will consist of three tabs. The first tab will contain the results including the quantity of materials to be ordered, the number of labor hours, and the costs. A preliminary layout for this tab is shown in Figure 33-2.

The second tab will contain the takeoff information, the plan view area of the roof, and the cost per square foot for the roof. A preliminary layout for this tab is shown in Figure 33-3.

The third tab will contain the pricing information and labor productivity rates. Two columns are used for the price and labor productivity. The left column will contain the numeric value and be used in the calculations. The right column will contain the units. A preliminary layout for this tab is shown in Figure 33-4.

Roof Information	
Roof Slope:	
Shingle Type:	
Underlayment:	
Ridge(s):	
No. of Ridges:	
Hip(s)/Valley(s):	
Horiz. Perimeter:	
Sloped Perimeter:	
Horiz. Counter:	
Sloped Counter:	
HVAC Flashings:	
Plumbing Flashings:	
Vent Type:	

Roof Area		
Area	Length	Width
1		
2		
3		
4		
5		
6		
7		
8		
9		
10		

Plan View Area:	
Unit Price:	

FIGURE 33-3 Second Tab of New Spreadsheet

Item	Price
Shingles	
20-year, Three Tab	$ / bundle
25-year Architectural	$ / bundle
30-year Architectural	$ / bundle
40-year Architectural	$ / bundle
⋮	⋮

Item	Labor Productivity
20-year, Three Tab	1hr / squ
25-year Architectural	1hr / squ
30-year Architectural	1hr / squ
40-year Architectural	1hr / squ

FIGURE 33-4 Third Tab of New Spreadsheet

SETTING UP THE SPREADSHEET

Now that the spreadsheet is planned it is ready to be set up. The setting up of a new form follows the process used to set up existing forms. The first step in setting up the spreadsheet is to create the form layouts, which is done in the next two exercises.

Exercise 33-1

In this exercise you will create the layout for the third tab to be used in the roof-estimating spreadsheet by completing the following steps:

1. Begin by opening a new workbook.
2. Save the workbook by clicking the **Save** button on the Quick Access toolbar or typing **Ctrl+S**, double clicking on **This PC**, selecting the **Excel Exercises** folder in the Save in: dropdown box, typing *Chapter 33* in the File Name text box, and clicking the **Save** button.

For this spreadsheet three worksheets are needed in the workbook.

3. If there are more than three sheets, delete the extra sheets by right clicking on the tabs of the extra sheets and selecting **Delete** from the popup menu.
4. If there are one or two sheets, add the needed sheets by clicking on the **New Sheet** button located to the right of the sheet tabs.
5. Name the left worksheet "Bid" by right clicking the tab of the worksheet, selecting **Rename** from the popup menu, and, with the sheet name highlighted, typing *Bid*.
6. Rename the center worksheet "Takeoff" and the right worksheet "Pricing Data" using the same procedures used to rename the left worksheet.

Next, you will format and add the pricing data to the Pricing Data worksheet.

7. Make sure that the Pricing Data worksheet is selected.
8. Change the width of Column A to 20 by right clicking on Column A, selecting **Column Width ...** from the popup menu to bring up the Column Width dialogue box, entering *20* in the Column Width: text box, and clicking the **OK** button.
9. Change the width of Columns B and C to 8.
10. Type the text shown in Table 33-3 into the specified cells.
11. Underline Cells A1 through C1 by highlighting Cells A1 through C1 and clicking the **Home>Font>Bottom Border** button (shown in Figure 3-47). If another border is shown in the Borders button, select the correct border by clicking on the small arrow to the right of the Borders button and selecting the **Bottom Border** from the Borders popup menu.
12. Underline Cells A30 through C30 using the same procedure used to underline Cells A1 through C1.
13. Bold the data in Cells A2, A8, A14, A18, and A26 by selecting the cells and selecting the **Home>Font>Bold** button (shown in Figure 3-36).
14. Format the numbers in Column B to have two decimal places by selecting Column B and selecting the **Home>Number>Comma Style** button (shown in Figure 3-42).
15. Center the text in Column C by selecting Column C and selecting the **Home>Alignment>Center** button (shown in Figure 3-32).
16. Merge and center Cells B1 and C1 by highlighting Cells B1 and C1 and selecting the **Home>Alignment>Merge & Center** button (shown in Figure 3-40).
17. Merge and center Cells B30 and C30.
18. Save the workbook.

The workbook should look like Figure 33-5.

TABLE 33-3 Data for Cells in Pricing Data Worksheet

Cell	Data	Cell	Data
A1	Item	B15	14.95
A2	Shingles	B16	13.95
A3	20-year Three Tab	B19	2.99
A4	25-year Architectural	B20	3.99
A5	30-year Architectural	B21	42.00
A6	40-year Architectural	B22	7.00
A8	Cap Shingles	B23	9.00
A9	20-year Three Tab	B24	4.00
A10	25-year Architectural	B27	1.39
A11	30-year Architectural	B28	1.00
A12	40-year Architectural	B30	Labor Productivity
A14	Underlayment	B31	2.50
A15	15# Felt	B32	2.80
A16	30# Felt	B33	3.00
A18	Flashings & Vents	B34	3.20
A19	Drip Edge	C3	$/bundle
A20	Counter	C4	$/bundle
A21	Ridge	C5	$/bundle
A22	Turtle	C6	$/bundle
A23	HVAC Pipe Flashing	C9	$/bundle
A24	Plumbing Flashing	C10	$/bundle
A26	Nails	C11	$/bundle
A27	Roofing Nails	C12	$/bundle
A28	Underlayment Nails	C15	$/roll
A30	Item	C16	$/roll
A31	20-year Three Tab	C19	$/ea
A32	25-year Architectural	C20	$/ea
A33	30-year Architectural	C21	$/ea
A34	40-year Architectural	C22	$/ea
B1	Price	C23	$/ea
B3	12.39	C24	$/ea
B4	15.98	C27	$/lb
B5	16.65	C28	$/C
B6	17.35	C31	lhr/squ
B9	15.75	C32	lhr/squ
B10	31.29	C33	lhr/squ
B11	33.29	C34	lhr/squ
B12	35.29		

	A	B	C
1	Item	Price	
2	**Shingles**		
3	20-year Three Tab	12.39	$/bundle
4	25-year Architectural	15.98	$/bundle
5	30-year Architectural	16.65	$/bundle
6	40-year Architectural	17.35	$/bundle
7			
8	**Cap Shingles**		
9	20-year Three Tab	15.75	$/bundle
10	25-year Architectural	31.29	$/bundle
11	30-year Architectural	33.29	$/bundle
12	40-year Architectural	35.29	$/bundle
13			
14	**Underlayment**		
15	15# Felt	14.95	$/roll
16	30# Felt	13.95	$/roll
17			
18	**Flashings & Vents**		
19	Drip Edge	2.99	$/ea
20	Counter	3.99	$/ea
21	Ridge Vent	42.00	$/ea
22	Turtle Vents (61 sq in)	7.00	$/ea
23	HVAC Pipe Flashing	9.00	$/ea
24	Plumbing Flashing	4.00	$/ea
25			
26	**Nails**		
27	Roofing Nails	1.39	$/lb
28	Underlayment Nails	1.00	$/C
29			
30	Item	Labor Productivity	
31	20-year Three Tab	2.50	lhr/squ
32	25-year Architectural	2.80	lhr/squ
33	30-year Architectural	3.00	lhr/squ
34	40-year Architectural	3.20	lhr/squ

FIGURE 33-5 Pricing Data Worksheet

SERIES

Excel will automatically create a series or list of consecutive numbers (for example, 1, 2, 3, 4, and 5), dates, days, weekdays, months, or other text (for example, #4, #5, #6, and #7). A series is created by entering a start value in a cell and dragging the little box in the lower right-hand corner of the cell down or to the right the appropriate number of cells. If Excel simply copies the cell, it can be changed to a series by clicking on the **Auto Fill Options** button shown in Figure 33-6 that appears at the lower right-hand corner of the filled cells and selecting **Fill Series** from the popup box shown in Figure 33-7.

The series may consecutively number the lists by 10s or other increments by entering the first two numbers in

FIGURE 33-6 Auto Fill Options Button

FIGURE 33-7 Auto Fill Options Popup Box

the series, selecting both cells, and dragging the little box in the lower right-hand corner of the cell down or to the right the appropriate number of cells. In the following exercise a series is used in the creation of the second tab of the roofing worksheet:

Exercise 33-2

In this exercise you will create the layout for the first and second tabs to be used in the roof-estimating spreadsheet by completing the following steps:

1. Make sure that Chapter 33, the workbook created in Exercise 33-1, is open.
2. Select the Takeoff worksheet by clicking on the **Takeoff** tab.
3. Change the width of Column A to 4, of Column B to 18, of Column C to 1, of Column D to 20, of Columns E through G to 5, and of Columns H and I to 8.
4. Type the text shown in Table 33-4 into the specified cells.

Next you will use the series command to create the numbers 1 through 10.

5. Select Cell G4.
6. Drag the little square at the lower right-hand corner of Cell G4 down to Cell G8 by placing the cursors over the square, clicking and holding the left mouse button while dragging the cursor down to the bottom of Cell G8, and releasing the left mouse but-

TABLE 33-4 Data for Cells in Takeoff Worksheet

Cell	Data
B2	*Roof Information*
B3	*Roof Slope:*
B4	*Shingle Type:*
B5	*Underlayment:*
B6	*Ridge(s):*
B7	*No. of Ridges:*
B8	*Hip(s)/Valley(s):*
B9	*Horiz. Perimeter:*
B10	*Sloped Perimeter:*
B11	*Horiz. Counter:*
B12	*Sloped Counter:*
B13	*HVAC Flashings:*
B14	*Plumbing Flashings:*
B15	*Vent Type:*
B17	*Plan View Area:*
B18	*Unit Price:*
E3	*:12*
E6	*ft*
E7	*ea*
E8	*ft*
E9	*ft*
E10	*ft*
E11	*ft*
E12	*ft*
E13	*ea*
E14	*ea*
E17	*sft*
E18	*/sft*
G2	*Roof Area*
G3	*Area*
G4	*1*
H3	*Length*
I3	*Width*

ton. Cells G4 through G8 should now contain the number 1.

7. Change these numbers to a series by clicking on the **Auto Fill Options** button (shown in Figure 33-6) at the lower right-hand corner of the cells and selecting **Fill Series** from the popup box (shown in Figure 33-7). Cells G4 through G8 should now contain the numbers 1 through 5.
8. Select Cells G7 and G8.
9. Drag the little square at the lower right-hand corner of Cell G8 down to Cell G13. Cells G4 through G13 should now contain the numbers 1 through 10.
10. Right justify the text in Column B by selecting Column B and selecting the **Home>Alignment>Align Right** button (shown in Figure 3-33).
11. Format the numbers in Cells D3 through D15 and D17 to the comma style with no numbers after the decimal place by selecting the cells, selecting the **Home>Number>Comma Style** button (shown in Figure 3-42), and clicking the **Home>Number>Decrease Decimal** button (shown in Figure 3-45) twice.
12. Format the number in Cell D18 to the comma style with two numbers after the decimal place by selecting Cell D18 and selecting the **Home>Number>Comma Style** button.
13. Center the text in Cells E3 through E18 by selecting the cells and selecting the **Home>Alignment>Center** button (shown in Figure 3-32).
14. Center the text in Cells G3 through G13, H3, and I3.
15. Merge and center Cells B2 through E2 by highlighting Cells B2 through E2 and selecting the **Home>Alignment>Merge & Center** button (shown in Figure 3-40).
16. Merge and center Cells G2 through I2.
17. Underline Cells B2 through E2 by selecting the cells and clicking the **Home>Font>Bottom Border** button (shown in Figure 3-47). If another border is shown in the Borders button, select the correct border by clicking on the small arrow to the right of the Borders button and selecting the **Bottom Border** from the Borders popup menu.
18. Underline Cells G2 through I2.
19. Place a thick box border around Cells B2 through E15 by selecting the cells, clicking on the **Home** menu tab, clicking on the small arrow to the right of the Borders button, and selecting the **Thick Outside Border** from the popup menu.
20. Place a thick box border around Cells B17 through E18.
21. Place a thick box border around Cells G2 through I13.
22. Save the workbook.

The workbook should look like Figure 33-8. Next, you will create the layout for the first tab.

23. Select the Bid worksheet by clicking on the **Bid** tab.

FIGURE 33-8 Takeoff Worksheet

24. Change the width of Column A to 2, of Column B to 35, of Columns C and D to 9, and of Columns E and F to 15.
25. Type the text shown in Table 33-5 into the specified cells.
26. Center the text in Cells D3 through D13, D16, E2, and F2.
27. Right justify the text in Cells E14 and E17.
28. Merge and center the text in Cells C2 and D2.
29. Format Cells C3 through C6 and C8 through C13 to whole numbers by selecting the cells, selecting the **Home>Number>Comma Style** button, and clicking the **Home>Number>Decrease Decimal** button twice.
30. Format Cells C7 and C16 to a number with one number after the decimal point by selecting the cells, selecting the **Home>Number>Comma Style**, and clicking the **Home>Number>Decrease Decimal** button once.
31. Format Cells E3 through F17 to a number with two numbers after the decimal point by selecting the cells and selecting the **Home>Number>Comma Style Home** button.
32. Underline Cells B2 through F2, B13 through F13, and B16 through F16.
33. Place a Thick Outside Border around Cells B2 through E17.
34. Save the workbook.

The workbook should look like Figure 33-9.

The next step in creating the roofing spreadsheet is to create the formulas and functions needed for the

TABLE 33-5   Data for Cells in Takeoff Worksheet

Cell	Data
B2	*Materials*
B5	*Roofing Nails*
B7	*Underlayment Nails*
B8	*10' Drip Edge*
B9	*10' Counter Flashing*
B10	*20' Ridge Vent*
B11	*Turtle Vents*
B12	*HVAC Flashing*
B13	*Plumbing Flashing*
B15	*Tax (6.5%)*
B16	*Roofing Crew*
C2	*Quantity*
D3	*bundle*
D4	*bundle*
D5	*lbs*
D6	*rolls*
D7	*C*
D8	*ea*
D9	*ea*
D10	*ea*
D11	*ea*
D12	*ea*
D13	*ea*
D16	*lhr*
E2	*Unit Price*
E14	*Subtotal*
E16	35.00
E17	*Total*
F2	*Total*

	A	B	C	D	E	F
1						
2						
3		Materials		Quantity	Unit Price	Total
4				bundle		
5				bundle		
6		Roofing Nails		lbs		
7				rolls		
8		Underlayment Nails		C		
9		10' Drip Edge		ea		
10		10' Counter Flashing		ea		
11		20' Ridge Vent		ea		
12		Turtle Vents		ea		
13		HVAC Flashing		ea		
14		Plumbing Flashing		ea		
15					Subtotal	
16		Tax (6.5%)				
17		Roofing Crew		1hr	35.00	
18					Total	

FIGURE 33-9 Bid Worksheet

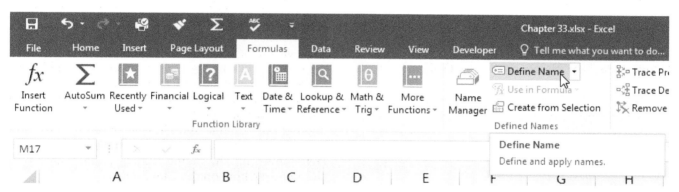

FIGURE 33-10 Formulas: Define Name

FIGURE 33-11 New Name Dialogue Box

spreadsheet to operate. In addition to the functions previously introduced, the roofing spreadsheet uses the naming, concatenate, and lookup functions. It will also use dropdown boxes for entering data. We will begin by looking at the naming function and the creation of dropdown boxes.

NAMING CELLS

In lieu of referencing a cell by its column and row number (for example, C5), one can reference a cell or group of cells by a name given to the cell or group of cells. Cells are given a name by selecting the cell or group of cells to be named, clicking on the **Formulas>Defined Names>Define Name** button (shown in Figure 33-10) to bring up the New Name dialogue box shown in Figure 33-11. The name for the cells is typed in the top text box. The name can be assigned to the entire workbook or the individual sheet in the Scope: dropdown box.

The location of the cells named is changed by changing the reference in the Refers to: box. This is done by typing the correct reference in the box or by

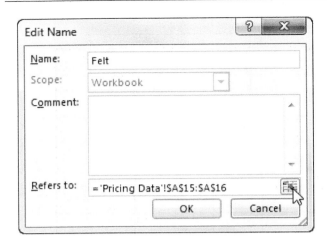

FIGURE 33-12 Click Here to Open New Name-Refers to: Dialogue Box

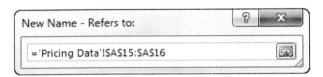

FIGURE 33-13 New Name – Refers to: Dialogue Box

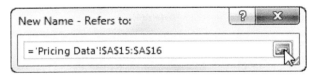

FIGURE 33-14 Click Here to Return to the New Name Dialogue Box

clicking the button at the right (shown in Figure 33-12) to go to the worksheet and bring up the New Name—Refers to: dialogue box shown in Figure 33-13. The user then selects the correct cells and clicks on the button to the right of the cell reference (shown in Figure 33-14) to return to the New Name dialogue box. The New Name dialogue box is closed by right clicking on the **OK** button.

Once a cell has been named, the cell may then be referenced by its name rather than its column and row number. For example, if Cell B12 were given the name "Area," the Excel formula

```
=B12
```

could be rewritten as follows:

```
=Area
```

A series or block of cells may also be named. When the named cell or the entire block of named cells is selected, the name appears in the Name Box located at the left of the Formula bar, which usually contains the cell reference by column and row. The Name Box is shown in Figure 33-15. Alternatively, cells may be named by selecting the cells, clicking on the Name Box, and typing their name.

The name is inserted into a formula by (1) typing the name in the formula, (2) clicking on the **Formulas>Defined Names>Use In Formula** (shown in Figure 33-16), and selecting the name from the popup menu shown in Figure 33-17, or (3) clicking on the **Formulas>Defined Names>Use, In Formula** and selecting **Paste Names...** from the popup menu shown in Figure 33-17, to bring up the Paste Name dialogue box shown in Figure 33-18. From the Paste Name dialogue box, the user selects the name to be used and clicks the **OK** button to close the dialogue box.

The advantage of naming cells is that it makes the formulas easier to follow because the cell can be named with a name that is representative of the data contained in the cells rather than a nondescriptive cell reference.

Names are edited and deleted by opening the Name Manager dialogue box (shown in Figure 33-19), which is opened by clicking on the **Formulas>Defined Names>Name Manager** button (shown in Figure 33-20). From the Name Manager the user may create another name by clicking on the **New** button, edit a name by highlighting a name and clicking the **Edit** button, or delete a name by highlighting a name and clicking the **Delete** button. The Name Manager dialogue box is closed by clicking on the **Close** button.

ADDING DROPDOWN BOXES

Dropdown boxes are created using the same data validation feature that was used to set limits on the data entered into a cell, which was discussed in Chapter 32. To create a dropdown box, the user selects the cell where the dropdown box is to be located, clicks on the **Data>Data Tools>Data Validation** button (shown in Figure 32-27) to bring up the Data Validation dialogue box shown in Figure 33-21. From the Setting tab the user selects **List** from the Allow: dropdown box. For the dropdown box to appear, the user must make sure the In-cell dropdown check box is checked.

The list of options or cells containing the list of options for the dropdown box is entered into the Source: text box. The list may be entered in a number of ways.

First, the cells containing the list of options are selected by clicking on the button to the right of the Source: text box and selecting the cell as was done with the naming function.

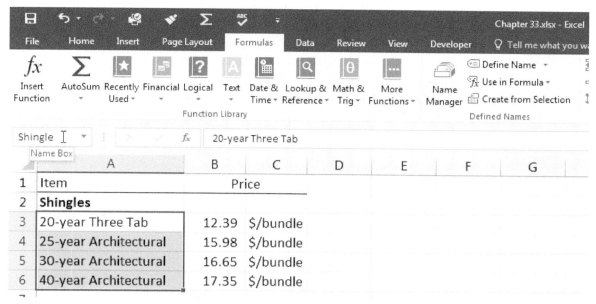

FIGURE 33-15 Name Box

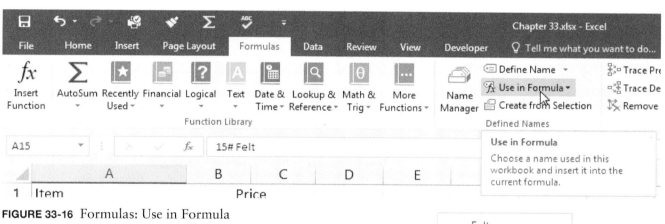

FIGURE 33-16 Formulas: Use in Formula

Second, the cells containing the list of options are named and the name of the cells is pasted into the Source: text box by clicking on the **Formulas>Defined Names>Use In Formula** (shown in Figure 33-16) and selecting the name from the popup menu (shown in Figure 33-17) or selecting **Paste Names...** from the popup menu (shown in Figure 33-17) to bring up the Paste Name dialogue box (shown in Figure 33-18), selecting the name of the cells, and clicking on the **OK** button. Alternatively, the name may be typed into the Source: text box by typing the equals sign and the name of the cells.

Third, the list of acceptable options is typed directly into the Source: box with a comma separating the items in the list.

Input messages and error alerts may be added to the dropdown box as was shown in Chapter 32. The Data Validation dialogue box is closed by clicking on the **OK** button. When the user selects the cell where the

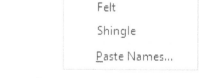

FIGURE 33-17 Paste Name Popup Box

FIGURE 33-18 Paste Name Dialogue Box

364 CHAPTER THIRTY-THREE

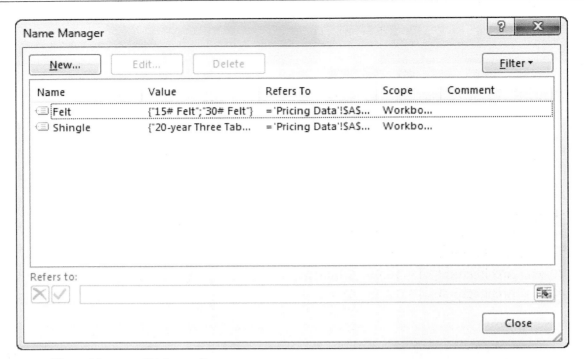

FIGURE 33-19 Name Manager Dialogue Box

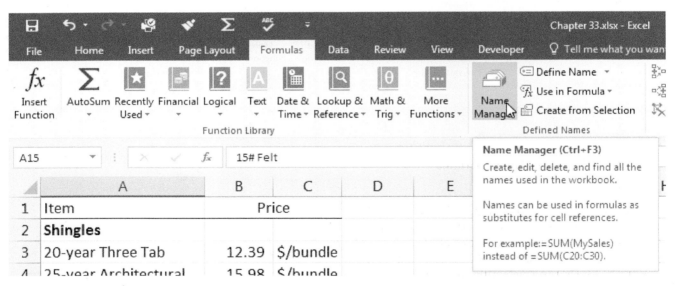

FIGURE 33-20 Formulas: Name Manager Button

dropdown box is located, an arrow appears to the side of the cell, and the user can select from the list of options by clicking on this arrow

Exercise 33-3

In this exercise you will add three dropdown boxes to the Takeoff worksheet, which will be used to select the shingle type, felt type, and vent type. You will begin by naming the cells containing the shingle and vent types by completing the following steps:

1. Make sure that Chapter 33, the workbook modified in Exercise 33-2, is open.
2. Select the Pricing Data worksheet.
3. Select Cells A3 through A6, select the Name Box at the left of the Formula bar (shown in Figure 33-15), type *Shingle*, and press the **Enter** key.
4. Select Cells A15 and A16, click on the **Formulas>Defined Names>Define Name** (shown in Figure 33-10) to bring up the New Name dialogue box, and type *Felt* in the top text box. The New Name dialogue box should look like Figure 33-11. Click the **OK** button to close the dialogue box.

5. Select the Takeoff worksheet by clicking on the **Takeoff** tab.

Next, you will restrict the values in Cell D4 to the values listed in the cells you named Shingle and add custom input and error messages to this cell.

6. Select Cell D4 and click on the **Data>Data Tools>Data Validation** button (shown in Figure 32-27) to bring up the Data Validation dialogue box (shown in Figure 33-21).

7. Select the **Setting** tab and select **List** from the Allow: dropdown box.

8. Place the cursor in the Source: text box, click on the **Formulas>Defined Names>Use In Formula** (shown in Figure 33-16), and select **Shingle** from the popup menu (shown in Figure 33-17).

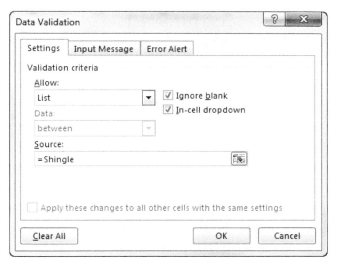

FIGURE 33-21 Settings Tab of the Data Validation Dialogue Box

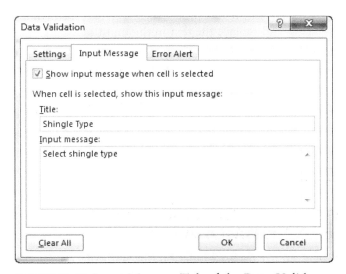

FIGURE 33-22 Input Message Tab of the Data Validation Dialogue Box

9. Make sure the **In-cell dropdown** check box is checked. The Settings tab should look like Figure 33-21.

10. Select the **Input Message** tab and make sure the **Show input message when cell is selected** check box is checked.

11. Type *Shingle Type* in the Title: text box and *Select shingle type* in the Input message: text box. The Input Message tab should look like Figure 33-22.

12. Select the **Error Alert** tab and make sure that the **Show error alert after invalid data is entered** check box is checked.

13. Type *Error* in the Title: text box and *Please select from dropdown list* in the Input message: text box. The Error Alert tab should look like Figure 33-23.

14. Click the **OK** button to complete the data validation.

15. Verify that the dropdown arrow and prompt appear when you select Cell D4 as shown in Figure 33-24.

16. Verify that when you click on the dropdown arrow the values in Cells A3 through A6 on the Pricing Data worksheet are available for selection as shown in Figure 33-25.

FIGURE 33-23 Error Alert Tab of the Data Validation Dialogue Box

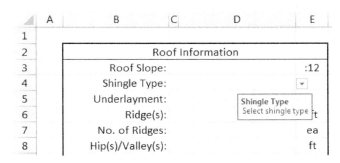

FIGURE 33-24 Dropdown Arrow and Prompt

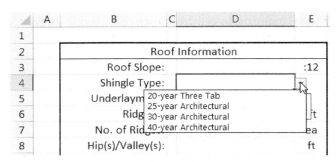

FIGURE 33-25 Dropdown List

FIGURE 33-26 Error Message

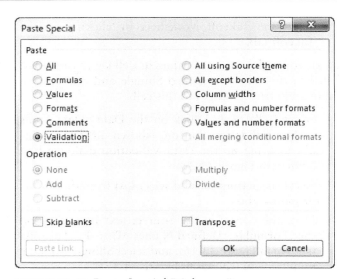

FIGURE 33-27 Paste Special Dialogue Box

17. Verify that the error message shown in Figure 33-26 appears when you enter unacceptable data by entering *shake* in the cell.

18. Click the **Cancel** button to close the error message.

Next, you will restrict the values in Cell D5 to the values listed in the cells you named Felt and add custom input and error messages to this cell. For this example the error message is copied from Cell D4 to Cell D5, and it does not have to be reentered.

19. Copy Cell D4 to Cell D5. When a cell is copied, the data validation properties are also copied.

20. From Cell D5, click on the **Data>Data Tools >Data Validation** button to bring up the Data Validation dialogue box.

21. Select the **Setting** tab and type *=Felt* in the Source: text box. Be sure to include the equals sign.

22. Select the **Input Message** tab.

23. Replace the text in the Title: text box with *Felt Type* and replace the text in the Input message: text box with *Select felt type*.

24. Click the **OK** button to complete the data validation.

25. Verify that the dropdown arrow, prompt, and error message are working, and that 15# Felt and 30# Felt are available for selection from the dropdown list.

Next, you will restrict the values in Cell D15 to gable end, ridge, or turtle and add custom input and error messages to this cell.

26. Right click on Cell D4 and select **Copy** from the popup menu.

27. Right click on Cell D15, select **Paste Special...** from the popup menu to bring up the Paste Special dialogue box (shown in Figure 33-27), click on the **Validation** radio button, and click on the **OK** button to copy the data validation properties from Cell D4 to Cell D15. If the paste command were used, the border at the bottom of the cell would be deleted and would need to be replaced.

28. From Cell D15, click on the **Data>Data Tools >Data Validation** button to bring up the Data Validation dialogue box.

29. Select the **Setting** tab and type *Gable End, Ridge,Turtle* in the Source: text box. Do not include a space between the commas and the words, before Gable, or after Turtle. The only space should be between the words Gable and End. Do not include an equals sign.

30. Select the **Input Message** tab.

31. Replace the text in the Title: text box with *Vent Type* and replace the text in the Input message: text box with *Select vent type*.

32. Click the **OK** button to complete the data validation.

33. Verify that the dropdown arrow, prompt, and error message are working and that Gable End, Ridge, and Turtle are available for selection from the dropdown list.

34. Save the file.

REFERENCING WORKSHEETS IN A FORMULA

When one writes a formula on one worksheet, cells on another worksheet may be used in the formula by clicking on the worksheet where the cell is found and clicking on the cell or by typing the name of the sheet followed by an exclamation point (!) followed by the cell reference. For example, Cell F17 on the Bid worksheet would be referenced as follows:

```
Bid!F17
```

When the worksheet name contains a space, the name of the worksheet must be enclosed in single quotes. For example, Cell B20 on the Pricing Data worksheet would be referenced as follows:

```
'Pricing Data'!B20
```

Exercise 33-4

In this exercise you will create the formulas for the Takeoff worksheet using the following steps:

1. Make sure that Chapter 33, the workbook modified in Exercise 33-3, is open.
2. Select Cell D17 on the Takeoff worksheet.

The area of the roof in plan view is calculated by multiplying the individual lengths and widths and adding the results together. The area may be written by the following formula:

$$\text{Area} = (\text{Length}_1)(\text{Width}_1) + (\text{Length}_2)(\text{Width}_2) + \cdots + (\text{Length}_{10})(\text{Width}_{10})$$

This formula is written in Excel as follows:

```
=H4*I4+H5*I5+H6*I6+H7*I7+H8*I8+H9*I9+H10*
I10+H11*I11+H12*I12+H13*I13
```

3. Type this formula into Cell C17, or select Cell C17 and type the equals sign (=), select Cell H4, type the asterisks (*), select Cell I4, type the plus sign (+), select Cell H5, and so forth until the formula is complete.

The unit price (found in Cell D18) is the total cost (found in Cell F17 on the Bid worksheet) divided by the sloped area of the roof. The sloped area of the roof is calculated from the area in plan view (found in Cell C17) by Eq. (9-7). The Excel formula for this is as follows:

```
=Bid!F17/(D17*(1+(D3/12)^2)^0.5)
```

4. Enter this formula into Cell D18 on the takeoff worksheet. When Cell D17 is zero, "#DIV/0!" will be displayed in Cell D18
5. Save the workbook.

CONCATENATE

The CONCATENATE function allows the user to link text, the results of formulas, and the contents of cells together into a single string of text. The contents of the cell may be text or the result of a formula. The CONCATENATE function is written as follows:

```
=CONCATENATE(text1,text2, . . . )
```

where text1, text2, and so forth are the references of the cells and text to be linked. The linking occurs without the addition of any spaces, so the user must add spaces as necessary. The spaces and text to be added must be enclosed in double quotes. For example, if Cell B1 were 17, the formula

```
=CONCATENATE("Cell B1 equals ",B1,".")
```

would result in the following text:

```
Cell B1 equals 17.
```

The space after the equals in the formula provides the space between the equals and the 17. Alternatively, this CONCATENATE function can be set up using the Function Arguments dialogue box shown in Figure 33-28. The Function Arguments dialogue box for the CONCATENATE function is opened by clicking on the **Formulas>Function Libaray>Text** button (shown in Figure 33-29) and clicking on CONCATENATE in the popup menu (shown in Figure 33-30).

Alternatively, text, the results of formulas, and the contents of cell can be linked together by the ampersand (&).

```
="Cell B1 equals "&B1&"."
```

produces the same result as

```
=CONCATENATE("Cell B1 equals ",B1,".") .
```

Exercise 33-5

In this exercise you will add material names for Cells B3, B4, and B6 and add the formulas that will calculate the quantities of materials. The type of materials ordered for the shingles, cap shingles, and underlayment depend on the type of material specified and will need to change on the Bid worksheet to match the items selected on the Takeoff worksheet. This is done using the following steps:

1. Make sure that Chapter 33, the workbook modified in Exercise 33-4, is open.
2. Click on the **Bid** tab and click on Cell B3.

Cell B3 on the Bid worksheet should be the same as Cell D4 on the Takeoff worksheet, unless the Cell D4 is blank, in which case Cell B3 should be "Select Shingle Type." This is done with an IF function, which checks to see if Cell D4 on the Takeoff worksheet is blank. If it is blank, the cell should contain Select Shingle Type; otherwise, the cell's contents should be the same as Cell D4 on the Takeoff worksheet. A blank dropdown box carries a numeric value of zero. The formula is written as follows:

```
=IF(Takeoff!D4=0,"Select Shingle Type",
Takeoff!D4)
```

3. Enter this formula into Cell B3 on the Bid worksheet.

Next we will use conditional formatting to change the font of Cell B3 to red when the shingle type has not been selected.

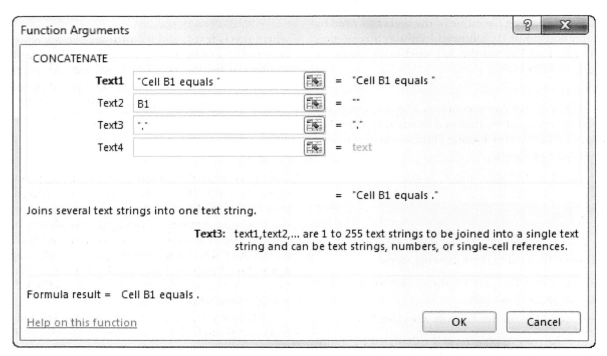

FIGURE 33-28 CONCATENATE Function Arguments Dialogue Box

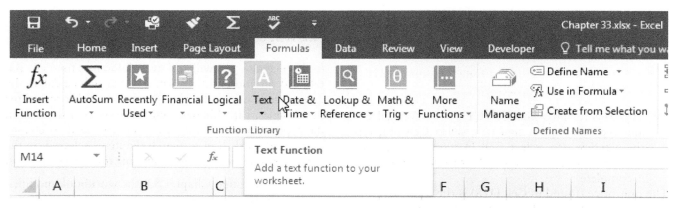

FIGURE 33-29 Functions: Text Button

4. Place the cursor in Cell B3, click on the **Home >Styles>Conditional Formatting** button (shown in Figure 32-19), place the cursor over the **Highlight Cell Rules** from the popup menu (shown in Figure 32-20), and click on **More Rules...** from the popup menu (shown in Figure 32-21) to bring up the New Rule Formatting dialogue box (shown in Figure 33-31).

5. Select **Cell Value** from the first dropdown box and **not equal to** from the second dropdown box and type =*Takeoff!D4* in the text box.

6. Click on the **Format . . .** button to bring up the Format Cells dialogue box (shown in Figure 33-32), click the **Font** tab, select the color red in the color dropdown box, and click the **OK** button.

7. The New Formatting Rule dialogue box should appear as shown in Figure 33-31, with red text in the Preview box.

8. Click on the **OK** button to close the New Formatting Rule dialogue box. "Select Shingle Type" should now be in red.

Cell B4 should combine the shingle type shown in Cell D4 of the Takeoff worksheet with the word "cap." This is done by using the CONCATENATE function. The Excel function is written as follows:

=CONCATENATE(Takeoff!D4," Cap")

9. Enter this function into Cell B4. Be sure to include a space before the word Cap.

Cell B6 on the Bid worksheet should be the same as Cell D5 on the Takeoff worksheet, unless Cell D5 is blank, in which case Cell B6 should be "Select Underlayment Type." This is done in the same manner as was done for Cell B3. The Excel formula is written as follows:

```
=IF(Takeoff!D5=0,"Select Underlayment Type",Takeoff!D5)
```

10. Enter this formula into Cell B6 on the Bid worksheet.
11. Add the same conditional formatting to Cell B6 as was added to Cell B3 except type =*Takeoff*!*D5* in the text box of New Formatting Rule dialogue box.
12. Save the workbook.

The quantity of shingles is determined by the sum of Eqs. (9-5), (9-7), and (9-8). Equation (9-5) for an exposure of 5 inches is written in Excel as follows:

```
=(5/12)*Takeoff!D9/100
```

Equation (9-7) is written in Excel as follows:

```
=Takeoff!D17*(1+(Takeoff!D3/12)^2)^0.5/100
```

Equation (9-8) for an exposure of 5 inches is written in Excel as follows:

```
=Takeoff!D8*((1+(Takeoff!
D3/12)^2)/2)^0.5*(5/12)/100
```

These equations will determine the number of squares of materials. The number of squares needs to be converted to bundles before the quantity is rounded up to

FIGURE 33-30 Text Popup Menu

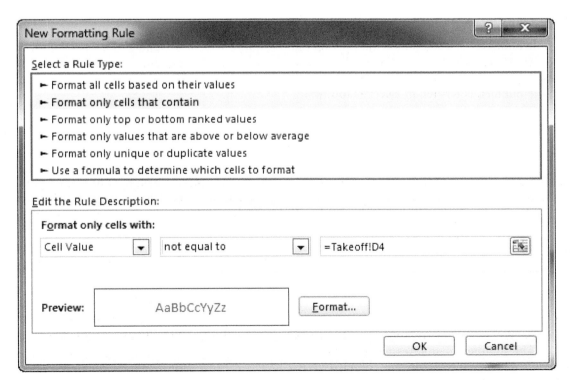

FIGURE 33-31 New Formatting Rule Dialogue Box

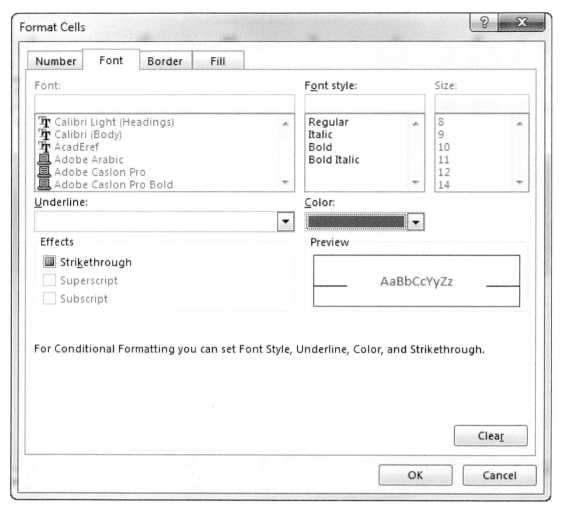

FIGURE 33-32 Font Tab of the Format Cells Dialogue Box

the next whole number. This is done by multiplying the sum by 3 because there are 3 bundles in a square. The sum of these equations, including the rounding up to the next whole number, is written in Excel as follows:

```
=ROUNDUP(3*((5/12)*Takeoff!D9/100+Takeoff!
D17*(1+(Takeoff!D3/12)^2)^0.5/100+Takeoff!
D8*((1+(Takeoff!D3/12)^2)/2)^0.5*(5/12)/
100),0)
```

13. Enter this formula into Cell C3 on the Bid worksheet.

The quantity of cap shingles is determined by the sum of Eqs. (9-9) and (9-10). Equation (9-9) is written in Excel as follows:

```
=Takeoff!D6/100
```

Equation (9-10) is written in Excel as follows:

```
=Takeoff!D8*(1+(Takeoff!D3/17)^2)^0.5/100
```

These equations will determine the number of squares of materials. The number of squares needs to be converted to bundles before the quantity is rounded up to the next whole number. This is done by multiplying the sum by 100 and dividing by 21, because there are 21 square feet in a bundle. The sum of these equations, including the rounding, is written in Excel as follows:

```
=ROUNDUP((100/21)*(Takeoff!D6/100+Takeoff!
D8*(1+(Takeoff!D3/17)^2)^0.5/100),0)
```

14. Enter this formula into Cell C4 on the Bid worksheet.

The number of roofing nails equals the number of bundles of shingles divided by 3 because 1 pound of nails is needed for every 3 bundles of shingles. The Excel formula for the number of nails is written as follows:

```
=ROUNDUP((C3+C4)/3,0)
```

15. Enter this formula into Cell C5 on the Bid worksheet.

The quantity of underlayment is calculated using Eq. (9-3). For a standard 3-foot width and a 6-inch lap, Eq. (9-3) is written in Excel as follows:

```
=(Takeoff!D17*(1+(Takeoff!D3/12)^2)^0.5*
 (3/2.5)+Takeoff!D8*((1+(Takeoff!D3/
 12)^2)/2)^0.5*(3+1)*(3/2.5)+(Takeoff!D6
 +Takeoff!D7)*3)
```

For 30-pound felt, the quantity is divided by 200; for 15-pound felt, the quantity is divided by 400. This decision is made using an IF function written as follows:

```
=IF(Takeoff!D5="30# Felt",200,400)
```

The IF function does not need to check to see if the cell contains 15-pound felt because the dropdown box for felt is only allowed to contain 30-pound felt or 15-pound felt; if it is not 30-pound felt, it must be 15-pound felt. Combining these two functions, we get the following Excel formula:

```
=ROUNDUP((Takeoff!D17*(1+(Takeoff!D3/
12)^2)^0.5*(3/2.5)+Takeoff!D8*((1
+(Takeoff!D3/12)^2)/2)^0.5*(3+1)*(3/2.5)
+(Takeoff!D6+Takeoff!D7)*3)/IF(Takeoff!
D5="30\# Felt",200,400),0)
```

16. Enter this formula into Cell C6 on the Bid worksheet.

From historical data, 125 nails are needed for 1 square of felt. The quantity of underlayment nails needed equals the number of squares of felt multiplied by 125, where there are 2 squares on a roll of 30-pound felt and 4 squares on a roll of 15-pound felt. The quantity is based on the number of rolls of felt ordered and is divided by 100 to express the quantity in 100s (C). The quantity is rounded to tens or 0.1 C. The Excel function is written as follows:

```
=ROUNDUP(C6*IF(Takeoff!D5="30#
Felt",2,4)*125/100,1)
```

17. Enter this function into Cell C7 on the Bid worksheet.

The quantity of drip edge equals the length of the horizontal perimeter plus the sloped perimeter of the roof, where the sloped perimeter is converted from a plan view to a length along the slope using Eq. (9-1). The length of the drip edge is converted to 10-foot-long pieces by dividing by 9.8 feet, thus allowing 0.2 feet for lap. The Excel function for the drip edge is written as follows:

```
=ROUNDUP((Takeoff!D9+Takeoff!D10*
(1+(Takeoff!D3/12)^2)^0.5)/9.8,0)
```

18. Enter this function into Cell C8 on the Bid worksheet.

Counter flashing is handled in the same manner. The Excel function for the counter flashing is written as follows:

```
=ROUNDUP((Takeoff!D11+Takeoff!D12*
(1+(Takeoff!D3/12)^2)^0.5)/9.8,0)
```

19. Enter this function into Cell C9 on the Bid worksheet.

If a ridge vent is specified, the length of the ridge will equal the quantity of ridge vent needed. If a ridge vent is not specified then the length is zero. The quantity of ridge vent needs to be converted to 20-foot rolls. The Excel function for the ridge vent is written as follows:

```
=IF(Takeoff!D15="Ridge",ROUNDUP(Takeoff
!D6/20,0),0)
```

20. Enter this function into Cell C10 on the Bid worksheet.

If turtle vents are specified, the quantity of turtle vents is calculated by dividing the area of the roof in plan view by 127. The number of vents must be rounded up. The Excel function for the turtle vents is as follows:

```
=IF(Takeoff!D15="Turtle",ROUNDUP(Takeoff
!D17/127,0),0)
```

21. Enter this function into Cell C11 on the Bid worksheet.

The quantities of HVAC and plumbing flashings are directly taken off of the plans and are equal to Cells D13 and D14 on the Takeoff worksheet.

22. Enter =*Takeoff!D13* into Cell C12 on the Bid worksheet.

23. Enter =*Takeoff!D14* into Cell C13 on the Bid worksheet.

24. Save the worksheet.

LOOKUP AND VLOOKUP

The LOOKUP function is used to select the correct material price based on the type of shingle or underlayment selected. This is done by comparing the type of material on the Takeoff worksheet to the list of materials and their prices on the Pricing Data worksheet. The LOOKUP function is written as follows:

```
=LOOKUP(lookup_value,array)
```

where the lookup_value is the value to be looked up (the type of material in our case) and the array is where the list of values is stored (the materials and their prices in our case). The array should be two columns wide with the lookup value in the left column and the associated value in the column to the right. The values in the left column of the array must be in ascending order (−1, 0, 1, 2, ... , A-Z) because the LOOKUP function will begin comparing the lookup value to the values in the left column of the array beginning with the top and working its way down, stopping when it finds a value equal to or greater than the lookup value. If the LOOKUP function finds a value in the left column of the array equal to the lookup value, it will return the value of the cell to its right. If the LOOKUP function finds a value in the left column of the array greater than the lookup value before it finds a value it is equal to, it will return the value of the cell one column to the right and one row up. For example, for the data in Figure 33-33, the formula

```
=LOOKUP(A1,A3:B6)
```

	A	B
1		
2		
3	a	1
4	b	2
5	g	3
6	e	4

FIGURE 33-33 LOOKUP array

is equal to 2 when cell A1 is equal to b. Alternatively, this LOOKUP function is equal to 2 when cell A1 is equal to e because the function finds g (which is greater than e) before it finds the e. In this case the LOOKUP function selects the value from the cell to the right and one row up (Cell B4) from Cell A5, where g is found.

Alternatively, this LOOKUP function can be set up using the Function Arguments dialogue box shown in Figure 33-34. The Function Arguments dialogue box for the LOOKUP function is opened by clicking on the **Formulas>Function Library>Lookup & Reference** button (shown in Figure 33-35), clicking on LOOKUP in the popup menu (shown in Figure 33-36), clicking on **lookup_value,array** from the Select Arguments dialogue box (shown in Figure 33-37), and clicking on **OK**.

The VLOOKUP function works in a way similar to the LOOKUP function, but it allows the user to select between multiple return values stored in different columns. The VLOOKUP function is written as follows:

=VLOOKUP(lookup_value,table_array,col_index_number)

where the lookup_value is the value to be looked up, the table_array is where the list of values is stored, and the col_index_number identifies which column is to be used for the return values. The table array should be at least two columns wide with the lookup values in the left column and the return values in the columns to the right. The lookup values in the left column must be listed in ascending order just as they were for the LOOKUP function. The VLOOKUP function follows the same rules as the LOOKUP function when looking for the lookup value. A column index number of 2 selects return values

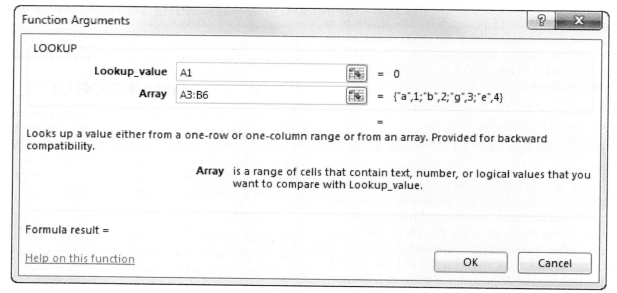

FIGURE 33-34 LOOKUP Function Arguments Dialogue Box

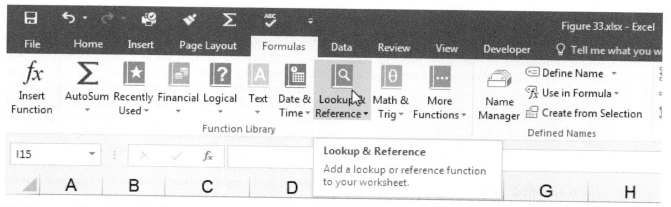

FIGURE 33-35 Functions: Lookup & Reference Button

FIGURE 33-36 Lookup & Reference Popup Menu

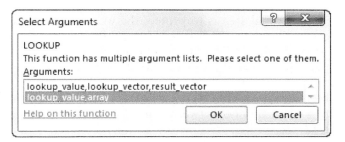

FIGURE 33-37 Select Arguments Dialogue Box

from the second column of the array, whereas a column index number of 3 selects return values from the third column. The column index number must be greater than 1 and less than or equal to the number of columns in the array. For example, for the data in Figure 33-38, the formula

```
=VLOOKUP(A1,A3:D6,2)
```

is equal to 2 when Cell A1 is equal to b. If the column index number is changed to 3 as follows:

```
=VLOOKUP(A1,A3:D6,3)
```

then the function returns 12 when Cell A1 is equal to b. The VLOOKUP function can be used for multiple pricing lists or skipping columns when looking up prices. The use of the LOOKUP function is shown in the following exercise:

	A	B	C	D
1				
2				
3	a	1	11	21
4	b	2	12	22
5	c	3	13	23
6	d	4	14	24

FIGURE 33-38 VLOOKUP array

Exercise 33-6

In this exercise you will determine the material, labor, and total costs of the roof by completing the following steps. The prices in Cells E3, E4, and D6 will need to change based on the type of materials specified. The prices in Cells E5, and E7 through E13 do not change.

1. Make sure that Chapter 33, the workbook modified in Exercise 33-5, is open.
2. Begin by entering the formulas in Table 33-6 into the associated cells on the Bid worksheet.

The materials pricing for the shingles must be selected from Cells B3 through B6 based on the shingles specified. This is done by using a LOOKUP function, which is written as follows:

```
=LOOKUP(Takeoff!D4,'Pricing Data'!A3:B6)
```

3. Enter this function into Cell E3 on the Bid worksheet. The cell will contain #N/A when a shingle type has not been selected.
4. Enter =*LOOKUP(Takeoff!D4,'Pricing Data'!A9: B12)* into Cell E4 on the Bid worksheet.
5. Enter =*LOOKUP(Takeoff!D5,'Pricing Data'!A15: B16)* into Cell E6 on the Bid worksheet.

The next step is to calculate the total cost for each type of material. This is done by multiplying Column E by Column C.

TABLE 33-6 Formulas

Cell	Formula
E5	='Pricing Data'!B27
E7	='Pricing Data'!B28
E8	='Pricing Data'!B19
E9	='Pricing Data'!B20
E10	='Pricing Data'!B21
E11	='Pricing Data'!B22
E12	='Pricing Data'!B23
E13	='Pricing Data'!B24

6. Enter =E3*C3 into Cell F3 on the Bid worksheet.

7. Copy Cell F3 to Cells F4 through F13 without copying the border by right clicking on Cell F3, selecting **Copy** from the popup menu, selecting Cells F4 through F13, right clicking on one of the selected cells, selecting **Paste Special…** from the popup menu to bring up the Paste Special dialogue box, selecting **All except borders**, and clicking on the **OK** button.

Next, we calculate the total cost of the materials without sales tax.

8. Enter =SUM(F3:F13) into Cell F14 of the Bid worksheet.

Next, we calculate the sales tax at a fixed rate of 6.5%.

9. Enter =F14*0.065 into Cell F15 of the Bid worksheet.

The next step is to determine the number of labor hours needed for the roof, which is determined by multiplying the labor hours per square for the specified type of shingle by the sloped area of the roof and adding two hours for mobilization. The labor hours per square is found in Cells B31 through B34 of the Pricing Data worksheet, and the LOOKUP function is used to select the appropriate production rate. The Excel formula for the sloped area is as follows:

```
=Takeoff!D17*(1+(Takeoff!D3/12)^2)^0.5/100
```

The Excel function to determine the labor hours is written as follows:

```
=ROUNDUP(LOOKUP(Takeoff!D4,'Pricing
Data'!A31:B34)*Takeoff!D17*(1+
(Takeoff!D3/12)^2)^0.5/100+2,1)
```

10. Enter this function into Cell C16 on the Bid worksheet.

The next step is to calculate the total labor costs by multiplying Column E by Column C.

11. Enter =E16*C16 into Cell F16 of the Bid worksheet.

The next step is to calculate the total costs by summing the material cost without tax, the sales tax, and the labor.

12. Enter =SUM(F14:F16) into Cell F17 of the Bid worksheet.

13. Save the workbook.

The next step is to automate with macros; however, macros are not needed in this workbook. The next step is to test the worksheet. This is done in the following exercise

Exercise 33-7

In this exercise you will test the workbook as follows:

1. Make sure that Chapter 33, the workbook modified in Exercise 33-6, is open.

2. Begin by entering the numbers and text in Table 33-7 into the associated cell in the Takeoff worksheet.

The Bid worksheet should appear as it does in Figure 33-39.

3. Exit without saving the worksheet.

The final step is to protect the worksheet against errors. For the roofing spreadsheet, this protection will be in two forms. First, data validation is used to restrict the values entered into Cells D3 through D15 and H4 through I13 on the Takeoff worksheet. This has already been done for Cells D4, D5, and D15. Cell C3 is limited to a whole number between 0 and 12. Cells D6 through D12 are limited to a whole number between 0 and 200. Cells D13 and D14 are limited to a whole number

	A	B	C	D	E	F
1						
2	Materials		Quantity		Unit Price	Total
3	25-year Architectural		51	bundle	15.98	814.98
4	25-year Architectural Cap		6	bundle	31.29	187.74
5	Roofing Nails		19	lbs	1.39	26.41
6	15# Felt		6	rolls	14.95	89.70
7	Underlayment Nails		30.0	C	1.00	30.00
8	10' Drip Edge		16	ea	2.99	47.84
9	10' Counter Flashing		2	ea	3.99	7.98
10	20' Ridge Vent		-	ea	42.00	-
11	Turtle Vents		12	ea	7.00	84.00
12	HVAC Flashing		2	ea	9.00	18.00
13	Plumbing Flashing		4	ea	4.00	16.00
14					Subtotal	1,322.65
15	Tax (6.5%)					85.97
16	Roofing Crew		46.1	lhrs	35.00	1,613.50
17					Total	3,022.12

FIGURE 33-39 Bid Worksheet

TABLE 33-7 Test Case

Cell	Data
C3	6
C4	25-year Architectural
C5	15# Felt
C6	12
C7	1
C8	91
C9	152
C10	0
C11	4
C12	8
C13	2
C14	4
C15	Turtle
H4	44
I4	32

between 0 and 20. Finally, Cells H4 through I13 are limited to a whole number between 0 and 200.

Second, the worksheet is protected, so only Cells D3 through D15 and H4 through I13 can be changed. In the following exercise you will protect the worksheets against errors:

Exercise 33-8

In this exercise you will protect the worksheet against errors by using the following steps:

1. Make sure that Chapter 33, the workbook modified in Exercise 33-6, is open.
2. Click on the **Takeoff** tab to move to the Takeoff worksheet.

First, the data validation for Cell D3 is set up.

3. Select Cell D3 and click on the **Data>Data Tools> Data Validation** button to bring up the Data Validation dialogue box (shown in Figure 33-40).
4. Select the **Settings** tab, select **Whole Number** from the Allow: dropdown box, select **between** from the Data: dropdown box, type *3* in the Minimum: text box, and type *12* in the Maximum: text box.
5. Select the **Input Message** tab and type *Enter a whole number between 3 and 12.* in the Input message: text box. The Title: text box will remain blank.
6. Select the **Error** tab, type *Error* in the Title: text box, type *Enter a whole number between 3 and 12.* in the Error message: text box, and click the **OK** button to close the Data Validation dialogue box.

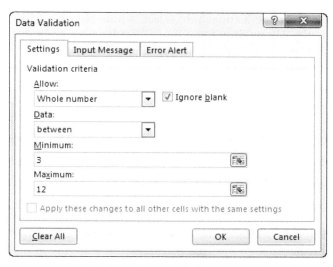

FIGURE 33-40 Settings Tab of the Data Validation Dialogue Box

Next, the data validation for Cell D6 is set up.

7. Select Cell D6 and click on the **Data>Data Tools> Data Validation** to bring up the Data Validation dialogue box.
8. Select the **Settings** tab, select **Whole Number** from the Allow: dropdown box, select **between** from the Data: dropdown box, type *0* in the Minimum: text box, and type *200* in the Maximum: text box.
9. Select the **Input Message** tab and type *Enter a whole number between 0 and 200.* in the Input message: text box.
10. Select the **Error** tab, type *Error* in the Title: text box, type *Enter a whole number between 0 and 200.* in the Error message: text box, and click the **OK** button to close the Data Validation dialogue box.

Next, copy the data validation from Cell D6 to Cells D7 through D12 and H4 through I13 using the following steps.

11. Right click on Cell D6 and select **Copy** from the popup menu.
12. Select Cells D7 through D12, right click on one of the selected cells, select **Paste Special . . .** from the popup menu to bring up the Paste Special dialogue box, select the **Validation** radio button, and click the **OK** button to close the Paste Special dialogue box.
13. Repeat this process for Cells H4 through I13.

Next, the data validation for Cells D13 and D14 is set up.

14. Select Cells D13 and D14 and click on the **Data>Data Tools>Data Validation** to bring up the Data Validation dialogue box.

15. Select the **Settings** tab, select **Whole Number** from the Allow: dropdown box, select **between** from the Data: dropdown box, type *0* in the Minimum: text box, and type *20* in the Maximum: text box.
16. Select the **Input Message** tab and type *Enter a whole number between 0 and 20.* in the Input message: text box.
17. Select the **Error** tab, type *Error* in the Title: text box, type *Enter a whole number between 0 and 20.* in the Error message: text box, and click the **OK** button to close the Data Validation dialogue box.

The final step is to protect the workbook. This is done in two steps. First, identify the cells that will be allowed to change. Second, protect each of the worksheets in the workbook.

18. Select Cells D3 through D15 and H4 through I13, right click on one of the selected cells, select **Format Cells . . .** from the popup menu to bring up the Format Cells dialogue box, select the **Protection** tab, uncheck the **Locked** check box, and click the **OK** button to close the dialogue box.
19. Protect the Takeoff worksheet by clicking on the **Review>Changes>Protect Sheet** button in the Changes group (shown in Figure 32-34) to bring up the Protect Sheet dialogue box (shown in Figure 32-35).
20. Make sure that the **Protect worksheet and contents of locked cells** check box is checked.
21. Type *1234* in the Password to unprotect sheet: text box.
22. Uncheck the **Select locked cells** check box.
23. Make sure that the **Select unlocked cells** check box is checked.
24. Make sure that all other check boxes are unchecked.
25. Click on the **OK** button to bring up the Confirm Password dialogue box.
26. Type *1234* in the Reenter password to proceed. text box and click on the **OK** button to close the dialogue box and protect the worksheet.
27. Check to make sure only Cells D3 through D15 and H4 through I13 can be selected by pressing the **Down Arrow** key to move through the unlocked cells.
28. Protect the Bid and Pricing Data worksheets using the same process.
29. Save the workbook.

PROPOSALS

In the previous sections, Excel was used to prepare an estimate for an asphalt shingle roof. It is a simple step to convert this estimate into a ready-to-sign proposal.

Exercise 33-9

In this exercise you will convert the spreadsheet from Exercise 33-8 into a ready-to-sign proposal by completing the following steps:

1. Open Chapter 33.xlsx, the workbook you created in the previous exercise.

Next, you will add a place on the Takeoff worksheet for the client and project information using the following steps:

2. Select the **Takeoff** tab.
3. Click on the **Review>Changes>Unprotect Sheet** button to bring up the Unprotect Sheet dialogue box (shown in Figure 32-37), type *1234* in the Password: text box, and click on the **OK** button to unprotect the worksheet.
4. Select Columns A through D and click on the **Home>Cells>Insert** button (shown in Figure 33-41) to insert four columns to the left of the worksheet.
5. Change the width of Columns A and C to 1, of Column B to 10, and of Column D to 25.
6. Type the text shown in Table 33-8 into the specified cells.
7. Underline Cells B2 through D2 by selecting the cells and clicking the **Home>Font>Bottom Border** button (shown in Figure 3-47). If another border is shown in the Borders button, select the correct border by clicking on the small arrow to the right of the Borders button and selecting the **Bottom Border** from the Borders popup.

TABLE 33-8 Data for Takeoff Worksheet

Cell	Data
B2	*Client Information*
B3	*Name:*
B4	*Address:*
B6	*City:*
B7	*State:*
B8	*Zip Code:*
B9	*Phone #:*
B11	*Project Information*
B12	*Name:*
B13	*Address:*
B15	*City:*
B16	*State:*
B17	*Zip Code:*
B18	*Phone #:*

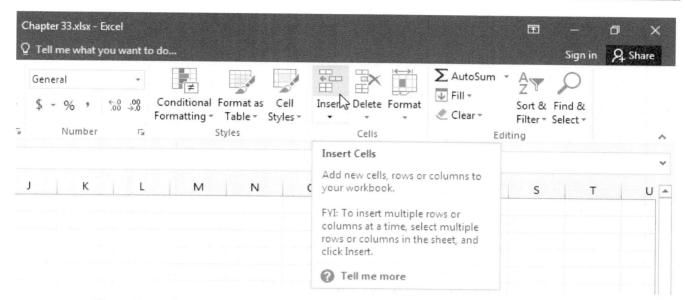

FIGURE 33-41 Home: Insert Button

8. Merge and center Cells B2 through D2 by highlighting Cells B2 through D2 and selecting the **Home>Alignment>Merge & Center** button (shown in Figure 3-40).
9. Underline Cells B11 through D11.
10. Merge and center Cells B11 through D11.
11. Right justify the text in Column B by selecting Column B and selecting the **Home>Alignment>Align Right** button (shown in Figure 3-33).
12. Draw a Thick Outside Border around Cells B2 through D9 by selecting Cells B2 through D9, clicking the **Home** menu tab, clicking on the small arrow to the right of the Borders button in the Font group, and selecting the **Thick Outside Border** from the Borders popup menu.
13. Draw a Thick Box Border around Cells B11 through D18.

The next step is to protect the worksheet using the following steps:

14. Select Cells D3 through D9 and D12 through D18, right click on one of the selected cells, select **Format Cells . . .** from the popup menu to bring up the Format Cells dialogue box (shown in Figure 32-33), select the **Protection** tab, uncheck the **Locked** check box, and click the **OK** button to close the dialogue box.
15. Protect the Takeoff worksheet by clicking on the **Review>Changes>Protect Sheet** button (shown in Figure 32-34) to bring up the Protect Sheet dialogue box (shown in Figure 32-35).
16. Type *1234* in the Password to unprotect sheet: text box.
17. Click on the **OK** button to bring up the Confirm Password dialogue box (shown in Figure 32-36).
18. Type *1234* in the Reenter password to proceed: text box and click on the **OK** button to close the dialogue box and protect the worksheet.
19. Check to make sure only Cells D3 through D9, D12 through 18, H3 through H15, and L4 through M13 can be selected by pressing the **Down Arrow** key to move through the unlocked cells.
20. Save the workbook.

The worksheet should now appear as shown in Figure 33-42.

Next, you will add a place on the Bid worksheet for the client, project, and other proposal information using the following steps:

21. Select the **Bid** tab.
22. Click on the **Review>Changes>Unprotect Sheet** button to bring up the Unprotect Sheet dialogue box, type *1234* in the Password: text box, and click the **OK** button to unprotect the worksheet.
23. Select Rows 1 through 9, click on the **Home>Cells>Insert** button in the Cells group (shown in Figure 33-41) to insert nine rows at the top of the worksheet.
24. Type the text shown in Table 33-9 into the specified cells.

Use the **Underline Key** to create the underlining. Press the **Underline Key** 34 times in Cell B30 and 18 times in Cell E30.

25. Merge and center Cells B2 and C2.
26. Merge and center Cells B3 and C3.
27. Merge and center Cells D2 through F2.
28. Merge and center Cells D3 through F3.

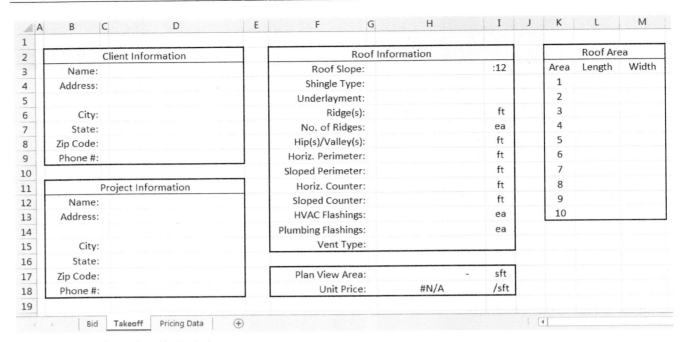

FIGURE 33-42 The Takeoff Worksheet

TABLE 33-9 Data for Bid Worksheet

Cell	Data
B2	*Bill To:*
D2	*Ship To:*
B28	*Half of the payment is due at delivery of materials. The remaining payment is due upon completion of the roofing.*
B30	*By: _____*
E30	*Date: _____*

TABLE 33-10 Formulas for Bid Worksheet

Cell	Data
B3	=IF(Takeoff!D3=" "," ",Takeoff!D3)
D3	=IF(Takeoff!D12=" "," ",Takeoff!D12)

29. Enter the formulas in Table 33-10 into the specified cells.
30. Copy Cell B3 to Cells B4 through B9.
31. Copy Cell D3 to Cells D4 through D9.
32. Left justify Cells B2 through F9 by selecting Cells B2 through F9, selecting the **Home>Alignment> Align Left** button (shown in Figure 3-31).
33. Draw a Thick Box Border around Cells B2 through C9.
34. Draw a Thick Box Border around Cells D2 through F9.
35. Select Cells A1 through F30 and change the font size to 14 point by selecting Cells A1 through F30, selecting the **Home** menu tab, and then selecting **14** from the Font dropdown box in the Font group (shown in Figure 3-35).
36. Merge and center Cells B28 through F28 and change the font size to 8.
37. If needed, adjust the underlining in Cell B30 so that it fits in Cells B30 to D30 and the underlining in Cell E30 so that it fits in Cells E30 and F30.
38. Merge and center Cells B30 through D30.
39. Merge and center Cells E30 and F30.
40. Protect the Bid worksheet.
41. Save the workbook.

The worksheet should now appear as shown in Figure 33-43.

42. Enter data into Cells D3 through D9 of the Takeoff worksheet and verify that the data appears in the Bill To: area on the Bid worksheet.
43. Enter data into Cells D12 through D18 of the Takeoff worksheet and verify that the data appears in the Ship To: area on the Bid worksheet.
44. Delete the data in Cells D3 through D9 and Cells D12 through D18.
45. Exit the workbook. There is no need to save it if you saved the workbook in Step 41.

In the next chapter, we will look at some estimating methods that may be used to prepare conceptual and preliminary estimates.

	A	B	C	D	E	F
1						
2		Bill To:		Ship To:		
3						
4						
5						
6						
7						
8						
9						
10						
11		Materials		Quantity	Unit Price	Total
12		Select Shingle Type	-	bundle	#N/A	#N/A
13		Cap	-	bundle	#N/A	#N/A
14		Roofing Nails	-	lbs	1.39	-
15		Select Underlayment Type	-	rolls	#N/A	#N/A
16		Underlayment Nails	-	C	1.00	-
17		10' Drip Edge	-	ea	2.99	-
18		10' Counter Flashing	-	ea	3.99	-
19		20' Ridge Vent	-	ea	42.00	-
20		Turtle Vents	-	ea	7.00	-
21		HVAC Flashing	-	ea	9.00	-
22		Plumbing Flashing	-	ea	4.00	-
23					Subtotal	#N/A
24		Tax (6.5%)				#N/A
25		Roofing Crew	#N/A	1hr	35.00	#N/A
26					Total	#N/A
27						
28		Half of the payment is due at delivery of materials. The remaining payment is due upon completion of the roofing.				
29						
30		By:_____			Date:_____	
31						
32						
33						

| Bid | Takeoff | Pricing Data |

FIGURE 33-43 The Bid Worksheet

CONCLUSION

The steps for planning a new form are to identify the desired results, identify input data, solve practice problems, identify steps and equations, and sketch spreadsheet layouts. The steps for setting up a new form are the same as for existing forms.

PROBLEMS

1. What are the steps for creating a new form?
2. How can a series be created?
3. What is the advantage of naming cells?
4. How are cells named?

5. How do you create a dropdown box?
6. What is the purpose of the CONCATENATE function?
7. What is the purpose of the LOOKUP function?
8. Make a copy of the spreadsheet developed in this chapter. Modify the copy to use the plan view area from a takeoff software package in lieu of calculating the area. Add the needed error protection and protect your worksheets.
9. Make a copy of the spreadsheet developed in this chapter and use it to calculate the cost of the roof shown in Figure 9-11. The roof is constructed of 30-year architectural shingles over 30-pound felt. Ridge vents are used and there are four plumbing flashings and one HVAC flashing. The roof slope is 4:12.
10. Make a copy of the spreadsheet developed in this chapter and use it to calculate the cost of the roof shown in Figure 9-12. The roof is constructed of 20-year, three-tab shingles over 15-pound felt. Turtle vents are used and there are six plumbing flashings and two HVAC flashings. The roof slope is 6:12.

CHAPTER THIRTY-FOUR

OTHER ESTIMATING METHODS

In this chapter you will learn to prepare estimates that may be used to check a detailed estimate or may be used to prepare an estimate during the design process. This chapter will cover the project comparison, square-foot, and assembly estimating methods.

Throughout the previous chapters, you have learned to prepare a detailed estimate for a project by including costs and productivity rates for all of the components needed to complete the building. This works well when you have a complete set of contract document that identify all of the components needed to complete the building. However, when working on design-build (DB) and construction manager general contractor (CMGC) projects, construction estimators need to prepare estimates for incomplete designs, which often require using other estimating methods. In this chapter we will look: (1) the role estimates play in the design process, (2) the role estimates play in the different delivery methods, and (3) three advanced estimating methods (project comparison, square-foot, and assembly estimating methods) that may be used to develop estimates during the design process or just provide a quick check on a detailed estimate.

DESIGN PROCESS

The design process for building project is typically divided into four phases: (1) conceptual, (2) schematic design, (3) design development, and (4) contract documents. Let's look at what occurs during each of these phases.

During the **conceptual** phase, which may be referred to as the feasibility or programming phase, the feasibility of the project's concept is tested to see if it is worthwhile to pursue. During this phase, enough of the design must be completed to establish a budget for the project and provide the information necessary to convince the management, investors, and financiers to support the project; or decide not to pursue the project. During this phase the building's general size and shape (square footage, perimeter, and height) are established, including how the building's square footage is divided among the building's uses. This design includes a programming statement (a written narrative describing the proposed building) and may include sketches, rough drawings, brochures, and conceptual models. The conceptual estimate is based on the broad parameters available at this stage of the design.

The **schematic design** phase focuses on the allocation of the space to different uses, the physical relationship between uses, and the aesthetics (look and feel) of the building. During this phase, the designer comes up with an achievable design, which shows the scale and arrangement of the various building elements (such as rooms, stairs, hallways, etc.), location of the building on the site, and the general materials to be used in the construction (such as exterior finish materials). The design consists of preliminary floor plans, elevations, and major sections without detail and may include a simple outline specification. The purpose of this level of design is to communicate intended design to the building's stakeholders and get their approval of the direction that the design is proceeding before continuing with a more detailed design. At this phase, the conceptual estimate needs to be updated to insure that the proposed design direction is staying within the budget constraints for the project.

During the **design development** phase the schematic design is further refined. During this phase the space requirements for building's structure, mechanical, plumbing, electrical, and other key systems are coordinated with the architectural design to insure that there is sufficient space for these systems and that, when these systems work together, the building will meet the project's performance requirements. At this phase the architect and other design professionals seek to establish dimensions for the building as the floor plans, elevations, and section become more refined. Estimates are often prepared to select between different buildings systems (e.g., steel frame versus reinforced concrete structure) and the project's estimate needs to be updated to insure that the project is staying within the budget constraints.

During the **contract documents** phase all documents needed to obtain approvals (such as building permits) from the local municipalities, to obtain bids for the project, and to construct the project are prepared. The contract document includes the working drawings—which have enough detail to construct the project—and the project manual—which includes the technical specifications. The level of detail of these documents may be affected by the delivery system used on the project. Estimates prepared at this phase are most often prepared using the methods described in the previous chapters.

DELIVERY METHODS

In this section we will look at the three most common delivery methods: design-bid-build (DBB), design-build (DB), and construction manager general contractor (CMGC).

The most common is the **design-bid-build** delivery method. In this method, the general contractor is hired, most commonly through a competitive bidding process, after the contract document has been prepared and the design is complete. During the design phase, the architect and other design professionals are responsible for preparing or contracting with a third-party to prepare the estimates for the project. The contractor must prepare an estimate during the bidding process. Most commonly, contractor estimates are prepared using the methods described in the previous chapters.

The second most common method is the **design-build** deliver method. In this method, the owner prepares a performance criteria for the proposed project and contracts with a single entity for the design and construction of the project. The design-build contractor may be hired at the completion conceptual phase or the owner may choose to hire an architect to prepare a partial design, which may include some of the schematic design and design development phases. In this delivery method, the contractor must prepare an estimate with enough accuracy to provide the owner with a price during the hiring process. Seldom is there enough detail at this point to prepare an estimate using the methods described in the previous chapters. Additionally, the contractor must prepare estimates at completion of each of the design phases, if not more often, to ensure that the project stays within the budget.

The third method is **construction manager general contractor** delivery method. In this method, the owner contracts with an architect to prepare the building design and then contracts with a general contractor to provide input during the design phase with respect to constructability, the availability and cost of material, and other construction-related concerns. Part of the contractor's responsibilities is to provide estimates at the various stages of design and cost comparisons between different buildings systems (e.g., steel frame versus reinforced concrete structure). At the completion of the contract documents, the contractor provides the owner with an estimate of the construction costs. The owner often compares the contractor's estimate to an independent cost estimate (ICE) prepared by a third party. If the estimate is acceptable, the owner can contract with the general contractor to complete the work; otherwise, the owner often retains the right to seek other bidders.

There are many times when a construction estimator needs to prepare estimates from incomplete designs. In the next section, we will look at some methods that may be useful in preparing estimates from incomplete designs. Care must be taken when using these methods because they often produce less accurate results than the methods described in the previous chapters; however, they may be accurate enough for the purpose of early estimates. The estimator should make sure that the estimating method used is appropriate for the purpose of the estimate. When ordering materials, there is no substitute for a good detailed estimate.

PROJECT COMPARISON METHOD

Project comparison estimates are made by comparing the proposed project to one or more completed projects that are very similar in scope (size, building use, construction type, etc.). When preparing a project comparison estimate, care must be taken to ensure that the projects used in the comparison are similar to the proposed project. The estimator needs to consider the following:

- **Size.** As the size of the building increases, the cost per square foot decreases. Conversely, as the size decreases, the cost per square foot increases. This is due to economies of scale.
- **Height between floors.** Increasing the height between floors increases the envelope cost without increasing the square footage of the building.
- **Length of perimeter.** Two buildings, a long skinny one and a square one, will have very different perimeters and envelope costs for the same square footage.
- **Project location.** Labor availability, materials availability, and government regulations vary by location.
- **When the project was built.** Inflation in labor, materials, and fuel costs; weather (summer versus winter).
- **Type of structure.** Concrete, steel, wood framed, block.

- **Level of finishes.** Quality of materials and workmanship.
- **Utilization of the space.** Some spaces cost more than others do. Bathrooms and kitchens cost more per square foot than bedrooms.
- **Union versus nonunion labor.**
- **Soil conditions.**

The estimate is prepared by starting with the costs of a completed project and adjusting those costs for differences in project specifications (square footage, project features, etc.), location, and time.

Bledsoe[1] recommended using a total cost multiplier (TCM) to adjust for difference in project size—usually measured in square feet for building construction—when the size of the proposed project differs from the comparison project by more than 10 percent. Bledsoe proposes that the following multiplier be used to adjust for size:

$$\text{TCM} = \left(\frac{\text{Proposed size}}{\text{Comparison size}}\right)^E$$

where E is about 0.9 for simple construction projects (buildings). The sizes may be in any units (square foot, parking stalls, tons per hour) as long as the same units are used for both the proposed project and the comparison project.

After adjusting for size, other adjustments for building features—such as the addition of extra elevators—are made. This is done by adding or subtracting the estimated cost of the features using the estimating methods from the previous chapters.

Construction costs vary by location based on location's conditions—such as the availability and cost of labor, transportation costs, and local regulations. For a project comparison estimate to be accurate, the comparison project must be from the same location as the proposed project. Because it is not always possible to find comparison projects in the same location, projects from other locations may be used by including an adjustment for the projects' locations. RSMeans—a major publisher of construction data—annually publishes a database that includes location factors for major cities in the United States and Canada. Figure 34-1 shows location factors for some U.S. cities. The following multiplier may be used to adjust for location:

$$\text{LM} = \left(\frac{\text{Proposed location factor}}{\text{Comparison location factor}}\right)$$

where LM is the location multiplier (location factor).

In general, construction costs increase over time, and thus inflation must be taken into account when preparing a project comparison estimate. There are two types of inflation corrections that may be required when preparing an estimate: (1) bringing historical data to the current year, and (2) projecting costs from the current year into the future.

RSMeans annually publishes a database that includes historical cost indexes (see Figure 34-2). The following multiplier may be used to bring costs from previous years into the current year:

$$\text{HCM} = \left(\frac{\text{Index for current year}}{\text{Index for year of historical cost}}\right)$$

where HCM is the historical cost multiplier.

For the estimate to be accurate, the estimated costs of the proposed project must be projected into the future to the expected time of construction. Because the future inflation rate of construction costs is not known, an estimate of the annual inflation rate must be made. The proposed project cost may be projected into the future using the following equation:

$$\text{Cost}_{t+n} = \text{Cost}_t (1 + \bar{f})^n$$

where

Cost_t is the estimated cost of the project in the current year

Cost_{t+n} is the projected future cost of the project n years from now

$\bar{f}$ is the average annual inflation factor

When the proposed project is compared to two or more projects, the cost of the proposed project is selected by looking at the adjusted cost for each of the comparison projects and selecting a cost or cost range. Statistical tools, such as the mean of the adjusted project costs, may be used to estimate the cost of the proposed project. Construction managers often add contingencies or increase the profit margin on the project to account for the uncertainties introduced by the estimating method. The preparation of a project comparison estimate is shown in the following example.

EXAMPLE 34-1 PROJECT COMPARISON

Two years ago, your company built a bank for a client in Temple, Texas, for $1,016,000. The client is looking at building a bank next year in Corpus Christi, Texas. The new branch is expected to be 20 percent larger. The cost indexes for this year is 100 and for two years ago is 97.3. Costs are expected to rise 3 percent during the next year. Using the project comparison method, prepare a cost estimate for the client.

Solution: First adjust for size using an E of 0.9 using Eq. 34-1 as follows:

$$\text{Cost} = \text{Cost} \times \text{TCM} = \$1,016,000(1.2)^{0.9} = \$1,197,173$$

Next, adjust for location. From, the location factor for Temple is 80.3 and for Corpus Christi is 84.6. Using Eq. 34-2 adjust the cost for location.

$$\text{Cost} = \text{Cost} \times \text{LM} = \$1,197,173\left(\frac{84.6}{80.3}\right) = \$1,261,280$$

Location Factors

STATE/ZIP	CITY	MAT.	INST.	TOTAL	STATE/ZIP	CITY	MAT.	INST.	TOTAL
NORTH CAROLINA (CONT'D)					**PENNSYLVANIA (CONT'D)**				
286	Hickory	98.8	53.9	79.2	177	Williamsport	93.8	83.0	89.1
287-288	Asheville	100.8	55.3	80.9	178	Sunbury	95.8	95.8	95.8
289	Murphy	99.6	48.3	77.2	179	Pottsville	94.9	98.5	96.5
NORTH DAKOTA					180	Lehigh Valley	95.8	114.0	103.7
580-581	Fargo	102.7	67.6	87.4	181	Allentown	97.9	108.6	102.5
582	Grand Forks	102.7	51.6	80.4	182	Hazleton	95.4	98.3	96.6
583	Devils Lake	102.0	58.6	83.1	183	Stroudsburg	95.3	101.4	97.9
584	Jamestown	102.1	45.2	77.3	184-185	Scranton	98.6	99.2	98.9
585	Bismarck	103.1	66.2	87.0	186-187	Wilkes-Barre	95.1	98.7	96.6
586	Dickinson	102.8	60.9	84.6	188	Montrose	94.7	96.9	95.7
587	Minot	102.7	71.3	89.0	189	Doylestown	94.9	120.6	106.1
588	Williston	101.2	60.9	83.7	190-191	Philadelphia	99.7	133.6	114.5
OHIO					193	Westchester	95.9	121.5	107.1
430-432	Columbus	98.6	89.2	94.5	194	Norristown	95.0	129.5	110.0
433	Marion	95.1	85.3	90.8	195-196	Reading	97.1	102.1	99.3
434-436	Toledo	98.7	98.1	98.4	**PUERTO RICO**				
437-438	Zanesville	95.6	86.5	91.6	009	San Juan	122.7	24.4	79.8
439	Steubenville	96.8	94.9	96.0	**RHODE ISLAND**				
440	Lorain	98.9	93.0	96.3	028	Newport	98.4	117.5	106.7
441	Cleveland	99.0	100.3	99.6	029	Providence	100.6	117.5	108.0
442-443	Akron	99.9	94.2	97.4	**SOUTH CAROLINA**				
444-445	Youngstown	99.2	89.0	94.7	290-292	Columbia	99.3	56.5	80.6
446-447	Canton	99.3	85.6	93.3	293	Spartanburg	98.6	56.7	80.3
448-449	Mansfield	96.7	87.3	92.6	294	Charleston	99.9	65.2	84.8
450	Hamilton	97.9	85.1	92.3	295	Florence	98.2	56.9	80.2
451-452	Cincinnati	98.3	84.5	92.2	296	Greenville	98.3	55.8	79.8
453-454	Dayton	98.0	84.4	92.1	297	Rock Hill	97.8	50.3	77.1
455	Springfield	98.0	84.7	92.2	298	Aiken	98.6	69.1	85.7
456	Chillicothe	96.8	93.0	95.1	299	Beaufort	99.3	44.9	75.6
457	Athens	99.6	80.4	91.2	**SOUTH DAKOTA**				
458	Lima	100.0	86.8	94.3	570-571	Sioux Falls	100.0	57.7	81.5
OKLAHOMA					572	Watertown	99.0	49.1	77.2
730-731	Oklahoma City	100.3	65.7	85.2	573	Mitchell	97.8	48.4	76.3
734	Ardmore	98.3	61.9	82.4	574	Aberdeen	100.5	49.6	78.3
735	Lawton	100.7	62.8	84.2	575	Pierre	101.7	58.4	82.8
736	Clinton	99.6	63.1	83.7	576	Mobridge	98.5	48.8	76.8
737	Enid	100.2	60.5	82.9	577	Rapid City	100.4	59.5	82.6
738	Woodward	98.3	63.4	83.1	**TENNESSEE**				
739	Guymon	99.4	59.7	82.1	370-372	Nashville	98.4	74.1	87.8
740-741	Tulsa	98.1	61.6	82.2	373-374	Chattanooga	100.0	66.2	85.2
743	Miami	94.8	67.5	82.9	375,380-381	Memphis	99.0	71.3	86.9
744	Muskogee	97.4	55.7	79.2	376	Johnson City	99.8	56.5	80.9
745	Mcalester	94.5	60.4	79.7	377-379	Knoxville	96.6	66.9	83.7
746	Ponca City	95.1	62.1	80.7	382	Mckenzie	98.2	59.2	81.2
747	Durant	95.1	63.5	81.3	383	Jackson	99.9	61.2	83.0
748	Shawnee	96.5	61.8	81.4	384	Columbia	96.9	66.1	83.5
749	Poteau	94.3	61.4	80.0	385	Cookeville	98.1	59.1	81.1
OREGON					**TEXAS**				
970-972	Portland	98.8	100.4	99.5	750	Mckinney	98.7	63.1	83.2
973	Salem	100.5	99.4	100.0	751	Waxahackie	98.6	65.9	84.4
974	Eugene	98.6	99.3	98.9	752-753	Dallas	99.5	67.4	85.5
975	Medford	100.0	97.1	98.8	754	Greenville	98.8	64.0	83.6
976	Klamath Falls	99.8	97.1	98.6	755	Texarkana	98.0	58.4	80.7
977	Bend	98.9	99.4	99.1	756	Longview	98.5	58.7	81.2
978	Pendleton	94.8	101.4	97.7	757	Tyler	98.5	65.0	83.9
979	Vale	92.9	98.1	95.2	758	Palestine	95.0	64.4	81.7
PENNSYLVANIA					759	Lufkin	95.6	67.8	83.5
150-152	Pittsburgh	99.9	104.6	102.0	760-761	Fort Worth	99.3	65.3	84.5
153	Washington	97.0	103.2	99.7	762	Denton	98.9	62.5	83.0
154	Uniontown	97.3	101.7	99.2	763	Wichita Falls	97.1	64.4	82.8
155	Bedford	98.2	91.3	95.2	764	Eastland	95.9	63.2	81.6
156	Greensburg	98.2	100.9	99.4	765	Temple	94.7	61.7	80.3
157	Indiana	97.1	99.3	98.0	766-767	Waco	96.5	64.1	82.4
158	Dubois	98.5	96.8	97.8	768	Brownwood	98.9	59.4	81.7
159	Johnstown	98.1	97.2	97.7	769	San Angelo	98.5	59.6	81.5
160	Butler	92.5	102.0	96.6	770-772	Houston	99.4	70.5	86.8
161	New Castle	92.5	97.9	94.8	773	Huntsville	98.0	66.7	84.4
162	Kittanning	92.9	103.3	97.4	774	Wharton	98.9	67.7	85.3
163	Oil City	92.4	96.9	94.4	775	Galveston	97.0	70.0	85.2
164-165	Erie	94.5	95.8	95.1	776-777	Beaumont	97.4	68.6	84.9
166	Altoona	94.6	91.9	93.4	778	Bryan	94.9	68.2	83.3
167	Bradford	95.7	96.4	96.0	779	Victoria	99.0	65.6	84.4
168	State College	95.3	92.8	94.2	780	Laredo	98.3	64.5	83.6
169	Wellsboro	96.3	92.6	94.7	781-782	San Antonio	98.5	65.1	84.0
170-171	Harrisburg	99.8	95.6	98.0	783-784	Corpus Christi	100.9	63.5	84.6
172	Chambersburg	96.2	88.0	92.6	785	Mc Allen	101.0	59.8	83.0
173-174	York	96.7	95.9	96.3	786-787	Austin	100.0	63.6	84.1
175-176	Lancaster	95.1	89.0	92.4					

FIGURE 34-1 Sample Location Factors (From RSMeans Building Construction Cost Data, 2015. Copyright RSMeans, Rockland, MA 781-422-5000; All rights reserved.)

Historical Cost Indexes

The table below lists both the RSMeans® historical cost index based on Jan. 1, 1993 = 100 as well as the computed value of an index based on Jan. 1, 2015 costs. Since the Jan. 1, 2015 figure is estimated, space is left to write in the actual index figures as they become available through either the quarterly *RSMeans Construction Cost Indexes* or as printed in the *Engineering News-Record*. To compute the actual index based on Jan. 1, 2015 = 100, divide the historical cost index for a particular year by the actual Jan. 1, 2015 construction cost index. Space has been left to advance the index figures as the year progresses.

Year	Historical Cost Index Jan. 1, 1993 = 100		Current Index Based on Jan. 1, 2015 = 100		Year	Historical Cost Index Jan. 1, 1993 = 100	Current Index Based on Jan. 1, 2015 = 100		Year	Historical Cost Index Jan. 1, 1993 = 100	Current Index Based on Jan. 1, 2015 = 100	
	Est.	Actual	Est.	Actual		Actual	Est.	Actual		Actual	Est.	Actual
Oct 2015*					July 2000	120.9	58.5		July 1982	76.1	36.8	
July 2015*					1999	117.6	56.9		1981	70.0	33.9	
April 2015*					1998	115.1	55.7		1980	62.9	30.4	
Jan 2015*	206.7		100.0	100.0	1997	112.8	54.6		1979	57.8	28.0	
July 2014		204.9	99.1		1996	110.2	53.3		1978	53.5	25.9	
2013		201.2	97.3		1995	107.6	52.1		1977	49.5	23.9	
2012		194.6	94.1		1994	104.4	50.5		1976	46.9	22.7	
2011		191.2	92.5		1993	101.7	49.2		1975	44.8	21.7	
2010		183.5	88.8		1992	99.4	48.1		1974	41.4	20.0	
2009		180.1	87.1		1991	96.8	46.8		1973	37.7	18.2	
2008		180.4	87.3		1990	94.3	45.6		1972	34.8	16.8	
2007		169.4	82.0		1989	92.1	44.6		1971	32.1	15.5	
2006		162.0	78.4		1988	89.9	43.5		1970	28.7	13.9	
2005		151.6	73.3		1987	87.7	42.4		1969	26.9	13.0	
2004		143.7	69.5		1986	84.2	40.7		1968	24.9	12.0	
2003		132.0	63.9		1985	82.6	40.0		1967	23.5	11.4	
2002		128.7	62.3		1984	82.0	39.7		1966	22.7	11.0	
2001		125.1	60.5		1983	80.2	38.8		1965	21.7	10.5	

Adjustments to Costs

The "Historical Cost Index" can be used to convert national average building costs at a particular time to the approximate building costs for some other time.

Example:

Estimate and compare construction costs for different years in the same city.

To estimate the national average construction cost of a building in 1970, knowing that it cost $900,000 in 2015:

INDEX in 1970 = 28.7

INDEX in 2015 = 206.7

Note: The city cost indexes for Canada can be used to convert U.S. national averages to local costs in Canadian dollars.

Example:

To estimate and compare the cost of a building in Toronto, ON in 2015 with the known cost of $600,000 (US$) in New York, NY in 2015:

INDEX Toronto = 110.9

INDEX New York = 131.8

$$\frac{\text{INDEX Toronto}}{\text{INDEX New York}} \times \text{Cost New York} = \text{Cost Toronto}$$

$$\frac{110.9}{131.8} \times \$600{,}000 = .841 \times \$600{,}000 = \$504{,}600$$

The construction cost of the building in Toronto is $504,600 (CN$).

Time Adjustment Using the Historical Cost Indexes:

$$\frac{\text{Index for Year A}}{\text{Index for Year B}} \times \text{Cost in Year B} = \text{Cost in Year A}$$

$$\frac{\text{INDEX 1970}}{\text{INDEX 2015}} \times \text{Cost 2015} = \text{Cost 1970}$$

$$\frac{28.7}{206.7} \times \$900{,}000 = .139 \times \$900{,}000 = \$125{,}100$$

The construction cost of the building in 1970 is $125,100.

*Historical Cost Index updates and other resources are provided on the following website. http://info.thegordiangroup.com/RSMeans.html

FIGURE 34-2 Historical Cost Indexes (From RSMeans Building Construction Cost Data, 2015. Copyright RSMeans, Rockland, MA 781-422-5000; All rights reserved.)

Next, adjust the cost to the current year using Eq. 34-3 as follows:

$$\text{Cost} = \text{Cost} \times \text{HCM} = \$1{,}261{,}280 \left(\frac{100}{97.3}\right) = \$1{,}296{,}280$$

Finally, adjust for inflation using Eq. 34-4 as follows:

$$\text{Cost}_{t+n} = \text{Cost}_t (1 + \bar{f})^n = \$1{,}296{,}280(1 + 0.03)^1 = \$1{,}335{,}168$$

Use $1,340,000.

When collecting data for use in project comparison, it is important that any unusual conditions that would skew the costs higher or lower be documented. It is simply not enough to record just the cost. The estimator must know whether the cost is typical for the type of project and understand why the cost might have been higher or lower than the average.

SQUARE-FOOT ESTIMATING

Square-foot estimates are prepared by multiplying the square footage of a building by a cost per square foot and adjusting for building features such as length of perimeter, building height, and other building components. Other units—such as number of parking stalls—may be used in lieu of square footage. RSMeans annually publishes a database of square-foot costs for a wide variety of buildings. Different data sets are provided for approximately 75 different commercial building uses. Figures 34-3 and 34-4 show the RSMeans square foot data for a parking garage. Using data from RS-Means, the estimated costs for a construction project may be determined by the following steps:

1. Find the correct page for the type of construction to be estimated. Figures 34-3 and 34-4 are for a parking garage.
2. Determine the base cost per square foot by finding the type of exterior wall and frame system in the left two columns and the area in the top row of Figure 34-3. For example, the base cost is $66.80 for a 145,000-square-foot parking garage with a face brick and concrete block backup exterior walls and a precast concrete frame. When the size falls between two sizes on the page, the estimator should interpolate between the two sizes. For example, the base cost per square foot for a 165,000-square-foot parking garage with a face brick and concrete block backup exterior walls and a precast concrete frame would be determined using linear interpolation as follows:

$$\text{Cost}(\$/\text{sf}) = \frac{165{,}000 - 145{,}000}{175{,}000 - 145{,}000}(\$66.10 - \$66.80) + \$66.80$$

$$= \$66.33$$

3. Next, the base cost per square foot needs to be adjusted for differences in the perimeters. This is done by determining the difference between the proposed building's perimeter and the perimeter specified in the second row of the table. The differences are measured in 100 feet or a fraction thereof. For example, for a 145,000-square-foot parking garage, the perimeter is 723 feet; and if the proposed building's perimeter is 700 feet, the difference in the perimeters would be 0.23 hundred feet, which would be a deduction because we are reducing the perimeter. This is multiplied by the perimeter adjustment, and the resultant is added or deducted from the base cost per square foot. The perimeter adjustment for a 145,000-square-foot parking garage is $1.10 per 100 feet.

4. Next, the base cost per square foot needs to be adjusted for differences in the heights of the stories. This is done by determining the difference between the proposed building's story height and the story height used to develop the square foot costs. The story height for a parking garage is 10 feet and is found in the top left-hand corner of Figure 34-4. This difference is then multiplied by the story height adjustment, and the resultant is added or deducted from the base cost per square foot. The story height adjustment for a 145,000-square-foot parking garage is $0.65 per foot of height.

5. The square footage of the building is multiplied by the cost per square foot, including adjustments for the perimeter and story height.

6. Other costs are then added or subtracted from this price to account for differences in the design.

7. Finally, adjustments for location and inflation are made using Eq. 34-2, Eq. 34-3, and Eq. 34-4. When using Eq. 34-2 to adjust for location, a comparison location factor of 100 is used. When using a current database, it is not necessary to use a historical cost multiplier (Eq. 34-3); however, the cost must still be projected into the future using Eq. 34-4.

Construction managers often add contingencies or increase the profit margin on the project to account for the uncertainties introduced by the estimating method. The preparation of a square-foot estimate using the data from RSMeans is shown in the following example.

EXAMPLE 34-2 SQUARE-FOOT ESTIMATE

Using Figures 34-3 and Figure 34-4, determine the cost for a four-story, precast parking garage that is to be used as a free park-and-ride. The exterior walls are to be precast and the average story height is 11 feet. The garage is expected to be 180 feet by 285 feet and have 650 parking stalls.

Solution: The area and perimeter of the parking garage is determined as follows:

Area = (180)(285 ft)(4 stories) = 205,200 sf

Perimeter = 180 ft + 285 ft + 180 ft + 285 ft = 930 ft

COMMERCIAL/INDUSTRIAL/INSTITUTIONAL — M.270 — Garage, Parking

Costs per square foot of floor area

Exterior Wall	S.F. Area	85000	115000	145000	175000	205000	235000	265000	295000	325000
	L.F. Perimeter	529	638	723	823	875	910	975	1027	1075
Face Brick with Concrete Block Back-up	Steel Frame	81.70	80.05	78.80	78.15	77.30	76.55	76.15	75.75	75.45
	Precast Concrete	69.70	68.00	66.80	66.10	65.30	64.50	64.15	63.75	63.40
Precast Concrete	Steel Frame	89.70	87.15	85.20	84.15	82.70	81.50	80.85	80.25	79.70
	Precast Concrete	76.25	73.90	72.15	71.20	69.95	68.85	68.20	67.65	67.15
Reinforced Concrete	Steel Frame	79.05	77.65	76.65	76.10	75.45	74.90	74.55	74.30	74.05
	R/Conc. Frame	70.40	69.00	68.05	67.50	66.80	66.25	65.95	65.65	65.35
Perimeter Adj., Add or Deduct	Per 100 L.F.	1.80	1.35	1.10	0.90	0.70	0.65	0.50	0.55	0.45
Story Hgt. Adj., Add or Deduct	Per 1 Ft.	0.85	0.75	0.65	0.60	0.55	0.50	0.45	0.45	0.45
Basement—Not Applicable										

The above costs were calculated using the basic specifications shown on the facing page. These costs should be adjusted where necessary for design alternatives and owner's requirements. Reported completed project costs, for this type of structure, range from $36.45 to $140.95 per S.F.

Common additives

Description	Unit	$ Cost
Elevators, Electric passenger, 5 stops		
2000# capacity	Ea.	167,000
3500# capacity	Ea.	173,000
5000# capacity	Ea.	179,500
Barrier gate w/programmable controller	Ea.	4150
Booth for attendant, average	Ea.	13,800
Fee computer	Ea.	16,000
Ticket spitter with time/date stamp	Ea.	7475
Mag strip encoding	Ea.	21,400
Collection station, pay on foot	Ea.	128,500
Parking control software	Ea.	27,500 - 118,500
Painting, Parking stalls	Stall	8.60
Parking Barriers		
Timber with saddles, 4" x 4"	L.F.	7.70
Precast concrete, 6" x 10" x 6'	Ea.	63
Traffic Signs, directional, 12" x 18", high density	Ea.	97.50

FIGURE 34-3 Parking Garage Square-Foot Costs (From RSMeans Square Foot Costs, 2015. Copyright RSMeans, Rockland, MA 781-422-5000; All rights reserved.)

Garage, Parking

Model costs calculated for a 5 story building with 10' story height and 145,000 square feet of floor area

			Unit	Unit Cost	Cost Per S.F.	% Of Sub-Total
A.	**SUBSTRUCTURE**					
1010	Standard Foundations	Poured concrete; strip and spread footings	S.F. Ground	2.05	.41	
1020	Special Foundations	N/A	—	—	—	
1030	Slab on Grade	6" reinforced concrete with vapor barrier and granular base	S.F. Slab	7	1.40	4.6%
2010	Basement Excavation	Site preparation for slab and trench for foundation wall and footing	S.F. Ground	.18	.04	
2020	Basement Walls	4' foundation wall	L.F. Wall	84	.46	
B.	**SHELL**					
	B10 Superstructure					
1010	Floor Construction	Double tee precast concrete slab, precast concrete columns	S.F. Floor	28.03	22.42	44.5%
1020	Roof Construction	N/A	—	—	—	
	B20 Exterior Enclosure					
2010	Exterior Walls	Face brick with concrete block backup 40% of story height	S.F. Wall	50	5	
2020	Exterior Windows	N/A	—	—	—	9.9%
2030	Exterior Doors	N/A	—	—	—	
	B30 Roofing					
3010	Roof Coverings	N/A	—	—	—	0.0%
3020	Roof Openings	N/A	—	—	—	
C.	**INTERIORS**					
1010	Partitions	Concrete block	S.F. Partition	33.80	1.30	
1020	Interior Doors	Hollow metal	Each	23,140	.16	
1030	Fittings	N/A	—	—	—	
2010	Stair Construction	Concrete	Flight	11,600	1.28	15.1%
3010	Wall Finishes	Paint	S.F. Surface	1.82	.14	
3020	Floor Finishes	Parking deck surface coating	S.F. Floor	4.73	4.73	
3030	Ceiling Finishes	N/A	—	—	—	
D.	**SERVICES**					
	D10 Conveying					
1010	Elevators & Lifts	Two hydraulic passenger elevators	Each	164,575	2.27	4.5%
1020	Escalators & Moving Walks	N/A	—	—	—	
	D20 Plumbing					
2010	Plumbing Fixtures	Toilet and service fixtures, supply and drainage 1 Fixture/18,125 S.F. Floor	Each	725	.04	
2020	Domestic Water Distribution	Electric water heater	S.F. Floor	.07	.07	3.2%
2040	Rain Water Drainage	Roof drains	S.F. Roof	7.50	1.50	
	D30 HVAC					
3010	Energy Supply	N/A	—	—	—	
3020	Heat Generating Systems	N/A	—	—	—	
3030	Cooling Generating Systems	N/A	—	—	—	0.0%
3050	Terminal & Package Units	N/A	—	—	—	
3090	Other HVAC Sys. & Equipment	N/A	—	—	—	
	D40 Fire Protection					
4010	Sprinklers	Dry pipe sprinkler system	S.F. Floor	4.28	4.28	8.7%
4020	Standpipes	Standpipes and hose systems	S.F. Floor	.09	.09	
	D50 Electrical					
5010	Electrical Service/Distribution	400 ampere service, panel board and feeders	S.F. Floor	.25	.25	
5020	Lighting & Branch Wiring	T-8 fluorescent fixtures, receptacles, switches and misc. power	S.F. Floor	3.07	3.07	6.9%
5030	Communications & Security	Addressable alarm systems and emergency lighting	S.F. Floor	.12	.12	
5090	Other Electrical Systems	Emergency generator, 7.5 kW	S.F. Floor	.05	.05	
E.	**EQUIPMENT & FURNISHINGS**					
1010	Commercial Equipment	N/A	—	—	—	
1020	Institutional Equipment	N/A	—	—	—	2.6%
1030	Vehicular Equipment	Ticket dispensers, booths, automatic gates	S.F. Floor	1.32	1.32	
1090	Other Equipment	N/A	—	—	—	
F.	**SPECIAL CONSTRUCTION**					
1020	Integrated Construction	N/A	—	—	—	0.0%
1040	Special Facilities	N/A	—	—	—	
G.	**BUILDING SITEWORK**	N/A				
				Sub-Total	50.40	100%
	CONTRACTOR FEES (General Requirements: 10%, Overhead: 5%, Profit: 10%)			25%	12.62	
	ARCHITECT FEES			6%	3.78	
				Total Building Cost	**66.80**	

FIGURE 34-4 Parking Garage Square-Foot Costs (From RSMeans Square Foot Costs, 2015. Copyright RSMeans, Rockland, MA 781-422-5000; All rights reserved.)

From Figure 34-3, the cost per square foot for a 205,000-square-foot parking garage with precast exterior walls and a precast structure is $69.95. From Figure 34-3, the base perimeter is 875 feet and the perimeter adjustment is $0.70 per square foot per 100 foot of perimeter. The perimeter adjustment is as follows:

$$\text{Perimeter Add} = \left(\frac{930 \text{ ft} - 875 \text{ ft}}{100 \text{ ft}}\right)\$0.70/\text{ft} = \$0.39/\text{sft}$$

From Figure 34-4, the base story height is 10 feet and, from Figure 34-3, the adjustment for the story height is $0.55 per square foot per foot of height. The story-height adjustment is as follows:

Story-Height Add = (1 ft)($0.55/ft − sft = $0.55 sft)

The base cost per square foot is as follows:

Cost = $69.95 + $0.39 + $0.55 = $70.89/sft

The base cost is calculated as follows:

Cost = 205,200 sft($70.89/sft) = $14,546,628

Add the following costs to the base cost: one 3500# elevator at $173,000; 650 painted parking stalls at $8.60 per stall; and 650 precast parking bumpers at $63.00 per stall.

Cost = $14,546,628 + $173,000 + 650($8.60 + $63.00)
     = $14,766,168

Use $14,800,000.

One of the assumptions of the square-foot method is that the average cost per square foot for a building is constant. Because one square foot of bathroom space costs more than one square foot of storage space, for this assumption to be true, the mix of expensive space (such as bathrooms) and inexpensive space (such as storage) must remain constant. A problem occurs when inexpensive space is decreased without decreasing the expensive space, which causes the cost per square foot to increase. Some of this uncertainty can be addressed by dividing a building's area up into different uses and developing a cost per square foot for the different areas.

Alternately, the square-foot method may be used for specific trades within the project. For example, a square-foot estimate may be prepared for the electrical to check the subcontractor's electrical pricing. In addition, the square-foot method may be applied to each cost code in a bid. This requires the estimator to think through each cost code and determine the best unit of measure to use for each cost code. Some items may be best based on the square footage (for example, floor slabs and roofing) or the length of the perimeter of the building (for example, exterior brick), and others may be nearly fixed for a building unless a large change in size occurs (for example, elevators).

ASSEMBLY ESTIMATING

An assembly is a group of building components that work together to perform a specific function. For example, an exterior wall assembly could consist of the wall framing, insulation, sheathing, interior finishes, and exterior finishes.

To create an assembly, the estimator determines the quantity of materials needed for one unit of the assembly, linear foot of wall in the case of the interior wall. From these quantities, the cost for materials, equipment, and labor to construct one unit of the assembly is then determined. This cost is then used to bid the assembly by multiplying the quantity of the assembly by the unit cost of the assembly. The most common units for assemblies include each (i.e., numeric count), lineal foot, and square foot. Determining the cost of an assembly is shown in the following example.

EXAMPLE 34-3 WALL ASSEMBLY UNIT COST

Determine the quantities of materials needed and the costs for an 8-foot-high interior partition wall assembly. The wall consists of track and metal studs, insulation, 1/2-inch drywall on both sides, and paint. The studs are spaced 16 inches on center. Openings in the partition will be handled as a separate assembly. The unit costs are as follows: track, $1.80 per foot; metal studs, $12.30 each; insulation, $0.76 per square foot; drywall, $0.91 per square foot; and paint, $0.50 per square foot. The costs include labor, materials, and equipment. What is the cost of 251 feet of an 8-foot-high interior wall?

Solution: The required materials for one lineal foot of wall are as follows:

$$\text{Length of track} = 2 \times 1 \text{ ft} = 2 \text{ ft}$$

$$\text{Number of studs} = \left(\frac{1 \text{ ft}}{16 \text{ in}}\right) = \left(\frac{12 \text{ in}}{16 \text{ in}}\right) = 0.75 \text{ each}$$

$$\text{Area of insulation} = (8 \text{ ft})(1 \text{ ft})\left(\frac{15 \text{ in}}{16 \text{ in}}\right) = 7.5 \text{ sf}$$

$$\text{Area of drywall} = (2 \text{ sides})(8 \text{ ft})(1 \text{ ft}) = 16 \text{ sf}$$

$$\text{Area of paint} = (2 \text{ sides})(8 \text{ ft})(1 \text{ ft}) = 16 \text{ sf}$$

The costs are as follows:

$$\text{Track} = (2 \text{ ft})(\$1.80/\text{ft}) = \$3.60$$
$$\text{Studs} = (0.75 \text{ each})(\$12.30/\text{ft}) = \$9.22$$
$$\text{Insulation} = (7.5 \text{ sf})(\$0.76/\text{sf}) = \$5.70$$
$$\text{Drywall} = (16 \text{ sf})(\$0.91/\text{sf}) = \$14.56$$
$$\text{Paint} = (16 \text{ sf})(\$0.50/\text{sf}) = \$8.00$$

The cost per lineal foot is as follows:

Cost = $3.60 + $9.22 + $5.70 + $14.56 + $8.00 = $41.08

The cost for the 251-foot wall is as follows:

Cost = (251 ft)($41.08) = $10,311.08

An assembly estimate is prepared by: (1) dividing the building up into assemblies, (2) determining the quantity of each assembly, (3) determining the cost of each assembly by multiplying its quantity by a cost per unit, and (4) summing the cost of the assemblies. RSMeans annually publishes a database of costs for a wide variety of assemblies. Figure 34-5 is a sample from the RSMeans database showing costs for a steel joist/deck roof assembly that is supported by walls and columns. Unlike the

B10 Superstructure

B1020 Roof Construction

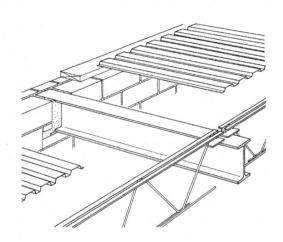

The table below lists the cost per S.F. for a roof system with steel columns, beams and deck using open web steel joists and 1-1/2" galvanized metal deck. Perimeter of system is supported on bearing walls.

Fireproofing is not included. Costs/S.F. are based on a building 4 bays long and 4 bays wide.

Column costs are additive. Costs for the bearing walls are not included.

Steel Joists, Beams and Deck on Bearing Walls

B1020 108		Steel Joists, Beams & Deck on Columns & Walls						
	BAY SIZE (FT.)	SUPERIMPOSED LOAD (P.S.F.)	DEPTH (IN.)	TOTAL LOAD (P.S.F.)	COLUMN ADD	COST PER S.F.		
						MAT.	INST.	TOTAL
3000	25x25	20	18	40		4.29	1.44	5.73
3100					columns	.39	.09	.48
3200		30	22	50		4.60	1.53	6.13
3300					columns	.52	.12	.64
3400		40	20	60		5.10	1.69	6.79
3500					columns	.52	.12	.64
3600	25x30	20	22	40		4.52	1.41	5.93
3700					columns	.44	.10	.54
3800		30	20	50		5.15	1.57	6.72
3900					columns	.44	.10	.54
4000		40	25	60		5.30	1.61	6.91
4100					columns	.52	.12	.64
4200	30x30	20	25	42		4.98	1.51	6.49
4300					columns	.36	.09	.45
4400		30	22	52		5.45	1.64	7.09
4500					columns	.44	.10	.54
4600		40	28	62		5.65	1.69	7.34
4700					columns	.44	.10	.54
4800	30x35	20	22	42		5.15	1.57	6.72
4900					columns	.37	.09	.46
5000		30	28	52		5.45	1.64	7.09
5100					columns	.37	.09	.46
5200		40	25	62		5.90	1.76	7.66
5300					columns	.44	.10	.54
5400	35x35	20	28	42		5.15	1.58	6.73
5500					columns	.32	.08	.40

FIGURE 34-5 Steel Joist/Deck Roof Assembly (From RSMeans Square Foot Costs, 2015. Copyright RSMeans, Rockland, MA 781-422-5000; All rights reserved.)

project comparison and square-foot estimating methods, there is little need to make adjustments for building features because most of the building's features have been incorporated into the assemblies. The final step is to make adjustments for location and inflation using the same procedures used for square-foot estimates.

Most specialized estimating programs—such as WinEst—incorporate assembly estimating into their software. Using broad assemblies (sometimes referred to as building systems) may be used to adjust for different types of building construction. Construction managers often add contingencies or increase the profit margin on the project to account for the uncertainties introduced by the estimating method. The preparation of an assembly estimate is shown in the following example.

EXAMPLE 34-4: ASSEMBLY ESTIMATING

Using the assembly estimating, prepare an estimate for a 50-foot by 100-foot warehouse. The exterior wall is 8-inch-thick CMU wall 25 feet high and rests on a 24-inch-wide by 12-inch-thick concrete footing. The roof structure consists of joist and deck supported by the exterior walls and a beam running down the center of the long axis of the building. The beam is supported by three wide-flange columns supported by 54-inch by 54-inch by 15-inch-deep footings. The floor consists of a 6-inch-thick, reinforced concrete slab. The warehouse has four personnel doors and eight 12-foot by 12-foot overhead doors. The roof consists of a membrane roofing over 3-inches of ridged insulation. Include 32 4-foot by 4-foot skylights and one roof hatch, one unisex bathroom, fire sprinklers (ordinary hazard), and lighting for the warehouse.

Solution: The following costs are from RSMeans Square Foot Costs 2015. Some assumption must be made, which will affect the cost of the building.

Area = (50 ft)(100 ft) = 5,000 sf
Perimeter = 50 ft + 100 ft + 50 ft + 100 ft = 300 ft
Footings = 300 ft × $40.25/ft = $12,075
Spread footings = 3 ea × $406/ea = $1,218
Floor slab = 5,000 sf × $8.05/sf = $40,250
Block wall = 25 ft × 300 ft × $11.03/sf = $82,725
Steel columns = 5,000 sf × $0.64/sf = $3,200
Joist, beams, and deck = 5,000 sf × $6.13/sf = $30,650
Personnel doors = 4 ea × $1,875/ea = $7,500
Overhead doors = 8 ea × $3,575/ea = $28,600
Roof insulation = 5,000 sf × $2.11/sf = $10,550
Roofing = 5,000 sf × $2.26/sf = $11,300
Skylights = 32 ea × 16 sf × $37.80/sf = $19,354
Roofhatch = 1 ea × $934/ea = $934
Bathroom = 1 ea × $3,400/ea = $3,400
Fire sprinklers = 5,000 sf × $5.43/sf = $27,150
Electrical service (200A) = 1 ea × $2,900/ea = $2,900
Lighting (HID) = 5,000 sf × $5.43/sf = $27,150
Total = $308,956–Use $310,000

AND BEYOND

During the course of this text you have been introduced to Excel as a powerful tool for estimating. You have seen sample worksheets that can be used for the quantity takeoff, developed a workbook for use in summarizing an estimate and documenting the details of the estimate, incorporated error checking into spreadsheets, automated repetitive tasks with macros, and set up a workbook to prepare the estimate and proposal for an asphalt shingle roof. With little effort, an estimator can develop a spreadsheet to prepare a conceptual estimate for specific types of construction projects or prepare more complex quantity takeoff spreadsheets. The possibilities are only limited by the mind and imagination of the estimator. Enjoy the productivity and time-saving features of using Excel for estimating.

CONCLUSION

Estimates are required at each stage of the design process. The design process is typically divided into four phases: (1) conceptual, (2) schematic design, (3) design development, and (4) contract documents. When dealing with a design-bid-build project, the architect prepares estimates during the design process and the contractor prepares a detailed estimate during the bidding process. For a design-build project, the contractor must prepare an estimate based on an incomplete design during the contractor selection process. If successful, the contractor completes the design for the project and prepares estimates at each of the design phases to make sure that the project stays within budget. For a construction manager general contractor (CMGC) project, the contractor is hired during the design process to provide estimates and input on other construction issues to the architect. Because there are many times when a contractor cannot prepare a detailed estimate, the contractor needs other estimating methods that can be used on incomplete designs. This chapter discussed the project comparison, square-foot and assembly estimating methods, which can be used on an incomplete design.

PROBLEMS

1. How would you determine the costs for a project using the project comparison method?
2. Determine the following using Figures 34-3 and 34-4:
    a. Base cost per square foot for a 235,000-square-foot, reinforced-concrete parking garage with a reinforced concrete frame.
    b. Base cost per square foot for an 85,000-square-foot precast parking garage with a precast exterior.

c. Base perimeter for a 295,000-square-foot parking garage.

d. Perimeter adjustment for an 115,000-square-foot parking garage.

e. Story height adjustment for an 85,000-square-foot parking garage.

f. The added cost for a 5000-pound elevator.

g. The added cost for an attendant booth.

3. What components would you include in an exterior wall assembly for a residence?

4. List the items that you would need to include in a concrete footing and foundation wall assembly.

5. Last year your construction company built a 2,400-square-foot home with a two-car garage for $278,400. Another client wants a similar home built, except they want a three-car garage. It is estimated that the garage will cost an additional $10,400 and that costs have risen 4 percent during the last year. Using this information, prepare a preliminary estimate for the house.

6. Your company has been asked to prepare a preliminary cost estimate for a 10,000-square-foot automotive repair garage, with four hydraulic lifts. From past projects, you have determined that the average cost per square foot to construct an automotive repair garage is $133.40 and excludes the cost of the hydraulic lifts. The lifts cost $10,200 each. Ignoring inflation, prepare an estimate for the repair garage.

7. Your company has been asked to prepare a preliminary cost estimate for a two-story, 16,000-square-foot office building. From projects you have constructed during the last year, you have determined that the average cost per square foot to construct an office building is $231.65. Using an inflation rate of 5 percent, prepare an estimate for the office building.

8. Using Figures 34-3 and 34-4, determine the cost for a 180-foot by 410-foot, four-story, precast parking garage with a precast exterior. The parking garage is expected to have 940 parking stalls and two parking attendant booths. The average story height is 11.5 feet.

9. Using Figures 34-3 and 34-4, determine the cost for a 240-foot by 300-foot, three-story, reinforced-concrete parking garage with a reinforced-concrete frame. The parking garage is expected to have 690 parking stalls and is to be used as a park-and-ride. The average story height is 10.5 feet.

10. Using the project comparison method, prepare an estimate for a two-story home in your local area. The home is to have three bedrooms, two bathrooms, a living room, kitchen and dining facilities, and a two-car garage. The square footage should be between 1,800 and 2,200 square feet. Exclude the lot, all site work, and all permits from your estimate. To prepare this estimate, you will need to find two or three projects to use as a comparison. These may be obtained via the Internet or talking to local builders. Be sure to adjust the prices as necessary to make the projects truly comparable, including adjustment for age and quality of materials. Be prepared to share the projects you used as a comparison and your calculations with the class, if your instructor chooses to do so.

REFERENCES

1. Bledsoe, John D., *Successful Estimating Methods . . . from Concept to Bid*, Kingston, RSMeans, 1992.

APPENDIX A

REVIEW OF ESTIMATING MATH

This appendix is provided as a review of the mathematical equations and concepts needed to perform the quantity takeoff for those who struggle with math or just want to review the necessary mathematical concepts before reading the chapters covering the quantity takeoff. Do not let the math scare you. The mathematical principles used in estimating are relatively simple and can be mastered with a little effort.

In estimating we often need to determine lengths, areas, and volumes in order to determine the quantities of materials needed to complete the project. In many cases we need to convert the quantities from one set of units to another set of units. Each of these topics is discussed in this appendix. Let's first look at determining lengths.

LENGTHS

In most cases lengths are determined by calculating the distances based on dimensions shown on the plans or by measuring the distances using a scale, plan measurer, digitizer, or takeoff software package; or by extracting the data from a building modeling software package. There are cases, such as in a roof-framing plan, where the distances shown on the drawings are not accurate because the building component is not parallel to the drawing plane, but instead is being projected to a horizontal plane. When this occurs the Pythagorean theorem needs to be used to determine the true distance. Let's look at ways to determine distances.

Calculating Distances from Dimensions

To determine distance from the dimensions, we may have to use two or more dimensions to determine the correct distance. For example, to find the distance from the corner of the building to the center of the door (A) in Figure A-1 one needs to add the distance from the corner of the building to the center of the window to the distance from the center of the window to the center of the door. Similarly, to determine the distance from the center of the window to the center of the closet wall (B) in Figure A-1, one needs to subtract the distance from the corner of the building to the center of the closet wall from the distance from the corner of the building to the center of the window.

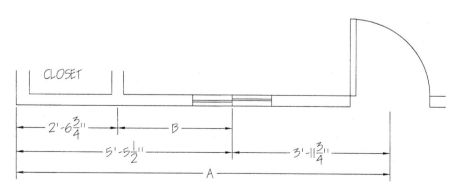

FIGURE A-1 Calculating Distance from Dimensions

When adding dimensions with feet and inches, the feet must be added to the feet, the inches to the inches, and the fractions of an inch to the fractions of an inch. Before adding the dimensions, one must first make sure that all the fractions of an inch have the same denominator (the bottom portion of the fraction). This is done by multiplying the numerator (the top portion) of a fraction and its denominator by the same number (which doesn't change the overall value of the fraction) in such a way that each fraction ends up with the same denominator. After you add the dimensions together, if the numerator of the fraction is greater than the denominator of the fraction, subtract the denominator from the numerator and add 1 inch to the inches portion of the distance. Repeat this until the numerator is less than the denominator. When the inches portion exceeds 12 inches, subtract 12 inches from the inches portion of the distance and add 1 foot to the feet portion of the distance. Repeat this until the inches portion of the distance is greater than or equal to 0 and less than 12.

EXAMPLE A-1

Determine distance between the corner of the building and the center of the door (*A*) in Figure A-1.

Solution: The distance is equal to the distance between the corner of the building and the center of the window (5 feet 5 1/2 inches) plus the distance between the center of the window and the center of the door (3 feet 11 3/4 inches). First, convert the fractions of an inch to a common denominator, in this case 1/4 inch, by multiplying the numerator (the top) and the denominator (the bottom) of 1/2 by 2 to get 2/4. The feet, the inches, and the fractions of an inch are added together as follows:

$$\text{Distance} = 5'\ 5\ 1/2'' + 3'\ 11\ 3/4'' = 5'\ 5\ 2/4'' + 3'\ 11\ 3/4''$$
$$= 8'\ 16\ 5/4''$$

Next, reduce the fraction of inches to less than 1 inch by adding 1 to the inches and subtracting 4 from the numerator of the fraction as follows:

$$\text{Distance} = 8'\ 16\ 5/4'' = 8'\ 17\ 1/4''$$

Finally, reduce the inches portion of the dimension to less than 12 by subtracting 12 from the inches and adding 1 to the feet as follows:

$$\text{Distance} = 8'\ 17\ 1/4'' = 9'\ 5\ 1/4''$$

The distance is 9 feet 5 1/4 inches.

When determining the distance between the center of the wall and the center of the window (*B*) in Figure A-1, one needs to subtract the smaller dimension (2 feet 6 3/4 inches) from the larger dimension (5 feet 5 1/2 inches). Before these dimensions are subtracted, the fractions of an inch must have the same denominator. This is done by multiplying both the numerator and the denominator of the fraction with the lower denominator by the same number. Next, one must make sure the fraction of an inch in the larger dimension is greater than the fraction of an inch in the smaller dimension. If this is not the case, one adds the number in the denominator to the number in the numerator of the fractions of an inch and subtracts 1 from the whole number of inches. One must also make sure that the number of inches in the larger dimension is greater than the number of inches in the smaller dimensions. If this is not the case, 12 must be added to the number of inches and 1 must be subtracted from the number of feet. Once this is accomplished, the feet are subtracted from the feet, the inches from the inches, and the fraction of an inch from the fraction of an inch, as shown in the following example.

EXAMPLE A-2

Determine the distance between the center of the closet wall and the center of the window (*B*) in Figure A-1.

Solution: The distance is equal to the distance between the corner of the building and the center of the window (5 feet 5 1/2 inches) minus the distance between the corner of the building and the center of the closet wall (2 feet 6 3/4 inches). First, make sure the fractions of an inch have the same denominator. Do this by multiplying the numerator (the top) and the denominator (the bottom) of 1/2 by 2 to get 2/4 as follows:

$$\text{Distance} = 5'\ 5\ 1/2'' - 2'\ 6\ 3/4'' = 5'\ 5\ 2/4'' - 2'\ 6\ 3/4''$$

Next, make sure that the fraction of an inch in the larger dimension is greater than the fraction of an inch in the smaller dimension. In this example, this is not the case, so in the larger dimension the denominator (4) must be added to the numerator (2) and 1 subtracted from the inches as follows:

$$\text{Distance} = 5'\ 5\ 2/4'' - 2'\ 6\ 3/4'' = 5'\ 4\ 6/4'' - 2'\ 6\ 3/4''$$

Next, make sure the number of inches in the larger dimension is greater than the number of inches in the smaller dimension. In this example, this is not the case, so 12 must be added to the inches of the large dimension and 1 subtracted from the feet as follows:

$$\text{Distance} = 5'\ 4\ 6/4'' - 2'\ 6\ 3/4'' = 4'\ 16\ 6/4'' - 2'\ 6\ 3/4''$$

Now we can subtract the feet from the feet, the inches from the inches, and the fractions of an inch from the fractions of an inch as follows:

$$\text{Distance} = 4'\ 16\ 6/4'' - 2'\ 6\ 3/4'' = 2'\ 10\ 3/4''$$

The distance is 2 feet 10 3/4 inches.

SCALING

Another method of determining the distance on the plans is scaling the distance. Plans are prepared at reduced scale. The length of components on the plans may be determined by scaling the length using a scale, plan measurer, digitizer, or takeoff package. When scaling drawings it is important to make sure that the correct scale is used and that the drawings are to scale. This may be done by scaling a known dimension and verifying that the scaled dimension is the same as the dimension shown on the drawing.

A triangular architect's scale includes the following scales: 1″ = 1′ (1 inch on the plans equals 1 foot of the building) 1/2″ = 1′, 1/4″ = 1′, 1/8″ = 1′, 3/16″ = 1′, 3/32″ = 1′, 3/4″ = 1′, 3/8″ = 1′, 1 1/2″ = 1′, and 3″ = 1′ along with a standard 12-inch ruler. Most of the sides have two scales, one reading left to right and the other reading right to left, with one scale being one-half of the other scale. For example, the architect scale shown in Figure A-2 includes two scales: 1/8″ = 1′ reading left to right and 1/4″ = 1′ reading right to left. For the 1/8″ = 1′ scale, the feet are read on the right side of the scale and the inches on the left side of the scale. The feet for the 1/8″ = 1′ scale are the numbers on the scale closest to the edge and increase in increments of 4 from left to right. For the 1/8″ = 1′ scale, both the long and short lines represent a foot. For the 1/4″ = 1′ scale, the feet are read on the left side of the scale and the inches on the right side of the scale. The feet for the 1/4″ = 1′ scale are the numbers on the scale furthest from the edge and increase in increments of 2 from right to left. For the 1/4″ = 1′ scale, the long lines represent a foot and the short lines represent a half of a foot.

The 1/8″ = 1′ scale is read by lining up the feet on the right side of the drawing such that the left side of the drawing falls within the inch marks on the left side of the scale. The feet are then read off the right side of the scale and the inches off the left side of the scale.

The 1/4″ = 1′ scale is read by lining up the feet on the left side of the drawing such that the right side of the drawing falls within the inch marks on the right side of the scale. When reading the 1/4″ = 1′ scale, care must be taken to line up with a whole foot mark—the longer marks—rather than the short marks, which indicate half a foot for this scale. The feet are then read off the left side of the scale and the inches off the right side of the scale.

EXAMPLE A-3

Determine the length of the wall shown in Figure A-3 using the scale at the bottom of the figure. The figure is drawn at a scale of 1/8″ = 1′.

Solution: The scale is lined up with a foot mark at the right side of the drawing, and the left side of the drawing is lined up within the inch marks at the left side of the scale. The feet and inches are then read off of the scale. The length of the wall is 28 feet 3 inches long. ■

A triangular engineer's scale typically has the following scales: 1:10 (1 inch equals 10 feet), 1:20, 1:30, 1:40, 1:50, and 1:60. Each side has only one scale and is read left to right. Inches are not included on the engineer's scale. The engineer's scale is read by lining up the zero on the left side of the drawing and reading the feet on the right side of the scale.

EXAMPLE A-4

Determine the length of the building shown in Figure A-4 using the scale at the bottom of the figure. The figure is drawn at a scale of 1″ = 20′.

Solution: The scale is lined up with the zero; therefore, the feet can be read off of the right side of the scale. The length of the wall is between 68 and 69 feet. ■

PYTHAGOREAN THEOREM

The Pythagorean theorem is useful when two dimensions at right angles are used to describe a construction component. This is commonly found in roof-framing plans, where the run of the roof joists is shown on the framing plan and there is a rise in the joist that is calculated from the slope of the roof. The Pythagorean theorem is also useful for walls that are constructed at odd angles to the rest of the building, as we will see in Example A-5.

The Pythagorean theorem equates the length of the hypotenuse (C) of a right triangle, a triangle with one 90-degree (right) angle, to the other two sides (A and B) of the triangle. The Pythagorean theorem states that the square of the hypotenuse equals the sum of the squares of the other two sides of the triangle and is written as

$$C^2 = A^2 + B^2 \qquad (A-1)$$

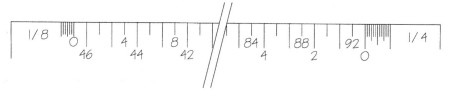

FIGURE A-2 Architect Scale

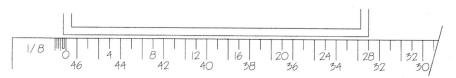

FIGURE A-3 Example A-3

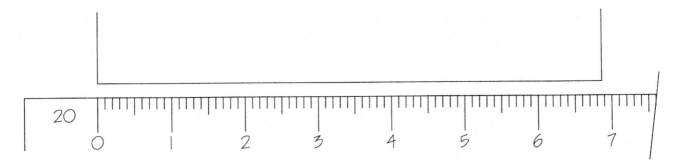

FIGURE A-4 Example A-4

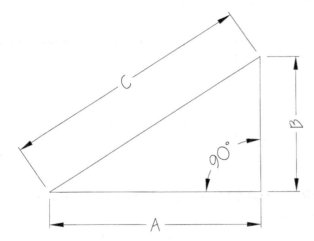

FIGURE A-5 Right Triangle

where A, B, and C are the lengths of the sides of the triangle as shown in Figure A-5. It does not matter which of the two sides adjacent to the 90-degree angle is designated A and which is designated B.

EXAMPLE A-5

Determine the length of the diagonal wall in Figure A-6.

Solution: Using Eq. (A-1), we get

$$C^2 = A^2 + B^2 = (13 \text{ ft})^2 + (9 \text{ ft})^2 = 250 \text{ ft}^2$$

Taking the square root of both sides to solve for C, we get

$$C = (250 \text{ ft}^2)^{0.5} = 15.81 \text{ ft} = 15 \text{ ft } 9 \text{ 3/4 in}$$

The wall is 15 feet 9 3/4 inches long.

AREAS

Estimators are often required to calculate areas in order to complete the quantity takeoff. The area of an object is determined by one of two methods: the geometric method and the coordinate method. Let's look first at the geometric method.

Geometric Method

The geometric method breaks complex shapes down into six simpler geometric shapes: the circle, the square, the

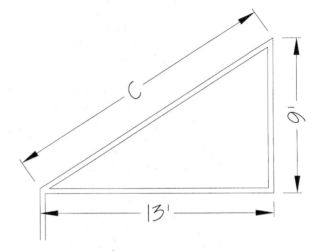

FIGURE A-6 Example A-5

rectangle, the triangle, the trapezoid, and the parallelogram. These shapes are shown in Figure A-7. The areas of these shapes are calculated as follows.

Circle   The area of a circle is calculated from the radius using the equation

$$\text{Area} = \pi R^2 \qquad (A\text{-}2)$$

or from the diameter using the equation

$$\text{Area} = \pi \frac{D^2}{4} \qquad (A\text{-}3)$$

where

π = 3.14159
R = Radius of the Circle
D = Diameter of the Circle

EXAMPLE A-6

Find the area of a circle whose radius is 12 inches.

Solution: Using Eq. (A-2), we get

$$\text{Area} = \pi R^2 = \pi (12 \text{ in})^2 = 452 \text{ in}^2$$

Rectangle   The area of a rectangle is calculated from its length and width using the equation

$$\text{Area} = LW \qquad (A\text{-}4)$$

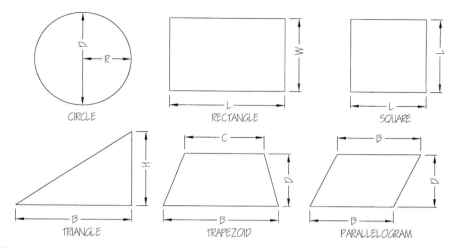

FIGURE A-7 Basic Shapes

where

L = Length
W = Width

EXAMPLE A-7

Find the area of a rectangle whose length is 24 inches and width is 12 inches.

Solution: Using Eq. (A-4), we get

Area = LW = (24 in)(12 in) = 288 in^2

Square A square is a special case of a rectangle in which the length and the width are equal; therefore, all of the sides are the same length. The area of a square is calculated from its length using the equation

$$\text{Area} = L^2 \quad (A\text{-}5)$$

where

L = Length

EXAMPLE A-8

Find the area of a square whose length is 12 inches.

Solution: Using Eq. (A-5), we get

Area = L^2 = (12 in)^2 = 144 in^2

Triangle The area of a triangle is calculated from its base and height using the equation

$$\text{Area} = \frac{BH}{2} \quad (A\text{-}6)$$

where

B = Base
H = Height

The height of the triangle must be measured perpendicular to the base. This is true for all of the triangles shown in Figure A-8.

EXAMPLE A-9

Find the area of a triangle whose base is 12 inches and height is 24 inches.

Solution: Using Eq. (A-6), we get

$$\text{Area} = \frac{BH}{2} = \frac{(12 \text{ in})(24 \text{ in})}{2} = 144 \text{ in}^2$$

Trapezoid A trapezoid is a four-sided figure with two parallel sides. The area of a trapezoid is calculated from the length of the two parallel sides and the perpendicular distance between the parallel sides using the equation

$$\text{Area} = \frac{(B + C)}{2}D \quad (A\text{-}7)$$

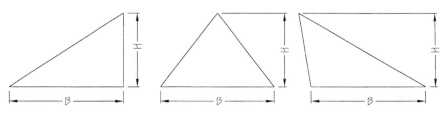

FIGURE A-8 Triangles

where

>B and C = Lengths of the Parallel Sides
>D = Perpendicular Distance between the Parallel Sides

EXAMPLE A-10

Find the area of a trapezoid whose parallel sides are 12 and 16 inches long with a distance between the parallel sides of 24 inches.

Solution: Using Eq. (A-7), we get

$$\text{Area} = \frac{(B + C)}{2}D = \frac{(12 \text{ in} + 16 \text{ in})}{2}(24 \text{ in}) = 336 \text{ in}^2$$

Parallelogram A parallelogram is a four-sided plane figure with opposite sides parallel. Because both pairs of opposite sides are parallel, the opposite sides have the same length. A parallelogram is a special case of a trapezoid in which B and C are the same length. The area of a parallelogram is determined from the length of one side and the perpendicular distance between that side and the side parallel to it, using the equation

$$\text{Area} = BD \qquad (A\text{-}8)$$

where

>B = Length of Each of a Given Pair of Parallel Sides
>D = Perpendicular Distance between the Two Sides That Have Length B

EXAMPLE A-11

Find the area of a trapezoid whose parallel sides are 12 inches long with a distance between the parallel sides of 24 inches.

Solution: Using Eq. (A-8), we get

$$\text{Area} = BD = (12 \text{ in})(24 \text{ in}) = 288 \text{ in}^2$$

Now that we have covered the basic geometric shapes, let's look at two complex shapes. The area of a complex shape is determined by breaking the shape down into its basic geometric shapes, determining the area of the basic geometric shapes, and adding their areas as shown in the following example.

EXAMPLE A-12

Determine the area of the shape in Figure A-9.

Solution: Divide the complex shape into two rectangles and a square as shown in Figure A-10. The area of the left rectangle is determined from Eq. (A-4) as follows:

$$\text{Area} = LW = (30 \text{ ft})(20 \text{ ft}) = 600 \text{ ft}^2$$

The area of the right rectangle is determined from Eq. (A-4) as follows:

$$\text{Area} = LW = (20 \text{ ft})(10 \text{ ft}) = 200 \text{ ft}^2$$

The area of the square is determined from Eq. (A-5) as follows:

$$\text{Area} = L^2 = (30 \text{ ft})^2 = 900 \text{ ft}^2$$

The area of the complex shape is determined by summing the areas of the smaller shapes as follows:

$$\text{Area} = 600 \text{ ft}^2 + 200 \text{ ft}^2 + 900 \text{ ft}^2 = 1,700 \text{ ft}^2$$

The area of the complex shape is 1,700 square feet.

Sometimes it is advantageous to make the shape a part of a larger shape and subtract the missing pieces as shown in the following example.

EXAMPLE A-13

Determine the area of the shape in Figure A-9.

Solution: Make the complex shape into a 60-foot by 30-foot rectangle less a 10-foot by 10-foot square as shown in Figure A-11. The area of the rectangle is determined from Eq. (A-4) as follows:

$$\text{Area} = LW = (60 \text{ ft})(30 \text{ ft}) = 1,800 \text{ ft}^2$$

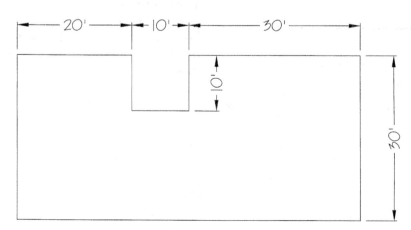

FIGURE A-9 Complex Shape

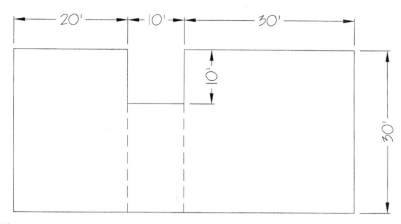

FIGURE A-10 Complex Shape

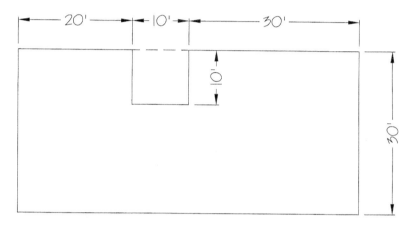

FIGURE A-11 Complex Shape

The area of the square is determined from Eq. (A-5) as follows:

$$\text{Area} = L^2 = (10 \text{ ft})^2 = 100 \text{ ft}^2$$

The area of the complex shape is determined by subtracting the area of the square from the area of the rectangle as follows:

$$\text{Area} = 1{,}800 \text{ ft}^2 - 100 \text{ ft}^2 = 1{,}700 \text{ ft}^2$$

The area of the complex shape is 1,700 square feet, which is the same as the answer from Example A-12.

Another common example of creating a large shape and subtracting the missing pieces arises in the case of an outside radius that creates a complex shape as shown in Figure A-12. The area of this shape is calculated by taking the area of a square of length R and subtracting the area of a quarter of a circle as shown in Figure A-13.

The area is calculated by using the equation

$$\text{Area} = R^2 \left(1 - \frac{\pi}{4} \right) \quad \text{(A-9)}$$

where

$\pi = 3.14159$
$R = $ Radius of the Circle

The use of this equation is shown in the following example.

FIGURE A-12 Complex Shape

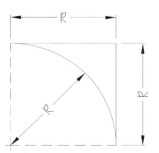

FIGURE A-13 Square-Less-a-Quarter-Circle

EXAMPLE A-14

Determine the area of the complex shape in Figure A-14.

Solution: Divide the shape into two rectangles and a square-less-a-quarter-circle as shown in Figure A-15.

The area of the left rectangle is determined from Eq. (A-4) as follows:

$$\text{Area} = LW = (70 \text{ ft})(30 \text{ ft}) = 2{,}100 \text{ ft}^2$$

The area of the right rectangle is determined from Eq. (A-4) as follows:

$$\text{Area} = LW = (40 \text{ ft})(30 \text{ ft}) = 1{,}200 \text{ ft}^2$$

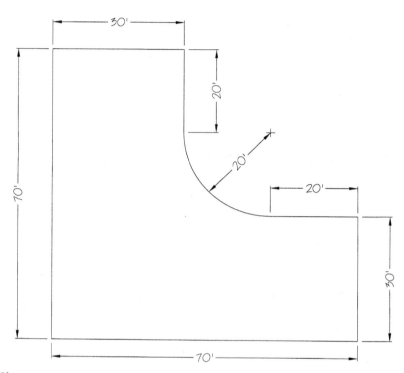

FIGURE A-14 Complex Shape

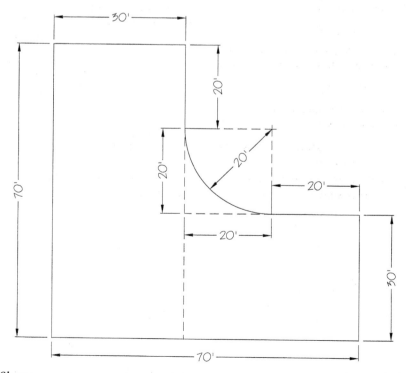

FIGURE A-15 Complex Shape

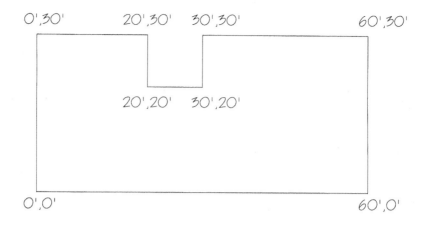

FIGURE A-16 Complex Shape

The area of the square-less-a-quarter-circle is determined from Eq. (A-9) as follows:

$$\text{Area} = R^2\left(1 - \frac{\pi}{4}\right) = (20 \text{ ft})^2\left(1 - \frac{\pi}{4}\right) = 86 \text{ ft}^2$$

The area of the complex shape is determined by summing the areas of the smaller shapes as follows:

$$\text{Area} = 2{,}100 \text{ ft}^2 + 1{,}200 \text{ ft}^2 + 86 \text{ ft}^2 = 3{,}386 \text{ ft}^2$$

The area of the complex shape is 3,386 square feet.

Coordinate Method

The coordinate method is used to calculate the area of a complex shape in which all the sides are straight by using the coordinates of the intersections of the sides. The area of an *n*-sided shape is determined by the equation

$$\text{Area} = 0.5[(X_2Y_1 + X_3Y_2 + \cdots + X_nY_{n-1} + X_1Y_n) \\ - (X_1Y_2 + X_2Y_3 + \cdots + X_{n-1}Y_n + X_nY_1)] \quad (A\text{-}10)$$

The use of this equation is shown in the following example.

EXAMPLE A-15

Determine the area of the complex shape in Figure A-16.

Solution: Using the intersection in the lower-left corner as Point 1, proceeding around the complex shape in a clockwise direction, and using Eq. (A-10), we get

$$\text{Area} = 0.5[(0 \times 0 + 20 \times 30 + 20 \times 30 + 30 \times 20 \\ + 30 \times 20 + 60 \times 30 + 60 \times 30 + 0 \times 0) - (0 \times 30) \\ + 0 \times 30 + 20 \times 20 + 20 \times 20 + 30 \times 30 + 30 \times 30 \\ + 60 \times 0 + 60 \times 0] \text{ft}^2$$

$$\text{Area} = 0.5[(0 + 600 + 600 + 600 + 600 + 1{,}800 + 1{,}800 + 0) \\ - (0 + 0 + 400 + 400 + 900 + 900 + 0 + 0)] \text{ft}^2$$

$$\text{Area} = 0.5[6{,}000 - 2{,}600] = 0.5[3{,}400] = 1{,}700 \text{ ft}^2$$

Because Figure A-16 is the same size as Figure A-9, they should have the same areas, which they do.

An alternate method for calculating the area using the coordinate method is to set up the coordinates as shown in Figure A-17, with the coordinates for the first point at both the top and bottom.

Starting with the second row, the X-coordinate is multiplied by the Y-coordinate from the previous row, and the resultant is entered in the second row of the first column as shown in Figure A-18. This is repeated for the third through the last rows. Next, starting with the second row, the Y-coordinate is multiplied by the X-coordinate from the previous row, and the resultant is entered in the second row of the fourth column as shown in Figure A-18. This is repeated for the third to the last rows. Next, the first and fourth columns are added as shown in Figure A-18.

To find the area, the sum of the fourth column is subtracted from the sum of the first column and the resultant is divided by two. For the coordinates in Figures A-17 and A-18, which correspond to the coordinates of Example A-15, the area is calculated as follows:

$$\text{Area} = 0.5(6{,}000 - 2{,}600) = 0.5(3{,}400) = 1{,}700 \text{ ft}^2$$

This gives us the same area as was found in Example A-15.

	X	Y	
	0'	0'	
	0'	30'	
	20'	30'	
	20'	20'	
	30'	20'	
	30'	30'	
	60'	30'	
	60'	0'	
	0'	0'	

FIGURE A-17 Table for Coordinate Method

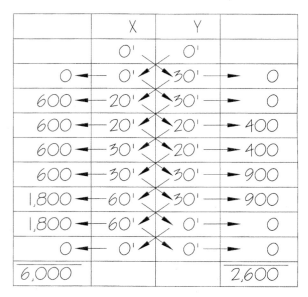

FIGURE A-18 Completed Table for Coordinate Method

VOLUMES

In addition to determining distances and areas, estimators often must determine volumes. Most volumes are created by extending a two-dimensional shape along a line perpendicular to the plane it occupies, as in the case of a cylinder (circle), a column (a square or rectangle), or a prism (triangle), or by extending the shape to a point, as is the case with a cone or a pyramid. The basic volumetric shapes are shown in Figure A-19. The volumes of these shapes are calculated as follows.

Cylinder

The volume of a cylinder is calculated by multiplying the area of a circle by the height of the cylinder, where the height of the cylinder is measured perpendicular to the area of the circle. The volume of a cylinder is calculated from the radius by using the equation

$$\text{Volume} = \pi R^2 H \qquad (A\text{-}11)$$

or from the diameter, which is twice the radius, by using the equation

$$\text{Volume} = \frac{\pi D^2 H}{4} \qquad (A\text{-}12)$$

where

π = 3.14159
R = Radius of the Circle
D = Diameter of the Circle
H = Height Measured Perpendicular to the Area of the Circle

EXAMPLE A-16

Find the volume of a cylinder whose radius is 1 foot and height is 10 feet.

Solution: Using Eq. (A-11), we get

$$\text{Volume} = \pi R^2 H = \pi (1 \text{ ft})^2 (10 \text{ feet}) = 31.4 \text{ ft}^3$$

Column

The volume of a column is calculated by multiplying the area of a rectangle by the height of the column, where the height of the column is measured perpendicular to the area of the rectangle. Remember, a square is a special case of a rectangle in which L and W are equal. The volume of a rectangular column is calculated from its length, width, and height using the equation

$$\text{Volume} = LWH \qquad (A\text{-}13)$$

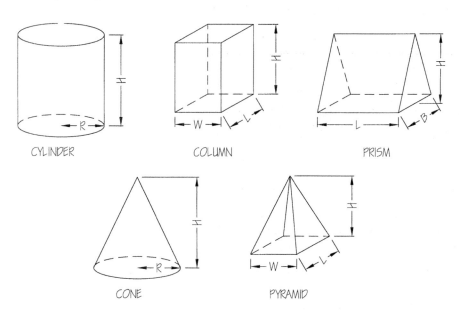

FIGURE A-19 Volumetric Shapes

where

    L = Length
    W = Width
    H = Height Measured Perpendicular to the Area of the Rectangle

EXAMPLE A-17

Find the volume of a column whose height is 10 feet, length is 2 feet, and width is 1 foot.

Solution: Using Eq. (A-13), we get

$$\text{Volume} = LWH = (2 \text{ ft})(1 \text{ ft})(10 \text{ ft}) = 20 \text{ ft}^3$$

Prism

The volume of a prism is calculated by multiplying the area of a triangle by the length of the prism (length is used because height has already been used to describe the triangle), where the length of the prism is measured perpendicular to the area of the triangle. The volume of a prism is calculated from its length, base, and height using the equation

$$\text{Volume} = \frac{BHL}{2} \quad (A\text{-}14)$$

where

    H = Height Measured Perpendicular to the Base of the Triangle
    B = Base
    L = Length Measured Perpendicular to the Area of the Triangle

EXAMPLE A-18

Find the volume of a prism whose height is 5 feet, length is 20 feet, and base is 3 feet.

Solution: Using Eq. (A-14), we get

$$\text{Volume} = \frac{BHL}{2} = \frac{(3 \text{ ft})(5 \text{ ft})(20 \text{ ft})}{2} = 150 \text{ ft}^3$$

Pyramids and Cones

For any volumetric shape that is created by extending the shape to a point, such as a cone or pyramid, the volume is one-third of the area of the base of the object times the height of the object, where the height is measured perpendicular to the area. For a pyramid with a rectangular base, the volume of the pyramid is calculated by using the equation

$$\text{Volume} = \frac{LWH}{3} \quad (A\text{-}15)$$

where

    L = Length
    W = Width
    H = Height Measured Perpendicular to the Area of the Rectangle

This is true for all of the pyramids shown in Figure A-20, even though their points are at different locations.

EXAMPLE A-19

Find the volume of a pyramid whose height is 12 feet, length is 2 feet, and width is 1 foot.

Solution: Using Eq. (A-15), we get

$$\text{Volume} = \frac{LWH}{3} = \frac{(2 \text{ ft})(1 \text{ ft})(12 \text{ ft})}{3} = 8 \text{ ft}^3$$

The volume of a cone is calculated from the radius using the equation

$$\text{Volume} = \frac{\pi R^2 H}{3} \quad (A\text{-}16)$$

or from the diameter using the equation

$$\text{Volume} = \frac{\pi D^2 H}{12} \quad (A\text{-}17)$$

where

    π = 3.14159
    R = Radius of the Circle
    D = Diameter of the Circle
    H = Height Measured Perpendicular to the Area of the Circle

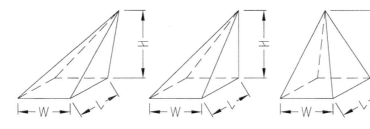

FIGURE A-20 Pyramids

EXAMPLE A-20

Find the volume of a cone whose radius is 1 foot and height is 10 feet.

Solution: Using Eq. (A-16), we get

$$\text{Volume} = \frac{\pi R^2}{3} = \frac{\pi (1 \text{ ft})^2 (10 \text{ ft})}{3} = 10.5 \text{ ft}^3$$

Volume of a Shape with a Constant Area

The volume of any shape with a constant area over a given length is calculated by multiplying the cross-sectional area of the shape by the length of the object over which the cross-sectional area is representative of the shape of the object, using the equation

$$\text{Volume} = AL \quad \text{(A-18)}$$

where
- A = Cross-Sectional Area of the Shape Measured Perpendicular to the Length
- L = Length

The length should be measured along the centerline of the shape, and the cross-sectional area should be measured perpendicular to the centerline of the shape. When the centerline of the shape forms a curved line, the volume is an approximation.

EXAMPLE A-21

Find the volume of a curb and gutter whose cross-sectional area is 1.7 ft² and whose length is 200 ft.

Solution: Using Eq. (A-18), we get

$$\text{Volume} = AL = (1.7 \text{ ft}^2)(200 \text{ ft}) = 340 \text{ ft}^3$$

Complex Volumes

Like the areas of complex two-dimensional shapes, the volumes of complex three-dimensional bodies can be calculated by breaking the complex body into simpler shapes for which volume can be calculated. For example, the complex volume created by a building excavation shown in Figure A-21 can be divided into four pyramids (one at each corner), four prisms (one along each side), and one column (located in the center), as shown in Figure A-22. The volume of the complex shape is calculated by determining the volumes for the pyramids, prisms, and column and adding them together. Additional methods of approximating the volume of complex bodies are discussed in Chapter 16.

CONVERSION FACTORS

So far all of the example problems have been in the same units, and we have not had to convert units. Typical unit conversions are given in Table A-1.

The two most common mistakes when converting from one set of units to another set of units are: (1) using the wrong conversion factor and (2) multiplying by the conversion factor when one should be dividing or vice versa. The best way to avoid these mistakes is to include the units in the calculations and make sure the answer is in the correct units by canceling out units.

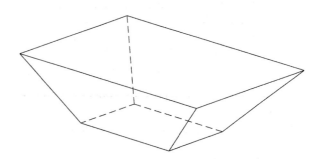

FIGURE A-21 Complex Shape

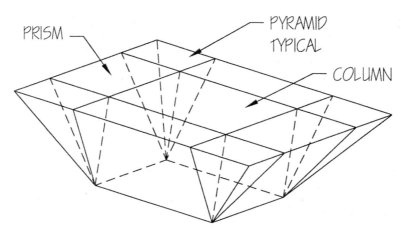

FIGURE A-22 Complex Shape

TABLE A-1 Conversion Factors

Length	Volume
1 mile = 5,280 feet (ft)	1 yd^3 = 27 ft^3
1 yard (yd) = 3 ft	1 ft^3 = 1,728 in^3
1 foot = 12 inches (in)	1 ft^3 = 7.48 gallons

Area	Weight
1 acre = 43,560 ft^2	1 ton = 2,000 pounds (lb)
1 yd^2 = 9 ft^2	1 hundredweight (cwt) = 100 lb
1 ft^2 = 144 in^2	1 lb = 16 ounces (oz)

EXAMPLE A-22

Convert 216 square inches into square feet.

Solution: To convert square inches to square feet, we need to divide the square inches by 144 in^2/ft^2 as follows:

$$216 \text{ in}^2 = \frac{216 \text{ in}^2}{144 \text{ in}^2/\text{ft}^2} = 1.5 \frac{\text{in}^2 \times \text{ft}^2}{\text{in}^2} = 1.5 \text{ ft}^2$$

Two hundred and sixteen square inches equals 1.5 square feet.

In Example A-22, had we tried to multiply by 144 in^2/ft^2, we would have ended up with the following:

$$216 \text{ in}^2 = 216 \text{ in}^2 \times 144 \frac{\text{in}^2}{\text{ft}^2} = 31,104 \frac{\text{in}^2 \times \text{in}^2}{\text{ft}^2} = 31,104 \frac{\text{in}^4}{\text{ft}^2}$$

In this case the inches would not cancel out, leaving us with in^4/ft^2 as the units for our answer, not the units we wanted. Likewise, had we used the wrong conversion factor and divided by 1,728 in^3/ft^3, we would have ended up with the following:

$$216 \text{ in}^2 = \frac{216 \text{ in}^2}{1,728 \text{ in}^3/\text{ft}^3} = 0.125 \frac{\text{in}^2 \times \text{ft}^3}{\text{in}^3} = 0.125 \frac{\text{ft}^3}{\text{in}}$$

In this case, the inches squared on the top reduce the inches cubed on the bottom to inches, leaving us with ft^3/in for our units—not the answer we wanted. I cannot emphasize this enough. Write your units down and cancel the units out to make sure that your answer is in the units you want!

Dimensional Analysis

In dimensional analysis one writes the equation without the numbers to make sure the units cancel out. This is useful for checking your units without having all of the numbers, which at times can be confusing. Dimensional analysis on Example A-22 is done as follows:

$$\text{in}^2 = \frac{\text{in}^2}{(\text{in}^2/\text{ft}^2)} = \frac{\text{in}^2 \times \text{ft}^2}{\text{in}^2} = \text{ft}^2$$

Board Feet

Lumber, especially random lengths and finish lumber, are often purchased by the board foot. A board foot (bft) is defined as the amount of wood contained in a 12-inch by 12-inch by 1-inch nominal board. Figure A-23 shows the definition of a board foot. The nominal dimensions are the dimensions used to describe the board, for example, 2 × 4 or 2 × 12. For dimensional lumber, the actual dimensions are always smaller than the nominal dimensions. A 2 × 4 typically measures 1 1/2 inches by 3 1/2 inches. The units of board feet are inches-feet-squared (in-ft^2).

The number of board feet in a piece of lumber is calculated by using the equation

$$\text{Board Feet} = \frac{(\text{Thickness})(\text{Width})(\text{Lineal Feet})}{12 \text{ in/ft}} \quad (A\text{-}19)$$

where the nominal thickness and the nominal width are in inches.

EXAMPLE A-23

Find the number of board feet in a 10-foot long 2 × 4.

Solution: Using Eq. (A-19), we get

$$\text{Board Feet} = \frac{(\text{Thickness})(\text{Width})(\text{Lineal Feet})}{12 \text{ in/ft}}$$

$$= \frac{(2 \text{ in})(4 \text{ in})(10 \text{ ft})}{12 \text{ in/ft}} = 6.67 \text{ in-ft}^2 = 6.67 \text{ bft}$$

There are 6.67 board feet in a 10-foot-long 2 × 4.

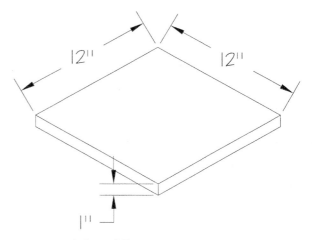

FIGURE A-23 A Board Foot

APPENDIX B

SAMPLE JOB COST CODES

The following is a sample list of standardized job cost codes for a commercial construction company along with a list of typical items that would be included in the cost code. The first two digits of each cost code correspond to the divisions of the 2014 edition of the MasterFormat®. The last three digits are company specific and are used to collect cost data in a way that matches the company's individual work practices. The cost codes should not be confused with the numbering of the MasterFormat®. For more information on the MasterFormat visit www.csinet.org/MasterFormat.

Code	Description	Typical Items
01-000	**GENERAL REQUIREMENTS**	
01-300	Supervision	Labor for supervision and supervisor's truck and tools.
01-500	Temporary Utilities	Temporary utility bills including electricity, water, sewer, natural gas, garbage, and propane. Temporary electrical, sewer, water, and gas lines and connection fees for temporary utilities.
01-510	Temporary Phone	Phone bills including long-distance service, mobile phones, and temporary phone lines.
01-520	Temporary Facilities	Temporary office trailer, office furniture, copier, computers, fax machine, telephone equipment, first-aid supplies, and office supplies.
01-700	Cleanup	Labor and supplies for general cleanup—cleanup that cannot be tied to a specific task.
02-000	**EXISTING CONDITIONS**	
02-400	Demolition	Demolition of existing structures.
03-000	**CONCRETE**	
03-200	Rebar	Rebar materials used in building concrete and masonry: and anchor bolts embedded in the concrete and masonry.
03-300	Footing and Foundation—Labor	Labor, forms, ties, tie wire, equipment, and all other items (except concrete, rebar, and concrete pump) needed to pour footings and foundations for buildings.
03-310	Footing and Foundation—Concrete	Concrete to pour footings and foundations for buildings.
03-320	Slab/Floor—Labor	Labor, forms, ties, tie wire, equipment, and all other items (except concrete, rebar, and concrete pump) needed to pour building slabs and floors.

03-330	Slab/Floor—Concrete	Concrete to pour building slabs and floors.
03-340	Concrete Pump	Concrete pump for placement of building concrete.
03-400	Precast Concrete	Precast concrete used in buildings.
03-500	Light-weight Concrete	Light-weight concrete and gypcrete.
04-000	**MASONRY**	
04-200	Masonry	Concrete Masonry Unit (CMU) and face brick used in buildings.
05-000	**METALS**	
05-100	Structural Steel	Structural steel materials including beams, girders, columns, lintels (not provided by mason), bracing, and bolts through the structural steel. Excludes installation and bolts not through the structural steel.
05-200	Joist and Deck	Steel open web joists and metal deck. Excludes installation.
05-500	Metal Fabrications	Stairs, handrail, decorative steel, pipe bollards, and other non-structural steel. Excludes installation.
05-900	Erection	Installation of structural steel, joist, deck, and metal fabrications, including equipment and welding supplies. Excludes installation of embeds.
06-000	**WOOD, PLASTICS, AND COMPOSITES**	
06-100	Rough Carpentry	Labor, nails, and equipment to install wood framing.
06-110	Lumber	Rough carpentry materials including lumber, engineered lumber (glue-laminated beams, wood I-joists, and so forth), wood siding, framing hardware, bolts, fasteners, and foam seal. Excludes nails.
06-120	Trusses	Pre-manufactured wood trusses.
06-200	Finish Carpentry	Labor, nails, and equipment to install metal doors, wood doors, hardware, wood trim, bathroom accessories, fire extinguishers and cabinets, and roof specialties when not installed by their suppliers.
06-210	Wood Trim	Wood trim, including base, casing, chair rail, and crown molding.
06-400	Cabinetry and Countertops	Furnish and install cabinetry and, if included in the bid, countertops.
06-410	Countertops	Furnish and install countertops when not included with cabinetry bid.
07-000	**THERMAL AND MOISTURE PROTECTION**	
07-100	Waterproofing	Furnish and install waterproofing, dampproofing, and drain board.
07-200	Insulation	Furnish and install batt, blanket, and blown insulation.
07-210	Ridged Insulation	Furnish and install ridged insulation.
07-220	Stucco	Furnish and install synthetic stucco and Exterior Insulation and Finish System (EIFS).
07-400	Siding	Furnish and install vinyl and aluminum siding, soffit, and fascia.
07-500	Roofing	Furnish and install asphalt, metal, built-up, and membrane roofing including metal trim, asphalt-impregnated felt, ice and water shield, and flashing not provided by other subcontractors. Install flashings provided by other subcontractors.

07-600	Sheet Metal	Furnish and install flashing (excluding roof flashings) and general sheet metal.
07-700	Roof Specialties	Furnish roof hatches, walkways, skylights, and so forth. May include installation.
07-710	Rain Gutters	Furnish and install rain gutters and down spouts.
07-800	Fireproofing	Furnish and install fire caulk and fire collars not provided by other subcontractors.
07-900	Caulking and Sealants	Furnish and install caulking not provided by other subcontractors.
08-000	**OPENINGS**	
08-100	Metal Doors and Frames	Furnish hollow-metal doors and frames.
08-110	Wood Doors	Furnish wood doors, wood frames, and prehung residential-grade doors. Prehung doors include butts and thresholds.
08-300	Overhead Doors	Furnish and install sectional and coiling overhead doors and coiling grilles, including all necessary hardware.
08-400	Storefronts	Furnish and install glass and aluminum storefronts including all necessary hardware (cylinders by hardware supplier).
08-500	Windows	Furnish vinyl, vinyl-clad, wood, aluminum, and aluminum-clad residential windows.
08-700	Hardware	Furnish hardware for hollow-metal and wood doors and cylinders for storefront doors.
08-800	Glass and Glazing	Furnish and install glazing in hollow-metal doors, wood doors, and window frames.
09-000	**FINISHES**	
09-200	Drywall	Furnish and install drywall, drywall draft stops, and, if included in the bid, nonstructural metal studs.
09-210	Metal Studs	Furnish and install nonstructural metal studs, if not included with drywall bid.
09-300	Ceramic Tile	Furnish and install ceramic tile and marble.
09-500	Acoustical Ceilings	Furnish and install acoustical ceiling with grid.
09-600	Flooring	Furnish and install carpet, vinyl, vinyl composition tile (VCT), wood flooring, laminate flooring, and rubber base.
09-700	Wall Coverings	Furnish and install fabric and paper wall coverings.
09-900	Paint	Furnish and install paints and stains for walls, ceilings, floors, doors, and trim.
10-000	**SPECIALTIES**	
10-100	Signage	Furnish and install interior building signage including room numbers, bathroom signage, Americans with Disabilities Act (ADA) insignias, and building directories.
10-200	Toilet Partitions	Furnish and install toilet partitions and screens.
10-210	Toilet and Bath Accessories	Furnish and install toilet and bath accessories, including toilet paper holders, soap dispensers, paper towel dispensers, built-in waste receptacles, sanitary napkin dispensers, sanitary napkin disposal receptacles, baby changing stations, mirrors, shower enclosures, and medicine cabinets.
10-400	Fire Extinguishers and Cabinets	Furnish and install fire extinguishers and fire extinguisher cabinets.

Code	Category	Description
11-000	**EQUIPMENT**	
11-300	Appliances	Residential appliances, including dishwashers, ranges, cooktops, ovens, refrigerators, ice makers, washers, and dryers.
12-000	**FURNISHINGS**	
12-200	Window Treatments	Furnish and install blinds and curtains.
14-000	**CONVEYING EQUIPMENT**	
14-200	Elevators	Furnish and install elevators, including excavation for hydraulic ram.
21-000	**FIRE SUPPRESSION**	
21-100	Fire Sprinklers	Furnish and install fire sprinkler systems.
22-000	**PLUMBING**	
22-100	Plumbing	Furnish and install building water supply and sanitary sewer systems to five feet outside the building, plumbing fixtures and garbage disposals. Provide flashing for plumbing roof penetrations.
23-000	**HEATING, VENTILATION, AND AIR-CONDITIONING**	
23-100	HVAC	Furnish and install heating, ventilation, and air-conditioning systems. Provide flashing for HVAC roof penetrations.
26-000	**ELECTRICAL**	
26-100	Electrical	Furnish and install electrical, including power, lighting, and control systems. Provide flashing for electrical roof penetrations.
27-000	**COMMUNICATIONS**	
27-100	Communications	Furnish and install communication (telephone and data). Provide flashing for communication roof penetrations.
31-000	**EARTHWORK**	
31-100	Clearing and Grubbing	Clearing and grubbing of site.
31-200	Grading and Excavation	Mass excavation and rough grading of site, including import or export of soils. Excavation, backfill, and import or export of soils for structures.
32-000	**EXTERIOR IMPROVEMENTS**	
32-100	Asphalt	Furnish and install asphalt pavement, slurry seals, base under asphalt, prime coat, and tack coat.
32-110	Site Concrete—Labor	Labor, forms, ties, tie wire, equipment, and all other items (except concrete and rebar) needed to pour site concrete, including sidewalks, driveways, curbs, gutters, waterways, and slabs outside the building.
32-120	Site Concrete—Concrete	Concrete to pour site concrete, including sidewalks, driveways, curbs, gutters, waterways, and slabs outside the building.
32-130	Rebar	Rebar materials used in the construction of site concrete and retaining walls.
32-300	Fencing	Chain link, masonry (including footings), wood, ornamental steel, and vinyl fencing, including gates.
32-310	Retaining Walls	Furnish and install reinforced concrete (excluding rebar materials), block, or rock retaining walls, including footings.
32-320	Dumpster Enclosures	Furnish and install dumpster enclosures, including footings and foundation, block wall, cap, and gates.

32-330	Signage	Furnish and install traffic signs, permanent project signs, and monuments.
32-340	Outside Lighting[1]	Furnish and install lighting mounted on poles and buildings to illuminate the site, including underground wiring.
32-900	Landscaping	Finish grading. Top soil import and placement. Furnish and install irrigation (sprinkler) systems, plants (trees, shrubs, bushes, and ground cover), sod, and bark.
33-000	**UTILITIES**	
33-100	Water Line	Furnish and install site culinary water lines beginning five feet outside the perimeter of the building, including gravel and bedding materials.
33-300	Sanitary Sewer	Furnish and install site sanitary sewer lines beginning five feet outside the perimeter of the building, including gravel and bedding materials.
33-400	Storm Drain	Furnish and install storm drain, including gravel and bedding materials.
33-500	Gas Lines	Furnish and install site natural gas lines beginning at the low-pressure side of the gas meter, including any fees charged by the natural gas provider.
33-700	Power Lines	Furnish and install electrical power lines to the meter base, including transformers located outside the building and any fees charged by the electrical power provider.
33-800	Telephone Lines	Furnish and install phone lines to the building, including any fees charged by the telephone company.

[1] Site lighting is placed here instead of Division 26000 to keep building costs separate from site costs.

APPENDIX C

SAMPLE LABOR PRODUCTIVITY RATES

Sample labor productivity rates are shown in the following table. These rates are for use in classroom exercises and are based on a review of a number of published productivity rates and personal experience. Where possible, estimators should use productivity rates based on the company's past performance. When this data is not available, the estimator should consult a current, more detailed source of productivity rates.

Item	Productivity
03-000 CONCRETE	
Forms, columns, square to 24" × 24"	0.13–0.20 lhr/sfca
Forms, footings, 2 × 12" high	0.06–0.09 lhr/lft
Forms, walls	0.09–0.15 lhr/sfca
Rebar, 20' continuous horizontal, to #6	0.20–0.30 lhr/ea
Rebar, dowel, to #6	0.05–0.07 lhr/ea
Rebar, ring tie	0.08–0.13 lhr/ea
Rebar, vertical, to 12' high, to #6	0.12–0.20 lhr/ea
Anchor bolt	0.05–0.10 lhr/ea
Embed	0.25–0.50 lhr/ea
Concrete placement, columns	0.40–0.60 lhr/cyd
Concrete placement, continuous footings	0.40–0.50 lhr/cyd
Concrete placement, slabs	0.40–0.60 lhr/cyd
Concrete placement, spot footings	0.60–1.00 lhr/cyd
Concrete placement, walls	0.50–0.70 lhr/cyd
Concrete finishing	0.010–0.015 lhr/sft
04-000 MASONRY	
Concrete masonry unit, including rebar	0.09–0.14 lhr/sft
Brick facing	0.16–0.20 lhr/sft
05-000 METALS	
Steel beams	0.07–0.10 lhr/lft
Steel columns	0.90–1.20 lhr/ea

Steel joists	0.04–0.07 lhr/lft
Metal deck	0.007–0.014 lhr/sft

06-000 WOODS, PLASTICS, AND COMPOSITES

Blocking	0.03–0.04 lhr/ft
Columns	0.60–1.00 lhr/ea
Fascia	0.04–0.06 lhr/lft
Girders and beams, to 20' long	1.0–1.5 lhr/ea
Joists	0.015–0.019 lhr/lft
Rafters	0.02–0.04 lhr/lft
Sheathing	0.011–0.016 lhr/sft
Sills and ledgers	0.02–0.05 lhr/lft
Soffit	0.04–0.06 lhr/sft
Trusses, to 40' span	0.60–1.00 lhr/ea
Walls, 16' O.C., to 10' high	0.13–0.16 lhr/lft
Add for door opening	0.25–0.50 lhr/ea
Add for window opening	0.50–0.75 lhr/ea
Trim	0.015–0.040 lhr/lft
Cabinetry	0.60–1.00 lhr/ea
Countertops, plastic laminate	0.25–0.33 lhr/lft

07-000 THERMAL AND MOISTURE PROTECTION

Building paper	0.20–0.40 lhr/squ
Insulation—batt or blanket	0.005–0.008 lhr/sft
Insulation—blown	0.010–0.016 lhr/sft
Exterior Insulation and Finish System	0.05–0.06 lhr/sft
Shingles, asphalt	1.4–2.2 lhr/squ
Fascia, aluminum or vinyl	0.04–0.05 lhr/lft
Fascia, wood	0.04–0.06 lhr/lft
Siding, aluminum or vinyl	0.03–0.04 lhr/sft
Siding, wood board	0.020–0.035 lhr/sft
Siding, wood sheet	0.011–0.025 lhr/sft
Soffit, aluminum or vinyl	0.04–0.05 lhr/sft
Soffit, wood	0.04–0.06 lhr/sft
Membrane roofing	0.010–0.015 lhr/sft
Sheet metal, drip edge	0.02–0.03 lhr/lft
Sheet metal, flashings	0.04–0.08 lhr/lft
Roof hatch	3.0–5.0 lhr/ea

08-000 OPENINGS

Doors, commercial exterior	2.0–4.0 lhr/ea
Doors, commercial interior	1.0–2.5 lhr/ea
Doors, residential exterior	1.0–2.0 lhr/ea
Doors, residential interior	0.5–1.0 lhr/ea
Sectional overhead door	2.0–4.0 lhr/ea

Add for opener	1.0–1.5 lhr/ea
Add for keyless entry	0.5–1.0 lhr/ea
Storefront	0.10–0.16 lhr/sft
Windows—residential	0.8–1.0 lhr/ea
Hardware, lockset or deadbolt	0.5–1.0 lhr/ea
Hardware, panic	1.0–2.0 lhr/ea
Hardware, stop	0.20–0.40 lhr/ea
Hardware, threshold	0.40–0.70 lhr/ea
Hardware, weather stripping	1.0–1.50 lhr/ea

09-000 FINISHES

Metal stud wall	0.018–0.020 lhr/sft
Drywall, taped and finished	0.015–0.020 lhr/sft
Ceramic tile	0.08–0.11 lhr/sft
Suspended acoustical ceilings	0.020–0.035 lhr/sft
Wood and laminate floors	0.04–0.08 lhr/sft
Sheet vinyl	0.03–0.04 lhr/sft
Vinyl composition tile	0.015–0.020 lhr/sft
Rubber base	0.25–0.30 lhr/lft
Carpet and pad	0.14–0.17 lhr/syd
Paint, walls, sprayed, per coat	0.004–0.007 lhr/sft
Paint, walls, rolled, per coat	0.006–0.010 lhr/sft
Paint, trim, brushed, per coat	0.008–0.015 lhr/lft
Paint, doors, brushed, per coat	0.5–1.0 lhr/ea

10-000 SPECIALTIES

Door signs	0.33–0.50 lhr/ea
Grab bar	0.33–0.67 lhr/ea
Mirror	0.50–1.00 lhr/ea
Paper towel dispenser	0.50–1.00 lhr/ea
Shower surround	3.0–4.0 lhr/ea
Toilet paper holder	0.25–0.50 lhr/ea
Towel bar	0.25–0.50 lhr/ea
Fire extinguisher cabinet	2.00–3.00 lhr/ea

21-000 FIRE SUPPRESSION

Fire sprinkler, to 1.5″ pipe, includes hangars	0.12–0.20 lhr/ft
Fire sprinkler, 2″ pipe, includes hangars	0.20–0.25 lhr/ft
Fire sprinkler, 2.5″ pipe, includes hangars	0.28–0.32 lhr/ft
Fire sprinkler, 3″ pipe, includes hangars	0.32–0.38 lhr/ft
Fire sprinkler, to 1.5″ fitting	0.14–0.33 lhr/ea
Fire sprinkler, 2″ fitting	0.33–0.50 lhr/ea
Fire sprinkler, 2.5″ fitting	0.40–0.60 lhr/ea
Fire sprinkler, 3″ fitting	0.50–0.75 lhr/ea
Fire sprinklers, sprinkler head and drop	0.50–0.80 lhr/ea

Fire sprinklers, standpipe	3.0–4.0 lhr/ea
Fire sprinklers, valve assembly	24–32 lhr/ea

22-000 PLUMBING

Clean-outs	0.8–1.0 lhr/ea
Drinking fountain, rough in supply and waste	4.0–5.0 lhr/ea
Drinking fountain, finish	2.0–2.5 lhr/ea
Floor drain	1.0–1.5 lhr/ea
Kitchen sink, rough in supply and waste	8.0–10.0 lhr/ea
Kitchen sink, finish	2.5–3.5 lhr/ea
Lavatory, rough in supply and waste	6.0–8.0 lhr/ea
Lavatory, finish	2.5–3.0 lhr/ea
Roof drain, including wye	3.0–4.0 lhr/ea
Shower, rough in supply and waste	7.0–8.0 lhr/ea
Shower, finish, including shower door	6.0–8.0 lhr/ea
Tub, rough in supply and waste	7.0–8.0 lhr/ea
Tub, finish	2.5–4.0 lhr/ea
Urinal, rough in supply and waste	6.0–8.0 lhr/ea
Urinal, finish	5.0–6.0 lhr/ea
Valve, to 2"	0.5–1.0 lhr/ea
Water closet, rough in supply and waste	6.0–8.0 lhr/ea
Water closet, finish	3.0–3.5 lhr/ea
Water heater	4.0–6.0 lhr/ea

23-000 HVAC

Duct, to 12" × 24"	0.10–0.13 lhr/lft
Duct, 12" × 30"	0.12–0.15 lhr/lft
Duct, 4"-diameter vent pipe	0.05–0.06 lhr/lft
Duct, 6" diameter	0.05–0.06 lhr/lft
Duct, 9" diameter	0.09–0.10 lhr/lft
Duct, 12" diameter	0.13–0.15 lhr/lft
Duct, 16"-wide sheet metal nailed to joists	0.05–0.07 lhr/lft
Duct, tee, elbow, transition, or cap, to 12" diameter	0.10–0.15 lhr/ea
Duct, tee, elbow, transition, or cap, to 12" × 12"	0.10–0.15 lhr/ea
Duct, tee, elbow, transition, or cap, 24" × 12"	0.15–0.20 lhr/ea
Diffuser, including drop	0.85–1.00 lhr/ea
Furnace and air conditioner, residential, includes plenum	6.0–8.0 lhr/ea
Package unit	20–26 lhr/ea
Register, including boot or 90-degree elbow	0.30–0.50 lhr/ea
Return air grill	0.30–0.50 lhr/ea
Thermostat	1.0–1.5 lhr/ea

26-000 ELECTRICAL

Conduit, to 1″, including fittings	0.08–0.10 lhr/lft
Junction or metal box	0.40–0.60 lhr/ea
Light fixtures, commercial	1.5–2.5 lhr/ea
Light fixtures, residential, including wiring	0.5–1.0 lhr/ea
Outlet, commercial	0.20–0.30 lhr/ea
Outlet, residential, 110V, including wiring	0.40–0.60 lhr/ea
Outlet, residential, 220V, including wiring	1.0–2.0 lhr/ea
Panel, commercial	8.0–14.0 lhr/ea
Panel, residential	6.0–8.0 lhr/ea
Switches, commercial	0.20–0.30 lhr/ea
Switches, residential, including wiring	0.40–0.60 lhr/ea
Wiring, commercial	0.02–0.03 lhr/lft

27-000 COMMUNICATION

Conduit, to 1″, including fittings	0.08–0.10 lhr/lft
Junction or metal box	0.40–0.60 lhr/ea
Phone jack, including wiring	0.40–0.60 lhr/ea
Television jack, including wiring	0.40–0.60 lhr/ea

31-000 EARTHWORK

Excavation, machine	0.08–0.10 lhr/cyd
Footings excavation, hydraulic excavator	0.08–0.10 lhr/cyd
Place and compact	0.15–0.20 lhr/cyd
Rough grading	0.02–0.03 lhr/sft

32-000 EXTERIOR IMPROVEMENTS

Road base, place and compact	0.15–0.20 lhr/cyd
Asphalt	0.02–0.03 lhr/sft
Forms, to 12″ high	0.06–0.09 lhr/lft
Forms, 13″ to 24″ high	0.09–0.15 lhr/sfca
Rebar placement	15–16 lhr/ton
Concrete placement, curb and gutter	0.40–0.50 lhr/cyd
Concrete placement, slab	0.40–0.60 lhr/cyd
Concrete finishing	0.010–0.015 lhr/sft
Finish grade	0.02–0.03 lhr/sft
Sod	0.005–0.008 lhr/sft

33-000 UTILITIES

Backfill	0.15–0.20 lhr/cyd
Bedding	0.15–0.20 lhr/cyd
Catch basin	5.0–7.0 lhr/ea
Connection to existing manhole or box	1.0–2.0 lhr/ea
Pipe, to 4″	0.15–0.20 lhr/lft
Trench excavation, hydraulic excavator	0.08–0.10 lhr/cyd

ABBREVIATIONS

ea = each
lft = linear feet
sfca = square-foot contact area
sft = square foot
squ = square (100 ft^2)
syd = square yard

APPENDIX D

SAMPLE EQUIPMENT COSTS

Sample equipment cost are shown in the following table. These rates are for use in the classroom exercises and are based on a review of published cost data. Estimators should use equipment costs obtained from local sources or company historical data.

The hourly operation cost includes fuel, lubrication and filters, tires, and repairs.

The rental cost includes the equipment ownership costs and does not include operation costs. The weekly rental cost is approximately 3 times the daily rental cost and the monthly rental cost is approximately 3 times the weekly rental cost.

All cost exclude the labor to operate the equipment.

Equipment	Hourly Operation Cost	Daily Rental Cost	Weekly Rental Cost	Monthly Rental Cost
Asphalt paver, 8 to 15 feet wide	100	1,365	4,100	12,300
Compactor, pad foot, 60 inch wide	30	695	2,075	6,225
Compactor, pneumatic, 3 ton	18	410	1,225	3,675
Compactor, single drum, 50 inch wide	33	205	625	1,875
Compressor, 100 cfm	15	55	175	525
Compressor, 150 cfm	23	80	250	750
Concrete pump	25	870	2,600	7,800
Dozer, 150 hp	55	525	1,575	4,725
Dozer, 240 hp	63	795	2,375	7,125
Generator, 150 kW	15	345	1,025	3,075
Generator, 300 kW	30	570	1,700	5,100
Hydraulic excavator, 1.25 cyd	60	540	1,625	4,875
Hydraulic excavator, 2.25 cyd	103	765	2,300	6,900
Light tower	15	60	175	525
Loader, 1 cyd	32	155	475	1,425
Loader, 2.25 cyd	67	270	800	2,400
Motor Grader, 165 hp, 12 foot wide	65	460	1,375	4,125
Office trailer	0	15	50	150
Pavement breaker, handheld	2	10	25	75

(continued)

(Contd.)

Equipment	Hourly Operation Cost	Daily Rental Cost	Weekly Rental Cost	Monthly Rental Cost
Pump, 18,000 gph	15	45	125	375
Pump, 40,000 gph	30	110	325	975
Skid steer, 49 hp	12	145	425	1,275
Sweeper	12	150	450	1,350
Telescopic excavators, 260 hp	67	1,575	4,725	14,175
Tractor-loader-backhoe	24	210	625	1,875
Truck, bottom dump, highway, 20 cyd	60	505	1,525	4,575
Truck, dump truck, highway, 12 cyd	72	390	1,175	3,525
Truck, light duty	14	75	225	675
Truck, water, 3000 gallon	60	205	625	1,875

APPENDIX E

MODEL SCOPES OF WORK

The following are sample scopes of work for multiuse developments (which includes commercial buildings, single-family homes, and twin homes) that may be used as model scopes of work for other projects. Not all provisions in these scopes of work will apply to all projects, and some projects will need additional provisions.

FOOTINGS AND FOUNDATIONS

Provide all labor, materials, equipment, tools, and supervision required to furnish and install all footings and foundations and all appurtenances required for a complete installation in accordance with the contract documents, applicable codes, and governing agencies. The work includes, but is not limited to, the following:

1. Installation of all footings, foundations, concrete walls, and concrete columns.
2. Installation of all embeds, anchor bolts, sleeves, hold-downs, and post bases that are to be cast in place. The subcontractor shall supply anchor bolts, hold-downs, and post bases. Embeds and sleeves to be provided by others.
3. Installation of all structural steel anchor bolts that are installed in the concrete footings, foundations, and columns. Construct plywood templates as required for proper bolt placement and alignment. All structural steel anchor bolts to be provided by others.
4. The subcontractor is to provide window bucks.
5. The subcontractor shall protect all concrete at all times during installation from weather damage, vandalism, and all other destructive elements.
6. The subcontractor shall provide layout for the work contained in this contract.
7. Concrete and rebar is to be provided by the contractor and installed by the subcontractor.
8. The subcontractor shall ensure that only approved concrete mixes from approved suppliers are poured. The subcontractor shall obtain a list of approved concrete mixes and suppliers from the project superintendent.
9. Concrete pumping is excluded. The contractor shall pay for any required concrete pumping. The subcontractor shall obtain approval to pump concrete prior to doing so. If the subcontractor fails to obtain approval prior to pumping concrete, the subcontractor shall pay for the cost of pumping the concrete.
10. All concrete placed under this contract shall be installed to avoid thin edges and voids.
11. Any concrete rejected by the owner, contractor, or building inspector, or concrete that is rejected because of failure to meet specified strength, shall be replaced by the subcontractor at its sole expense.

FRAMING

Provide all labor, equipment, tools, and supervision required to furnish and install all framing and all appurtenances required for a complete installation in accordance with the contract documents, applicable codes, and governing agencies. The work includes, but is not limited to, the following:

1. The subcontractor shall install all rough carpentry members, including, but not limited to, sills, studs, plates, sleepers, columns, posts, beams, girders, floor joists, header joists, stringers, risers, treads, trimmers, backing, headers, blocking, lintels, bridging, firestopping, rafters, purlins, trusses, floor sheathing, wall sheathing, roof sheathing, and drops.

2. The subcontractor shall install all exterior drywall and all drywall sandwiched between wood walls.
3. The subcontractor shall install all framing hardware, including but not limited to, nails, shots, pins, expansion bolts, machine bolts, lag screws, post-to-beam connections, joist and beam hangers, and straps.
4. The subcontractor shall provide all nails required for the completion of this contract. All other framing hardware is to be provided by the contractor.
5. The subcontractor shall install all exterior doors, sliding glass doors, and windows. All doors and windows shall be hung true and plumb. Windows that line up shall be hung such that the vertical and horizontal lines are straight and true. Doors shall be hung such that they do not bind or swing open or closed by themselves.
6. The subcontractor shall furnish the forklift and/or crane for use in transporting and placing materials installed as a part of this contract.
7. The subcontractor shall furnish all tools, saws, cords, and miscellaneous equipment required for the completion of this contract.
8. Contractor shall supply temporary power for the subcontractor.
9. The subcontractor shall coordinate with the project superintendent to:
   a. ensure that correct material sizes and quantities are ordered;
   b. ensure prompt and timely material deliveries;
   c. coordinate material distribution and storage;
   d. coordinate installation procedures with other trades;
   e. set up and maintain inventory control for receiving and distributing materials;
   f. spot-check rough openings and backing requirements.

FINISH CARPENTRY

Provide all labor, materials, equipment, tools, and supervision required to furnish and install all finish carpentry and all appurtenances required for a complete installation in accordance with the contract documents, applicable codes, and governing agencies. The work includes, but is not limited to, the following:

1. Installation of all interior doors. All doors shall be hung true and plumb. Doors shall be hung such that they do not bind or swing open or closed by themselves.
2. Installation of mirrored wardrobe doors and associated hardware.
3. Installation of all interior trim, including, but not limited to, base, casing, chair rail, crown mold, wood handrail, wood guardrail, and window seats.
4. Installation of fireplace mantels.
5. Installation of wood closet shelving and rods with associated hardware.
6. Installation of door hardware and door viewers.
7. Installation of house numbers.
8. Installation of windowsills.
9. Installation of bath hardware, including, but not limited to, medicine cabinets, grab bars, towel bars, towel rings, toilet paper holders, and shower rods. All bath hardware shall be fastened into wood backing or installed with molly bolts.
10. The subcontractor is to provide nails, shims, and glue. All other materials are to be provided by others.
11. The subcontractor shall coordinate the location of required backing with the project superintendent and the framing subcontractor.

DRYWALL

Provide all labor, materials, equipment, tools, and supervision required to furnish and install all drywall and all appurtenances required for a complete installation in accordance with the contract documents, applicable codes, and governing agencies. The work includes, but is not limited to, the following:

1. Provide and install all interior drywall and associated components, including, but not limited to, all trim, accessories, corner and edge trims, and reveals as required by the contract documents and required for a complete job.
2. Provide and install fire walls and draft stops as shown on the contract documents.
3. Provide and install fire-rated and moisture-resistant board and compounds where required by code and the contract documents.
4. Furnish and install all rated enclosures, chases, soffits, shafts, and so forth for plumbing, electrical, HVAC ducting and piping, elevators, and other items as required by the contract documents.
5. The subcontractor shall pre-rock rated enclosures, chases, soffits, shafts, behind the tub, furnace rooms, and so forth prior to rough plumbing, rough HVAC, and rough electrical as directed by the project superintendent.
6. All walls and ceilings are to be taped and textured and ready for paint or wall coverings.
7. The subcontractor is to provide a complete drywall finish in the furnace room.

8. The subcontractor is to provide a complete drywall finish in the garage.
9. Materials shall be stocked by the subcontractor at right angles to the floor trusses/joists and distributed so the materials' weight will not concentrate in any areas that may cause damage to the structure.
10. The subcontractor shall not cover any areas that require insulation prior to installation of insulation and will at all times advise the project superintendent of any deficiencies in the insulation.
11. All drywall shall be screwed into place.
12. The subcontractor shall patch around openings of pipes, ductwork, electrical, and other openings in a workmanlike manner. Patches in fire-rated walls shall comply with the applicable fire and building codes.
13. All vertical and horizontal external corners shall be reinforced with corner bead trim applied in strict accordance with the manufacturer's recommendations. Vertical corners shall be straight and plumb. Horizontal corners shall be straight and level.
14. Joints and screw depressions shall be invisible to the naked eye when surface is painted. The subcontractor shall verify that no shadows occur at joints by shining a lamp on the face of the finished surface to check for shadows and bad joints.
15. After the initial coat of primer has been applied, the subcontractor shall touch up all areas for screw pops, shadows, rough paper, faulty areas, and areas unacceptable to the contractor.
16. Care shall be taken and exercised to minimize dropping joint compound on other surfaces. Droppings shall be scraped loose before they set up in all areas, including closets.
17. The subcontractor shall be responsible for the disposal of all drywall scrap materials. The subcontractor shall not dispose of drywall scrap materials in the dumpster supplied by the contractor.
18. Exterior drywall and drywall between the wood framing of the party walls are excluded from this subcontract.

FLOOR COVERINGS

Provide all labor, materials, equipment, tools, and supervision required to furnish and install all floor coverings and all appurtenances required for a complete installation in accordance with the contract documents, applicable codes, and governing agencies. The work includes, but is not limited to, the following:

1. Furnish and install all resilient flooring, including sheet vinyl and vinyl tile.
2. Furnish and install all carpeting.
3. Furnish and install all rubber base.
4. Furnish and install all underlayment.
5. Furnish and install trim metal where carpet or vinyl meets other flooring materials or the tub.
6. The subcontractor shall prepare and submit seam diagrams for the contractor's review. Installation of carpet shall strictly follow the approved seaming diagram.
7. The subcontractor shall not combine different lots of materials within a home or building. This applies to both carpet and vinyl materials.
8. The subcontractor shall sweep the floors prior to installation of carpet or vinyl.
9. The subcontractor shall be responsible for any minor float leveling (1/4 inch or less) that may be necessary prior to the installation of floor coverings. The subcontractor shall provide and install all items as required for floor preparation.
10. The subcontractor shall caulk the trim metal to the tub.
11. The subcontractor shall rehang doors after installing flooring.

PAINTING AND STAINING

Provide all labor, materials, equipment, tools, and supervision required to furnish and install all painting and staining and all appurtenances required for a complete installation in accordance with the contract documents, applicable codes, and governing agencies. The work includes, but is not limited to, the following:

1. Furnish and install all paint and stain for the walls, ceilings, doors, interior handrails, interior guardrails, and interior trim.
2. Prime and seal under wall coverings.
3. Furnish and install concrete floor sealer to interior concrete floors as listed on the finish schedule.
4. Furnish and install all painting required to paint the garage and furnace/utility room.
5. Furnish and install all painting required to paint the exterior wood siding, trim, soffit, and fascia.
6. Furnish and install all painting required to paint the shutters.
7. Furnish and install all painting required to paint the exterior stairs, stair treads, stair stringers, railings, and handrails.
8. Furnish and install all painting required to paint the power meters, gas meters, telephone boxes, cable TV boxes, and disconnects.
9. Furnish and install all painting required to paint the vents, roof vents, and roof flashings to match the roof. The said items are to receive one coat of

zinc chromate primer where needed and one coat of paint.

10. Seal and finish the exposed edges of all metal and wood doors including the top and bottom when they are visible from stairs, balconies, and so forth.
11. Paint shall be applied to all drywall surfaces by roller.
12. All surfaces of building components (which are installed by others) shall be inspected and accepted by the paint subcontractor prior to start of any work. Application of subcontractor's work shall constitute acceptance.
13. Mask, protect, and cover surfaces and items that do not receive paint, including, but not limited to, door hinges, fire-rating labels, exposed concrete and masonry, fixtures, furnaces, equipment, and so forth.
14. Properly clean and prepare all surfaces before applying all material and finishes.
15. Fill gouges, fill nicks, fill nail holes, caulk joints, and prepare all surfaces, including spot priming, before proceeding with the work.
16. The subcontractor shall rehang doors after painting.
17. The subcontractor is to include touch-up of paint and finishes prior to owner's walkthrough as required by normal damage, splatters, and marring caused by the process of construction work. Touch-up is to include minor drywall patching. This touch-up is in addition to the touch-up that occurs after the owner's walkthrough.
18. The subcontractor is to include four hours of touch-up per residence for touch-up after the owner's walkthrough. This touch-up is in addition to the touch-up that occurs before the owner's walkthrough.

FIRE SPRINKLERS

Provide all labor, materials, equipment, tools, and supervision required to furnish and install all fire sprinklers and all appurtenances required for a complete installation in accordance with the contract documents, applicable codes, and governing agencies. The work includes, but is not limited to, the following:

1. Furnish and install all piping, fittings, heads, supports, seismic bracing, anchors, backflow preventer stations, post indicator valves, fire sprinkler risers, fire department connections, supervisory tamper/flow switches, alarm bell/gong, roof manifold, and so forth.
2. The subcontractor shall be responsible for the design and adequacy of the fire sprinkling system. The subcontract price includes obtaining approved design drawings, professional engineer's certification or stamp, and related permit from the local municipality.
3. The subcontractor shall provide trenching, backfill, piping, fittings, and connections to 5 feet outside the building line. Site utilities are provided by others.
4. Furnish and install all access doors required for the fire sprinkling system.
5. Install all heads in the center of the ceiling tiles.
6. Provide all testing, flushing, and so forth required to comply with the fire marshal's requirements.
7. The subcontractor shall furnish and install fire stops, fire-retardant compound, and/or safing to seal all of the subcontractor's penetrations through fire-rated assemblies.
8. The construction of the fire sprinkling system shall closely follow the construction of the building and shall be placed in service as soon as applicable laws permit following completion of each story.

PLUMBING

Provide all labor, materials, equipment, tools, and supervision required to furnish and install all plumbing and all appurtenances required for a complete installation in accordance with the contract documents, applicable codes, and governing agencies. The work includes, but is not limited to, the following:

1. Furnish and install all fixtures and piping, including, but not limited to, water heater, water softener, toilets, sinks, vanities, shower pans, tubs, washer hookups, icemaker boxes, faucets, floor drains, hose bibs, drain piping, vent piping, and supply piping.
2. The subcontractor shall provide a floor drain by laundry facilities and water heaters.
3. The subcontractor is to connect the water heater to the gas line. The gas line is to be provided by others.
4. The subcontractor is to connect the dishwasher to the plumbing.
5. The subcontractor shall provide all trenching required for their work.
6. The subcontractor shall provide all block-outs and sleeves through floors, walls, and ceilings required for their work.
7. The subcontractor shall coordinate with the framing subcontractor for the location and size of all required backing for the plumbing fixtures.
8. The subcontractor shall furnish and install fire stops, fire-retardant compound, and/or safing to

seal all of the subcontractor's penetrations through fire-rated assemblies.

9. The subcontractor shall furnish all pipe flashings and sheet metal jacks for its work. Pipe flashings and sheet metal jacks are to be installed by others.

HVAC

Provide all labor, materials, equipment, tools, and supervision required to furnish and install all HVAC and all appurtenances required for a complete installation in accordance with the contract documents, applicable codes, and governing agencies. The work includes, but is not limited to, the following:

1. Furnish and install all equipment required to complete the HVAC system, including, but not limited to, furnace, fan coil, and condenser.
2. Furnish and install all required ducting, including, but not limited to, ducting, supply registers, return-air grilles, combustion air ducting, and weather caps.
3. Furnish and install smoke and fire dampers as required by code and by the contract documents.
4. Furnish and install ducting for clothes dryers.
5. Furnish and install ducting to all exhaust fans. The exhaust fans are provided and installed by others.
6. Furnish and install a setback thermostat and all required low-voltage wiring. Provide temporary thermostats for use during construction.
7. The subcontractor is to connect electrical power to furnace and air-conditioning units from the outlet box or disconnect provided by others.
8. Furnish and install all refrigerant lines. Provide protection from nail punchers for all lines.
9. Furnish and install all condensate drain lines and piping for all mechanical equipment.
10. Furnish and install all natural gas lines including gas line service to gas ranges, gas dryers, gas fireplaces, and water heaters.
11. The subcontractor is to connect gas dryers and gas ranges to gas lines after appliance installation.
12. Connect water heater to gas lines. Water heater to be provided by others.
13. The subcontractor shall furnish all pipe flashings and sheet metal jacks for its work. Pipe flashings and sheet metal jacks are to be installed by others.
14. The subcontractor shall provide all block-outs and sleeves through floors, walls, and ceilings required for its work.
15. The subcontractor shall coordinate with the framing subcontractor for location and size of all required backing for the HVAC.
16. The subcontractor shall furnish and install fire stops, fire-retardant compound, and safing to seal all of the subcontractor's penetrations through fire-rated assemblies.
17. The subcontractor shall work with the contractor to provide an early operation of the furnace to be used as temporary heat in the building.
18. The subcontractor shall replace all furnace filters just prior to the building being turned over to the owner.

ELECTRICAL

Provide all labor, materials, equipment, tools, and supervision required to furnish and install all electrical and all appurtenances required for a complete installation in accordance with the contract documents, applicable codes, and governing agencies. The work includes, but is not limited to, the following:

1. Furnish and install all electrical fixtures, switches, meter bases, breaker boxes, and wiring.
2. Provide one meter base per unit and one house meter per apartment building.
3. Furnish and install all exhaust fans. Ducting of the exhaust fans is to be provided by others.
4. Furnish, install, and provide power to the disconnect for the air-conditioner condenser.
5. Furnish, install, and provide power to an electrical box for the furnace.
6. Provide electrical hookup of the dishwasher and disposal.
7. Install range hood or space-saver microwave.
8. Install pigtails on range and electrical clothes dryers.
9. Furnish and install wiring, jacks, boxes, a terminal box, and other items required to prewire the telephones.
10. Furnish and install wiring, jacks, boxes, a terminal box, and other items required to prewire the cable television.
11. Furnish and install wiring, door and window contacts, and an outlet required to prewire the security system. The key pad, motion detectors, and equipment are provided and installed by others.
12. Furnish and install wiring, jacks, boxes, equipment, and other items required to complete the intercom/sound system.
13. Furnish and install wiring, jacks, boxes, and other items required to prewire the home theater system.

EARTHWORK AND UTILITIES—ROADS AND PARKING LOTS

Provide all labor, materials, equipment, tools, and supervision required to furnish and install all earthwork and utilities and all appurtenances required for a complete installation in accordance with the contract documents, applicable codes, and governing agencies. The work includes, but is not limited to, the following:

1. Furnish and install sanitary sewer lines, associated manholes, and so forth to complete the sanitary sewer system.
2. Furnish and install sanitary sewer laterals for all lots. The subcontractor shall furnish and install one sewer lateral per lot. The sewer laterals shall be constructed from the sanitary sewer in the road to a point 10 feet inside the lot boundaries. The locations of the sewer laterals are to be coordinated with the site superintendent. The subcontractor shall mark the end of each sewer lateral in such a manner that it can be located for future connection.
3. The subcontractor shall make all required connections to connect the sanitary sewer lines to existing utilities.
4. Furnish and install the water lines, fittings, valves, hydrants, washouts, boxes, covers, and so forth to complete the water system.
5. Furnish and install water laterals, including, but not limited to, the water meter yoke and meter box for all lots. The subcontractor shall furnish and install one water meter yoke per lot. The water laterals shall be constructed from the water line in the road right-of-way to the water meter locations shown on the plans. Water meters are excluded from this contract.
6. The subcontractor shall make all required connections to connect the water lines to existing utilities. Utility company's connection, impact, and review fees are excluded from this subcontract.
7. Furnish and install storm drain piping, catch basins, clean-out boxes, manholes, and so forth to complete the storm drain system.
8. The subcontractor shall construct the detention and retention ponds shown on the plans.
9. The subcontractor shall excavate and backfill the trench required for the placement of power lines. The power lines shall be furnished and installed by others.
10. Furnish and install conduits for phone and cable television. The phone and cable television conduits may be placed in a joint trench with the power lines. The phone and cable television wiring shall be furnished and installed by others.
11. Furnish and install conduits for site lighting. The site lighting conduits may be placed in a joint trench with the power lines. The site lighting and its associated wiring shall be furnished and installed by others.
12. The subcontractor shall clear and grub the site.
13. The subcontractor shall provide the rough grade for the roads and parking lots and their associated gutters, sidewalks, and curb walls within the road right-of-way.
14. Furnish and install base for all gutters, sidewalks, and curb walls required for completion of all concrete within the road right-of-ways and parking lots.

LANDSCAPING

Provide all labor, materials, equipment, tools, and supervision required to furnish and install all landscaping and all appurtenances required for a complete installation in accordance with the contract documents, applicable codes, and governing agencies. The work includes, but is not limited to, the following:

1. Furnish and install the landscaping including hydroseeding, sod, trees, shrubs, and bedding plants.
2. Furnish and install an irrigation system with automatic controls and pop-up heads. The electrical supply to control locations shall be furnished by others. The subcontractor shall coordinate with the project superintendent and electrical subcontractor to ensure proper location of the automatic controls and electrical supply.
3. The subcontractor shall adjust all sprinkler heads and set automatic controllers prior to completion of each building or residence.
4. The subcontractor shall place all sleeves, pipes, and conduits required to complete the sprinkling system under roads, driveways, walks, curbs, and other hard surfaces prior to the construction of the said surfaces.
5. Preparation of the soil.
6. Finish grade to ± 0.1 foot. Rough grade to ± 0.2 foot to be provided by others.
7. Disposal of rock uncovered by installation of the irrigation system.
8. The subcontractor is to warranty all trees and shrubs for one year.
9. The subcontractor is to warranty all hydroseeding and sod for 90 days.
10. The subcontractor shall maintain the landscape for 90 days after final acceptance.

GLOSSARY

A

Absolute Reference All or part of an Excel cell's reference that does not change as the cell is copied, which is designated by a dollar sign ($) in front of the row and/or column designation for the cell.

Addenda A construction document, issued before the bid, that changes the scope of work for a construction project.

Add-On Costs that are added to the cost of the bid, such as bonds, building permit, and profit and general overhead. Also known as a markup.

AGC Associated General Contractors of America.

AIA American Institute of Architects.

Air Handler An HVAC component that combines a heat exchanger and a fan.

Alternate Bid Item A bid in which the contractor provides alternate or optional pricing for specific bid items in addition to the base bid. Also known as an optional bid item.

American Standard Beam An I-beam designated with the prefix S.

American Standard Channel A steel channel designated with the prefix C.

Anchor Bolt A bolt used to connect steel or wood to concrete or masonry.

Angle or Angle Iron An L-shaped steel member designated with the prefix L.

Annual Contract A contract for materials that is set up for a period of time (often 1 year), which covers multiple construction projects.

ANSI American National Standards Institute.

Architect's Scale A scale or ruler that allows the user to read dimensions from plans drawn at common architectural scales.

ASTM American Society for Testing and Materials.

Avoidable Waste Waste that can be avoided by careful ordering, storage, and use of materials.

B

Bank Cubic Yards The volume of soil in cubic yards before it has been excavated in its natural or in situ condition.

Beam A horizontal member that supports a load and transfers the load to girders or columns.

Bedding Gravel or sand that provides a place for underground pipe to rest and may surround the pipe to protect it.

Bid The amount that the contractor charges the owner for the work.

Bid Bond A bond issued by a surety guaranteeing that the bidder, if successful in winning the bid, will execute the contract and provide payment and performance bonds.

Bid Documents Forms used for the submission of the bid, which may include bid forms, bid-bond forms, a schedule of values, and contractor certifications.

Bid Form The construction document on which the bid or price is submitted.

Bid Instructions A construction document that provides bidders with a set of instructions that must be followed to prepare a complete bid.

Bid Package The construction documents that define the scope of work for a construction project, on which bids are prepared.

Billable Hours The number of hours that an employee can be billed to projects during a year.

Block Concrete masonry unit used to construct structural walls.

Blocking A small piece of wood used to strengthen structural members or fill the space between members, as is the case with fire blocking.

Boiler A closed vessel used to heat a liquid.

Bond *See* Bid Bond, Payment Bond, and Performance Bond.

Bond Beam A horizontal row of concrete blocks or bricks reinforced with rebar and grout, used to add structural strength to a wall.

Bonding Limit The maximum dollar value for a single job and total work in progress a contractor may be bonded for.

Bottom Plate The bottom framing member in a wood wall.

Brick A masonry unit made from clay or shale.

Brick Facing A nonbearing layer of brick on the face of a block or wood wall. Also known as brick veneer.

Brick Ties Corrugated metal ties used to connect brick veneer to masonry or wood walls.

Brick Veneer A nonbearing layer of brick on the face of a block or wood wall. Also known as brick facing.

Budget The target cost set for a project or portion of a project.

Building Paper Asphalt-impregnated felt, often used as a vapor barrier.

Bullnose A rounded outside corner.

Burden The costs that must be added to an employee's wages to determine his or her total cost, which includes cash equivalents and allowances paid to the employee, payroll taxes, unemployment insurance, workers' compensation insurance, general liability insurance, insurance benefits, retirement contributions, union payments, vacation, sick leave, and other benefits paid by the employer. Commonly referred to as labor burden.

Butts A common type of hinge.

Buyout The process of hiring subcontractors and procuring materials and equipment for a construction project.

C

Cell (1) A location on an Excel spreadsheet designated by a row and column. (2) The hollow space in a concrete masonry unit.

Certifications A construction document that the contractor signs certifying that it meets specified criteria.

Change Order A construction document that is issued after the bid and changes the scope of the work and/or the contract amount for a construction project.

Channel American Standard Channel, a steel channel designated with the prefix C.

Check Valve A valve that allows a fluid to flow in one direction only.

Chiller An HVAC unit used to cool water.

Close-Out Audit A review of all activities completed during the construction process, including the estimating process, which occurs after the project is complete.

CMU Concrete masonry unit.

Coil An HVAC component consisting of a coil of pipes used to transfer heat from the air to a refrigerant or from the refrigerant to the air. Also known as a heat exchanger.

Collar Ties A wood member connected to two rafters used to prevent the wall supporting the rafters from spreading apart due to the forces on the rafters. Also known as rafter ties.

Column A slender vertical structural member.

Compacted Cubic Yards The volume of soil in cubic yards after it has been excavated, placed, and compacted, and is used to measure the volume of fill.

Competitive Bidding The bid process in which two or more contractors compete for the bid based on price and/or other criteria.

Composite Slab A structural slab that derives its strength from structural steel beams in combination with a concrete deck.

Concrete A mixture of cement, large aggregate, fine aggregate, and water.

Concrete Masonry Unit Blocks made of concrete.

Condenser An HVAC component consisting of a compressor and a heat exchanger, used to cool and compress a refrigerant.

Continuous Footing A long, narrow footing used to support walls and columns.

Contract A legally binding agreement between the project's owner and a contractor for the work done on a construction project.

Cost Codes A code for the logical grouping of items separated by supplier or installer type. Cost codes are often based on the MasterFormat®.

CPVC Chlorinated polyvinyl chloride, a pipe used for water lines.

Crew Rate The hourly or daily cost of labor and equipment for a construction crew.

Cripple Stud A short stud used below windowsills and in other areas.

Cycle Time The time it takes to complete one cycle of a repetitive task.

D

Damper A blade or louvers in an air duct used to control the airflow.

Dampproofing A water-resistant material applied to foundations.

Davis-Bacon Wages Wage rates that must be used on contracts with federal funding.

Design Professional The architect or engineer responsible for the design of a project.

Dialogue Box A small box in Excel that allows the user to make choices.

Diffuser An air distribution component located on the ceiling used to direct the airflow into a room.

Digitizer A device used to measure distances and areas off of plans.

Dimensional Analysis A mathematical procedure in which only the units of an equation are written to verify that the equation produces an answer with the correct units.

Direct Costs Costs that can be specifically identified with the completion of a task. For example, for the task of installing windows the direct cost includes the material cost for the windows (including sales tax and delivery), the labor cost (including labor burden), and the cost of any equipment used to install the windows.

Drain Board A material used to allow water to freely drain vertically down the face of a wall.

Drawings A graphical representation showing the dimensions of the project and where different construction materials are used on the project. Also known as plans.

Dropdown Box A box in Excel that allows the user to select among a list of preset choices with only one choice being visible until the user selects the arrow to the right side of the box.

Dry Unit Weight The density of the solid portion of a soil, usually measured in pounds per cubic foot.

Drywall Preformed sheet of gypsum.

Drywall Compound The material used to tape the joints of the drywall.

Durable Wood Wood that is naturally protected against decay and termites; for example, redwood.

E

EIFS Exterior insulation finish system.

End Condition The need to add a repetitive member to the beginning of a wall.

Engineer's Scale A scale or ruler that allows the user to read dimensions from plans drawn at common engineering scales.

Escutcheon A decorative cover around plumbing and fire sprinkler pipes.

Estimated Cost The projected cost of the work including the cost of labor, materials, equipment, and markups. The estimated cost should not be confused with the bid or selling price.

Estimating Estimating is the process of determining the expected quantities and costs of the materials, labor, and equipment for a construction project.

Estimating Software Package High-end software packages that combine a spreadsheet with a database.

Estimator A person responsible for preparing cost estimates regardless of his or her job title.

Ethics (1) A set of principles or rules by which members of a group agree to abide. (2) A set of values or a guiding philosophy of a person.

Excel A spreadsheet program developed by Microsoft.

F

Fascia A vertical board running perpendicular at the end of the rafters or trusses.

FICA Federal Insurance Contributions Act, which requires the collection of Social Security and Medicare taxes.

File A uniquely named storage unit containing contents of a computer spreadsheet, drawing, or other computer document.

Fillet Weld A triangular weld used to connect two steel members together at a right angle.

Fire Department Siamese A wye fire hose connection located outside a building and connected to a fire sprinkler system for use by the fire department.

Flow Alarm An alarm that goes off when flow occurs in a pipe.

FOB Free on board. The point where ownership and responsibility for materials transfers from the supplier to the contractor.

Folder A location on the computer used to organize files and subfolders.

Footing A structural member used to transfer the loads from a column or wall to the ground.

Forms Temporary structures used to support concrete until it has cured and can support its own weight.

Foundation The concrete or masonry structure below the first floor of a building, including the footing.

Foundation Wall The wall portion of the foundation, which often acts as a retaining wall.

Framing Anchor A light-gage metal connector used to connect two or more wood members together or connect wood members to concrete and masonry.

Function A computer instruction that selects a value based upon a specified set of requirements and the values located in other cells. For example, the MAX function selects the largest value from a series of cells.

Fusible Link A metal link that melts when exposed to fire conditions, opening a fire sprinkler head or closing a fire damper.

FUTA Federal Unemployment Tax Act, which requires employers to provide unemployment insurance for workers.

G

Gable-End Truss A nonbearing truss installed over walls to construct the gable end of the building.

General Conditions A construction document that identifies the relationships among the owner, design professionals, and the contractor and addresses provisions that are common to the entire project. Typically, this is the standard document used for many different projects.

General Decision A document containing the Davis-Bacon labor rates, which is published by the U.S. Department of Labor.

General Overhead Costs that cannot be identified with a project, for example, main-office expenses and accounting costs. Also known as indirect costs.

Girder A horizontal structural member used to transfer loads from beams to columns.

GLB Gluelam beam.

Gluelam Beam A beam constructed by laminating with glue a number of smaller wood members parallel to each other into a larger beam.

Grille An HVAC component consisting of horizontal slots used to cover an air duct.

Grout A mixture of cement, fine aggregate, and water used to fill the cells of blocks or bricks.

H

Header A horizontal structural member used to transfer vertical loads from above an opening to the vertical members located at the sides of the opening.

Heat Exchanger An HVAC component consisting of a coil of pipes used to transfer heat from the air to a refrigerant or from the refrigerant to the air.

Hip Rafters The rafters located at the junction of two inclined roof planes where the exterior surfaces of the planes intersect at an angle greater than 180 degrees.

Hold-Down A structural metal connector used to provide resistance against uplift.

HVAC Heating, ventilation, and air-conditioning.

I

Icon A computer image used to represent a command, file, or folder.

I-Joist An engineered wood joist with an I-shaped cross section.

Included by Reference Documents that are included in the contract documents by referencing them in the contract documents rather than physically attaching them.

Indirect Costs Costs that cannot be identified with a project, for example, main-office expenses and accounting costs. Also known as general overhead.

Indirect Project Costs Costs that can be identified with a construction project but are not identifiable with a specific task. Most Division 1 (for example, supervision, jobsite trailer, and temporary utilities) costs fall into this category. Also known as job overhead.

In Situ Soil in its natural condition.

Invitation to Bid A construction document that invites contractors to bid on a project.

J

Jack Rafters A shortened rafter in a hip roof that is connected to a hip or valley rafter.

Jack Stud A short stud used to support a header.

Jamb The exposed side of a door or window.

Job Overhead Costs that can be identified with a construction project but are not identifiable with a specific task. Most Division 1 (for example, supervision, jobsite trailer, and temporary utilities) costs fall into this category. Also known as indirect project costs, which are different than indirect costs.

Joist A horizontal wood or steel member used to frame a floor or ceiling.

Joist Girder A joist used to transfer loads from joists to columns.

Joist Hangar A light-gage metal connector used to connect joists to joist headers and rim joists.

Joist Header A joist used to transfer loads from joists intersecting an opening in a floor or ceiling at a right angle to the joist trimmers.

Joist Trimmer The joists running parallel to a side of an opening.

K

King Rafters A full-length rafter that runs from the eave of the roof to the peak of the roof.

King Stud A full-length stud located at the side of an opening.

L

Labor Burden The costs that must be added to an employee's wages to determine his or her total cost, which includes cash equivalents and allowances paid to the employee, payroll taxes, unemployment insurance, workers' compensation insurance, general liability insurance, insurance benefits, retirement contributions, union payments, vacation, sick leave, and other benefits paid by the employer.

Labor Hour One employee working 1 hour.

Labor Hours The total number of labor hours it takes to complete a task.

Labor Productivity A measure of how fast construction tasks can be performed, which is reported as the number of units performed during a labor hour or crew day (output) or the number of labor hours required to complete one unit of work.

Labor Rate The hourly cost of an employee per billable hour including wages and labor burden, which is calculated by dividing the annual cost of the employee by the number of billable hours.

Landscape When specifying a page layout for printing, the paper orientation that has the long side horizontal.

Left Click Pressing the button on the left side of the mouse.

LHR Labor hours.

Line Set The copper lines used to connect a condenser with a coil in an HVAC system.

Lintel A structural member used to support the wall above an opening. *See also* Header.

List Box A box in Excel that allows the user to select from a list of preset choices, with multiple choices being visible in the box.

Lookouts Short wood members supporting a roof overhang.

Loose Cubic Yards The volume of soil in cubic yards after it has been excavated but before it has been placed and compacted, which is used to measure the volume of soil being transported.

Lump Sum Bid A bid in which the contractor provides a single price for the construction work.

M

Man Hours *See* Labor hours.

Markup Costs that are added to the cost of the bid, such as bonds, building permit, and profit and general overhead. Also known as an add-on.

Membrane Roofing A roof constructed of an impermeable flexible plastic membrane.

Metal Deck Formed sheet metal used to support a roof or floor.

Mono Truss A truss with a single pitch or slope.

Mortar A mixture of cement, lime, fine aggregate, and water used to adhere brick or block together.

Mullion A member that separates the glazing panels in a window.

N

Nominal In name only. Lumber is often specified in nominal dimensions, which are larger than the actual dimensions.

Nonresponsive Bid An incomplete bid, which is disregarded.

O

Optional Bid Item A bid in which the contractor provides alternate or optional pricing for specific bid items in addition to the base bid. Also known as an alternate bid item.

Oriented Strand Board A panel of wood created by gluing wood chips together.

OSB Oriented strand board.

Output Labor productivity measured in the number of units performed during a labor hour or crew day.

Overbuild A portion of a roof built over the top of a roof structure, often used to create the appearance of a dormer.

P

Package Unit An HVAC unit that incorporates the heating and cooling equipment into a single unit.

Paid-When-Paid Clause Language in a contract that ties the payment of the subcontractor to the receipt of payment from the owner.

Path The computer location of a file that includes the drive designation and all folders from the drive to the file.

Payment Bond A bond issued by a surety guaranteeing that the contractor will pay for the labor, materials, and equipment used on the project.

Performance Bond A bond issued by a surety guaranteeing that the contractor will complete the work and that, should they fail to do so, the surety will step in and complete the work.

Pex A plastic pipe used for water lines.

Pipe Hanger A device used to support a pipe from a structural member.

Plan Measurer A device that allows the user to measure distances from plans by rolling the wheel of the device along a line.

Plans A graphical representation showing the dimensions of the project and where different construction materials are used on the project.

Plate Steel Flat sheet of structural steel used to make connectors and bearing plates.

Plenum (1) The space between a dropped ceiling and the structure above. (2) The ductwork above a residential furnace.

PO Purchase order. A legally binding agreement used to order materials or work on a project.

Portrait When specifying a page layout for printing, the paper orientation that has the short side horizontal.

Post A secondary column.

Pressure-treated Wood Wood that has been treated to prevent rot and insect infestation and/or fire by forcing chemicals into the wood under pressure.

Progress Payment A payment made from the owner to the contractor or from the contractor to the subcontractor before the project is complete as payment for the work completed to a specified date.

Project Manual A bid document that contains the invitation to bid, bid instructions, bid documents, bond and contract forms, general and special conditions, technical specifications, and other contract documents.

Proposal A letter or document from a contractor to an owner (or from a subcontractor to a contractor) in which the contractor (subcontractor) proposes to perform a specified scope of work for a set price and set conditions.

PRV Pressure-reducing valve.

P-Trap A P-shaped pipe used to prevent sewer gases from escaping from the sewer system.

Purchase Order A legally binding agreement used to order materials or work on a project.

PVC Polyvinyl chloride.

Q

Quantity Survey or Quantity Takeoff The process of converting the building dimensions and details into estimated quantities.

R

Rafter A sloped wood member used to frame a roof.

Rafter Ties A wood member connected to two rafters, used to prevent the wall supporting the rafters from spreading apart due to the forces on the rafters.

Raised Slab A slab that does not bear on the ground.

Random Length A bundle of lumber with differing lengths.

Rate of Progress The speed at which a linear task (for example, paving roads, installing pipes, painting striping on roads, installing curbs, and so forth) proceeds.

Rebar Steel bars used to reinforce concrete and masonry.

Register A diffuser combined with a damper to direct and control the flow of air into a room.

Relative Reference All or part of an Excel cell's reference that changes as the cell is copied.

Request for Material Quote A written request for pricing on materials.

Retention Funds withheld from a payment to ensure that a contractor completes a construction project.

Ridge Board A board located at the ridge of a roof used to support the rafters.

Right Click Pressing the button on the right side of the mouse.

Rim Board or Rim Joist The board or joist located at the perimeter of a wood-framed floor or ceiling.

Ring Shank Nail A nail with rings around its shank to increase its resistance against pullout.

Ring Ties Round rebar ties used to tie the vertical rebar together in a round column.

Rise (1) The change in height from the supporting members to the ridge for a truss or rafter. (2) The vertical distance between stairs.

Roof-Top Unit An HVAC unit that incorporates the heating and cooling equipment into a single package and is designed to be placed on the roof of a structure.

RTU Roof-top unit.

Run The distance between the outside of an exterior wall and the ridge or peak of a roof.

Runner A metal channel used at the top and bottom of a wall to support the metal studs.

S

Schedule of Values A list of values used to determine the amount of the progress payments.

Scope of Work "The construction and services required by the Contract Documents . . . and includes all other labor, materials, equipment and services provided or to be provided by the Contractor to fulfill the Contractor's obligations" (*General Conditions of the Contract for Construction*, American Institute of Architects, AIA Document A201-1999, p. 9.)

Second-Tier Subcontract A subcontractor to a subcontractor to the general contractor. The contractual relationship of second-tier subcontractors passes through a subcontractor on the way to the general contractor.

Selling Price The amount that the contractor charges the owner for the work. More commonly known as the bid or bid price.

Shear Panel A structural panel used to resist horizontal (shear) forces.

Sheathing A panel of wood used to cover walls, floors, and roofs.

Shrink The relationship between the bank and compacted condition of soil.

Siamese A wye fire hose connection located outside a building and connected to a fire sprinkler system for use by the fire department.

Sill The bottom horizontal member of a window.

Sill Plate The bottom framing member in contact with concrete or masonry.

Slab on Grade A slab supported by earthen materials.

Slope The relationship between the rise and run of a rafter, truss, or side of an excavation.

Soffit The underside of a roof overhang.

Soils Report A document that describes the soil conditions and water table at the construction site.

Special Conditions A construction document that identifies conditions, in addition to the general conditions, that apply to a specific project.

Specifications (1) The technical specifications. A construction document that identifies the quality of materials, installation procedures, and workmanship to be used on the project. (2) Incorrectly used to refer to the project manual, which contains the invitation to bid, bid instructions, bid documents, bond and contract forms, general and special conditions, and other contract documents, in addition to the technical specifications.

Spread Footing A square or rectangular footing used to support a column.

Sprinkler Head The portion of a fire sprinkler system that sprays water to suppress a fire and often includes a fusible link or liquid-filled vial to release the water.

Stirrups U-shaped rebar located near the end of concrete beams to resist shear and diagonal tension.

Structural Brick A masonry unit made from clay or shale that may be used as structural support.

Structural Tee A steel shape made by cutting an I-beam in half, which is designated with the prefix T.

Structural Tubing A hollow square or rectangular steel shape.

Stud (1) A wood member used to frame walls, most commonly 2 × 4s. (2) A metal bolt with one end securely fastened to a deck or steel member, for example, a Nelson stud.

Subcontract A legally binding agreement between the contractor and a subcontractor for the work done on a construction project.

SUTA State Unemployment Tax Act.

Sweating The process of connecting copper pipe using heat, flux, and solder.

Swell The relationship between the bank and the loose condition of soil.

T

T&G Tongue and groove.

Tail Joist A short joist that ends at an opening.

Takeoff Software A software package used to measure distances, areas, and volumes from a set of digital plans.

Task A discrete segment of the scope of work that consumes time and resources.

Technical Specifications A construction document that identifies the quality of materials, installation procedures, and workmanship to be used on the project. Often referred to as specifications.

Tempered Glass A form of safety glazing in which the glass has been heat-treated to resist breakage.

Text Box A box in Excel where text is typed.

Thickened Slab An area of a slab-on-grade where the thickness has been increased to provide greater strength or to act as a footing.

Ties Round, rectangular, or square rebar ties used to tie the vertical rebar together in a column.

Tongue and Groove Sheeting that has a tongue on one side designed to fit into a groove on the opposite side of an adjoining sheet.

Top Plate The top horizontal framing members in a wood wall.

Transom A window located above a door.

Truss A wood-framing member composed of multiple structural members, which often form smaller triangles, and which is used to support a floor or roof.

U

Unavoidable Waste Waste that cannot be avoided because of the design of the project.

Unit Pricing Bid A bid in which the contractor provides unit pricing for each bid item.

Unit Weight of Water The weight of the water in the soil, expressed in pounds per cubic foot.

V

Valley Rafters The rafters located at the junction of two inclined roof planes where the exterior surfaces of the planes intersect at an angle of less than 180 degrees.

Vapor Barriers Building papers and plastics used to prevent the infiltration of water vapor.

VAV Box Variable-air-volume box.

VCT Vinyl composition tile.

V Weld A V-shaped weld used to weld the end of structural steel members together.

W

Waste Materials lost during the construction process, which equals the difference in the quantities of materials ordered and the calculated quantities of materials to be placed on the project.

Water Content The relationship between the solids and the water in soils, which is the weight of the water (not volume) expressed as a percentage of the weight of the solids.

Waterproofing A water-resistant material applied to foundations.

Web Stiffener A piece of steel or wood used to stiffen the webs of an I-beam or I-joist; prevents the beam's or joist's web from buckling.

Welded Wire Fabric A series of wires welded at right angles in a grid to form rolls or sheets of reinforcement for use in concrete slabs.

Wet Unit Weight The density of soil, both the solids and the water, usually measured in pound per cubic foot.

Wheel Plan measurer. A device that allows the user to measure distances from plans by rolling the wheel of the device along a line.

Wide-Flange Beam An I-beam with wider flanges than an American Standard Beam, which is designated with the prefix W.

Work "The construction and services required by the Contract Documents . . . and includes all other labor, materials, equipment and services provided or to be provided by the Contractor to fulfill the Contractor's obligations" (*General Conditions of the Contract for Construction,* American Institute of Architects, AIA Document A201-1999, p. 9.)

Workbook An Excel file containing one or more worksheets.

Worksheet The pages within an Excel workbook or file.

WWF Welded wire fabric.

X

xlsx File extension for a 2007 or later Excel file.

xlsm File extension for a 2007 or later Excel file that contains macros.

APPENDIX G

INDEX OF DRAWING SETS

(The loose project drawings are provided in a separate package shrink-wrapped to this text.)

Residential Garage

Sheet	Title	Rev. Date
1	Title Page	3/24/06
2	Site Plan	3/24/06
3	Plan Views	3/24/06
4	Elevations	3/24/06
5	Sections	3/24/06
6	Details	3/24/06
7	Details	3/24/06

Johnson Residence

Sheet	Title	Rev. Date
T	Title Page	3/24/06
G	Specifications	3/24/06
A1	Site Plan	3/24/06
A2	Floor Plan	3/24/06
A3	Exterior Elevations	3/24/06
A4	Exterior Elevations	3/24/06
A5	Building Sections	3/24/06
A6	Finishes and Int. Elevations	3/24/06
A7	Details	3/24/06
A8	Details	3/24/06
S	Structural	3/24/06
P	Plumbing Plan	3/24/06
M	HVAC Plan	3/24/06
E	Electrical Plan	3/24/06

West Street Video

Sheet	Title	Rev. Date
T	Title Page	3/24/06
G1	Specifications	3/24/06
C1	Site Plan	3/24/06
C2	Grading Plan	3/24/06
A1	Floor Plan	3/24/06
A2	Exterior Elevations	3/24/06
A3	Exterior Elevations	3/24/06
A4	Building Sections	3/24/06
A5	Sections	3/24/06
A6	Sections	3/24/06
A7	Reflected Ceiling	3/24/06
A8	Interior Elevations	3/24/06
A9	Interior Finishes/Doors & Wind.	3/24/06
A10	Architectural Details	3/24/06
S1	Foundation Plan	3/24/06
S2	Roof-Framing Plan	3/24/06
S3	Structural Details	3/24/06
S4	Structural Details	3/24/06
P1	Plumbing Plan	3/24/06
P2	Plumbing Details	3/24/06
FP1	Fire Sprinkler Plan	3/24/06
FP2	Fire Sprinkler Details	3/24/06
M	Mechanical Plan	3/24/06
E1	Lighting Plan	3/24/06
E2	Power Plan	3/24/06
E3	One-Line/Panel Boards	3/24/06

INDEX

401(k) 270

A
Absolute reference 337, 425
Accident history 269
Accounting
  style 38
  system 65, 250, 257, 269, 288
Accrued cost 264
Accuracy 309, 328
Acoustical
  ceilings 184–5, 201
  tile 184
Addenda 425
Add-on 16, 425, 429
Adjustment factor 258
Admixtures 85
AGC 321, 324, 425
Aggregate 85
AIA 10, 321, 324, 425
Air-conditioning 201–6, 409
Air
  chamber 196
  entrainment 85
  handler 425
  infiltration 126
Alarm
  check-valve assembly 192
  flow 192
Align left button 37
Align right button 37
Allowances 264
Alternate bid item 425
Alternatives 317
Aluminum siding 162
American Institute of Architects (AIA) 2, 10, 321, 425, 430
American National Standards Institute (ANSI) 7, 425
American Society for Testing and Materials (ASTM) 7, 425
American Society of Professional Estimators (ASPE) 9, 10, 329
American standard beam 116, 425
American standard channel 116, 425

Anchor bolts 74, 76, 95, 96, 103, 107, 108, 136, 406, 425
Angle 107, 111, 116, 118, 122, 425
Annual contract 425
ANSI 7, 425
Appliances 209
Arbitration 323
Architect 184, 187, 287
Architect's
  estimate 12
  scale 7, 394, 425
Architectural-grade shingles 160
Area
  calculating 396–401
  constant 404
  method 72
Asphalt 236–7
  impregnated felt 154–6
  roofing 352
Assemblies 8–9
  Associated General Contractors of America (AGC) 321, 425
Assumptions 9
ASTM 7, 425
Athletic surfaces 236
Attachments 315
Attic 201
Attorney 321
  fees 323S
Auto fill 358
Autosum button 56
Average-end method 220, 224–5
  modified 225–6, 231
Average function 56–7
Average-width-length-depth method 224
Avoidable waste 425

B
Backcharge 322
Backer board 184
Backfill 231–2, 247
Backing 132, 181, 419, 420

Bank 220–1, 231
   cubic yards 425
Bankruptcy 3
Base 236–7, 420
   plate steel 116
   rubber 187
Batten strip 165
Beam 97, 116–18, 425
   American standard 116
   wide-flange 116
   wood 382
Bearing walls 128, 136
Bedding 245–7, 425
   plants 424
Benefits 264, 270
Bid 425
   best 286
   bond 4, 11–12, 314, 425
   checklist 314
   competitive 11
   courtesy 293
   date 286, 293
   deadline 15
   documents 4, 232, 314, 425
      preparation 15
   form 4–5, 314, 425
   instructions 4, 425
   nonresponsive 314
   package 4–7, 425
   planning 11
   preliminary 315
   review 17–8
   rigging 329
   scheduling 11
   selection 288
   shopping 329
   subcontractor 192, 315
   submission 17, 314–7
   time 286
Bid-day activities 15–7
Bidding
   fair 328
   practices 328–9
   rhythm 293
   warming up 293
Billable hours 425
Billing disputes 321
Block 107, 425
   bond-beam 107, 108
   bullnose 109
   lintel 109
Blocking 126, 136, 419, 425
Block-out 422
Board feet 126, 128, 131, 405
Bodily injury 270
Boiler 426
Bold 37
   button 37
Bolts 116, 118
   expansion 365
   molly 365

Bond 426
   beam 89–90, 426
   bid 4, 11, 314, 425
   capacity 317
   cost 16, 334
   limits 12, 426
   payment 4, 263, 270, 374
   performance 4, 291, 323, 429
   schedule 335
Borders 33, 40–2, 333
   button 44
   popup menu 40
Bottom plate 426
Branch lines 192–3
Brick 426
   facing 111426
   lintels 111
   structural 107
   ties 426
   veneer 107, 111–12, 426
Bridges 236
Budget 4, 426
Build in mind 72, 75, 84, 325
Building
   code 85, 134, 290
   paper 153–7, 160, 426
   permit 15, 290, 292, 299, 317, 334
Bullnose 426
Burden 264, 426
Butt joint 77–9, 164
Butts 426
Buyout 17, 321–4, 426

C

Cabinets 143
   base 201
   medicine 420
Cable TV 421
Calculations 310
Calculator 8
Caret (^) 51
Carpentry finish 420
Carpet 181, 421
Cash equivalents 264
Casing 420
Catalogue cut sheets 315, 317
Catalogue of Standard Specifications and Load Tables for
   Steel Joists and Joist Girders 119
Catch basins 424
Caulk 421
Ceiling
   dropped 205
   function 55, 427
   joists 123, 139
   painting 188
Cell 426
   combining 38
   formatting 36–40
   merging 39
   naming 361–2
   unmerge 39

Cement powder 85
Center button 36–40, 44–6
Centerline method 93
Ceramic tile 184–5
Certifications 4, 314, 426
Certified payroll 322
Chair rail 420
Chair rebar 85
Change order 4, 11, 287, 322, 426
Channel American standard 426
Chases 420
Check
  joint 322
  valve 192, 426
Chiller 426
Chimney 160
Circle 396
  quarter of 399
Circular reference 334, 335
Clarifications 287, 315, 318
Cleanouts 197
Clean-out boxes 424
Cleanup 5, 323
Clipboard 30, 31
Close-out audit 18, 426
Closet rod 420
CMU 426
Coil 426
Collar ties 137, 139, 426
Color 341
Column 87–9, 116–18, 223, 402–3, 426
  headings 60
  widths 33, 35
    dialogue box 36
    minimum 35
  wood 419
Combining cells 38
Combustion air 423
Comma style button 39
Comments 33
Commercial storefront 174–5
Commitment 328
Communication 17, 307, 409
  productivity 415
  verbal 3
  with field 74–6
  written 3
Compacted 220, 231
  cubic yards 426
Competitive bidding 426
Competitors tracking 293
Completeness 288, 309
Complex
  shape 398–402
  volumes 393
Composite slab 426
Composites 407
Compression fitting 196
Computers 8
Concatenate 367
Conceptual estimate 3–4

Concrete 84–103, 203, 247, 406, 426
  admixtures 85
  chilled water 86
  columns 419
  curb 236–7, 258
  curing 97
  design-based specification 85
  finishing 84, 86, 97
  footings 419
  forms 84, 237
    life of 85
    reuse 84
  foundations 419
  gutter 236, 237
  hot water 86
  ice 85
  masonry unit 407
  material 85–6
  mixes 419
  performance-based specification 85
  placement 84, 86
  productivity 411
  protection 86
  pump 86, 419
  short loads 86
  site 237
  slump 85
  stairs 100–1
  strength 85
  truck 86
  walls 419
  waste 86
  weight 100
Condenser 423, 426
Conditional formatting 343, 345, 351
  button 343
  rules manager dialogue box 343
Cone 402
Conferences 328
Confirm password dialogue box 348
Conflict of interest 329
Connectors steel 116, 120
Constant area 404
Construction
  cost 292
  documents 252, 321, 328
  methods 2, 74, 325
Construction Specifications Canada 9
Construction Specifications Institute 9
Contact information 317
Continuous
  footing 426
  improvement 17
Contour lines 227
Contract 4, 426
  alternatives 315
  annual 250, 252
  assignment 321
  change order 322
  documents 287, 293
  general provisions 323

Contract (*Contd.*)
   inspections 322
   materials 250, 324
   parties 321
   payment 322
   price 322
   project identification 321
   quality 322
   quantities 321
   retention 322
   schedule 322
      of values 321–2
   submittals 322
   termination 321
   testing 322
   voiding 321
Convection 205
Conversion factors 310, 354, 404–5
Conveying equipment 409
Cooktop 209
Coordinate method 401–2
Coordination 322
   special 326
Copper pipe 196
Copy 31–3
   button 31
   worksheet 27
Corner
   piece siding 162
   trim 165
Cost
   accrued 264
   actual 250
   code 11, 18, 67, 73, 296, 298, 309, 311, 406–10, 426
   construction 292–293
   controls 4, 17
   equipment 73, 276–80
   escalation 252–3
   labor 2, 9, 13, 14, 73, 76, 283, 297, 355
   large 309
   material 2, 12, 13, 73, 76, 296–7, 354
   operating 276, 278s
   ownership 276, 277
   shipping 251, 296
   square foot 311
   storage 252, 296
   total 298
   tracking 250
Counted items 76
Countertops 143
Coverage rate 80, 159
CPVC 196, 426
Crane 420
   sizing 116
Crawl space 201
Crew 14
   day 255
   in-house 286
   makeup 284, 326
   rate 283–5, 353–4, 426
   size 256
      minimum 322

Cripple stud 426
Cross bracing steel 116
Cross section 242
   area 100
Cross-sectional method 226–31
Cross tees 184
Crown molding 407
Ctrl key 27
Curb 424
   walls 424
Current
   date button 48
   time button 48
Cushioning 186
Cut 30
   button 30
Cwt. 251
Cycle time 257–8, 426
Cylinder 402

D

Damage 252
Damper 426
   fire 422
   smoke 423
Dampproofing 153–4, 426
Data
   entry 30–3
   historical data 80, 81, 158–9, 182, 184, 187, 250, 257, 260, 286, 288, 354
   manufacturer's 80, 128, 158, 182, 187, 278
   validation 346–7, 362–3, 374
      button 345
      dialogue box 345–6, 363–4
Database 9, 18
Davis-Bacon wages 273, 426
Deck
   metal 120
   wood 131
Decrease decimal button 34, 43
Deflection 100
Delays 74, 76, 256
Delete
   button 27
   popup menu 27
Delivery 251
Depreciation 276–7
Design
   information 256
   professional 4, 85, 132, 232, 314, 318, 426
Design-build 315
   subcontract 192
Desired results 352–3
Detail worksheet 33, 296–8, 301–7, 332, 336, 338
Detailed
   estimate 3–4, 11, 309–10
   oriented 3
Detention ponds 424
Dialogue box 426
   column 36
   conditional formatting rules manager 343
   confirm password 348

data validation 346–7, 347
format cells 38, 39, 344, 348
function argument 53–5
insert 27
   function 53
launcher 38
macro 338
move or copy 29
name manager 362
new
   formatting rule 369
   name 361
open 22
options 26, 335
page setup 60
paste
   name 363
   special 33
picture 48
print 59–65
protect worksheet 348
recent files 24
record macro 337
save as 20–4
security options 336
trust center 336
unprotect worksheet 350
warning 27, 30
zoom 42
Diffuser 185, 472
Digitally signed 336
Digitizer 7, 77, 93, 96, 393, 426
Dimensional analysis 405, 426
Dimensions
   adding 394
   subtracting 394
Direct
   costs 427
   overhead defined 72
Disconnects 421
Dishwasher 422, 423
Disposal 423
Dispute resolution 323
Dollar sign 40
Door 172, 422, 423
   access 422
   cylinder 174, 175
   exterior 420
   frame 176
      wood 181
   jamb 135, 172
   mortising 172
   opening 111, 131, 133
   painting 188
   predrilled 172, 176
   prehung 72, 176
   schedule 172, 176
   sliding glass 172, 420
   stop 175, 176
   swings 172
   wood 172
Double check 310, 311

Dowels 86
Draft stops 420
Drain
   board 153, 427
   pipe 196, 197
   valve 192
Drainage system 153
Drawings 4, 11, 76, 287, 301, 312, 422
   as-built 322
Drip edge 160
Driveway 203, 424
   excavation 231
Dropdown box 362–4, 427
Dry
   density 220
   unit weight 220, 427
Dryer 209, 423
Drywall 182, 183, 420–1, 427
   compound 182, 183, 427
   fasteners 182
   fire-rated 182, 420
   moisture-resistant 420
   patching 422
   scrap 421
   tape 182
   trim 182
   water-resistant 182
Duplication 309
Durable wood 427
DWV 196–9

E
Earthquakes 136
Earthwork 9, 220–33, 409, 424
   productivity 415
Economy 252
Effective
   length 77
   width 78
Efficiency system 258
EIFS 159, 427
Electrical 209–16, 381, 389, 391
   box 210
      junction 210, 211
   breaker 209, 212
      box 423
   commercial 210–3
   conduit 209–12
   controls 209
   disconnect 421
   distribution 211
   fixtures 422
   fluorescent light 210
   light fixture 185, 209–11
   meter 211
      base 421
   motors 209
   outlet 209–10
   panel 209–10
   panelboards 211–13
   plan 209–11,

Electrical (*Contd.*)
  productivity 413
    raceway 210
    Romex 209
  starters 209
  switches 209–10, 423
  transformer 209
  wire nuts 210
  wireway 212
  wiring 211, 423
    commercial 210–3
    diagram 211
    residential 201–4
Elevators 420
E-mail 318–20
Embed 419
Employee
  annual cost 264
  class 264
  cost 264
  exempt 267
  injury 270
  occupational illness 269
End condition 76, 77, 90, 130, 132, 427
Engineer 287
Engineer's
  estimate 12
  scale 7, 395, 427
Environmental inspections 7
Equations identifying 352, 354–5
Equipment 325, 326, 409, 421
  bucket teeth 278
  costs 2, 9, 13, 73, 276–80, 283, 297–8
  cutting edges 276
  damage 279
  depreciation 276–7
  different 74
  expected life 276
  filters 276, 279
  fuel 276, 279
  insurance 276, 277, 279
  interest 276–8
  leased 276, 279
  license 276, 279
  life 276
  lubricants 276, 279
  maintenance 279
  overhaul 279
  owned 276
  price 14–15, 276
  procurement 18, 321
  quantities 2, 13
  rented 276, 279
  repair 276, 279
    reserve 279
  sales tax 276
  salvage value 276
  setup 276
  storage 276, 277, 279
  taxes 276, 277, 279
  tire 276, 277, 279
    repair 278

  transportation 276
  use 74
  value 277
  wear 279
    items 276, 277, 279
Error
  alert 347
  canceling 310
  checking 9, 17
  identification of 18
  protection 332, 341–5
Escalation 296
Escutcheon 194, 427
Estimate
  accuracy 18
  architect's 12
  complete 15
  conceptual 3–4
  detailed 3–4, 11, 309–10
  engineer's 12
  errors 17
  final 3–4
  preliminary 3–4, 11
  review 311
  too high 2
  too low 2
  types of 3–4
Estimated cost 427
Estimating 427
  art of 3
  computerized 8–9
  database 9, 18
  department 2
  fundamental principles 3
  inaccuracies 4
  math 2, 393–405
  poor 14
  practice 3
  process 11–19
  software 8
    package 42
  tools 7–8
Estimator 2–3, 427
  skills of 2–3
Ethics 328–9, 427
  dilemmas 329
Ethylene propylene diene monomer (EPDM) 165
Excavation
  average-end method 224–5, 233, 242
    modified 225–6, 231, 242, 244
  average-width-length-depth method 224
  backfill 231–2, 247
  base 424
  basement 225
  building 230, 232
  clear and grub 424
  comparison of methods 231
  contour lines 227
  costs 7
  cross sections 224
  cross-sectional method 226–31
  cut 224–30

driveway 232
fill 224–9
footing 231, 232
foundation 231–2
geometric method 223–4, 231, 244, 396
grade
    existing 226
    proposed 226
grading data 231
gravel underslab 232
grids 226
rough grade 424
sidewalk 232
site 226
trench 242
zero line 226–8
Excel 3, 8, 21–68, 427
    advanced options 31
    asphalt 237
    average-end method 244
        modified 244
    base 237
    batt insulation 158
    bedding 246
    block 110
    board foot 128
    bond cost 292
    brick 112
    building permit 290
    ceramic tile 184
    column
        rectangular 88
        round 89
    competitor tracking 294
    continuous footing 94
    crew rates 284
    cycle time 258, 262
    decking 131
    detail worksheet 298
    drywall 183
    equipment costs 280
    floor sheathing 130–1
    foundation wall 96
    icon 21
    insulation batt 158
    joists 130
    labor rate 272
    metal deck 120
    rafters 140
    rate of progress 258–60
    request for material quote 298
    roofing
        felt 157
        shingles 161
    siding 163
    slab
        on grade with rebar 98
        on grade with wire mesh 99
        raised 100
    spread footings 87
    structural steel 118
    studs 135

    summary worksheet 300
    trench excavation 244, 245
    waterproofing 154
Exclusions 16, 287, 315
Executable code 336
Exhaust fan 204, 423
Existing conditions 406
Expense additional 74
Experience modifier 270
Export 231, 247
Exterior improvements 236–8, 409
    productivity 415
Exterior insulation finish system 159–60, 427

F
F4 key 52
Fan 205
    coil 423
Fascia 141, 162–5, 421, 427
Faucets 199, 422
Federal Insurance Contributions Act (FICA) 268, 427
Federal unemployment tax (FUTA) 268
Fence 236
Fiberbond 419
Field measure 174
File 427
    name button 48
    path button 48
    type 21
Fillet weld 427
Final estimate 3–4
Finish carpentry 143
Finishes 181–9, 408
    productivity 413
Fire
    blocking 129, 136, 419
    department
        connections 192
        Siamese 193, 427, 430
    prevention 129, 136
    sprinkler 419
        drop 193
        heads 194
    suppression 192–4, 409
        productivity 413–4
    treated wood 181
    walls 140, 420
Fireplace 423
    gas 204
    mantels 420
Fire-rated labels 421
Fit selection 43
Fittings 424
    compression 196
Fixtures 422, 423
Flashing 160, 165, 421, 423
    HVAC 204
    pipe 424
    plumbing 199
Floor
    covering 419
    drains 199, 422

Floor (*Contd.*)
   function 55, 56
   opening 128
   painting 188
   sheathing 130–1
   wood 126–45, 185–6
Flooring
   carpet 188, 421
   laminate 185–6
   pad 188
   trim 188
   vinyl 186–7, 421
      composition tile 187
   waste 185
   wood 126–45, 185–6
Flow alarm 192, 427
Fly ash 85
FOB 251, 427
Folder 427
Font 46, 341
   dialogue box launcher 38
   dropdown box 37
   size 37
      dropdown box 37
   style 37
Foolproof 65
Footers 46–8
Footing 419, 427
   centerline method 93
   continuous 90–5
   corners 90, 93
   excavation 231
   intersection 90, 93
   spot 86
   spread 86–7
Forklift 420
Form layout 332
Format
   button 27, 36
   cells 36
      dialogue box 38, 39, 309, 312
   picture button 48
   popup menu 26
Formatting 33
Forms 84–5, 90, 95, 97, 100, 427
   converting 332–51
   new 352–79
   planning 352–6
   supporting 97, 100
Formula 9, 33, 333–4
   bar 51
   copying 336
   editing 51
   selecting between two 354
   testing 311
   wrong 310
Foundation 419, 427
   end condition 90
   wall 95–6, 427
      corners 95
Framing 419–20
   anchor 128, 427

plan 76
   floor 127
   joist 118–20
   roof 141
   steel 117
Free on board 251
Freight 16
Fringe benefits 267
Frost wall 102
Function 53–8, 427
   argument dialogue box 55
   library 53, 57
Furnace 201–2, 420–3
   filters 423
Furnishings 409
Fusible link 194, 427
FUTA 268–9, 427

G
Gable
   ends 137, 140, 161
      truss 427
   roof 160
Garage 420, 421
Gas
   line 203, 209, 422
   meters 421
Gasket 247
General
   conditions 5, 427
   decision 264, 427
   liability insurance 264, 270
   overhead 428
   requirements 406
Geographical area 293
Geometric method 223–4, 231, 244, 396
Girder 116–18, 428
   wood 128
Glass 175
Glazing 175–6
   safety 173, 175
   tempered glass 173, 175, 431
Glue
   carpet 188
   lam beam 128, 428
   laminated beam (GLB) 128, 428
   thinset 184
   VCT 187
Go-footer button 49
Go-header button 50
Grab bars 136, 420
Grade finish 237
Grading 86
Grammar 320
Gravel 86, 245
Gridlines 60
Grille 165, 181, 185, 367, 372
Groups in Excel 22
Grout 107–10, 184, 428
Gutters 236, 424
Gypcrete 203
Gypsum board 182–4

H

Handrail 420
  steel 120
Hardibacker 184
Hardiplank 164
Hardware 420
  butts 176
  closers 176
  cylinder 176
  doorstop 174–6
  hinges 174–6, 422
  locksets 176
  panic 174–6
  pulls 176
  schedule 174
  silencers 176
  threshold 174–6
  weather stripping 174, 176
Hanger pipe 193, 429
Head storefront 174, 175
Header 46–50, 428
  metal studs 181
  wood 127, 136, 419
Header and footer
  button 46
  tools design menu 46
Heat
  exchanger 425
  loss 153
  temporary 420
Heating, ventilation and air-conditioning (HVAC) 201–6, 207, 409, 428
  air
    conditioners 209, 423
    ejector 203
    filtration 201
    handler 205
  boiler 203, 205
  cap termination 203
  chillers 205
  coil 201–2
  combustion gasses 201
  commercial 204–6
  condenser 201
  controls 204
  damper 205
  dehumidifying 201
  diffusers 185, 427
  ductwork 201, 205
  electric baseboard 204
  exhaust fans 204, 423
  expansion tank 203
  fan 205
  fin-tube convector 205, 206
  flashing 203
  flexible duct 205
  furnace 201, 420–3
  grille 185, 201, 205, 423, 428
  heat
    exchanger 205
    pump 204
  heaters 209
  humidifying 201
  line sets 201, 428
  package unit 205, 429
  plenum 202, 205
  productivity 414
  pump 203
  radiant heat 203, 204
  registers 185, 201
  residential 201–4
  return
    air 201–3, 205
    manifold 203
  rooftop 204
  supply 201, 202
    manifold 203
  thermostat 203, 205, 423
  trunk line 201
  valves 203
  variable-air-volume box (VAV box) 205, 431
  vents 201
Heavy Construction Systems Specialists Inc. 9
Heavy timber construction 131
Heavybid 9
Height dropdown box 60
High dollar items 13
Highlighter 310
Hinges 174, 176, 422
Hip
  rafters 428
  roof 139, 140, 155, 160–2
Historical data 80, 81, 158, 250, 251, 257, 260, 286, 288, 354
Hold down 134, 135, 419, 428
Holidays 264
Hollow metal 172, 175
Honesty 329
Hose bibs 422
Hours
  billable 264, 271, 277
  unbillable 271
Hundredweight 251
Hydrants 424
Hydroseeding 424

I

Ice and Water Shield 155, 157, 160
Ice-maker boxes 422
Icon 428
If function 53, 57–9
  nested 58
I-joist 428
Illegal behavior 329
Imbeds 107, 108
  steel 108
Import 231, 247
Improvement constant 3
In situ 220, 428
Included by reference 7, 428
Increase decimal button 39
Indirect
  costs 428
  project costs 428
Inflation rate 252

Information proprietary 329
Input data 354
   identifying 354–5
   message 347
Insert
   button 25
   dialogue box 26
   function
      button 53
      dialogue box 53
   picture dialogue box 33
   popup menu 27
   worksheet tab 19, 25
Inspections 322, 326
Instructions to bidders 4, 425
Insulation 107, 157–9, 420
   batt 157
   blown 157
   fiberglass 157
   paperbacked 157
   ridged 159
Insurance
   benefits 264, 270
   certificates 322
   dental 270
   disability 270
   equipment 276
   health 270
   life 270
   requirements 322
   workers' compensation 322
Insured additional 322
Intercom 423
Intermediate mullion storefront 175
Internal Revenue Service 268, 275
International Building Code 171, 175, 180
International Code Counsel 171, 175, 180
Interpolation 227
Interruptions 258
Inventory control 420
Invitation to bid 4, 428
Invoice 250, 251
Irrigation systems 236, 424
IRS 268, 275
Isometric drawing 197
Iteration 291, 292, 334

J
Jack
   rafters 428
   stud 428
Jamb 428
   storefront 174
J-molding 160, 183
Job
   overhead 428
   title 317
Jobsite overhead defined 72
Joist 201, 428
   girder 118–20, 428
   hangar 428
   header 428
   steel 188–20
   trimmer 428
   wood 127, 422

K
Key
   shortcut 337
   word search 53
Keyless entry 176
King
   rafters 428
   stud 428

L
Labor 321, 325–6
   breaks 258
   burden 264, 426, 428
      markup 271–3
   class 256
   cost 2, 9, 13, 73, 76, 183, 287, 355
   hour 255–61, 260–1, 283, 297, 326, 352, 355, 428
      tracking 257
   pricing 14
   productivity 9, 255–7, 283, 411–6, 297, 309, 353, 428
      factors affecting 255–7
      field observations 257–60
   quantities 2, 13
   rates 9, 264–73, 297, 309, 353, 428
   skill of 2
Laminate floors 185–6
Laminated lumber 135
Landscape page orientation 60, 428
Landscaping 236, 424, 428
   plan 238
   plants 237–8
Lap joint 78, 107
Lateral forces 136
Lavatory 196, 199
Layout 419
Learning curve 256
Ledger steel 111, 116
Left click 428
Length 341
   effective 61–2
Letter
   business 317–8
   formats 318
   proposal 314–7
Lhr 428
License requirements 322
Lien waivers 322
Light fixtures 185, 209–10
Limitations spreadsheet 65
Line set 201, 428
Linear components 77–8
Link 33
Lintel 428
   masonry 107
   steel 107, 111, 116
Liquid damages 293
List box 428
Lists 358

INDEX

Loan 276
Locksets 176
Logical operators 57
Long-lead 326
Lookouts 141, 429
Lookup 371–6
Loose 220, 231
   cubic yards 429
Lost cost factor 269
Loyalty 329
Lumber 126–45, 407
   backing 136, 181, 419
   beam 419
   blocking 128, 136, 425
   ceiling joists 137, 139
   collar ties 137, 139, 426
   columns 419
   decks 131
   doors 172
   fascia 141
   fire blocking 129, 136, 419
   fire-treated 181
   floor 126–31, 185–6
     joist 128–30, 419
     sheathing 130–1
     squeaks 130
     trim 186
   girder 128, 419
   gluing 130
   header 128, 132, 135–6, 419
   I-joist 128
   joist 128
     ceiling 139
     header 128, 129
     trimmers 128
   laminated 135
   lintels 419
   naturally durable 131
   package 126
   paneling 143
   plate 131, 419
   plywood 130, 135
   post 128, 419
   pressure-treated 126, 131, 429
   productivity 412
   purlins 419
   rafter 137–40, 419, 430
     common 139
     hip 139–40
     jack 139
     king 139
     valley 139–40
   random lengths 126
   ridge 156, 160
     board 140
   rim board 128
   ring-shank nail 130, 430
   risers 419
   roofs 137–43
   shear panels 136, 430
   sheathing 130–1, 419
     roof 141
     waste 130
   shelving 143, 420
   sill 126, 419
     plate 131
     window 134
   sleeper 419
   slip-resistant nail 130
   stingers 419
   stud 131, 132, 181, 419
     cripple 134
     jack 132
     king 132
     precut 132
   treads 419
   trim 143, 165
   trimmers 419
   trusses 137, 140–1, 419
     tails 140
   walls 131–7
     corners 132, 134
     ends 132
     hold-downs 132, 135
     intersections 132, 133
     openings 132
   web stiffeners 128, 431
Lump sum 56
   bid 428

M

Macro 310, 332, 335–40
   button 337
   dialogue box 338
   recording 337
   security
     level 336
     warning 336
   settings 336
   stop recording 338
Main
   lines 192
   tees 184
Maintenance equipment 279
Man hours 429 see also labor hours
Manholes 424
Mantels 420
Manual of Steel Construction 115
Manufacturer's data 80, 128, 158, 182, 187, 278
Markups 16–7, 290–4, 425, 429
Masonry 107–12, 407
   bond beam 107
   cells 107
   insulation 107, 108
   lintels 107
   openings 108
   productivity 411
   ties 107, 111
   waste 109
Masterformat 6, 10, 406
Material 2, 324
   contract 250, 324
   costs 2, 14, 73, 76, 206–8, 354–5
   damage 252
   different 74
   fob 251, 427

Material (*Contd.*)
   free on board 251
   improper use 81
   installation 6
   invoice 250, 251
   long-lead 326
   orders 3–4
   pricing 14, 16, 250–3, 353
   procurement 18
   protection 252
   purchase 321
   quantities 2, 4, 13
   safety data sheets 323
   scrap 81
   testing 5
   theft 252
   use 4, 74, 81
   wrong 74, 76
Math 2, 393–405
Mediation 323
Medicare 268
Meeting pre-bid 11
Membrane roofing 165, 429
Menu tab 25
Merge and center button 39
Merging cells 39
Metal
   deck 99, 120, 429
   grid 184
   productivity 411
   stud 81
     header 181
     jack 181
     king 181
     opening 181
Metals 115, 407
Method of construction 74
Mhr 429
Microwave 423
Miscellaneous steel 122–5
Moisture 182, 185
   productivity 411
   protection 153–67, 412
Moldings wall 184
Mono-truss 429
Mortar 107, 111, 429
   thickset 184
Move after typing 31
Move or copy
   dialogue box 26
   sheet 26
MSDS 323
Mullion 429
Multiple cells selecting 27

N

Nails 420
Name
   box 362
   cells 361–2
   manager
     button 362
     dialogue box 362

Naming cells 413
National Council on Compensation Insurance (NCCI) 269
National standard books 251
Naturally durable wood 126
Nested if functions 58
New
   folder button 25
   formatting rule dialogue box 344
   name dialogue box 361
Nominal 429
Noncontiguous cells 56
Nonresponsive bid 419
Normal view 63
   button 48, 62
Notice-proceed 322
Number
   dialogue box launcher 40
   of pages button 46
Numbers consecutive 358

O

Occupational Safety and Health Act 323
Office trailer rental 72
On Center Software 7
On-Screen Takeoff 3, 7, 9
Open dialogue box 22
Opener overhead door 176
Opening 111, 172–6
   door 131, 135, 162, 164
   floor 128
   productivity 412
   window 135, 162, 164
Operation and maintenance manuals 322
Operating cost 276, 278
Operators 51
   logical 57
Optional bid item 429
Options dialogue box 26, 336
Order of operation 51
Orientation button 59
Oriented-strand board 130, 429
OSB 130, 142, 429
OSHA 322
Output 255, 429
Overbuild 137, 429
Overhead 292–4
   checklist 15, 16
   direct defined 72
   door 176
   general 17, 292
   jobsite defined 72
   main office 292
   markup 17, 299
   project 15, 288
Overtime 256, 322
Owner 2, 4, 314
Ownership cost 276–8

P

Package unit 205, 429
Pad 188
Page
   break

button 62
preview 62
setting 62
view button 62
layout 59
button 46
view 46
number button 46
orientation 59
setup dialogue box 60–3
launcher 60, 61
Paid-when-paid 322, 429
Paint 188, 420–2
Paneling wood 143
Panic hardware 174
Paper forms 8
Parallelogram 306, 308, 401
Parapet wall 165
Parenthesis 51
Parking lots 424
Partitions
fire-rated 182
non-bearing 181
Password 348
Paste 30–1
button 30–1
name dialogue box 362
special 31
dialogue box 33
popup menu 30
Path 429
Pattern 341
match 186, 187
Paving 236, 258
Paydirt 7
Payment
bond 4, 291–2, 299, 429
final 322
progress 322
terms 315, 318, 324
Payroll taxes 264
Peer review 311
Penetrations 160, 165
Pennies dropping 311
Pension plan 270
Percent style button 39
Performance bond 4, 291–2, 299, 322, 429
Pergo 185
Perlite 108
Permit 422
Personal
injury 270
title 317
Pex 196, 429
Phone quote 16, 250
Picture
adding 33
button 33, 48
dialogue box 48
Pipe 258, 423
copper 196
CPVC 196, 426
drain 196–9

hanger 193, 429
Pex 196, 429
steel 116
vent 160, 196–9
waste 196–9
Plan
and profile 242
measurer 7, 77, 96, 210, 303–4, 354, 429, 431
view 354
Plans 4, 12, 76, 287, 293, 310, 317, 353, 419, 427, 429, 432
framing 75
Plastic
laminate 143–5
sheet 157
tubing 203
Plastics 407
Plate
wood 127, 131, 419
steel 116, 118, 429
Plenum 429
Plumbing 196–9, 287, 409, 420, 422
air chamber 196
bushing 199
cleanouts 197, 198
equipment 199
faucet 199, 422
flashing 199
floor drain 199, 422
flush valve 199
lavatory 196, 199
pressure reducing valve 196, 429
productivity 413–4
p-traps 196, 199, 429
sanitary tee 199
shutoff valve 196,
sink 196, 199, 422
stub-out fitting 196
test cap 199
toilet 196, 199, 422
flange 199
urinal 196, 199
water
closet 196, 199
heater 196
Plywood templates 419
Polyvinyl chloride 165, 429
Portrait page orientation 59–60, 429
Post 429
base 128, 419
cap 128
wood 128, 419
Post-bid activities 17–18
Post-it 15, 314
Power 51
lines 242, 423
meters 421
temporary 420
Pozzolan 85
Practice problems 354
Preliminary estimate 3–4, 11
Prequalification package 11
Presentations 3
Pressure test 196

Pressure-treated wood 126, 131, 429
Price 288, 309, 315, 318
   breakdown 315
   combining 16
   extensions 296–307
   too high 18
   too low 18
Pricing
   database 18
   materials 14, 250–3, 353
Primer 366
Print 48, 59–60
   button 59
   dialogue box 59
   entire workbook 59
   gridlines 607
   landscape 60
   number of copies 59
   portrait 60
   range 59
   scale 60
     to fit 60
   setting
     page breaks 62–63
     scale 60–61
   titles button 60, 62
Printer 8
   properties 59
   selection 59
Prism 223, 402
Production
   daily 260
   time 258
Productivity
   communication 415
   concrete 411
   earthwork 415
   electrical 415
   exterior improvements 415
   finishes 413
   fire suppression 413
   HVAC 415
   labor 9, 255–61, 283, 411–16, 297, 309, 353, 428
     factors affecting 255–7
     field observations 257–60
   masonry 411
   metals 411–12
   openings 412–13
   plumbing 414
   rates 14
   specialties 413
   thermal and moisture protection 412
   utilities 415
   woods 412
Profit 292–4, 299, 309
   margin 2
   markup 16
Progress payment 4, 315, 429
Project
   buyout 18, 321–4, 426
   layout 257
   location 321

   management team 18
   managers 2, 18
   manual 4–7, 12, 287, 293, 320
   overhead 15, 288
   size 256, 288, 293
   type 293
Property damage 270
Proposal 286, 314–17, 429
Protect worksheet
   button 348
   dialogue box 348
Protection of work 257, 322, 419
PRV 196, 429
P-trap 196, 199, 429
Public agencies 4
Purchase order 3, 18, 323–4, 429
   log 324
Purlins 419
PVC 165, 429
Pyramid 223, 402–4
Pythagorean theorem 138, 141, 142, 393, 395–402

Q

Quality 4, 288, 322, 328
   actual 250
   survey 72, 429
   Quantity
   takeoff 4, 9, 250, 309, 429
     defined 72
     fundamentals 72–82
     package 7–8, 77, 93, 96, 210
     performing 72–3
     quick-and-dirty 310
     software package 354, 393, 431
Quantity-from-quantity goods 64
Quick
   access toolbar 22, 59
     customizing 60
   print 59
Quotes
   e-mailed 16
   faxed 16
   phone 16, 17, 250

R

Radiant heat 203
Rafter 137–40, 419, 430
   ties 430
Raised slab 430
Random length 430
Range 209, 423
   hood 423s
Rate of progress 258–60, 430
Rebar 84, 87, 90, 95, 96, 107, 108, 430
   dowels 86
   epoxy coated 85
   shop fabricated 85
   size 85
   ties 86, 87
   weight of 85
Recent files dialogue box 22
Record macro dialogue box 337

Rectangle 396–7
Redlining 75
Redwood 126, 131
Reference
   absolute 51–2, 337, 425
   books 80
   relative 52, 337, 429
Register 429
Reimbursements 268
Reinforcing 85
   lap 85
Relationships working 328
Relative reference 52, 337, 338, 430
Repair multiplier 278
Repetitive members 76
Request
   for materials quote 250, 251, 297, 323, 429
   for proposal 286
   for quote 12–3, 286
Resilient floor 421
Retaining walls 236
Retention 322, 429
   ponds 424
   rate 315
Retirement contributions 264, 270–1
Return line 205
RFP 286
Request for quote (RFQ) 12–13, 286
Ridge board 430
Right click 430
Rim
   board 430
   joist 430
Ring-shank nail 130, 430
Ring ties 430
Rise 138, 430
Riser pipe 192, 193
Risk 13, 293, 309
Roads 424
Rock ballast 165
Rod threaded 214
Roll goods 78–80
Romex 209
Roof 395
   framing plan 142
   hip 139, 155, 157
   joist 395
   lookouts 141, 429
   overhang 138, 140
   ridge 156, 160
   sheathing 141–2
   shingle 160–2
      cap 160–1
   exposure 160
   slope 138, 143
   span 139
   starter row 160, 161
   valley 155, 157
   waste 142, 155
Roofing
   asphalt shingle 352
   membrane 165

Roof-top unit 420
Rough carpentry 126–145
Round function 53, 54
Rounddown function 54
Rounding 77, 79
Roundup function 54
Row and column method 79–80
Row heights 36
Rows repeating at top of page 60
RTU 420
Rubber base 187, 421
Run 138, 420
Runner 181, 420

S
Safety 257, 270, 323
   glazing 173, 175
   meetings 322
   plan 323
   regulations 242
Sag 99, 100
Sales tax 16, 251, 252, 296
Sample
   durations garage 326
   takeoff garage 101–3, 143–5, 165–7, 176, 188–9, 213–6, 232–3, 238, 301–7, 326
Samples 314
Sand 245
Sanitary sewer 242
Save as dialogue box 22
Scale 7, 77, 210, 393–5, 425
   to fit 60
Scaling 394–5
Schedule 73, 193, 288, 314, 320, 322, 325–6
   durations 325–6
   of values 4, 6, 15, 314–15, 321–2, 425
Scope of work 2, 11, 15, 286–8, 315–17, 321, 425
   model 419–24
Scrap 81
Screws 420
Seaming diagram 421
Search key word 53
Second-tier subcontract 425
Security 252, 277
   options dialogue box 335
   systems 423
Seismic straps 74
Selecting multiple cells 27
Self-adhering bitumen 155
Selling price 425
Series 358–61
Setting page breaks 62–3
Severe weather 315
Sewer
   gas 196
   lines 423
Shafts 420
Shear panel 136, 420
Sheathing 137, 420
   wood 127, 130, 419
Sheet
   goods 78–80

Sheet (*Contd.*)
    name button 48
    tabs 22
    vinyl 186–7, 421
Shelving 143, 420
Shift key 27
Shimming 135
Shingles 160–2
Shipped lapped 164
Shipping cost 251–2, 296
Shop drawings 115
Shortcut key 337
Shower pans 422
Shrinkage 220–3, 247, 425
    percentage 221, 231
Shrubs 424
Shutoff valve 196
Shutter 421
Siamese connection 193, 427, 430
Sick leave 264, 271
Sidelight 175
Sidewalk 424
    excavation 232
Siding 162–5, 421
    aluminum 162
    board 164
    sheet wood 165
    vinyl 162
    waste 162
    wood 164
Significance 55
Sill 126–7, 430
    plate 430
Seal 126, 131, 430
    storefront 174
Sink 199, 432
Siphoned 197
Site lighting 424
Slab
    composite 100
    on grade 97–9, 430
    raised 99–100
Sleeper 419
Sleeves 419, 422–4
Slope 138, 142, 430
Social security 268
Sod 424
Soffit 162–5, 420, 421, 430
    vented 162
Soil
    bank 220, 221, 231
    characteristics 220
    compacted 220, 221, 231
    compaction 220
    conditions 242
    dry
        density 220
        unit weight 220, 427
    excavation 220
    export 231, 247
    import 231, 247
    in situ 220, 428

    loose 220, 221, 231
    report 6, 231, 430
    shrinkage 220–223, 247, 430
        percentage 220, 221, 231
    swell 220–3, 431
        percentage 220, 231
    transportation 220
    unit weight of water 220, 431
    water content 220
    wet
        density 220
        unit weight 220, 431
Soldered 196
Solvent weld 196
Sonotube 84
Sound
    attenuation 186
    system 423
Spaced joint 79
Spacing 76
Special conditions 5, 430
Specialization 256, 293
Specialties 408
    productivity 413
Specifications 5, 251, 317, 322, 353, 430
    sample 314, 315
Spelling 317, 320, 328
Splice 79
Spot footings 76
Spread footings 86–9, 430
Spreadsheet 8–9, 43, 310
    layout 352, 355–6
    limitations 65
    testing 9, 65–68
Sprinkler head 430
Sprinkling system 424
Square 395, 396
    root 51
Stain 421–2
Stairs 421
    concrete 99
    nosing 186
    steel 120
    stringers
        steel 116
        wood 137
Standardized form 309
Starter strip 162
State unemployment tax 268, 269
Status bar 46
Steel 115–22
    angles 107, 111, 116
    base plate 116
    bolts 116
    connections 116–18, 119
    dimensioning 115
    fabrication 115
    framing plan 117
    handrails 121–2
    imbeds 122
    Joist Institute 118, 119
    joists 118–20

ledger 111, 116
lintels 107, 111, 116
miscellaneous 122
pipe 116
plate 116, 118, 429
prime 115
shapes 115–16
stairs 121–2
structural 108–11
timber brackets 122
truss 120–1
types of 115
welds 120
Steps identifying 352, 354–5
Stick framed 137, 140
Stirrups 97, 430
Storage costs 252–4
Storefronts 172, 174–5
Storm drain 242, 424
Striping 258
Structural
  brick 430
  engineer 85, 115
  steel 108
  tee 116, 430
  ties 107, 111
  tubing 116, 430
Stucco 159
Stud 131, 181, 201, 419, 430
Subcontract 3, 18, 287, 321–3, 431
  design-build 192
  pricing 286–8
Subcontractor 5, 12–13, 18, 318
  bid 192, 298
  past experience 16
  performance 288
  pricing 15–16
  selection 15–16
Subgrade 86
Subject line 317
Submittals 6, 322
Sum function 56
Summary worksheet 13, 14, 65, 68, 296, 298–300, 308, 332
Superintendents 18
Suppliers 250, 251, 318
Supply
  line 205
  registers 423
Surety 5, 12
SUTA 268, 430
Swamp cooler 204
Sweated 196
Sweating 431
Swell 220–3, 247, 431
  percentage 220, 231
Synthetic stucco 159
System efficiency 258–260

T

T&G 130, 141, 431
T1-11 162
Tackless fastening strips 188
Tail joist 431
Takeoff 4, 9, 296, 353
  defined 72
  fundamentals 72–82
  package 7–8, 77, 93, 96, 210
  performing 72–3
  quick-and-dirty 310
  software package 354, 393, 431
Task 325, 431
  assignment 11
  linear 258
  relationships 325
  repetitive 257
Tax
  advisor 268
  equipment 276, 279–80
  property 277
Technical specifications 5, 251, 317, 322, 353, 431
Tee
  suspended ceilings 184–5
  steel 116
Telephone 423
  boxes 421
  lines 242
Tempered glass 173, 175, 431
Templates 419
Temporary fencing 252
Testing 322
  spreadsheets 65–8
Text
  box 431
  button 367
  string 367
  wrapping 38
Theater system 423
Theft 322
Thermal protection 153–67, 407–8
  productivity 412
Thermostat 203, 205, 423
Thickened slab 431
Thickset mortar 184
Thinset glue 184
Threshold 174–6
Ties 431
Tile
  acoustical 184–5
  ceramic 184
  grout 184
Timber brackets 122
Time curing 326
Toilet 196, 199, 422
  paper holder 420
Tongue and groove 131, 185, 431
Top plate 431
Topsoil 237
Towel
  bars 420
  rings 420
Tracking
  competitors 293
  worksheet 46

Trade
    journals 328
    magazines 252
    publications 328
Transom 175, 431
    profile 175
Transpose 33
Trapezoid 396–8
Travel time 258
Trees 424
Trenching 242, 422
Triangle 395, 397
Trim 143, 165, 420, 421
Trimble 7
Trimble WinEst 8
Truss 431
    wood 127, 419
    steel 120–1
Trust 328
    center dialogue box 336
Trusted sources 336
Tubing steel 116
Tubs 422
Typographical errors 31, 320
Tyvek 107

U
U. S. Department of Labor 264
Unavoidable waste 431
Uncertainty 293
Underground service 192
Underlayment 160, 186, 421
Unemployment
    claims history 268
    insurance 273
Unethical behavior 329
Union 270
    contracts 264
    dues 270
    payments 264, 271
Unit 310, 352
    pricing bid 431
    weight of water 220, 431
Unmerge 39
Unprotect worksheet
    button 348
    dialogue box 348
Use relative reference button 337
Utilities 192, 242–7, 410, 424
    productivity 415
Utility
    boxes 242, 246
    conduit 247
    lines 247
    manhole 242, 246
    pipe 246
    room 421
    valves 247
    wire 247

V
V weld 431
Vacation 264, 271
Validation 33
Valley 155, 161
    rafters 431
Valve 203, 424
    check 192, 426
    drain 192
    flush 199
    pressure reducing 196, 429
    shutoff 196
    utility 247
Vandalism 322, 419
Vanities 422
Vapor barriers 154–7, 185, 431
Variable-air-volume box 205
Variables 51
VAV box 205, 431
Vendors 5, 13
Vent 421
    pipes 160, 196
Ventilation 201–6
Vinyl
    composition tile (VCT) 187, 431
    flooring 186, 421
    siding 162–5
Viruses 336
Vlookup 371–379
Volumes 402–4
Volumetric goods 80

W
Wage rate 264
Wages 264
    bonus 264
    overtime 264
    raise 264
    weekends 264
Walks 424
Wall 421
    bearing 128, 131, 140
    coverings 420
    fire 140
    foundation 95–6
    moldings 184
    painting 188
    wood 131
Warning dialogue box 27, 30
Warranty 322
Washer hookups 422
Washout 423
Waste 78, 81, 86, 431
    actual 86
        percentage 81
    avoidable 4
        control of 81
        definition 81
        tracking 81

controlling 86
factor 81
flooring 185
pipe 196
roof 141, 155
unavoidable 4, 79
   definition 79
Water
  closet 196, 199
  content 220, 431
  heater 196, 422,
  infiltration 153
  lines 242, 423
  main 192
  meter 423
  softener 422
  supply 196
  table 6
Waterproofing 153–4, 431
Weather 256
  barrier 107, 111
  caps 419
  stripping 174, 175
Web stiffener 128, 431
Weld field 115
Welded wire fabric (WWF) 85, 431
Welds 118, 120
Wet
  density 220
  pipe fire sprinklers 236
  unit weight 220, 431
Wheel 7, 429, 431
Wide-flange beam 116, 431
Width
  dropdown box 60
  effective 78
Wind forces of 136
Window 172–4, 420
  bucks 419
  fixed panels 173
  opening 111, 133
  operable 173
  safety glazing 173, 175
  schedule 172, 173
  seats 420
  sills 420
  sliding residential 173
  tempered glass 173, 175, 430
WinEst 8
WinEstimator Inc. 8
Wire mesh 85, 97, 99, 431
Wiring
  diagram 211
  low voltage 203, 423
Wood 126–45, 407,
  backing 132, 181, 419, 420
  beam 419
  blocking 126, 136, 419, 425
  ceiling joists 137, 139
  collar ties 137, 139, 426
  columns 425
  decks 126
  doors 172
  fascia 141
  fire blocking 129, 136, 425
  fire-treated 181
  floor 126–8, 185–6
    joist 128–30, 419
    sheathing 130
    squeaks 130
    trim 186
  framing 175
  girder 128, 419
  gluing 130
  header 126, 132, 135–6, 419
  I-joist 128
  joist 126
    ceiling 139
    header 126
    trimmers 128, 129
  laminated lumber 135
  lintels 419
  naturally durable 131
  package 126
  paneling 143
  plate 126, 131–2, 419
  plywood 135
  post 128, 419
  pressure-treated 126, 131, 429
  productivity 412
  purlins 419
  rafter 137–40, 419, 430
    common 139
    hip 139
    jack 139
    king 139
    valley 139
  random lengths 126
  ridge 156, 160
    board 140
  rim board 128
  ring-shank nail 130, 430
  risers 419
  roofs 137
  shear panels 136, 430
  sheathing 126, 130–1, 419
    roof 141
    waste 130
  shelving 143, 420
  sill 126, 419
    plate 126
    window 134
  sleeper 419
  slip-resistant nail 128
  stingers 419
  stud 126, 419
    cripple 134
    jack 126, 132, 134
    king 126, 132
    precut 132

treads 419
trim 141, 163
trimmers 419
trusses 126, 137, 140–1, 419
    tails 140
walls 131–7
    corners 132, 134
    ends 132
    hold-downs 132, 134
    intersections 132
    openings 132
web stiffeners 128, 431
Word processing 3, 9
Work 431
    complexity 288
    duplication 287
    ethic 328–9
    package 73–4, 286
    protection of 322
    scope of 2, 4–5, 286, 315, 316, 321, 430
        model 419–24
Workbook 431
    creating 22
    macro-enabled 338
    management 21–5
    new 23
    opening 22
    saving 22
Workers' compensation insurance 264, 265, 269–70
Workmanship 6

Worksheet 431
    adding 25–6
    copying 26–7
    deleting 27
    detail 29
    formatting 35–44
    organizing 27–9
    protecting 348–350, 374
    referencing 366–7
    renaming 27
    summary 29, 65, 68
    testing 340–1
    working with 25–9
Wrapping 38
Wrap text button 38
Writing formulas 50–2

X

xls 21
xlsm 21, 338, 431
xlsx 21, 431

Z

zoom 42–4
    button 43
    dialogue box 43
    level 42
    selection button 43
    slider 42
    to fit 42